Cognitive and Linguistic Aspects of Geographic Space

NATO ASI Series

Advanced Science Institutes Series

A Series presenting the results of activities sponsored by the NATO Science Committee, which aims at the dissemination of advanced scientific and technological knowledge, with a view to strengthening links between scientific communities.

The Series is published by an international board of publishers in conjunction with the NATO Scientific Affairs Division

A Life Sciences **B Physics**	Plenum Publishing Corporation London and New York
C Mathematical and Physical Sciences **D Behavioural and Social Sciences** **E Applied Sciences**	Kluwer Academic Publishers Dordrecht, Boston and London
F Computer and Systems Sciences **G Ecological Sciences** **H Cell Biology** **I Global Environmental Change**	Springer-Verlag Berlin, Heidelberg, New York, London, Paris and Tokyo

NATO-PCO-DATA BASE

The electronic index to the NATO ASI Series provides full bibliographical references (with keywords and/or abstracts) to more than 30000 contributions from international scientists published in all sections of the NATO ASI Series.
Access to the NATO-PCO-DATA BASE is possible in two ways:

– via online FILE 128 (NATO-PCO-DATA BASE) hosted by ESRIN, Via Galileo Galilei, I-00044 Frascati, Italy.

– via CD-ROM "NATO-PCO-DATA BASE" with user-friendly retrieval software in English, French and German (© WTV GmbH and DATAWARE Technologies Inc. 1989).

The CD-ROM can be ordered through any member of the Board of Publishers or through NATO-PCO, Overijse, Belgium.

Series D: Behavioural and Social Sciences - Vol. 63

Cognitive and Linguistic Aspects of Geographic Space

edited by

David M. Mark

National Center for Geographic Information and Analysis,
Department of Geography,
State University of New York at Buffalo, Buffalo, New York, U.S.A.

and

Andrew U. Frank

National Center for Geographic Information and Analysis,
Department of Surveying Engineering,
University of Maine, Orono, Maine, U.S.A.

Kluwer Academic Publishers

Dordrecht / Boston / London

Published in cooperation with NATO Scientific Affairs Division

Proceedings of the NATO Advanced Study Institute on
Cognitive and Linguistic Aspects of Geographic Space
Las Navas del Marqués, Spain
July 8–20, 1990

ISBN 0–7923–1537-5

Published by Kluwer Academic Publishers,
P.O. Box 17, 3300 AA Dordrecht, The Netherlands.

Kluwer Academic Publishers incorporates the publishing programmes of
D. Reidel, Martinus Nijhoff, Dr W. Junk and MTP Press.

Sold and distributed in the U.S.A. and Canada
by Kluwer Academic Publishers,
101 Philip Drive, Norwell, MA 02061, U.S.A.

In all other countries, sold and distributed
by Kluwer Academic Publishers Group,
P.O. Box 322, 3300 AH Dordrecht, The Netherlands.

Printed on acid-free paper

Printed in the Netherlands

Table of Contents

Table of Contents (continued)

PREFACE

This book contains twenty-eight papers by participants in the NATO Advanced Study Institute (ASI) on "Cognitive and Linguistic Aspects of Geographic Space," held in Las Navas del Marqués, Spain, July 8-20, 1990. The NATO ASI marked a stage in a two-year research project at the U. S. National Center for Geographic Information and Analysis (NCGIA). In 1987, the U. S. National Science Foundation issued a solicitation for proposals to establish the NCGIA—and one element of that solicitation was a call for research on a "fundamental theory of spatial relations". We felt that such a fundamental theory could be searched for in mathematics (geometry, topology) or in cognitive science, but that a simultaneous search in these two seemingly disparate research areas might produce novel results. Thus, as part of the NCGIA proposal from a consortium consisting of the University of California at Santa Barbara, the State University of New York at Buffalo, and the University of Maine, we proposed that the second major Research Initiative (two year, multidisciplinary research project) of the NCGIA would address these issues, and would be called "Languages of Spatial Relations" The grant to establish the NCGIA was awarded to our consortium late in 1988.

The project included three meetings, each involving geographers, mathematicians, computer scientists, cognitive scientists, and others. The first was held in Buffalo, New York, in June 1988, before the NCGIA grant was awarded (see Mark, 1988). The second was the NCGIA "Specialist Meeting" held in Santa Barbara, California, in January 1989 (see Mark *et al.*, 1989, and Mark, 1989). And the third was the NATO ASI which produced this book. The second and third meetings built on the results of the earlier ones, as a core group of participants was invited back. This facilitated the breaking of the barriers to interdisciplinary work that are often erected by differing terminologies, philosophies, and approaches. Through these meetings and the research activities between them, we have gained a great deal of personal insight into the topic of human conceptualization of space, and the influence of that on technology. And we believe that other participants in the Initiative have been similarly enriched.

It was an enormous pleasure to share our ideas on a theory of geographic space with the participants in the workshop; we thank all the participants for their openness, and for their contributions to the good humor and camaraderie that developed at the Castillo-Palacio "Magalia" during the two weeks of the workshop. The fact that the village of Las Navas del Marqués was having its annual fiesta during our first week there, with events ranging from bull fights to bumper cars, contributed substantially to the atmosphere. Maria Cátedra, of the Universidad Complutense de Madrid, acted as the local organizer and discovered the Castillo-Palacio "Magalia" in Las Navas del Marqués; she participated in the organizing committee and contributed to the smooth preparation and operation of the workshop. Amilcar Sernadas and Hans Dieter Ehrich also were members of the organizing committee, but unfortunately were unable to attend the ASI. We are grateful to the Scientific Affairs Division of the North Atlantic Treaty Organization (NATO), and the Programme Director L. V. da Cunha, for the grant that made the ASI and this book possible. The NCGIA (supported by U. S. National Science Foundation grant SES-88-10917, Geography and Regional Science program) and its staff at Buffalo and Maine contributed in numerous ways to the organization of this workshop, and provided travel support for some participants. NSF also supported the travel to the ASI of five young U. S. scholars through the Division of Research Career Development. Roberta Balstad Miller encouraged us at various stages along the way. Scott Freundschuh, Max Egenhofer, and Werner Kuhn assisted with various aspects of the organization of the meeting, and Robert Scicogna provided some editorial assistance. Michael Gould contributed countless hours to the editing and reformatting of the papers in this book, and Connie Holoman assisted with the final manuscript production. Last but certainly not least, we must thank the Spanish Ministerio de Cultura, Instituto Nacional del las Artes Escénicas y de la Música that operates "Magalia" as a conference center, and the Castillo's director Mario García Bardaji, the bartender Jesús, and the rest of the staff, who arranged the meeting rooms and kept us happy with good food and drink, creating an atmosphere of well-being that permits intense intellectual work.

This book is dedicated to the memory of David S. Simonett, founding Director of the National Center for Geographic Information and Analysis. We learned immensely from him, and miss him deeply.

David M. Mark and Andrew U. Frank
Buffalo, New York (USA) and Orono, Maine (USA), August 1991

References:

Mark, D. M., 1988, 'Cognitive and Linguistic Aspects of Geographic Space'. Report on a Workshop. Santa Barbara, California: National Center for Geographic Information and Analysis.

Mark, D. M., 1989, 'Languages of Spatial Relations: Researchable questions and NCGIA research agenda.' Santa Barbara, California: National Center for Geographic Information and Analysis, Report 89-2A.

Mark, D. M., Frank, A. U., Egenhofer, M. J., Freundschuh, S. M., McGranaghan, M., and White, R. M., 1989, 'Languages of spatial relations: Initiative two specialist meeting report'. Santa Barbara, California: National Center for Geographic Information and Analysis, Report 89-2.

COGNITIVE AND LINGUISTIC ASPECTS OF GEOGRAPHIC SPACE: AN INTRODUCTION

This volume contains papers presented or discussed at a NATO Advanced Study Institute on "Cognitive and Linguistic Aspects of Geographic Space," held in Las Navas del Marqués, Spain, July 8-20, 1990. We organized the meeting as a place to discuss people's understanding of space, specifically geographic space. An important subsidiary goal was to try to understand how such cognitive concepts of space relate to Geographic Information Systems (GISs). We invited researchers from a wide spectrum of disciplines, from geography and linguistics, psychology and anthropology, to mathematics and engineering. Each of the disciplines deals with some aspect of understanding space. Each discipline, and indeed each individual, views the problem in a different way and from a unique vantage point. The papers collected in this volume represent much of the breadth of this spectrum. Any such a collection of papers is of course influenced by the limitations of the size of the meeting, the availability of participants, and their ability to contribute to the topic. Despite this, we believe that the papers in this volume present a valuable perspective on our current understanding, and open5 a host of research avenues.

We believe that this book presents an important and diverse collection of papers. However, we must note that it does not closely reflect what happened at the NATO ASI. The majority of the papers included were presented at the meeting by participants during the last 4 days of the 12-day ASI. Most of the time during first 8 days of the meeting was devoted to background lectures by a distinguished set of lecturers—Mark Blades, Maria Cátedra, Jean Gabriel Ganascia, John Herring, George Lakoff, Zenon Pylyshyn, and Len Talmy. Unfortunately, time and other writing commitments prevented four of these lecturers from submitting papers for this book. Their lectures and discussions, however, contributed tremendously to the meeting, and many of their ideas influenced the post-institute revisions of the papers included in this book. Papers related to the lectures by Maria Cátedra, Mark Blades, and John Herring appear in this book, and are introduced in the opening parts of Sections 2, 3, and 5, respectively. But to place the book in the broader context of the meeting, we believe that it is worth giving an overview of the topics that the other lecturers presented in the first part of the ASI.

On the first Tuesday morning, **George Lakoff,** from the Department of Linguistics, at the University of California, Berkeley, USA, presented a lecture on "Categories and Cognition". Much of the material was drawn from his recent book, *Women, Fire, and Dangerous Things: What Categories Reveal About the Mind* (University of Chicago Press, 1987). First, Lakoff discussed objectivism, which is his term for a traditional philosophical position in the sciences, related to realism and positivism, that asserts that a "God's eye" view of truth and reality does exist, and science attempts to discover it. In the objectivist view, language involves the manipulation of symbols; language has meaning through the correspondence between those symbols and the external reality. In objectivism, categories are equivalent to sets, and can be modeled using set theory. However, empirical research by Eleanor Rosch and others provided many counter examples to this generalization: categories often exhibit best examples, graded membership, and

D. M. Mark and A. U. Frank (eds.), Cognitive and Linguistic Aspects of Geographic Space, 1–6.

"basic level" effects. And categories often show a radial structure, with category members related to a central member through a variety of kinds of similarity.

To accommodate this evidence, Lakoff (working at times with Mark Johnson, a philosopher) has developed a philosophical framework that they call "experiential realism". Cognitive models are shaped by bodily and sensory experiences with the world. Then, further sensory stimuli are interpreted with respect to existing cognitive models, which may be modified through use. The meanings of words are always defined with respect to cognitive models—they include a variety of background assumptions, schemata, etc. Metaphor is a central part of experiential realism. Lakoff and Johnson use the term "metaphor" to refer to mappings from one situation to another preserving image-schematic structure. People often understand novel situations through metaphorical relations to familiar ones, and structure abstract domains based on concrete domains. Spatial experiences, such as journeys, are common source domains for metaphors. Image-schemata play a central role in experiential realism. Lakoff proposed that image schema exist at a neurophysiological level, and expressed the opinion that connectionist models (including "neural networks") are the most appropriate computing architecture for cognitive science.

Lakoff also raised several questions relating to geography and GIS. What kind of metaphors are behind certain kinds of geography? If a culture uses a metaphor such as "land is a possession" or "land is a commodity", this may profoundly influence the land tenure system and land management under that culture. But land tenure systems are not the same everywhere. Can different systems be traced to different metaphors for understanding man-land relationships? This is an interesting and open research question. Land management in the developed world is often based on a "nature is a resource" metaphor. In the early 1980s, the United States Geological Survey published a professional paper entitled "Nature—To be Commanded". A culture which holds a metaphor that "the Earth is our mother" might administer land-related laws and customs quite differently. Language and culture are not neutral or value-free, and neither is technology. A study of pervasive metaphors for geographical information in different cultures might give valuable clues regarding the role and value of geographic information in decision making. A practical implication is the potential impediment to cross-cultural technology transfer.

Jean Gabriel Ganascia, of the Université Pierre et Marie Curie, France, presented two lectures with his colleague Denis Blamont (CNRS, France). The first, on the first Friday morning, was entitled "Machine Learning and Knowledge Acquisition Applied to Cartography." The lecture was primarily a tutorial on knowledge acquisition. The more competent experts are, the less able they are to explain their knowledge. It also is important to recognize that knowledge acquisition is a social process as well as a technical one. Denis Blamont then continued the lecture by describing a study in which Ganascia and Blamont applied these methods to remotely-sensed imagery. French geography is based on landscape concept, and includes a "symbolic understanding of the world". Like other researchers, geographers approach the world with preconceived notions. These principles were further developed in a second lecture by Ganascia and Blamont, entitled "A Symbolic of Geographic Space Through Remote Sensing." A program called CHARADE was used to determine rules that could account for given, field-determined land-use and land-cover classes for an area in the Himalayas, based on satellite images.

Zenon Pylyshyn is Director of the Center for Cognitive Science at the University of Western Ontario, Canada. On Tuesday afternoon, he presented the second major lecture of the ASI, on the relation between computation and cognition. How is it possible that people can operate in a physical world? Behavior is determined by relationships to things which are not existent in that world, but are themselves mental states. The formalist program in mathematics provided solutions in the form of truth preserving transformations of statements. But observations of behavior are not sufficient to explain the internal structure of a black box. Pylyshyn asserted during the lecture and throughout the meeting that the details of brain physiology are not important—the architecture is relatively unimportant, compared to the representations of knowledge at a more abstract and conceptual level. This view stands in sharp contrast with Lakoff's opinion that brain architecture should influence computational models, favoring connectionist models over propositional models.

Pylyshyn proposed that three levels are important: a semantic level (knowledge); a syntactic level, encoded physically; and a physical model, mapped to physical action. There is a strong equivalence among things that are computable—but implementations are useful in that their properties of operation, such as response (reaction) times, may indicate whether a particular model has the appropriate form. Most proposed systems have the wrong computational properties. For example, the more rules and facts that are put into the computer model, the slower the operation is, whereas for people, it is often the case that the more they learn, the faster they can solve problems.

Pylyshyn went on to talk about reasoning with the aid of imagery, a topic that he has worked on for some time. He described Kosslyn's experiments about a "mind's eye" model, in which the mind uses the visual cortex or something very much like it for spatial inference. Many experiments by Kosslyn and his colleagues found essentially the same results, reaction times, etc., for processing of remembered images, as were found in the presence of visual stimuli. Is that telling us something about the architecture for spatial reasoning? Pylyshyn reported on performing similar experiments with different instructions to the subjects, and getting the opposite results. However, the reaction time effects cannot be reversed if the subject is looking at an actual visual stimulus, no matter what instructions are given. Pylyshyn concludes that the mind does not use the same architectures for reasoning about perceived images and for reasoning about remembered ones. Pylyshyn proposed that most reasoning is propositional, based on "sentences" in a "language of thought".

Leonard Talmy, who was a member of the Cognitive Science Program at the University of California, Berkeley, USA at the time of the ASI, gave two lectures during the second week of the conference. The first was entitled "Fictive Motion in Language and Perception", and discussed an extremely widely used concept in language—many times, static objects or situations are talked about as if they are in motion. In English and many other languages, we say things like "the road runs through the mountains", when in fact the road is not moving at all, or at least not in the ways that these examples suggest. Other more abstract examples of fictive motion are the conceptualization of vision as an activity in which something moves beyond the observer's body—"he gazed into the valley" or "she ran her eyes over the painting." Talmy also notes that the concept of radiation moving away from a source such as the sun, or a light bulb, is also fictive, because we cannot observe the motion; this concept of "emanation" of light or heat may be a universal across all people. Another interesting point is that motion in language seems to be rather independent from, and more basic than, concepts of time itself. It seems as if motion, and dynamic concepts, may be more fundamental in our mental models of the world than are static concepts. These examples clearly reveal how the links among language, thought, and reality are not as clear cut as we often assume.

Later, Talmy presented another lecture, this one based on his well-known paper, "How Language Structures Space". Talmy's research has mainly focused on "closed class elements" of language—in English, prepositions are examples, as is "s" for plural. It is very rare for a language to add a new member to a "closed class" (hence the term "closed"). In contrast, new members of open classes such as nouns, adjectives, and verbs, can be coined or borrowed from other languages very easily. Closed class elements could (in theory) be used for almost anything, but are used just for some. Almost all languages have a closed-class indicator of number (singular, plural), and most languages represent spatial relations this way. At the other extreme, a few languages have closed class elements to extract units from mass nouns—whereas in English one must say "a grain of rice", "a drop of water", or "a blade of grass". Yiddish and a few other languages can extract a unit by adding a closed-class suffix to the mass nouns such as rice or grass. Most languages encode spatial relations with a closed grammatical class. Situations can be described as involving a figure (thing being located), a ground (which provides a reference frame against which the location is specified), and a spatial relation. An important point is that, in the range of their meanings, the prepositions (or equivalents) in almost all languages and situations are magnitude-neutral (that is scale-independent), material-neutral, and shape-neutral. One of Talmy's example is "the ant crawled across my palm" and "the bus drove across the country". It seems that the language is based on topology rather than geometry, but another example contradicts this–a strict mathematical topology would see little similarity between a bird cage, a box, and a bowl, whereas most languages treat these all as containers and use the same preposition to express containment "in" them.

After the initial background lectures, more specific topics were presented by some of the participants. Most of these presentations are included in this book. Looking over the table of contents of the book, one may ask what these very different contributions have in common? In fact, before the meeting, even some of the participants wondered why some of the topics and speakers had been included. However, we believe that all of these topics are essential to studies of how people relate to geographic space, and to how computer-based systems can be developed to assist in those interactions. The common topic is human understanding of space and how people interact with space. But each author and discipline cast this common topic in a different light. Each of the sciences that has an interest in space and spatial thinking uses its own definition of space and of spatial concepts. Euclidean geometry, a marvelous abstraction from many aspects of our experience with real space, is one of the best-known spatial models. But every school child, after marveling at the parallel lines of railway tracks as they appear to converge, intuitively grasps the difference between the abstract, formal explanation in school and the "realities" of our perceptual worlds. Christian Freksa points out in his paper in this book that space is exceptional, because we experience space through several senses: visual, sensorimotor, auditory, haptic (tactile), and others. He asserts that this is why spatial configurations and relations seem to be among the most *real* of all our experiences, and form the source domain for so many metaphorical structures for abstract concepts and phenomena.

Turning to the behavioral side of the topic, it appears likely that people use different concepts of space, depending on the task they are trying to solve. This is not surprising, but conflicts with the idea of simple and universal formal models of space. Mark Blades surveys the extensive studies on human wayfinding behavior, and discusses different wayfinding strategies and the spatial understanding that would be necessary for each. Wayfinding and navigation are among the most fundamental aspects of spatial cognition, since they are needed for our very survival. They are also activities that almost all higher animals must conduct, and so at least some of the underlying principles might be found in comparisons of human spatial abilities to those of other species.

Understanding space, and influencing and controlling the cultural understanding of space, is a very political topic. Many conflicts have been based on land, access, and other aspects of geographic space. A culture's understanding of space can be influenced through technology—

while at the same time technologists may suggest that the technology they promote is an objective, value-free interpretation. Recently, geographer Brian Harley applied the post-modern "deconstructionist" method to maps, pointing out that maps do not directly reflect reality, but show the map maker's understanding of spatial reality—and thus is a cultural statement. Irene Campari points to the dangers of computerized technology, with its aura of "truth" and the lack of awareness for the subtle cultural and value statements built into software. "Garbage in, *gospel* out" is the new danger, as people abandon their mistrust of computers, and begin to trust them too much. Specifically, GIS software, mostly designed in the United States, but used around the world, has the potential to "export" with it the Anglo-American concepts of space, land tenure, urban planning, nature, resources, etc., to other countries. An immediate problem is the difficulty in translating menus, terminology, and manuals, but other, less obvious problems are likely to emerge, especially if the technology is extended beyond the technicians and technocrats.

As Joan Nunes reports, geographers have long tried to define regions, in a sometimes-futile attempt to identify exactly the objects of their studies. They have invariably found the fundamental "neighborhood", "adjacency", and "contiguity" properties of space, as formalized by mathematicians. Spatial interaction is another universal: Waldo Tobler has stated this as one of the fundamental geographic "laws": Everything interacts with everything, but close things interact more than distant things. Cartographers have faced the same problem when they have analyzed the structure of maps, and asked how maps communicate spatial knowledge. Grant Head gives an overview describing the different approaches cartographers have taken to analyze maps using a "linguistic" paradigm, while Hansgeorg Schlichtmann examines the semiotic aspect of maps and and the relations of map space to geographic space.

The topics discussed at the ASI and reflected in this volume have an engineering or technical side as well as a scientific one. The construction of geographic information systems requires knowledge of human cognition throughout. The data structures, the procedures and functions, and especially the user interfaces all must take human cognition and user needs and goals into account. Many commercial companies and government agencies (mostly in the United States, but also in Europe) develop GISs—and the engineers encounter all the scientific questions discussed above, and more, while under pressure to produce solutions now. There are many GIS users, and many more to come, as GIS is rumored to be a multi-billion dollar industry, and is expected to continue to grow rapidly. Progress is impeded by a lack of a theoretical base upon which software can be built. A general theory of spatial relations, consistent with human cognition is required. This must then be formalized by the mathematicians and implemented by the software engineers and programmers–John Herring's paper presents a partial solution. The engineers also need methods to translate the data stored in the databases of a GIS into cartographic form, and thus ask for formalized rules for cartographic design. Large amounts of spatial information, such as plans of public utility lines, are stored in graphical form. Such data could be scanned and automatically analyzed if we could automatically understand the cartographic language and how it communicates spatial information.

It is at first surprising that the experiential viewpoint, presented by Lakoff and supported by many participants at the ASI, is completely compatible with the new discipline of software engineering, specifically the object-oriented design method. This new approach to computer implementations uses multi-sorted algebra as a base, where a mathematical structure is defined by operations and rules for the effects of these operations. There are, however, major problems with the translation of cognitive concepts into formal, mathematical language, primarily due to the non-Aristotelian nature of cognitive categories, which by and large cannot be modeled as sets, or even as fuzzy sets. Nevertheless, the contributions from cognitive and other social sciences are extremely relevant for the engineers that design future geographic information systems. We agree with Irene Campari's view that the so-called "soft sciences" are important in the development of technology, and that the dialog between the social sciences and the engineering disciplines is valuable.

The discussion about space needs to be multi-disciplinary and we think that the dialog that is documented in this volume is an important step in this direction. We believe that none of the sciences involved has the complete answer, or is likely to find it alone. The full truth of what geographic space is, and how a GIS should handle it, is probably unattainable—but if we can bring together the different viewpoints represented in this book, and others, we may make significant progress.

SECTION 1: GEOGRAPHIC SPACE

This section includes three papers that take a philosophical view of geographic space. It is not surprising that all three papers have been written by geographers. Geographers have long been concerned with the nature of their discipline and its subject matter. However, the need to design better geographic information systems has brought new attention to the topic, with new needs for formal models, and for new tools to develop them.

In the first paper in this section, **Joan Nunes** sets the scene for much of what follows: Nunes tries to answer a substantive and central question: what do we know about "geographic space"? In doing so, he clarifies the topic of this volume. The paper is a very extensive essay, partially historical, in which he draws the major lines of the development of concepts of space and the major problems on this topic. On first consideration, it appears that space is a simple concept and not in need of further study. However, even two thousand years ago, the classic Greek philosophers found two different concepts of space, namely a space that exists independently of the objects embedded in it, contrasted against a space that is a fundamental property of the objects. This same dichotomy continues through to the fundamental discussion about geographic space and the question about the objects of geographic study, and then to the spatial data models in modern GIS, where we find the same difference between two technically substantially different solutions, often used for different tasks. It is somewhat surprising to find such a long and famous history of concern (and even preoccupation) with this topic. Nunes' paper concludes by separating the "traditional" GIS, based on a computer representation of a map, from the GIS as a independent model of reality, part of the field of database management systems. This idea will be followed more extensively in the fourth Section of this volume, where cartographic representations of space are discussed. The idea also introduces the crucial question regarding the spatial facts that are useful in order to represent space and spatial relations. In the past, GISs have sometimes answered this question with a digital reproduction of a map, but this is not the solution that responds to the multi-purpose requirements placed on many GISs. It is necessary to identify the most general, basic spatial relations, from which any other relation can be deduced, whatever the system is used for. Lastly, Nunes points to the attempts to understand geographic space based on human perception. It appears as if the "scientific" concept of space—Euclidian geometry in most cases—may not be the most relevant for the problems GIS or geography deals with. Cognitive science and its constituent disciplines are an important source of alternative models of geographic space.

The second paper in this section, by **Irene Campari**, contrasts the deep scientific questions in the design of GIS with the mundane difficulties encountered due to the technology. She starts by asserting that GIS is a widely used tool, and then examines its lack of a coherent theory. The apparent absence of theoretical investigations may perhaps be explained by the different time horizons for the software industry and for basic science. She then gives an extensive and very practical account of the current shortcomings of GIS, some of which are due to linguistic problems. Campari offers the example of the problems of building a GIS in Italy, where GIS may be the first contact with the English language for many users. The technical terms of GIS, such as "plot" or "digitize", are often used without any understanding of their meaning outside the GIS.

But this seems to be only a minor problem, compared with the contrast in the conceptual bases of a GIS between system users and software engineers, especially when the former are in Italy and the latter in the United States. Translation of technical terms alone will not be sufficient, because the "feel" of space in Italy is different from that in an Anglo-Saxon world. And there are even problems within Italian—Campari goes on to stress that the official language (e.g. "standard" Italian, in contrast to local dialects) expresses most efficiently the "official" spatial objects, such as parcels and electoral districts, but lacks the sensitivity to capture more traditional cultural arrangements.

The concluding short paper by **Alan Philbrick** points to the specific properties of space, primarily its continuity. Geography has a long history of attempting to divide the world into uniform regions, reified in the political hierarchy, in categorical coverages, or thematic layers used in many GIS applications. However, there is rarely if ever a single regionalization that fulfills all requirements. The basic property of topology, contiguity, which in turn gives rise to the neighborhood, is our way to subdivide space into isolated pieces to serve as regions in a feature-based view of the world. Philbrick advocates a nodal hierarchy, where higher level units are composed of lower level ones, over many orders of magnitude of sizes, from the subatomic to the universe. In his model, the boundary between the inside and outside of a single human being is the most significant geographic boundary.

In three different ways, the three papers point toward the relativity of space and spatial concepts, as dependent on human cognitive systems. Space may seem entirely subjective, depending on the individual, but the individuals communicate and cooperate, and thus the concepts used are agreed upon in a community, either explicitly (by laws or rules) or implicitly (through customs, traditions, or moralities), as much in a profession with its jargon as in the local community of a small town in Tuscany with its local dialect. The issue of culture and cultural relativity of spatial concepts is taken up even more directly in the section which follows (Section 2).

GEOGRAPHIC SPACE AS A SET OF CONCRETE GEOGRAPHICAL ENTITIES

JOAN NUNES
Departament de Geografia
Universitat Autònoma de Barcelona
08193 Bellaterra, Spain

ABSTRACT. This paper address the need for a theory of geographic space that handles space through the concrete objects or entities that actually create and form the space. Several trends seem to show the need for such a theory, which in fact would be a subset of a general spatial theory provided that it would specify what spatial relationships are possible and in effect amongst geographical (real world) entities, and what their concrete forms are. It also would have to establish their significance or influence in the concrete geographic processes into which real world entities enter and act. The paper will attempt to contribute to the development of such theory by extending revisions of the approaches to space from physics and philosophy, geography and cognitive science; and by proposing for discussion some basis for a conceptual model of geographic space suitable for GIS and built around the concept of the geographical entity: a part of the Earth surface defined by virtue of one or several spatial and/or not spatial properties, which make it a space in itself, physically or functionally different from its surrounding and therefore able of carrying out an individual existence persistent in time and of being perceived by human beings as such individual space with particular properties and relationships.

1. Introduction

The lack of a comprehensive and widely accepted conceptual model of the geographic space is now quite often recognized as a major hindrance for further development of geographic information systems. Although reflections on the conceptualizations of space implied by GIS have arisen from time to time since the beginning, it has not been until recently that this question has received substantial attention and effort. This notwithstanding, it has been implicitly present in much of the permanent debate about data structures and data models, as well as in the claims for improving GIS analytical and modeling capabilities and for adding more (or at least some) intelligence to these systems.

Looking at the concepts of space developed in the disciplines for which GIS has become a common tool, especially geography, has not helped much to fill the gap mainly because the same lack has been found, despite of a larger and more explicit debate. In fact, it is acknowledged that geographic theory is fragmentary, poorly structured and unsatisfactory.

The broadening of perspectives for geographical inquiry since the mid-seventies, by adding new focus, methodologies and underlying philosophies, has not provided geography with a more consistent and well-developed body of theory or basis for it. Instead, although highly beneficial in showing how many different ways of looking at spaces and places are possible and worthwhile, all

D. M. Mark and A. U. Frank (eds.), Cognitive and Linguistic Aspects of Geographic Space, 9–33.

this enormous flow of ideas has lead to further uneasiness and difficulty to define the discipline itself, or simply to find communalities in what geographers do.

That geography be able to define geographic space is important at least for two reasons. Leaving aside whether or not it will help geography to structure itself, it is important because if geography does not provide such definition perhaps no other discipline will do it, since for most disciplines dealing with geographic information the term 'geographic' is something assumed, that accounts for little more than being referenced to a position on the Earth. Secondly, and more crucial, it is important because in the event that a general theory of space can be devised —and many of its elements seem to be already available from modern physics and mathematics— it will be still necessary to establish what is specific in geographic space, what makes it a particular kind of space, where some of the general rules probably will not be pertinent or will have a particular meaning.

Both aspects are best illustrated by the current situation in GIS development. On the users side —from whichever discipline— most of the analytical tasks have to be undertaken outside the systems by the users themselves, leaving the systems little more than a spatial referencing tool. As long as the analysis is carried on in a pragmatic basis this situation seems to be endurable and everyone adapts to the structures and capabilities provided by the systems, even in many cases without much dissatisfaction, with the exception of ease-of-use.

On the research and development side much effort is spent in the development of precise formal models for geographic data, mainly based on mathematics and logic. Those data models are —at least potentially— able to be implemented, which is a highly desirable goal, but which can have very little reference to the real world phenomena for which they are supposed to represent. This has several negative consequences both for the research an the results. First, the research becomes in this way tentative and most likely larger than needed. Second, although the main concern is formalism and completeness —when not on computational efficiency— judgements about the adequacy of the models are many times unavoidable and in such situations, given the lack of a conceptual model, the criteria for judgements are either adequacy with respect to some geometric representations —which are not the phenomena themselves— or with respect to some set of common sense or metaphysical believes about the nature of space implicitly assumed.

Thus, a conceptual model of geographic space that supports current scientific thought about space, which will not necessarily solve the many unknowns that geography has as a scientific discipline, could help in the development of data models for GIS which fulfill the need to increase the semantic content of geographic data.

2. Geographic Space vs Space: The Domain of Geography

It seems clear that some kind of specificity is implied by the use of the term geographic. Otherwise, the distinction between geographic space and space (any kind of space), or between geographical and spatial, would be superfluous and geography would study any kind of space. Because the latter is not true, it should be apparent from the subject matter of geography what geographic space is, and what subdomain it occupies in the general spatial domain.

Unfortunately, however, this is not the case, as the practice of geography shows no agreement on its subject matter. Moreover, it is often argued that geography has not a distinct subject matter compared to other sciences but a distinctive point of view from which study is conducted: usually space (or the spatial dimension).

Leaving this point for a later discussion, it must be said that even if successful in preserving the discipline identity —or in reassuring geographers— this kind of definition leaves the essential question —what is geography about?— unanswered. Also, it appears in the light of geographical practice that not all kinds of spaces or, better, not all kinds of phenomena are studied in relation to space. In fact, geographers seem to have developed —although more unconsciously than formally— a particular sense or scale of measure to recognize what phenomena lay in the geographic domain and what do not.

Thus, through time, geography has been seen quite differently as a science whose subject matter has been:

- places
- landscapes
- regions
- the relationships between the human beings and the environment
- the areal differentiation of the Earth's surface as it is relevant for human beings
- the analysis of spatial distributions
    - spatial patterns
    - spatial relationships
    - spatial structures
    - spatial systems
- the relationship between social processes and spatial structures
- human behavior upon and influenced by geographic space

The list is far from being completed but long enough to show how diverse the interests of geography can be, to the point that from this list it is difficult to look at geography as a single discipline unless some communalities can be found. The quite early separation between human and physical geographies is just the most extreme case of such disparity, but some of the current approaches in human geography can be as far apart from each other as human from physical geography.

Also, from the list it is clear that no particular primary objects of study, specific to geography, are defined or implied, apart from those highly abstract like places, landscapes, regions and spatial structures or systems, which are in fact organizational mental constructs. The importance of this lack of definition can hardly be over-exaggerated, as it will be seen.

Finally, considering the whole propositions of each school of geography built around each of those subjects, it must be noted that in geography as soon as one moves from description to explanation (the ultimate goal of scientific work), the sources of explanation lay outside the 'objects' of geographic inquiry, in the nature of the phenomena studied either environment or society. The only exception, as will be discussed below, was the attempt to assign causal power to space itself by the early proponents of geography as spatial science.

Looking positively at communalities, however, shows a substantial agreement on several questions. First, on the domain of study, which is almost without exception restricted to the Earth's surface and thus to whatever might be, and was, referenced on the Earth surface. Second, there is agreement that the focus of geography is in the spatial aspects of phenomena which occur in that domain and only on the spatial aspects pertinent to that domain. Thirdly, both phenomena and spatial aspects are of interest to geography in so far as they are relevant to the life of human beings, which greatly restricts and specifies the scope and domain of geographical inquiry, though this assumption can be and usually is weaker in physical geography. Fourth, it follows from the previous that geography's main ability is to integrate —to relate— the phenomena over and to the space where they occur —take place—; whether this must lead to a stress on phenomena or places is far from general agreement. Similarly, it is far from clear the explanatory status of such integrative task or, in other words, how far one must proceed in search of explanation of those relationships and where to look for them.

Although a formal definition of the general subject matter of geography is still needed, these common traits provide the framework or context of geographic study. Strictly speaking, they do not yet provide the contents of the geographic domain —that is, its objects— but they allow us to define its bounds, upper (the whole Earth's surface) and lower (the smallest piece of the Earth's surface) in which human beings carry out some activity or which can influence those activities. At the same time they give some indications of the nature of such domain: it must be seen as a space or a set of spaces meaningful to human's activities.

This last point allows, of course, a wide range of interpretations and it could explain the astonishing diversity of scopes that geographical work shows. However, even if it seems difficult to cope with such a wide range of criteria of meaning, and the number of objects of geographical inquiry can be seen as an almost infinite or open-ended set, nothing prevents that such domain be specified and handled unambiguously and formally, provided that the criteria of relevance to humans is specified in each situation.

To go any further regarding the actual contents of the geographic domain —or geographic space— two crucial questions must be examined: the concept of space (what is it, or what humans think it is), both in general and in the context of the geographic domain; and the definition of geography's objects of study in light of such discussion.

3. Space as Geometry

Perhaps the most widely accepted conception of space is that of a container or framework where things exist. It is so accepted because it agrees with what people have been taught in school and with what people expect in every day experience —both in common life and in technical work such as engineering, architecture or planning. People can put things in some space or place, can move them along that space from place to place, and thus got the idea that space is something that is there before things begin to fill it. Something that exists regardless of things and previously to their existence: it exists even if empty.

One of the formal representations of such conception of space —again the most widespread and most commonly taught— is that of euclidean geometry. Euclidean space is empty of any physical concretion of matter, it must be seen as the set of all possible places —the whole container— and, because it has to provide for all possible places, it is understood and defined as infinite and continuous, allowing for infinite places even in a bounded range (i.e., between any two given places). In this view the objects that populate and constitute euclidean space are dimensionless points. Particularly arranged clusters (such as lines, surfaces, etc.) can be made out of these points, which can be judged to a certain degree as useful and appropriate representations of real world (material) objects.

Traditionally, euclidean space has been thought of as being three-dimensional in order to represent the three spatial dimensions commonly identified and experienced in the physical world. In classic euclidean space, therefore, points are defined by three numbers or coordinates each one expressing —or measuring— a point's position along each one of the three dimensions with respect to an absolute origin or center (a point in itself). Mathematically, euclidean space is said to be the result of the Cartesian product of the real numbers set by itself a certain number of times, over which a particular function of distance is defined. This generalizes the classic three-dimensional euclidean space to *n* dimensions, since the properties arising from that function of distance hold well beyond any number of dimensions and given the fact that a space with more than three dimensions may not have physical real world meaning is irrelevant in mathematics.

For centuries classic euclidean geometry was the only geometry available and imaginable. With the development of modern geometry, since the 19th Century, a number of spaces other than euclidean have emerged, among which were hyperbolic (Lobachevski), elliptic (Riemann), space-time (Minkowski), non-metric spaces such as topologic spaces, and others. Classic "Euclid's space" has become just one of a family of euclidean spaces, that defined on R^3. Up to this point geometry was seen as the description of physical space, from there on the definition of geometry had to be restated, as Klein did in 1872 in the *Erlangen program*, as the study of invariances of a set of objects —a space— subject to a group of transformations, in order to cope with all possible known geometries. Thus, it was made explicit the rejection of any sense of description or intuition of real world space in the nature of geometry, just like in any formal system like mathematics or formal logic, where the notions of truth and correctness are only internal, considered in relation to the axioms and rules upon which the system is built.

Some of the implications about the concept and nature of space arising from the new concept of geometry are extremely important. First of all, it should be clear that the objects that form the spaces studied by geometry, whether they be dimensionless —indefinitely small— points, vectors of three or more dimensions of real numbers or of any other set, or simply elements of abstract sets, have no real, physical, sensible existence.

Geometry, therefore, has never been a description of physical space, even if in the past that was seen explicitly as one of its main purposes. What geometry provided and still provides are formal structures that can be used as representations or tools for representing physical space, provided that a satisfactory mapping between both domains can be established. Such mappings should ensure correspondence between the set or sets of real world objects with the set or sets of objects underlying a particular geometry, and the operations that structure each of them respectively. Finally, and most important, this kind of proposition is made possible precisely because of the evolution followed by modern geometric thought on the construction of the concept of abstract space, which is considered as one or more sets of objects together with systems of relations between these objects (Alexandroff, 1961).

At this point, the question of whether the nature of the objects matters for the definition of a space, or rather if it depends only on the relations among them can have different answers. For geometry, as Alexandroff (1961) points out, the nature of objects is irrelevant because a space is actually constructed by devising particular kinds of relations regardless of their potential meaning and are those relations —the structure they provide— that gives a space its distinct properties. On the other hand, for disciplines dealing with real world phenomena, the nature of objects really matters because the kind of relations present among objects will depend on the properties of those objects. However, leaving aside what the answer should be, and it seems that both answers are appropriate depending on the purpose, the most important is that throughout the maturing of geometry a new concept of space, as a mental category, emerged: one that does not see it as an empty container —a sort of preexistent medium— but as a set of objects with their relations.

The fact that the concept of space as a void is still so accepted, and consequently the acceptance of euclidean geometry as the description of physical space, has to be related to the prevalence of certain philosophical views of old physical theories since, after quantum mechanics and relativism, modern physics offer a view of space that does not agree with that conception either.

4. Space in Physics

The number of reviews of the concept of space in physics (e.g. Jammer, 1954; Reichenbach, 1957; von Fraassen, 1970; Hawking and Ellis, 1973; Nerlich, 1976; Friedman, 1983; Sklar, 1985; Bergmann and Sabbata, 1986; Penrose and Isham, 1986) is large enough to try to add a new one. However, as some key conceptions of space in physics and philosophy are often quoted in geographical and GIS literature as foundations of concepts of geographic space and data models for such space, they need to be briefly discussed.

As Couclelis and Gale (1986) pointed out, the discussion about the nature of space organizes frequently around dichotomous oppositions. Particularly important in the history of philosophical and physical concepts of space, and most often used in geographic debate, is the opposition between absolute and relative space, and between space as attribute of things or as existent in itself so that things can be seen as attributes of space, which is a consequence or a particular form of the absolute vs relative opposition.

These views of space, absolute and relative, were already sustained as opposite by ancient Greek philosophers. Noteworthy examples are Democrit (the void must exist to allow for movement) and Parmenides (since what is nothing can not exist, the void does not exist). Aristotle's conception of space, often seen as one of the most influential in western thought, was in a sense relativistic by according to space the quality of attribute of objects, throughout the distinction between substance and form —space being one of the forms to be considered— but reintroduced the absolute notion by giving to space (and all other forms) a metaphysical existence, given that only substances had physical existence.

The most influential concept of space ever developed, to the point that it is still the prevalent one, has to be attributed to both Newton and Kant. This concept supposed the definite reification of absolute space as a reality, both physical (the universal empty container independent of matter and time, infinite and continuous set of points) and philosophical or logical (the *a priori* condition of any experience, not a concept derived from experience, nor something attached to matter). To discuss which enunciate had a stronger effect in subsequent scientific and philosophic thought, or to what extent have they links to Aristotelian ideas, is out of the scope of the paper. Most likely both reinforced each other and fused as one of the most common assumptions in scientific culture, since both provided the same disembodied, immaterial, timeless view of the nature of space, and of its *a priori* existence. Given that the transcendent character of Kantian space was never fully supported by the later idealist philosophy, Kant's absolute space remained as some sort of logical *a priori* super-category that had its 'material' counterpart in physical immaterial Newtonian space. Thus, as far as physics is concerned, it is quite obvious why euclidean geometry might be taken literally as the description of such immaterial, however physical, space.

As is well known, since the late 19th Century a series of empirical discoveries and theoretic elaborations had almost completely changed the face of physics, not to mention the whole science. To simplify, and with regard to the concepts of space and time, after Einstein's theories of relativity, both special and general, space and time can not longer be considered as absolute and separate. They are relative in that there cannot be absolute measurements of space and time, because any observation depends on the speed of movement of the observer. For some authors (e.g., Hawking, 1988) the concept of relative space was already implied by Newtonian physics because of the assumption of no prefered state of repose, and therefore the concept of absolute Newtonian space was due to Newton's reluctance to accept the notion of relative space because of its religious implications, ultimately a 'metaphysical' argument. But space and time are not only relative to movement —because of the constant value of light speed— but also to matter and energy (because of the influence of matter in the shape of the space-time).

The general theory of relativity was far more influential than special relativity in changing the well-accepted concepts of absolute space and time because it forced us to assume that energy and matter create their own space and time; only physically possible, since no space or time exists out of the material universe. Thus, space-time can no longer be separated from the objects that constitute it and any modern definition of physical space has to consider space as a collection of material objects (since the objects themselves are the space).

The notion of objects is particularly important at this point because it suggests clearly that space should be conceived as discrete, something that agrees with quantum mechanics, given that matter and energy are ultimately discrete.

5. Geography's View of Space

The concern about the conceptualization of space seems to be undoubtedly at the root of geography. Many authors have stressed its importance as one of the fundamentals of geographic research. Perhaps one of the best statements about its significance for geography is that of Harvey (1969, 191): "The whole practice and philosophy of geography depends upon the development of a conceptual framework for handling the distribution of objects and events in space". Harvey's point is remarkable, because it makes clear that such conceptualization is far from reached, even in its basic sense. It could be argued that what lacks is a consensus about which one of the many concepts of space is best for geography (or for geographic space). However, as the following review aims to show, it is not only a matter of agreement but of development, since none of the concepts has proven really fruitful.

As Couclelis and Gale (1986) indicate, there is not only one space, but a multitude, or at least a multitude of ways to look at space, and they list a hierarchy including pure euclidean, physical, sensorimotor, perceptual, cognitive and symbolic space. Probably all them are relevant to geography but here discussion will be confined to the lower 'more objective' levels, particularly because most of the past reflection on space in geography remained at those levels; and also because higher levels are perhaps not fully defined or differentiated yet.

Without doubt the most complete earlier reflection on space and its significance to the nature of geography is due to Richard Hartshorne (1939, 1959). Hartshorne's strict adherence to Kantian absolute space is explicit and paradigmatic. Even more, Hartshorne's definition of the nature of geography agrees with Kant's definition of geography, as Hartshorne insisted to view Geography as a science of space from Kant and Humboldt to Hettner. Hartshorne's contribution to the settlement of geography as a scientific discipline will never be appreciated enough, since it led the way to theoretic development. Otherwise, the concern about space in the nature of geography, though well understood in German and French early schools of geography (the strength of Ritter and Vidal on geography as the science of places and landscapes, not of man, must be remembered), would not have been systematically developed. However, because of the transcendent concepts of space and time, separated from matter, in Kant's thought, which lead to a different —ideographic— status of geography —and history— compared with that of proper sciences —nomothetic— dealing with substantive subject matters, Kantian and subsequently Hartshornian geography was deprived of generalization and explanation power, just concerned with the differentiation and description of spatial arrangements —chorology— considered to be unique, as much early regional geography did.

Despite this, the basis for theoretic geography was already set. In fact, as several authors have shown (e.g., Johnston, 1979, Gregory, 1982a), the critique of Schaefer (1953) just inverted the goals of geography from the previous ideographic 'orthodoxy' to the production of laws about spatial patterns, but relied on the same concept of absolute space, as proved by the morphological —not explanatory— character of such laws, and in the same —though reinforced— idea of space as the subject matter of geography.

The merits of Schaefer's critique of exceptionalism and those of early development of the 'new geography' are undeniable in that they introduced and consolidated the scientific method in geography, although, as many have criticized it, was introduced hand-in-hand with positivism leading to the identification of one to the other (e.g., Johnston, 1986). More important for what is discussed here, they contributed decisively to put space at the center of geographical research and debate, even though with a Kantian view or, in many cases, with a Newtonian one because of its stronger sense of a physical and empirical reality (a quality highly desirable for scientific and/or positivistic approaches).

Thus, as soon as concern about space grew, critiques of absolute space in geography began to arise, even parallel to some of the main formal enunciates of geography as spatial or locational science, heavily morphological and geometric in nature (Bunge, 1962; Haggett, 1965), when not deterministic (Stewart, 1950; Stewart and Warntz, 1958). With no doubt the best, probably still unmatched, of those criticisms was Blaut's (1961) celebrated article on space and process. His demolishing critique to both Newton and Kant's concepts of absolute space in the light of modern physics and philosophy, and his claim for relative space fused with time and matter into what are ultimately processes, has been widely acknowledged, at least nominally, even from positivistic standpoints (e.g., Harvey, 1969).

However, what seems to have escaped subsequent debate is that there are many more questions than answers in Blaut seminal work. To use his own words: "Deprived of the glamour of metaphysics and the false imagery of Newtonian atoms-in-the-Void, space becomes a most unprepossessing word. It labels certain purely mathematical notions, certain operations of measurement, and certain aspects of or relations among processes. Mathematical spaces and distance-measurements are the property of all science. If, therefore, the word space is to designate any peculiarly geographic concepts, these must be concepts of process. *But can any such qualify as our basic organizing concept?*" (Blaut, 1961, 5, emphasis added). After examination of some possible concepts —space as scale (or resolution or extent) of integration; space as intervening ground-processes (those between objects forming an integration); and space as cross-sectional relation of simultaneity within and between integrations— the conclusion is that "*...we are still without a spatial notion which qualifies as a suitable basic organizing concept.* The writer can think of no other candidates." (Blaut, 1961, p. 6, emphasis added)

Subsequent revisions of the concepts of space in geography started with the acceptance of relative space, but it has not been substantially further developed. Some authors (e.g., Bunge, 1962; Harvey, 1969; Abler *et al*, 1971; Haggett, 1979) tried to accommodate relative space into the morphological or geometric pursuit by looking for development of metrics and scales of distance measurement appropriate to the phenomena at hand (e.g. travel time, cost distance, etc). The insufficiency of such approach is quite evident since, without a full definition of the objects considered and of the significance for those objects of the space measured, it easily leads to a treatment of space separately of the objects that constitute it, no matter if with a different metric (one of the best examples is the distinction between spatial structure and spatial process (Bunge, 1962). After this, absolute space is more a matter of separation of space from matter than of a particular geometry (e.g., there is nothing intrinsically wrong in euclidean geometry; what is wrong is to take it as a physical reality). Spatial constructs are thus again reified by taking them as objects with attached 'thematic or substantive properties'.

To many others, taking seriously the need for explanation in geographical work, and consequent with the view of space as process, geographic inquiry had to be (and was) redirected to the processes underlying spatial structures; particularly to the social processes (e.g., Harvey, 1973; Sack, 1974, 1980; Gregory, 1978, 1982b), but also to those of human perception and behavior (e.g., Hägerstrand, 1953; Pred, 1967, 1969; Olsson, 1969; Golledge *et al*, 1972; Bunting and Guelke, 1979). In many cases, however, this did not move away from any specific geographic enquiry as no proper objects or subject matter for geography can be identified. This could be a reasonable point, but since it seems that space makes a difference there is need to look for ways to cope with it.

Very few have taken the abstract concept of relative space —a system of relations amongst a set of objects— literally to the point of considering that "any relation defined on a set of objects creates a space, with the further corollary that space cannot be regarded as the exclusive preserve of the geographer" (Gatrell, 1983, 4). Although this corollary is absolutely true, and the NATO ASI (upon which this book is based) is a good example of it, it is not less true that the geographer is not concerned with any kind of space, or with any kind of relationships between objects as Gatrell aims to show. Despite of the undeniable correctness of the starting premise about the nature of a space, the key point for geography —as well as for any discipline dealing with any space— is the definition of the universe or the set of objects that create this space. In this regard no answers are provided except a warning about the need for being careful in defining the sets of objects geographers would wish to study.

6. The Geographic Individuals

It seems that the ever-present need for defining the specific objects that geography studies is even more pressing after having looked at space as the prime subject matter of geography. All hopes put in space in order to avoid a search for specific geographic objects seem to have finally come down, as the debate on concepts of space in geography shows that no further specification of geographic space is possible unless the geographic objects be defined. The number of geographers who have addressed that question is, not surprisingly, far less than those who have written about space. David Harvey concluded his *Explanation in Geography* by recognizing that the crucial question still unsolved was the definition of the domain of geography: what are the geographic individuals, what the geographic populations or sets, what are the appropriate scales of measurement. The explicit treatment of this question is still relatively ignored, even after we acknowledge previous surveys devoted to the search of formal languages appropriate to determine spatial distributions, the morphometric laws that rule them, and the underlying processes and the laws that explain such processes. One may even wonder how these goals could be achieved without taking into account the nature of the objects spatially distributed and what the processes are which produce such distributions. Anyway, as a survey, it gives a measure of how few authors had previously addressed the question of geographic individuals (mainly Wilson, 1955; Dacey 1964; Grigg, 1967).

Harvey's treatment of the question tended to avoid explicit answers but rather to clarify the different criteria for defining individuals. Faced with the dilemma of what is a geographic individual (a farm, a farmer, a temperature measurement at some point?), he delays the answer by pointing out that if all may be legitimately considered geographic individuals then the problem is that there are individuals of different nature: some are discrete, and some continuous. Subsequently, he extends a little further in showing some other differences of this kind, namely that some individuals can be identified substantively (by a set of attributes) while others spatially (by a set of spatial, or space-time, coordinates); some are natural (like farms, countries, lakes or rivers) while others artificial (just spatial units of measurement, especially well suited for continuously varying phenomena); some singular and some collective. However there is no further discussion to elucidate all these alternatives (quite familiar in GIS literature about data models, by the way) nor to integrate them in a comprehensive framework. Not surprisingly, the question is considered mainly when dealing with data collections, and as such its treatment is essentially pragmatic, taking all supposed geographic individuals as possible and warning about the need of being aware of such differences in nature in order to choose the appropriate methods to handle each case and in order to not mix them inadvertently.

The only unambiguous statement about geographic individuals in Harvey (1969) is that they can be points, lines, areas, volumes or four-dimensional (space-time) volumes; and that they are best ordered and expressed by means of the so-called geographical data matrix (Berry, 1964; Chorley and Haggett, 1967), which according to Harvey combines substantive and spatial criteria in a single ordering. In fact, it must be said that this elements-observations matrix (or cube, if time is added) can only be called geographical in that their elements are assumed to be geographical, since for the rest it is just a plain data matrix as used in any other scientific domain. Actually, what makes it different is precisely the fact that their elements —the geographic individual— are ill-defined. As it has been seen, they were and are mainly conceived as generic locational entities or places (no matter if points, lines or areas), most commonly as areal *spatial units*, a sort of artificial and neutral 'geographic atoms' from which to build spatial structures. On the other hand, despite the claims that the geographic data matrix reflects the very nature of geographic data, it does not express any spatial or other relationship between the individuals, thus not conveying any spatial information and therefore failing to accomplish what was supposed to be its ultimate purpose. Even if their elements were genuinely geographic and correctly specified, the nature of this data matrix would remain substantive. To alleviate what has been seen as static information another kind of data matrices, complementary to the former, have been proposed (e.g., Johnston, 1978): those of interactions or flows, which can be generalized to any kind of data about relationships between pairs of individuals, although no attempts to combine both have been made. However, no matter if alone or integrated, matrix arrangements are just a registering or record keeping tool. There is nothing in them that grants a correct or reasonably good representation of the nature of data. Matrix or tabular arrangements are well suited for both statistical analysis and relational processing of substantive data, but because geographic data is not only substantive they do not fit well in tabular data structures alone.

It would be comforting to say that this is past history. Perhaps it really is; since the late sixties many geographers have concerned themselves with particular real processes (social, economic, perceptual, etc.) and hopefully have found their relevant objects and spaces. Maybe even they are not necessarily geographical, since perhaps there is no need for geography as a science but rather for a geographic education, as some authors have suggested (e.g. Gould, 1979; Bartels, 1982). There are good reasons, however, to keep looking at the problem of geographic objects, not the least important the debate in the GIS field about data models, since it revolves again and again around most of the questions mentioned. On the other hand, no geographic education could be provided if there is not a minimally consistent understanding of what geographic means, even if it changes over time.

As already said, there have been very few substantial contributions to the definition of geographic objects in the last twenty years. One of the outstanding exceptions, however, has been Chapman's (1977) explicit claim about the lack of entification in geographic work. Starting with the concept of object —"something which can be observed as a discrete entity by some perceptual mechanism, operated by man"— of which real world is made, and considering the benefits for any discipline to have defined —or attempt to define— its objects of study, as it often is the case, Chapman differentiates three kind of objects: the first order object of study or structurally integrated object -the proper object-; the areal aggregate object -often perceivable, as for instance wood-; and the non-areal aggregate object -a class intellectually constructed.

According to this author, proper objects must have the properties of wholes -structural integration-, and must be given in order to be experienced. Thus, first order objects are the only true objects. They have properties encompassed by taxonomy, morphology, physiology, ecology, chorology, chronology and composition. If something does not have all those properties then it is not an object. Objects may be made of more elemental objects and may also enter to form more complex ones. Heavily inspired by ecology and systems theory, Chapman views real world objects arranged along a hierarchy of levels of complexity or integration (from atoms to ecosystems) and points out that the difficulties for identifying objects lay at the intervening levels, above organisms -plants or animals- and below ecosystems.

In this way it is clear that space or pure spatial 'objects' —such as locations, places, areal units or regions— cannot be the objects of geography, since they are not objects at all. Clearly, according to the concept of relative space, Chapman states that space is a property of the objects, one that must have a different value for two objects of the same kind to be different.

The range of the objects of geography, given its concern with human beings, will become valid after the persons level, with households, farms, factories, shops and the like, as they are control points in the social human system. Settlements (towns, cities, etc.) could be the next higher level of geographical objects, but at this point the author expresses serious doubts about the morphological and performance properties of such candidates in relation to the whole economic and social system, which is -at least- nation-wide if not international or even global.

Chapman's work was not concluding at this point, since its main purpose was to show that "object and system come first and space comes second", somehow paraphrasing Hoffmann (1959): "These electrons and other fundamental particles, they do not exist in space and time. It is space and time which exist because of them." Because Chapman conceives of geographical objects as systems, it focuses further attention to the formal specification and study of systems.

To summarize, even if they are too restricted in the human domain (e.g. environmental systems are not included, perhaps judged as the domain of other disciplines), or too demanding in the requisites for qualifying as an object, or too biased by a systems approach, Chapman's proposals on the objects of geography should be considered a major breakthrough, in that they make full sense of the concept of relative space. It would be worthwhile to expand upon his ideas. Ironically, his work is much more known as one on systems theory in geography than on the subject of geography. A significant example of how far Chapman's ideas have been further developed is a list found in a quite recent work by Wilson and Bennett (1985). According to these authors, the entities that make up the subject matter of human geography are: people; organizations; goods and services; commodities; land; and physical infrastructures.

7. The Space Humans Interact with and Think of

A note of caution must be voiced about the following comments, since they are merely personal impressions born of very limited knowledge of the field of cognitive science, and particularly of cognitive semantics.

Interest in the cognitive approach in geography is not a new one as it came soon after interest in spatial or environmental behavior (e.g., Gould, 1973; Gould and White, 1974; Downs and Stea, 1973, 1977; Moore and Golledge, 1976; Kuipers, 1978). Whether this leads to the concept of a cognitive space different from that of a perceptual space (Couclelis and Gale, 1986) is not the point here. In the GIS literature it has not been until recently (Mark and McGranaghan, 1986; Mark *et al*, 1987; Mark, 1987, 1989; Mark and Frank, 1989, 1990; Frank and Mark, 1990) that it has received substantial attention.

The concern here with cognitive approach is rather *sui generis*, since it will be used as additional evidence of the need to deal with space in terms of discrete, embodied, objects. Therefore, the point discussed is not what concepts humans produce about space or how they produce them, but what general outlook of space arises from those concepts and from the way spatial concepts are used. Quite strange indeed, since it seems contradictory to try to gain evidence about an objective, real world throughout something that could be termed subjective (in that it involves subjects, not of course understood as arbitrary). It may be argued even that this is a futile enterprise since what can be known about real world, assuming it exists, will be always mediated by human, or other sensitive subjects, and then what ultimately will be known is what those subjects believe —in the most broader sense— to be the real world. The whole scientific enterprise has precisely no aim but knowing about the real world, and since quite long ago it has accustomed itself to the shortcomings due to the fact that humans —scientists or not— are not God. Thus, the mediated, provisional, and defective character of any knowledge is assumed from the beginning. Hopefully experience comes every day to help in correcting former ideas although the truly right ones will never exist.

Therefore, the contradiction between goal and means in the present case is only apparent. Rather, it expresses a note of realism by recognizing the importance of taking into account the way subjects view their real world. This does not imply, on the other hand, that any knowledge about real world must stem from the study of subjects' experience and thinking, and that this is the only matter worth being studied. Since any knowledge is mediated by subjects, then it is rather a matter of degree of abstraction to talk of knowledge about any real world or of subjects' knowledge about a real world; and it will depend on the topics of interest to the researcher to place his/herself toward to one or another end of the abstraction range.

So far in this paper, abstraction has been made of human cognition of space. Now it seems useful to look at it in order to see if very primary low-level concepts or mental constructs about space, common to all human beings, prove to be consistent with those highly abstract of scientific thought (nothing surprising after all because science is human made and it may well be that it simply reflects basic human thinking or basic human abilities to think about things). If both agree some more evidence will be added in support of scientific theory. If they do not agree, a theory will hold depending on whether scientific ideas are based on observations unaccessible to common human experience, as it happens in the case of physics versus naive physics for instance, or otherwise shall have to be revisited.

The case of findings in experientialist cognition (Talmy, 1983; Lakoff and Johnson 1980; Lakoff, 1987; Johnson, 1987; Langacker, 1987), however, is particularly valuable for this purpose because it shows that human thinking about space is to a great degree grounded in the bodily and sensory-motor nature of human beings, and therefore the notions of space arising from these basic capacities are most likely unbiased by existing elaborated ideas about space.

Of course the experientialist cognition approach to cognitive semantics goes well beyond establishing a number of notions of space. It accounts for what meaning is to human beings, and for the kind of mental constructs and mechanisms that humans use for reasoning and producing knowledge, many of which are innate and in many cases based on spatial-like notions and operations. However, since the issue here is not how human beings acquire knowledge and reason about space or using space, but how they characterize space in doing so, these further implications will not be treated here.

A final caution has to be voiced concerning to the irreconcilable criticism that experientialist cognition makes to what is called objectivist metaphysics (also objectivism or metaphysical realism). This critique has as its main sense a rejection of a particular theory of meaning (as association of arbitrary meaningless symbols with real things) and of reasoning (as algorithmic manipulation of symbols). Still, some of its aspects are relevant to all scientific and philosophical thought. The assumption is that reality has a mind-free structure which can be modeled by set theory. It will not be discussed here so as not to proceed endlessly, but rather it can be argued that in contexts other than contrasting theories of cognition a complete rejection of such assumption is not so necessary nor convenient, despite how valuable it can be to explore alternative ways to model reality. Moreover, as with using any formal system, use of set theory to model reality has to be decided in the light of the results achieved, since no identity between reality and the formal system is, by definition, predicated. In any case, experientialist cognition criticism is timely to remind us that not everything can be solved by using today's omnipresent set theory, at least it has to be proven in any particular case.

Although perhaps not completely according with experientialist cognition philosophical view, its findings about space can be brought in a rather speculative way in support of a conception of space as made of discrete objects. Among those 'evidences' there is the fact that human beings as sensorial bodies interact with objects perceived as bodies, that is discrete material concretions. Secondly, it is from that bodily nature —or perception of nature— of both people and objects that many image-schemata, regardless of their image nature and ways of being operated, have a strong spatial flavor or component, either static or dynamic (process-oriented) such as container, path, link, part-whole, contact, entity-multiplicity-mass, near-far, figure-ground, force dynamics, full-empty. Conversely it follows also that image-schemata refer to space as relationships between embodied objects, essentially topological —even in a broader extent than it is formally understood— as no metrics or quantification is usually involved. Thus there are not direct notions of infinite or continuous or dimensionless, nor is there a generic notion of location, since it is implied by each particular relationship amongst objects. Particularly interesting are notions like reference frame or the figure-ground contrast, as they express a sense that every object constitutes a space, and the sense that space is not something to be exhausted since it allows for other objects to be in it. Also interesting is the possibility of building complex objects as the transition from entity to multiplicity and to convex hulls.

It seems that as a consequence of the embodied nature of objects there is no notion of overlap, and this is a particularly interesting point since the concept of GIS as analytical tool relies heavily in the ability of systems to successfully perform spatial overlays, thus showing that what has been judged one of their major strengths could be in fact a major shortcoming due to the way they model geographic reality.

Finally, the whole experientialist cognition theory supports a relative concept of space not only because of its actual spatial content but, also, and perhaps more important, because it suggests that identifying or building spaces is one of the most basic and fundamental mechanisms and abilities to produce and structure knowledge, since it provides for developing systems of relationships. Most significant is not how cognition helps us to understand space but how space helps in cognition.

To conclude then, a generalized concept of space could be given by stating that space is a sense of relationship and identity. A way to assess the existence, identity and change of matter and energy, which is itself discrete and embodied.

8. What GISs Know about Geographic Space

GIS is at least as much a matter of technology than of science and therefore theoretical issues have emerged only after a long time. Most of the relative merits of data structures were discussed in the past in relation to computational efficiency, either for storage and retrieval or data processing (see for instance, Dutton, 1978), as well as in relation to their suitability for particular operations, which were not trivial problems in early times.

The term data model itself appeared quite recently (e.g., Peuquet, 1984) and it has subsequently coped with the attention of researchers marking a progressive shift of concern from implementation to conceptual issues: from how to arrange data to what is represented in data and what should be represented and how it has to be done.

Since GISs were born in the process of automation of cartography their first goal was to reproduce inside the computer the graphical or spatial nature and contents of maps, and to provide for some of the operations of analysis or presentation commonly carried out with maps. They were set in particular contexts of particular organizations to serve different needs of map data processing and according to different professional and scientific ways to think of geographical information and maps. Moreover, the earlier state of technological development made it difficult, either technically, economically, or both, to represent and process data of such nature, and it not was until the late seventies that substantial results were reached (see for a description of technological development see Tomlinson, 1984).

Two divergent ways to represent the graphical rendering of maps appeared almost since the beginning: square grid cell partitioning at a fixed resolution of the surface to be represented, such that each cell is given a coded value corresponding to the cartographic feature over which the cell lies; and vector encoding of series of pairs of coordinates corresponding to points that taken successively as starting and ending points of straight line segments approximate the shape of the boundaries of a given cartographic feature which is identified by means of a label or coded value.

Both methods, respectively known as raster (or grid) and vector, have a number of relative advantages and disadvantages for practical purposes, as well as some conceptual implications. Both had, and still have, a large number of proponents and have led to two separate families of data structures and to a number of what has been commonly named data models. At the risk to oversimplification, it can be said that in the beginning each of those methods reflected well the needs of two major groups of professionals.

Grid based GISs gained much support from the group best defined as that of 'map users' —like planners, geographers, earth scientists in general— that looked for a rather powerful analytical tool where generation of data was simple, easy and cheap, while sacrificing geometric accuracy and high quality graphic representations. In fact most early grid based GIS grew out of planning organizations or schools (e.g., Sinton and Steinitz, 1969; Corbin *et al.* 1972; Tomlin, 1975; see Tomlinson, 1984 and Tomlin 1983 for detailed surveys). This not withstanding, vector based systems are as old as grid or raster based systems, and they arose out of the group that can be labeled 'map makers', that is from cartographic or data gathering organizations, very much concerned with accuracy —both geometric and thematic— of data. Thus, some of the most conspicuous examples of early GIS were vector based, as is the case of Canadian GIS, which appeared in 1964 for land use inventory and is considered the first GIS in history; and the DIME system created as a geo-referencing tool for population census (Cooke and Maxfield, 1967).

Debate surrounding data structures and data models based on both kind of geometric representations occupied much attention until quite recently, but it has obscured the very nature of the problem that required discussion. This was, firstly, because it was conducted upon technical or pragmatic basis rather than a conceptual one. Secondly, the discussion centered around methods of digital representation of geometry or, to be precise, of cartographic elements instead of centering on data models properly. Therefore, even when implementation issues were progressively less considered, since optimized storage methods are largely independent of what is actually stored, it leaves the issue that each method of geometric representation constitutes in itself a data model based in a different conceptualization of geographic space, which is not necessarily true. Thirdly, representation of geometry has been taken as representation of geographic space, which is actually the relevant problem.

Conceptual reflection was almost nonexistent in early stages of GIS development. Some work, however, by Chrisman (1975, 1978) pointed very early to the center of the problem. Chrisman, by considering the development of concepts of space in scientific thought, showed that there are two opposite conceptualizations of geographic space in GIS, and in general: one that views geographic space as a set of objects having spatial properties, a view that relates to a concept of relative space; and another that considers geographic space a set of locations having thematic properties, which follows from a concept of absolute space. However, the identification he made of raster data structures with a location based concept of space, and of vector data structures with an object based concept can be considered misleading (also his claim of topological structures as an overcoming solution that leads to a true concept of relative space, should be considered not only misleading but unfounded since it reintroduces absolute space so far as objects are still abstract geometric objects).

Thus raster and vector data models have acquired, progressively, a status of conceptual models of geographic space that they do not really have, although their use implies some conceptual model, even unconsciously, as Chrisman (1978) pointed out.

Further reflection upon those questions by Peuquet (1984, 1988), although it gave an extended survey of data structures and data models used in GIS, has not provided significant advances. Rather, it has provided the consolidation of this false conceptual nature of geometric representation, since she presented both families of data structures and data models as logically dual —thus carriers of some conceptual meaning— and therefore with no possibility of a definite choice between both.

At this point we should mention the notion of data model, a term introduced to the GIS literature by Peuquet (1984). As it was proposed and widely used it refers to a model of what existing data are or how they are used, not to a more recent meaning of a model of what data should be in order to reflect a conceptual model of reality, as Ullman (1982, 12-17) notes: "A conceptual data representation may be defined as a general description of specific sets of entities and the relationships between those sets of entities. An entity is a thing which exists and is distinguishable, i.e. we can tell one entity from another." Influence of conceptual development of fields of information technology like DBMS has been decisive in the development of GIS theory, not only by the adoption of techniques and system architectures, but most important by using those concepts to clarify the kind of problems to handle. A view of GIS as a DBMS of a special (spatial) sort (Frank, 1984) has proven most fruitful as will be seen later.

In a later work Peuquet (1988) moves on to that direction but the actual contents of the model proposed does not in the slightest change her former conceptualization of geographic space, as the model presented is explicitly predicated as dual: with a set of entities and a set of locations, each one with their specific relations and scarcely integrated through a functional mechanism of mutual feeding. Furthermore, the necessity of such duality is predicated not only on the same basis as former work but also upon the dual nature of maps as images and structures, a claim that contradicts the nature of maps as representations of geographic space and as such their nature as structures. Thus, its somewhat cognitive approach through perceptual mechanisms and the way to reproduce them within computer vision looks inappropriate for their unfounded premises.

A move from map to database, from a representation of a document —itself a representation— toward a representation of the reality, follows from the view of GIS as a particular DBMS. One of the first conceptualizations of geographic space in GIS field not mediated by a previous representation of it, such as maps, nor by a digital representation of the geometry of such previous representation, arises from the work of Frank (1987, 1988), where "The notion of space... is essentially a finite or infinite collection of discernable objects together with a notion of neighborhood".

By this time a new, major influence from computer science came into GIS field, the adoption of object-oriented data models (Herring, 1987), which is a sort of consequent evolution from theory about data models in DBMS. Since then it can be certainly claimed that there exists an object —or feature— based model in GIS contrasting with a location based model one, since what was obscured along much of previous debate was that both raster and vector representations, which are only geometric representations, were used to convey the same conceptual model of geographic space, that location based, and particularly using the mediation of an intervening representation, the cartographic one.

It would be tempting to call this data model cartographic-based instead of location-based, because of such mediation, but the term location-based reflects best its nature of a concept of absolute space. Moreover, to term it cartographic would do a considerable injustice to cartographic representation of geographic space, which is far away from being based on an absolute concept of space. Cartographic representation of geographic space is intrinsically relative, as it models space with space, and has as one of its main aims to provide for an inventory of real world objects that includes their spatial properties and relationships.

It is not surprising that due to the nature of cartographic representation and to the cartographic origin of GIS, the development of these systems has provided new stimuli and fruitful insights to the question of geographic individuals, to a large extent stagnated in geography. First it cut across geographical debate —or lack of debate— since main concern of GISs with geographic data necessarily leads to deal, consciously or not, with geographic individuals and thus linked to the cartographic tradition of data gathering quite reasonably leading to a concern for accurate specifications of such individuals, if only for data collecting purposes. This has had a major effect of accustoming at least two generations of geographic data users to think of geographic space more and more in terms of objects, and to collect data for those objects, although still mediated by a notion of cartographic feature or geometric feature (many times objects are thought to be the points, lines or areas used to represent them on maps, and they effectively do have that punctiform, linear or areal nature). DBMS played an important role to reinforce the sense of having to deal with real world objects, since cartographic manipulation inside the computer could have maintained the sense of dealing just with maps or geometric representations. However, as DBMS became part of GIS, no matter how completely, it made unavoidable that dealing with collections of things and data, spatial or geometric data included, had to be seen as attributes of those things. Secondly, as this trend consolidated the need for a further and definite conceptualization of geographic space it has lead to review again the former and incomplete conceptual development that took place in geography, and thus GIS now provides geography with a hope for answers to its questions about the geographic domain.

Thus is why GIS are (or can be seen as) important for geography. Since the appearance of maps it is the first time that geography has had an instrument which is largely felt to be genuinely geographic, both in nature and in history of contribution to the development. The GIS strengthens the care for, and awareness of, the suitability and quality of data (regardless even of further theoretical implications); especially as they provide the opportunity to design and tailor the data, and to increase the detail or degree of 'entification' of geographical data, contrasting with its former predominant aggregate character —something that has been called the case of 'pointillism' vs 'percentagism' (Abler, 1987). This had a side effect already undergone of general diffusion and awareness of the value of geographic information, that is of geo-referencing of data, across disciplines and over all society, thus overcoming the conception of things without space, symmetrical to the absolute space conception. Finally, as mentioned, it can lead potentially to a further theoretic development about geographic space by acting as a theory proving tool.

Therefore, debate on the use of raster or vector representation should be conducted on the basis of their adequacy to represent the geometry that a conceptual model of space will require, and not on the basis of their supposed conceptual content. Said in passing, it seems rather clear that while vector geometric representation can hold in both location-based and object-based models, raster representation can only be used in systems grounded on location-based models because raster representation imposes a space too rigid to approximate real objects' geometric properties.

Current situation can be summarized as clarified by the agreement on the existence of two conceptual data models —location vs object (sometimes also called feature, a term to be avoided because of its mediative connotations)—, even if there is not a decision made upon which to choose because of the adherence to either the concept of absolute (e.g., Goodchild, 1990) or relative space (e.g., Frank, 1988, 1990). By now attempts to conciliate both is more a matter of pragmatism than of consistent conceptual development. In that respect, the proposals available are only apparently integrative since they are ultimately location-based. The model of dual functional components (Peuquet, 1988) actually restricts the spatial relations to the location side, which is viewed a a set of points —or, properly, of pixels— where thematic information providing from the objects' side can be anchored. A recent proposal by Goodchild (1990) is based on the complementarity between field view and object view of geographic space, both judged necessary though the only one empirically verifiable is the field view, and therefore the nature of geographic space is best stated as a set of tuples of thematic attributes corresponding to locations. Both views are called to form an object model, that has as objects those of field view (regions and segments) and those of object view (points, lines and areas). In this 'object' model, therefore, the best that could be achieved is the integration —by the way, a simple coexistence or juxtaposition— of all primitive geometric objects, and thus it is no more than location-based.

The two conceptual data models encompass former data models —understood as only data relevant—, such as regular tessellated —raster, grid and others—, irregular tessellated, unstructured vector ("spaghetti"), polygon-structured, topologically structured, geo-relational, topologically integrated, object-oriented (see Peuquet 1984; Rhind and Green, 1988; Frank and Mark, 1990). There is progressively more agreement on the presence of a cartographic mediation in the representation of geographic space in GIS, either considered as an unnecessary hindrance to be removed or as necessary (e.g., Peuquet, 1988; Goodchild, 1990).

It also seems clear that some authors (Frank, 1990; Frank and Mark, 1990; Mark and Frank, 1990; Herring 1989, 1990) increasingly would consider that geometric representation or any formal data model is not in itself part of the conceptual model or models of geographic space, but rather tools that will serve to formalize and subsequently implement the conceptual model. Quite obviously, however, not just any formal model will accommodate adequately any conceptual model.

Research is addressed, on one hand, to developing, and refining further, current or new formal models that can support current concepts of geographical space, which are somehow scattered or even conflicting given that there is not a well defined conceptual model. Foundations for it are provided mainly by what has been termed the object data model, but it has still to be fully specified and more decidedly accepted by the GIS community as the basic conceptualization of geographic space.

From another point of view, a search for new concepts of space and formalisms or models is actively being undertaken throughout the cognitive approach. This search aims to find means of modeling geographic space in a way closer to natural language or to common human understanding and interaction with space. Its direct application to the design of natural user interfaces and to the removal of linguistic barriers or biases for using GIS is immediately apparent (e.g., Mark *et al.* 1989). However, as has been suggested very recently (Mark and Frank, 1990) results from this search will be significant beyond the user front-end level and would enter in the structuring of GIS data models, either in substituting more traditional ones or in adding to the current collection.

A strong claim arising from the formal side of the research is the possibility of combining different formal models into an integrated system. Something workable and already feasible by use of multi-sorted algebras, most commonly known as object-oriented software construction techniques. Full adoption of the object-oriented approach in GIS also will allow us to go some steps ahead in the way of semantic data modeling, that is, in the way of actually modeling real world systems from inside the information systems instead of simply having a static representation, no matter if conceptually correct, of reality.

To conclude, the current state of development of general formal models is far beyond the current ability to fill them with meaningful content in order to model geographic space in GIS. Of course there is need for devising specific formalisms for geographic concepts, but this could be fairly straightforward on the basis of general formalisms once the concepts be known and fully specified. It can be argued that most likely a full specification of geographic space could require the construction of a particular formalism that allow us to do it, i.e. as Newtonian physics needed a new kind of mathematical calculus, or as cognitive semantics has need the development of new categories like image-schema (Lakoff, 1987). Therefore, it could be that both searches — conceptual and formal— must go hand-in-hand. However, it has to be clear that concepts should come first in order to make explicit what kind of problems the researchers are facing and what the formalisms are needed for. It is a personal impression, and thus the aim of this paper and of the proposal below, that conceptual research is lagging behind that of formalisms.

9. Geographic Space as a Set of Geographic Entities

Following the previous discussion, geographic space must be a relative space, and because the concept of relative space essentially means that objects are the space, the problem of defining geographic space or a conceptual model for it is actually a problem of defining and studying the geographical objects, their attributes and relationships.

According to modern physics —and also cognitive processes— reality, space, is made of discrete bodies. Therefore, geographic objects are, at least at the physical basic level, material discrete objects, not abstract locations nor primitive geometric objects like points, lines or areas. Objections usually made to the adoption of a relativistic physics point of view on space, or to the rejection of euclidean geometry are, to some extent, misguided. The point is not if gravitation effects are too weak at human scale for curving our space-time, nor if euclidean geometry might not give adequate measures of distances. It is, rather, a matter of changing the way of thinking of space; i.e. to cease thinking of it as an empty, infinite framework, something that unconsciously is already done in every day cognitive processes. It is to stop taking euclidean geometry for an actual description of physical space, since no geometry describes physical space. They help a lot in doing that, admittedly, provided that a mapping between physical and abstract geometric space can be established, a mapping that must reflect adequately the structure of physical space. But, precisely because there is nothing but a mapping between both spaces, one must not be substituted for the other.

There is something like a paradox in the fact that discrete objects are best measured using a continuous scale while continuous varying phenomena are best measured with discrete measurement procedures like sampling. But this is only apparent. In empirical work there is no such thing as infinitely continuous scales of measurement; they only exist in geometry. Empirical scales of measurement seem to be continuous because they allow for varying resolution by recursive partitioning of measurement units, but to a finite point only: that of atoms and elementary particles, as in the most refined measurement tools, like cesium clocks. However, when this point is reached the object or phenomena being measured will have disappeared and the observer will be faced with something completely different. Conversely, so-called continuously varying phenomena might be ultimately measured in terms of their elementary objects like water drops, and even better water molecules, in the case of rainfall. It must be kept in mind, therefore, that empirical measurement, just like any other cognitive action, is mediated by subject's abilities to interact and sense, and therefore no empirical input must be readily taken as an exact description of the phenomena itself. What cannot be seen with the naked eye may be seen with microscopes or telescopes. Thus, comparison of experiences and procedures often produce better results and lead to descriptions that are never literal observations. Otherwise, humans would still be thinking that the Earth does not move.

The same applies to geographic space and a fair draft definition of geographical entities could be the following: A part of the Earth surface defined by virtue of one or several spatial and/or non-spatial properties, which make it an space in itself, physically and/or functionally different of its surrounding environment, and therefore able of carrying out an individual existence persistent in time, and able of being perceived by human beings as such individual space and entity.

Use of the term entity is just a matter of convenience since it may well be termed object, but because of the wide use of the term object, especially after the appearance of the object-oriented programming methods, the term entity tells better that it is not only an object for implementation convenience, but a real one, and can serve to avoid confusion. Moreover, entity conveys best the sense of a real existent individual.

Therefore, geography and GIS have to deal with entities, all having spatial and non-spatial properties and relationships, but a fundamental distinction must be made between those (like people and other living organisms, organizations, or man-made artifacts) that in relation to Earth have location as the only spatial property, and those that allow for location of the former because of being a substantial extension (like rivers, islands, landforms, parcels, roads). Geographic phenomena (processes that occur over the Earth surface relevant to human beings) involve both non-geographic entities and geographic entities, and observations about both enter into what is known as geographic information. Geographic entities, those that form geographic space, those that allow for 'invariants' under geographic processes, have to be themselves pieces of the Earth surface (so far, Earth is the domain or space for such processes and the ultimate geographic entity), and they have to be distinguishable by their morphological properties and their actual or potential use —in its broader sense— by humans, since this is the criteria of pertinence in geography (mentioned earlier).

Geographic space can then be defined as a finite —but not fixed— set of such geographic entities having a recursive structure of partition and composition. The kind of formalisms, therefore, needed for modeling geographic space in GIS, and particularly for semantic data modeling, are algebras or geometries dealing with such entities. Clearly, what is needed are not formal models of spatial relationships among ideal polygons or lines, or topologic cells or simplices, but among rivers and lakes, or streets and buildings. Even if, admittedly, this may be a never-ending task, each step ahead will render better results than a never-ending search for ideal geometries always imperfectly matched by reality. It is assumed that semantic modeling is, by principle, a never-ending task (Codd, 1982), just as it is any scientific working, but nevertheless worthwhile to be undertaken.

Advantages in the practical domain would not be the least. Today GIS can identify a limited number of spatial relations like neighborhood, containment or overlap relations, and only after very expensive, blind searches and computations, even to produce trivial results quite well presumable or known in advance (David Rhind discussed in a recent lecture the example of how much effort a DEM-based GIS would need to provide a trivial answer for a query such how many points are higher than 400 meters above sea level in the Netherlands). More complex, or conversely very intuitive, queries simply cannot be asked.

Further steps in the analytical direction require first of all what could be termed a taxonomic approach, that is a systematic collection and specification of entities, their properties and relations. This will lead perhaps to a splitting of what is proposed now as geographic space in a number of interrelated sub-spaces, each one relevant to a kind of processes. It is too early to assess whether those entities will best be handled by set theory or not but, as biology shows, this is something just realizable after careful examination of a large number of instances, what is in essence a taxonomic enterprise, not in purpose but in procedure.

10. References

Abler, R.F. (1987) 'The National Science Foundation National Center for Geographic Information and Analysis', International Journal of Geographic Information Systems 1(4), 303-326.

Abler, R.F., Adams, J.S. and Gould, P.R. (1971) Spatial organization: the geographer's view of the world, Englewood Cliffs, NJ: Prentice Hall.

Alexandroff, P. (1961) Elementary concepts of topology, New York: Dover Publications.

Bartels, D. (1982) 'Geography: Paradigmatic change or functional recovery? A view from West Germany', in Gould, P. and Olsson, G. (eds.), A search for common ground, London: Pion.

Bergmann, P.G. and Sabatta, V. de (1986) (eds.) Topological properties and global structure of space-time, (Proceedings of a NATO Advanced Study Institute), New York: Plenum Press.

Berry, B.J.L. (1964) 'Approaches to regional analysis: a synthesis', Annals of the Association of American Geographers 54, 2-11.

Blaut, J.M. (1961) 'Space and process', The Professional Geographer 13(4), 1-7.

Bunge, W. (1962) Theoretical Geography, Lund: C.W.K. Gleerup.

Bunting, T.E. and Guelke, L. (1979) 'Behavioral and perception geography: a critical appraisal', Annals of the Association of American Geographers 69, 448-462.

Chapman, G.P. (1977) Human and Environmental Systems: A Geographer's Appraisal, London: Academic Press.

Chorley, R.J. and Haggett, P. (1967) Models in Geography, London: Methuen.

Chrisman, N.R. (1975) 'Topological information systems for geographic representation', Proceedings of Second International Symposium on Computer-Assisted Cartography (Auto-Carto II), Falls Church, VA: ASP-ACSM.

Chrisman, N.R. (1978) 'Concepts of space as a guide to cartographic data structures', in Dutton, G. (1978) (ed.), Proceedings of the First International Advanced Study Symposium on Topological Data Structures for Geographic Information Systems, Cambridge, MA: Laboratory for Computer Graphics and Spatial Analysis, Harvard University.

Codd, E.F. (1982) 'Extending the database relational model to capture more meaning', ACM Transactions on Database Systems 4(4), 397-434.

Cooke, D.F. and Maxfield, W.H. (1967) 'The development of a geographic base file and its uses for mapping', Proceedings of the Fifth Annual Conference of URISA, Washington, DC: Urban and Regional Information Systems Assoc.

Corbin, C.C. et al. (1972) IRIS/NARIS user's manual, Urbana, IL: Center for Advanced Computation, University of Illinois.

Couclelis, H. and Gale, N. (1986) 'Space and spaces', Geografiska Annaler 68 B, 1-12.

Dacey, M.F. (1964) 'Individuation in substance and space-time languages', Memo GSI, 1, Regional Science, University of Pennsylvania.

Downs, R.M. and Stea, D. (1973) (eds.), Image and the Environment. Cognitive Mapping and Spatial Behaviour, Chicago: Aldine.

Downs, R.M. and Stea, D. (1977) Maps in Minds. Reflections on Cognitive Mapping, New York: Harper and Row.

Dutton, G. (1978) (ed.), Proceedings of the First International Advanced Study Symposium on Topological Data Structures for Geographic Information Systems, Cambridge, MA: Laboratory for Computer Graphics and Spatial Analysis, Harvard University.

Fraassen, B. von (1970) An Introduction to the Philosophy of Time and Space, New York: Random House.

Frank, A.U. (1984) 'Requirements for database systems systems suitable to manage large spatial databases', Proceedings of the International Symposium on Spatial Data Handling, Zurich: IGU, Commission on Spatial Data Sensing and Processing.

Frank, A.U. (1987) 'Towards a spatial theory', in Aangeenbrug, R.T. and Schiffman, Y.M. (eds.), Proceedings of the International Geographic Information Systems (IGIS) Symposium, Washington, DC: NASA.

Frank, A.U. (1988) The concept of space and geometry in geography, Technical paper, Orono, ME: Department of Surveying Engineering, University of Maine.

Frank, A.U. (1990) 'Spatial concepts, geometric data models and data structures', in Proceedings of the Conference on GIS Design Models and Functionality, Leicester: Midlands Regional Research Laboratory University of Leicester.

Frank, A.U. and Mark, D.M. (1990) 'Language issues for Geographical Information Systems', in Maguire, D., Rhind, D.W. and Goodchild, M.F. (eds.), Geographic Information Systems: Principles and Applications, London: Longman. [in press]

Friedman, M. (1983) Foundations of space-time theories, relativistic physics and philosophy of science, Princeton, NJ: Princeton University Press.

Gatrell, A.C. (1983) Distance and Space: A Geographical Perspective, Oxford: Clarendon Press.

Golledge, R.G.; Brown, L.A. and Williamson, H. (1972) 'Behavioural approaches in geography: an overview', Australian Geographer 12, 59-79.

Goodchild, M.F. (1990) 'A geographical perspective on spatial data models', in Proceedings of the Conference on GIS Design Models and Functionality, Leicester: Midlands Regional Research Laboratory University of Leicester.

Gould, P. (1973) 'On mental maps', in Downs, R.M. and Stea, D. (eds.), Image and the Environment. Cognitive Mapping and Spatial Behaviour, Chicago: Aldine.

Gould, P. and White, R. (1974) Mental maps, London: Penguin.

Gould, P. (1979) 'Signals in the Noise', in Gale, S. and Olsson, G. (eds.), Philosophy in Geography. Dordrecht, Holland: Reidel.

Gregory, D. (1978) Ideology, science and human geography, London: Hutchinson.

Gregory, D. (1982a) 'Solid geometry: notes on the recovery of spatial structure', in Gould, P. and Olsson, G. (eds.), A search for common ground, London: Pion.

Gregory, D. (1982b) Social theory and spatial structure, London: Hutchinson.

Grigg, D.B. (1967) 'Regions, models and classes', in Chorley, R.J. and Haggett, P. (eds.), Models in Geography, London: Methuen.

Hägerstrand, T. (1953) [English translation (1968)] Innovation diffusion as a spatial process, Chicago: University of Chicago Press.

Haggett, P. (1965) Locational analysis in human geography, London: Edward Arnold.

Haggett, P. (1979) Geography: a modern synthesis, London: Harper and Row.

Hartshorne, R. (1939) The Nature of Geography: A critical survey of current thought in the light of the past, Lancaster, PA: Association of American Geographers.

Hartshorne, R. (1958) 'The concept of geography as a science of space, from Kant and Humboldt to Hettner', Annals of the Association of American Geographers 48, 97-108.

Hartshorne, R. (1959) Perspective on the Nature of Geography, Chicago: Association of American Geographers.

Harvey, D.W. (1969) Explanation in Geography, London: Edward Arnold.

Harvey, D.W. (1973) Social Justice and the City, London: Edward Arnold.

Hawking, S.W. and Ellis, G.F.R. (1973) The large scale structure of space-time, Cambridge: Cambridge University Press.

Hawking, S.W. (1988) A brief history of time: From the big bang to the black holes, New York: Bantam Books.

Herring, J.R. (1987) 'TIGRIS: Topologically integrated geographic information system', in Chrisman, N.R. (ed.), Proceedings of the Eighth International Symposium on Computer-Assisted Cartography (Auto-Carto 8), Falls Church, VA: ASPRS-ACSM.

Herring, J.R. (1989) 'A fully integrated geographic information system', in Anderson, E. (ed.), Proceedings of the Ninth International Symposium on Computer-Assisted Cartography (Auto-Carto 9), Falls Church, VA: ASPRS-ACSM.

Herring, J.R. (1990) 'The mathematical modeling of spatial and non-spatial information in geographic information systems', in Mark, D.M. and Frank, A.U. (eds.), Cognitive and Linguistic Aspects of Geographic Space, Kluwer Academic Publishers.

Hoffmann, B. (1959) The Strange Story of the Quantum, New York: Dover.

Jammer, M. (1954) Concepts of Space. The History of Theories of Space in Physics, Cambridge, MA: Harvard University Press.

Johnson, M. (1987) The Body in the Mind: The Bodily Basis of Reason and Imagination, Chicago: University of Chicago Press.

Johnston, R.J. (1978) Multivariate Statistical Analysis in Geography, London: Longman.

Johnston, R.J. (1979) Geography and Geographers: Anglo-American Human Geography since 1945, London: Edward Arnold.

Johnston, R.J. (1986) On Human Geography, London: Basil Blackwell.

Kuipers, B. (1978) 'Cognitive modelling of the map user', in Dutton, G. (ed.), Proceedings of the First International Advanced Study Symposium on Topological Data Structures for Geographic Information Systems, Cambridge, MA: Laboratory for Computer Graphics and Spatial Analysis, Harvard University.

Lakoff, G. and Johnson, M. (1980) Metaphors We Live By, Chicago: University of Chicago Press.

Lakoff, G. (1987) Women, Fire, and Dangerous Things: What Categories Reveal About the Mind, Chicago: University of Chicago Press.

Langacker, R.W. (1987) Foundations of Cognitive Grammar, Stanford: Stanford University Press.

Mark, D.M. and McGranaghan, M. (1986) 'Effective provision of navigation assistance to drivers: a cognitive science approach', in Blakemore, M. (ed.), Proceedings Auto-Carto London, London: Auto Carto London, Ltd.

Mark, D.M., Svorou, S. and Zubin, D. (1987) 'Spatial terms and spatial concepts: Geographic, cognitive, and linguistic perspectives', in Aangeenbrug, R.T. and Schiffman, Y.M. (eds.), Proceedings of International Geographic Information Systems (IGIS) Symposium, Washington, DC: NASA.

Mark, D.M. (1987) 'On giving and receiving directions: cartographic and cognitive issues', in Chrisman, N.R. (ed.), Proceedings of the Eighth International Symposium on Computer-Assisted Cartography (Auto-Carto 8), Falls Church, VA: ASPRS-ACSM.

Mark, D.M. (1989) 'Cognitive image-schemata for geographic information: Relations to user views and GIS interfaces', Proceedings, GIS/LIS'89, Falls Church, VA: ASPRS-ACSM.

Mark, D.M. and Frank, A.U. (1989) 'Concepts of space and spatial language', in Anderson, E. (ed.), Proceedings of the Ninth International Symposium on Computer-Assisted Cartography (Auto-Carto 9), Falls Church, VA: ASPRS-ACSM.

Mark, D.M. and Frank, A.U. (1990) 'Experiential and Formal Models of Geographic Space', submitted to Annals of the Association of American Geographers.

Mark, D.M., Gould, M.D. and Nunes, J. (1989) 'Spatial language and geographic information systems: Cross-linguistic issues', *Proceedings, II Conferencia Latinoamericana sobre la Tecnología de los Sistemas de Información Geográfica (SIG)*. Mérida, Venezuela: Universidad de los Andes.

Moore, G.T. and Golledge, R.G. (1976) (eds.), Environmental Knowing, Stroudsburg, PA: Dowden, Hutchinson and Ross.

Nerlich, G. (1976) The Shape of Space, Cambridge: Cambridge University Press.

Olsson, G. (1969) 'Inference problems in locational analysis', in Cox, K.R. and Golledge, R.G. (eds.), Behavioural problems in geography: A symposium, Evanston, Ill.: Northwestern University Press.

Penrose, R. and Isham, C.J. (1986) (eds.), Quantum concepts in space and time, Oxford: Clarendon Press.

Peuquet, D.J. (1984) 'A conceptual framework and comparison of spatial data models', Cartographica 21(4), 66-113.

Peuquet, D.J. (1988) 'Representations of geographic space: Toward a conceptual synthesis', Annals of the Association of American Geographers 78, 3, 375-394.

Pred, A. (1967) Behavior and location. Foundations for a geographic and dynamic location theory, Vol. I. Lund: C.W.K. Gleerup.

Pred, A. (1969) Behavior and location. Foundations for a geographic and dynamic location theory, Vol. II. Lund: C.W.K. Gleerup.

Reichenbach, H. (1957) The Philosophy of Space and Time, New York: Dover.

Rhind, D.W. and Green, N.P.A. (1989) 'Design of a geographical information system for a heterogeneous scientific community", International Journal of Geographical Information Systems 2(2), 171-189.

Sack, R.D. (1974) 'The spatial separatist theme in geography', Economic Geography 50, 1-19.

Sack, R.D. (1980) Conceptions of Space in Social Thought. A Geographic Perspective, Minneapolis: University of Minnesota Press.

Schaefer, F.K. (1953) 'Exceptionalism in geography: a methodological examination', Annals of the Association of American Geographers 43, 226-249.

Sinton, D.F. and Steinitz, C.F. (1969) GRID: a user's manual, Cambridge, MA: Laboratory for Computer Graphics and Spatial Analysis, Harvard University.

Sklar, L. (1985) Philosophy and space-time physics, Berkeley: University of California Press.

Stewart, J.Q. (1950) 'The development of social physics', American Journal of Physics 18, 239-253.

Stewart, J.Q. and Warntz, W. (1958) 'Macrogeography and social science', Geographical Review 48, 167-184.

Talmy, L. (1983) 'How language structures space', in Pick, H. and Acredolo, L. (eds.), Spatial Orientation: Theory, Research and Application, New York: Plenum Press.

Tomlin, C.D. (1975) The Tomlin subsystem of IMGRID, Masters Project, Department of Landscape Architecture, Harvard University. [unpublished]

Tomlin, C.D. (1983) Digital Cartographic Modeling Techniques in Environmental Planning, Ph.D. Dissertation, Yale University. [unpublished]

Tomlinson, R.F. (1984) 'Geographic information systems: A new frontier', Proceedings of the International Symposium on Spatial Data Handling, Zurich: IGU, Commission on Spatial Data Sensing and Processing.

Ullman, J. (1982) Principles of Database Systems, Rockville, MD: Computer Science Press.

Wilson, A.G. and Bennett, R.J. (1985) Mathematical methods in Human Geography and Planning, London: John Wiley and Sons.

Wilson, N.L. (1955) 'Space, time and individuals', Journal of Philosophy 52, 589-598.

SOME NOTES ON GEOGRAPHIC INFORMATION SYSTEMS: THE RELATIONSHIP BETWEEN THEIR PRACTICAL APPLICATION AND THEIR THEORETICAL EVOLUTION

IRENE CAMPARI
CNUCE-Italian National Research Council
via Santa Maria 36, 56100 Pisa, Italy
and
IREM-Italian National Research Council
viale Gramsci 5, 80122 Napoli, Italy

ABSTRACT. GISs have recently begun a new course: after a period of technical refinement, they are now passing through a phase in which their theoretical-cultural foundations are being re-examined. A debate on the construction of a new theoretical basis for GIS, begun by a group of experts in the field, now involves many new, as well as the long-standing scientific and technological, aspects of this discipline. Among the new-comers now demonstrating interest are the cognitive sciences. Efforts aimed at developing expert systems and user interfaces have stimulated in-depth consideration of the cognitive implications of the whole of GIS activity. Linguistics, non-Euclidean geometry and topology are becoming objects of such interest that GISs may act as springboards for a science of spatial information. Parallel to these new approaches however, there remains the current, widespread and purely operative view of GIS utilization, which, perhaps for solely incidental reasons, is at odds with the newer approaches. This paper presents some considerations of the two ways of viewing and working with GISs. The author, basing notes on experience in the field, hopes that it will contribute to the understanding of how the new courses and the variety of issues involved have come to be perceived by those far from where it has been set.

1. Introduction

The immediate practical usefulness of traditional GIS tools has contributed toward making some of the inadequacies in their theoretical base innocuous, especially given the fact that these inadequacies have not limited their original primary function of generating maps. The peculiar nature of GIS has also contributed to the slow realization of these short-comings and their consequences on what has to date been carried out, or attempted, using these tools. In fact, during the last few years work in the field has been directed at meeting the ever increasing market demand for instruments for the rapid solution of limited, well-defined technical problems. Symptomatic of this approach is the subordinate position which basic research in information science and the other disciplines involved have assumed in the development and application of current GISs.

Until now only opinions have been advanced as to what precisely occupies the pivotal position in which GIS functions are rooted. Many of these opinions have been directed at justifying the many technical, pseudo-scientific or scientific activities and investments, undertaken for and through GIS tools. Therefore, those justifications regarded some aspects considered relevant and

D. M. Mark and A. U. Frank (eds.), Cognitive and Linguistic Aspects of Geographic Space, 35–44.

beneficial for the adoption of GIS tools (i.e., attention to the user, to some application, to techniques, etc.). They were, however, only that, justifications; they could not, therefore, furnish any more than schematic palimpsests for technical method orientation for traditional GIS tools. In the few cases which went beyond pure justification, the theoretical base for territorial-data-processing technology was formulated from the "pure" confluence of theories and practices originated in the individual disciplines, or pseudo-disciplines, involved in cartographic applications.

The interdisciplinary nature of a GIS, in both its internal structure and applicative outcomes, requires efforts to construct an adequate theoretical base which is comparable to the creation of a new discipline. At the moment there seems to be a return to the basic elements underlying the field and a demand for in-depth scientific study in the fundamental sectors responsible for software development and its conceptual frameworks.

A number of questions have been suggested by the directives of this new approach, which has only recently and effectively been formulated and indicated by the issues raised at the NATO Advanced Study Institute upon which this book is based.

2. In Search of a Scientific Method

The need to establish a scientific method in the context of a GIS arose with the need for greater scientific dignity for a sector whose usefulness was simply too immediate, and at the same time, with the considerable financial involvement on the part of both the public and private sectors for the applications of a traditional GIS tool. Both aspects interest me (the tool, and the application, not to mention the theoretical consistency) and not being able to give priority to either, all of the factors involved must be simultaneously provided for, when any form of technological re-assessment is undertaken, particularly considering that it is to take place within only a few years' time from a massive market diffusion of the technology. The point should be emphasized because, as soon as the make-up of a new discipline has been established, space may be made for a tacit form of discrimination which could in the future be used to determine, a priori, what constitutes scientific activity, and what does not. In this sort of process, it is quite likely that the fields of some application of a GIS will fall once again within the realm of the non-scientific. The establishment of a scientific method must provide for the in-depth analysis of these aspects as well as others, while at the same time keeping in mind that in the field of geography, as well as others, there already are a number of basic theories in search of applications using a GIS tool.

It appears as if the objectives of some recent theoretical research directives are actually headed toward the formulation of a more coherent scientific method for activities involving GISs, in a broad sense. The definition of "scientific method" seems correct to me because all the conceptualizations and praxes which have come to constitute a tradition in the use of a GIS tool can be engaged. Organizing the efforts to renew GISs in this manner, all the positive aspects of the work done up to now can be conserved. Thus, the many activities initiated in the absence of clearly formulated theoretical premises (and often even of practical goals) should be re-framed within the correct methodological context, rather than disavowed. By choosing to define a scientific method, we may take this anomaly, incorporate it into the current transitional process, and then proceed. Method is, in fact, constituted at the end of empirical experimentation (Morin, 1984). In this way a research approach leaves an inheritance of select methodological techniques and some forms of conceptualization which have contributed to its success.

The definition of scientific method has another advantage: within its range the basic theories of the application's disciplines are present, simultaneously and to the same extent as the theories of the directly involved discipline in both the logical and technical fulfillment of tools. Therefore, the scientific method for a GIS inserts itself within a distinct tradition of research and study, whose scientific premises it can appropriate; the scientific method of a GIS may start to work at a high level of theoretical elaboration, controlling at the same time the degree of application obsolescence which is sometimes imposed by technological constraints. The new scientific method shall involve, for instance, the theories applicable to geographic phenomena such as complexity theory (Le Moigne, 1984; Morin, 1984b; Zanetto, 1988), systems theory (Chapman, 1977), fractal theory (Mandelbrot, 1982), and catastrophe theory (Thom, 1980). In this way the final object of technological interest, the territory, becomes as active a subject of method as the "space", while at the same time avoiding a gap that might be created between different interests involved with a GIS - those who deal with its "theoretical object", space, and those having to do with its "temporal aspect", territory.

There are, however, certain distinctions which must be considered, not to discriminate, but in order to be able to clarify in what scientific methodological context applications and basic research in scientific method for a GIS are to be inserted.

3. GIS and Epistemology Issues: Concrete Consideration

Efforts to bring some aspects of the disciplines of different applications, along with a new approach to the logical structure of GIS tools, toward the theoretical focal points solicited by a greater attention to recent scientific research methods originating in computational science (specifically the cognitive sciences) inserts themselves in the context of the scientific method. At this moment, this is a "natural" outcome of the debate concerning the scientific base of a discipline dealing with computer science.

We currently find ourselves in a period of definition of a new epistemology: from the laws of nature we have moved on to statements of bounds and possibilities; from research on the universal order we find ourselves studying chaos; from linear processes we have moved on to catastrophes theory; from scientific determinism to positivism and then to possibilism; and from mathematical certitude to scientific uncertainty. Every scientific field has recently recognized that the scientific method, in order to be such, must open itself to the age-old concepts of the humanities. On this theme, European and American scientific cultures are in the process of dialogue without vindictiveness or feelings of inferiority on the part of either. One needs only think of Gregory Bateson (1976, 1989), Edgar Morin, Von Foerster (1980, 1982) and the theoreticians of the new epistemology (Atlan, 1983; Maturana, 1978; Maturana and Varela, 1985; Prigogine and Stengers, 1979) and the scientific-philosophical current called "neoconnectionism".

In the new epistemology some fields of computer science, such as Artificial Intelligence and Logic, play a fundamental role; above all, the aspects of these fields that are reconsidering the processes of human knowledge. Computer science fields and humanistic disciplines such as cognitive psychology and linguistics have found a very fertile common terrain for integrating their new and old experimental acquisitions. This common terrain constitutes the cognitive science; within each of the above mentioned disciplines it maintains its sovereignty.

The convergence between cognitive sciences and the GIS field, which had its debut in articles published in the eighties (Mark and McGranaghan, 1986; Mark, 1987), would give that evolutionary approach a very current methodological character, that is, not dictated by the belief that a dutiful scientific method for GIS is unavoidable, but rather guided by the conviction that it is precisely this experimental practice which represents the future of GIS.

The cognitive sciences, which should have had primary roles in the development of cartography (Robinson, 1984), have some conceptual and strategic points of natural interaction with the GIS field: the attention paid by GIS sectors to causes of uncertainty of basic spatial conceptualizations, to the qualitative character of spatial reasoning, to the cultural differentiation of GIS users having heavy consequences on the effective use of tools, and so on. Whether the GIS field wants expert systems or optimization of human-GIS tool interaction as the final outcome, it must consider all these issues.

But there are other aspects of the relationship between GIS and cognitive science which are not altogether explicatory. The cognitive sciences, despite their having a philosophical tradition thousands of years old, have grown on the crest of the wave of computational science. If it is true that the computational sciences represent the search for all the computations necessary for various cognitive efforts (Pask, 1976; Fodor, 1979, Morin, 1989), then the cognitive sciences have a practically unlimited range of action and experimentation. Hoping to map out the field of competence of GIS, they should become the framework for the computation necessary for spatial analysis and territorial knowledge, methodologically obtained by means of a conceptual framework adequate to the recent themes so dear to some sectors of the cognitive sciences (value of metaphor, imagistics, different kinds of geometry tied to spatial human perception, etc.). There are different schools of scientists which tell us indirectly that the way is long and difficult. The more they explain their motives, the more shrill will be their laments at the difference between that which GIS tools are at present, and that which computational spatial science wishes they were.

For example, the current GISs, produced "ready to roll", have achieved great success because, despite their complex appearance, they represent the results of a process of simplification and standardization of the man and the territory relation carried out in a completely artificial environment. Paradoxically, this has been the main reason for both their commercial success and practical utility. Looking at GISs as that which they are now and have up to now been considered as being, that is, an international business, it is difficult to believe that the long experimentation times required in the cognitive sciences will be respected. These sciences are not much help in resolving the problems in this field at the moment, but they do aid in defining them, and, in any case, simple solutions which are immediately and practically transformable into a final product do not yet figure among the goals of cognitive science. The inter-relationship between GISs and cognitive science could also have important long-term implications. For example, will the topics left open at this time by the cognitive sciences, such as the images suggested by perception and formed by the mind (the mechanisms of which are as yet totally unknown) in turn become open questions in the GIS field? Or will they be dismissed by embracing only those parts which are the most understandably near current GIS, that is, the association between image and metaphor. Does this not seem like a form of operative reductionism aimed solely at giving a wider meaning to activities which revolve around GIS user interfaces?

In order to assure that the relationship between GIS and the cognitive sciences realizes its full potential, the latter should develop a scientifically broader perspective of GIS, such as might be provided by a new scientific method.

Applications to GIS of concepts deriving from the cognitive sciences have had, and continue to have, limited diffusion. Much of the explanation for this stems from the process of selection and reduction of cognitive science to GIS. Their basic roles have themselves been exchanged. Is it not the sciences which should develop new conceptualizations from experimentation on the interactions of man with machine during the use of a GIS? And why is it that this has not yet come about, nor is it in the process of being undertaken? This would be of little use, because through a GIS a type of consciousness is reproduced which has very little to do with intimate human spatial perceptions, as will be seen later.

In order to avoid disequilibrium, the relationship must be reversed. Although this will certainly complicate, rather than simplify the problems, current GIS must be considered only one of the possible fields for experimentation in the cognitive sciences.

At a level closer to our current interests, there is the fact that, in the parlance and destiny of GIS, all these acquisitions are involved from the point of view of theory as well as applications. One should be able to find suitable unifying factors between this age-old quest to understand man's thought and the systems which process spatial data which, all things considered, have for the last few years been largely given over to large-scale business, a fact which has provoked a certain incompatibility between the objects which are, in theory, to be unified.[1]

4. The Reality of GIS Tools

A consideration which comes to mind at the end of these first sections is in the form of a rhetorical question - everything is useful to a GIS, but what is a GIS useful for? There is an existing GIS reality and a population which uses them, some more correctly than others. I shall begin with an assumption based on the facts of what exactly a GIS has been until now and what it is becoming, paying particular attention to the private and public service facilities employees for whom a GIS represents their "daily bread". I do not believe that it would be a reduction of the topic to take up a discussion of GISs, starting with the current reality and practice in the field. The GISs available on the market are used for the production of maps, thematic more often than topographic. Moreover it seems that it is this function which has allowed GISs to spread rapidly into a large network of users. If we are still discussing GISs, it is precisely because there is a base of public and private agencies which have continually applied them to their originally projected function. This reality should certainly, then, not be underestimated considering the quantity of work aimed at rendering GISs usable, or the useful results stemming from it. In light of these affirmations, I would like to consider some issues advanced during the NATO Advanced Study Institute. The cognitive sciences are still involved.

Consider the example of linguistics. Language is one expression of spatial cognition; every language expresses unique syntactic differences, which in turn express semantic differences in the perception of space. Often the use of linguistic metaphors in daily communication stimulates the recognition of one single, exploitable, "common cultural element", making the communication possible too. Given that the main interface between a GIS and the user is language, the need was felt to create an interface language which was particular and appropriate to the culture of all those speaking a given language. In and of itself, this consideration is quite fascinating, in a cultural sense. Furthermore, since interfaces originated in English-speaking countries, by people who seem to have forgotten that in order to use GISs, or any advanced technology for that matter, people in non-English-speaking countries have had to learn and use English. This phenomenon is not restricted to particular intellectually or scientifically qualified environments, nor to those in which, despite the lack of these qualifications, the workers have set about learning just enough English in order to be able to read manuals and understand the commands which they must send the GIS. For many workers in the territorial sector, at least in Italy a few years ago, GISs represented their first contact with this language. In fact, a GIS is so inextricably linked to English that many terms used in digital map-making can be translated accurately and succinctly (that is without a rather wordy or cumbersome phrase) into Italian only with difficulty, if at all. Many employees in the Italian public offices devoted to the administration and control of the territory have had to learn typical GIS terms such as "editing, sorting, generating, plotting, drawing, overlay," etc. Their meanings in everyday English remain more than likely a mystery, and it would therefore be impossible to adopt them into more generalized use. They do not enhance the user's ability to understand English and are therefore, for them, part and parcel of a GIS. I know that learning a language does not mean getting hold of the conceptual structure which that language expresses. But until now the GIS goal has not been to represent the cultural conceptual structure of space (i.e., sense of territoriality) but just pieces of territory characterized by some quantitative variables.

Recuperating the specific linguistic connotations of the Italian language in order to construct a GIS interface language more appropriate to its structure, as well as to the culture of the Italian space, might be considered a forced enterprise. In the Italian school system, English is taught as a foreign language, and recently its teaching has been instituted in the first years of elementary school as well. In many areas of Italy, however, children will be learning Italian in the same way that they will learn English - in school from books - because their everyday language, the one they use to express spatial perception and behavior, is a local dialect, one of the hundreds of such dialects existing throughout Italy (thoroughness in such a discussion is difficult because its limits are not clear). The geography of language (Breton, 1976) informs us that national borders are the fundamental boundaries for the formation of a national language which will extend as far as the national culture and education. However a national language is quite different from the languages actually used by the population to express their particular and effectively different culture.

Concerning the current everyday reality of GISs, it would be enough to institute the process of standardization anxiously awaited for years. All that would be required in order to carry out the functions for which a GIS is acquired, is to adopt a standard terminology made up of a set of key words with unambiguous definitions. Despite the best intentions, this seemingly simple undertaking has not been done, no matter how willingly or unwillingly the process has been subjected to the linguistic constructs imposed by the various producers of GIS systems.

During a meeting on linguistic problems held at the NATO Advanced Study Institute, a number of curious considerations were expressed by computer scientists. They expressed the need to trace the linguistic archetype or archetypes of various languages with the aim of understanding how a particular perception of space, expressed through the language, has entered into a certain people's culture. The curiosity of this fact resides in the fact that it presumes the search for a differentiated archetype, when in fact the archetype is unified. De Santillana, working in the sixties on linguistic archetypes, reconstructed the history of various myths common to the different populations throughout the world, specifically those regarding the heavens, and built up a sort of planetary wayfinding by means of considerations of the archaic structure of time. It seems to me that this is a path beginning too far removed and which has, in any case, already been traversed and has produced relevant results. However, there seems to be little else for us to do other than make careful note of them, especially considering our fields of competence. Knowledge of the discoveries made by De Santillana Von Dechend (1983) and other authors are useful because they allow us to progress in our research with more current objectives in mind, in an arc of time some thousands of years nearer to us. The fact that the age of Pisces will give way to that of Aquarius does not have much of an effect upon us; our perception of time has changed, just as our perception of space. If four thousand years before Christ the planetary movements were determined by means of the knowledge of terrestrial space, now it is from planetary space that we regard terrestrial movements. Before long we will find the new archetypes in the technological terminology which at present is only a couple of decades old, transforming language of Remote Sensing into the virtual interface for the representation of terrestrial territory.

5. The Territory of GIS

What sorts of territories are, in the ultimate analysis, most easily represented by a GIS and what are the standard representations for these territories? The answer is official territories, those ratified by law; not those perceived by the people living there, but rather the myriad nations, regions, provinces and townships created by decree or other legal procedure. The agencies and institutions which can most easily afford a GIS are most usually public research or administrative offices. The use to which public agencies put a GIS is largely the cataloguing of the official territories under their jurisdiction, either by means of, or within a GIS. I do not think that the intentions of those who have promoted the widespread use of commercial GISs were any different from what has actually come to pass, which is the standardized processing and recognition of territorial representations of standardized and recognized regions of land. Even the collection of territorial representations particular to certain nations, such as the cadastral, for example, is official, established by law and carried out with the aim of taxation. What could be further from man's perception of space? In order to find the least resemblance to man's perception in these collections, one would have to look back to the days of the historical 19th century Napoleonic cadastral collection, in which the ability of the draughtsman was given free rein in the coloring (Campari, 1984) and pictorial descriptions of urban and agricultural plots of land.

On the other hand, the research institutes which utilize GIS for territorial representation base their work on the administrative divisions, because it is to these divisions that official data are referred. In this way a representation of the territory can be constructed without ever leaving the office. How much these differences actually count in our field is something which remains to be established. The inflexed arching streets of medieval Italian towns have nothing in common with the roads of Porterville. Yet both are represented by means of the exact same geometric primitives, the same topological properties, the same color set, and the same hierarchy of pre-defined relations among the elements which make up the urban network. In brief, whether it is the urban data for Cortona which is processed in California, or the data for Porterville processed in Tuscany, the results will be the same - an approximation of the reality with conceptual generalizations at a level which will necessarily eliminate any form of cultural identification. The real difference resides, instead, in the human-determined spatial relationships consequent to the layout of villages; this could be a main point of anthropological interest, welcome in the GIS field. But because the GIS future is not clear yet, I also question if interests are directed at maintaining these differences in the handling of territorial information or if, in any event, the goal is to reduce them to supra-cultural common denominators through the contributions of the Geographic Information Science. In the latter case the doubt arises that we are not seeking cognitive or cultural support for GIS, but a forced interdisciplinary character in which roles can be determined for those disciplines generally little associated with computer science.

6. Geographers Dealing with GIS

Mention was made earlier of the role which current geographic theories could play in future cultural adaptations to GISs. At this time, however, new conceptualizations in certain disciplines remain in the best of cases only fundamental theoretical points of departure for the renewal of consciousness. There is a certain difficulty in transferring new theoretical concepts to research methodology. Examples are numerous. The first article on complexity was written in 1947 by Warren Weaver, but it was only in the second half of the sixties that methodological discussion began referring to complexity as a subject or object for interdisciplinary study. Geography followed even later. Systemic theory is even older: Chapman (1977) wrote a splendid work on geography aimed at this, but there have been very few studies or in-depth research efforts based on his theories, and even less known still are the institutions in which research in this direction can be carried out with the necessary consensus. Despite all these theoretical fits, starts and interruptions, GIS technology has been continually utilized, largely because it has a precise referent, territory. Until new epistemological characteristics have been rendered explicit through effective research (also in the ecological sciences) it will remain difficult to justify efforts to advance a new scientific research method through the development of a technology whose supporters themselves relegate the very object of their study to a secondary position.

In order to accomplish this, not only must instruments be technically capable of following such a development, but there must, above all, be geographers available to carry it out. Actually, recently geographers involved with GISs have been occupied with mega-projects of (official) data acquisition and organization with the rather heroic ultimate goal of one day attaining real time GIS processing. Finding the logical link which joins the disparate national collections of territorial data is certainly not the job that a computer scientist would care to take on, but nor would a linguist, mathematician or statistician who has contributed to realizing the heroic enterprise. Usually it is the geographer to whom this task falls, taking years and years to find, as we say in Italian, the hank's end in an often inextricable tangle. When this logical connection has at last been located scientifically and technically, his or her work has very little value because it is an integral part of the technical routine. As anyone who has been dealing with GISs for even a short time knows, in order to obtain any high-level performance the information base must be at least above a standard threshold level of accuracy, at least as far as data such as administrative boundaries, population figures, and so on, are concerned. Personal experience has taught me that attaining a data base accuracy which surpasses the confines of a single nation takes between two and three years, and then only if the geographic areas considered are not excessively bureaucratized, in which case twice as long is needed, with all the consequences which this time period entails. I shall offer an example carried out in the Mediterranean area. A research project was undertaken for the development of a data base for several Mediterranean countries. Immediately problems arose the solution to which would require months, if not years, of application. Portugal, for example, has two agencies for the collection of population census data, the Commisao Nacional do Ambiente and the Instituto Nacional de Estatistica. The data amassed by these two agencies are thematically afferent, that is to say, in order to construct a useful data base for the evaluation of environmental impact in a certain "distrito", one must assemble the data from both. Furthermore, they both contain some of the same information, for example the results of the 1971-81 and 1951-61 censuses. During the early stages of a project of this type, the data from either source might seem equally reliable, and the choice of one or the other might be made with indifference. But it is soon discovered that the sources are not at all equivalent because the methods for administrative coding adopted by the two agencies are not in the least similar. They are in fact so different that arriving at the optimal methodology for reconstructing the reference territory for Portugal cannot be absolute, but must take into account the different paths chosen by the INE and the CNO in their data. The situation in Turkey is even worse. In order to establish a territorial structure for Turkey in which to enter the data, even only census data, one must work from original documents and then reconstruct by hand the current administrative situation, as well as the various changes brought about by the 1955 census (Campari, 1990). In effect, one must build an appropriate data base from scratch. Faced with problems of this kind, which admittedly must be solved before the GIS even enters the scene, but which at the same time are posed exclusively in function of the GIS, the geographer is supposed to be enthusiastic about the chance to broaden the understanding of complex space representation, if for no other reason than the fact that it frees him from the complications found only at lower levels.

It seems that at this point it is quite clear that difficult problems exist in the phases prior to the use of a GIS; problems which not only are still in need of solution, but which are not even being faced squarely. These problems also involve the time required for the quantity and quality of work to bring concrete experience in the use of GISs to conclusion. It might be objected that once an experience of this kind has been finished, a final, ready-to-use product results. An assertion of this sort cannot, however, be fully upheld. There are innumerable factors concerning accuracy and uncertainty which render the actual use of GISs within wide ranges and in non-homogeneous areas far from a foregone conclusion.

7. Conclusion

The definition of a scientific method in the GIS field will obviously take a long time and much competence. This activity must be rooted in a current context both theoretically and practical-operatively. Only in this manner can the confusion about goals and means be avoided.

8. Notes

1. The Piaget study, used by psychologists as a sort of vade-mecum for the study of human spatial orientation, certainly has a justifiable current relevance to discussions and clarifications regarding the approach to spatial representation through a GIS. But I ask myself if the descriptive uses to which Piaget's studies and school of thought are put is enough to justify recourse to them. Certainly a higher quality contribution would result if Piaget's work could be used on an analytical level as well, a level which should also be applied with regard to GIS.

The author may be contacted via electronic mail at IRENE@ICNUCEVM.CNUCE.CNR.IT

9. References

Atlan, H. (1983) L'emergence du nouveau et du sens, in P. Dumouchel and J. P. Dupuj (eds.), L'auto-organisation, Colloque de Cerisy, Seuil, Paris.

Bateson, G. (1976) Steps to an ecology of mind, Ballantine, New York.

Bateson, G. (1989) Mente e natura, Adelphy, Milan.

Breton, R. (1990) Geografia delle Lingue, Marsilio, Padova.

Campari, I. (1984) Insediamenti rurali. Note di filologia del territorio, Pacini, Pisa.

Campari, I. (1990) 'Accuracy vs spatial statistical data. The Mediterranean region. The MEDASE Project', Proceedings, European Geographical Information Systems Conference (EGIS'90), Amsterdam.

Chapman, G. P. (1977) Human and environmental systems: a geographer's appraisal, Academic Press, London.

Fodor, J. A. (1975) The language of thought, Harvard University Press, Cambridge, Mass.

Le Moigne, J. L. (1984) Progettazione della complessità e complessità della progettazione, Bocchi, Ceruti, La sfida della complessità, Feltrinelli, Milan.

Mandelbrot, B. B. (1982) The fractal geometry of nature, Freeman, San Francisco.

Mark, D. M. (1987), 'On giving and receiving directions: cartographic and cognitive issues', Proceedings, International Symposium on Computer-Assisted Cartography (Auto-Carto 8), Baltimore, 562-571.

Mark, D. M. and McGranaghan, M. (1986) 'Effective provision of navigation assistance to drivers: a cognitive science approach', Proceedings, Auto Carto London, 399-408.

Maturana, H. R. (1978) The biology of language: the epistemology of reality, in D. Rieben (ed.), The biology and psychology of language, Plenum Press, New York.

Maturana, H. and Varela, F. (1985) The three of knowledge, New Science Library, Boston.

Morin, E. (1984) Il Metodo. Ordine Disordine Organizzazione, Feltrinelli, Milan (o.t. La Méthode)

Morin, E. (1984b) Le vie della complessità, in G. Bocchi and M. Ceruti (eds.), La sfida della complessità, Feltrinelli, Milan

Morin, E. (1989) La conoscenza della conoscenza, Feltrinelli, Milan (o.t. La connaissance de la connaissance)

Pask, G. (1976) Conversation, Cognition and Learning, Elsevier, New York.

Prigogine, J. and Stergers, I. (1979) La nuovelle Alliance. Metamorphose de la Science, Gallimard, Paris.

Robinson, A. H, Sale, R. D. and Morrison, J. L., Muehrcke P. C. (1984) Elements of Cartography, Wiley, New York.

Santillana, G. and Von Dechend, B. (1983) Il mulino di Amleto, Adelphi, Milano.

Thom, R. (1980) Stabilità strutturale e morfogenesi, Einaudi, Turin.

Von Foester, H. (1982) Observing Systems, Inter-Systems Publications, Seaside, CA.

Von Foester, H. (1980) Epistemology of Communication, in K. Woodward (ed), The Myths of information : technology and postindustrial culture, Routledge, London.

Weaver, W. (1947) Science and Complexity, American Scientist 36, 1948, 536-544 (1947 first version of this paper).

A HAND-IN-GLOVE PARADIGM FOR GEOGRAPHY

ALLEN K. PHILBRICK
Department of Geography
University of Western Ontario
London, Ontario N6A 5C2 Canada

ABSTRACT. A two-dimensional mind-set hampered the search by geographers for the ideal set of 'natural regions'. Multi-dimensional perspectives provide a more useful approach. For, there are regional entities in volumetric timespace where boundaries are not "arbitrarily imposed". They are surfaces that separate inside from outside. The paper cites a hierarchical set of nested necessary volumetric natural regions from approximately the negative to the positive power of ten. In this context people are the 'atoms' of human geography. Joining forces with cognitive and linguistic sciences, inside/outside study of human behavior creates a hand-in-glove paradigm for geography.

1. Search for 'Natural Regions'

There was a time when geographers searched for the ideal set of "natural regions" (Herbertson, 1905). These were proposed to be the archetypal subdivisions of the earth's global surface into scientifically defined parts which everyone could accept. Success would have required natural regions to have been defined on the basis of "something inherent and not arbitrarily imposed" (Hartshorne, 1939, 296). The effort over a period of at least a century has witnessed the demise of regional geography as a primary thrust of the discipline. For many of us regional geography has become little more than personalized description.

Still, more than half a century ago, Hartshorne concluded his treatment of the problem with this paragraph, (1939, 300):

> "Is there then no possible way by which we may divide the world into regions that will give due account to the whole complex of natural elements involved and yet be independent of the human point of view? Such a 'natural division of the world into regions' would, if it were even theoretically possible to construct it, provide a framework independent of man against which we could study the development of cultural regions. The great value which a framework of this sort would have for regional geography requires us, in spite of all the logical arguments which make it seem impossible, to consider any serious suggestion that appears to provide it."

It is just barely possible that there is a message in this eternal quibble over the drawing of linear boundaries subdividing the earth's surface into regions of invariant significance. We could be approaching the problem incorrectly. Could it be that the phenomena we wish to classify and analyze cannot be contained in linearly bounded areas without destroying their inherent natures to

D. M. Mark and A. U. Frank (eds.), Cognitive and Linguistic Aspects of Geographic Space, 45–49.

some degree? Isn't it obvious that a line drawn around a set of criteria drawing attention to their perceived homogeneity must have been imposed arbitrarily? Are we not forgetting the truism that nature knows no linear boundaries, that only man draws boundaries?

In recent decades mathematics has made great contributions to geographical thinking. Geographers pay lip service, at least, to thinking dynamically, in three, even, four dimensions. For our quantitative wing, "hyper space" is multi-dimensional; but still for all too many geographers our underlying common sense thinking is still two dimensional.

The purpose of this brief paper is to show how that multi-dimensional perspective may lead to a more useful geographical regional approach. If geography is concerned with relative locations of things in earth space, then what we are talking about is not a two dimensional world. Whatever we do about simplifying things for convenience, we must start treating the surface zone of the planet and all it parts and the objects/energies on it as three dimensional, dynamically, which means we must recognize that time is object/energies in motion. In a word, we should think time-spatially. Even though we may continue picturing the objects of analysis two dimensionally on paper using the traditional convention of the map, we must accelerate our use of the electronic medium to depict in motion and in momentarily arrested static positions as well multidimensional time-spatial relationships graphically and conceptually in our ongoing development of Geographical Information Systems, GIS.

Even so, many people when faced with a four dimensional viewpoint simply shrug their shoulders and go right on thinking two dimensionally as before. But consider this. There are regional entities in volumetric timespace which have boundaries which are not arbitrarily imposed. The boundaries of such entities are surfaces which separate the inside from the outside. The earth, itself, as a planetary body is such a region. It is subdividable into "natural regions, not arbitrarily imposed" which are its core, plastic zone, brittle zone, so called 'Moho' discontinuity, biosphere, troposphere, stratosphere, etc. Air masses within the earth's atmospheric circulation are another example. They are distributed dynamically, volumetrically as well as concentrically. As well, tectonic plates responsible for continental drift over geologic time, override along some of their margins, so that they do not neatly subdivide the subsurface planimetrically. Nevertheless, they are part of the concentric inside-outside surface bounding of the planet. Even more important than the fact that they are volumetric surfaces, however, the interfacing of these plates represents their dynamic motion in timespace.

2. Hierarchical Nodality

The structure of the universe, to paraphrase Claude W. Heaps (1960), is characterized as "aggregates of smaller parts". There is a theory in geography called "hierarchical nodality" (Philbrick, 1982). According to it reality is at once nested, particulate, and discontinuous, as well as dynamic. Any appearances of continuity we may think we perceive are, like mathematics, mental constructs. A hierarchy of nodal entities exists which comprises a very complex set of nested necessary volumetric natural regions. This set of sets extends from approximately the negative power of ten to the positive power of ten, in quantum scale leaps, from sub-atomic particles, to the atom, to molecules, clusters of molecules, organic molecules, cells, other living entities, viruses, plants, animals, including humans, larger nodal regions of human organization, the planet earth, other astronomical bodies, solar/star systems, galaxies, clusters of galaxies, the universe, and other possible universes.

The boundaries of each of these entities throughout the immensity of its hierarchical magnitude are surfaces which separate an inside from an outside. Since this set of sets of entities exists in relative locations (which is what defines space), there is a corresponding set of sets of distances between and among them. In the work in which I have further outlined the theory of hierarchical nodality (Philbrick, 1988, 54), I referred to these distances "in the timespace mode as a distance/duration hierarchy". To avoid confusion in the mind of the reader, let me summarize this concept by referring to one brief paragraph.

> "There is a corollary 'uniformity' or accordance in the orders of magnitude of units at each level. Galaxies --and probably solar systems, planets, animals, cells and molecules, atoms, and particles-- all exist within particular size ranges, within their respective orders of magnitude. The same observation applies to the orders of magnitude of the distances/durations among them."

At the earth surface scale, in both physical and living realms, there are various kinds of entities which can be dealt with appropriately in such a dynamic volumetric geography. No claim is made that hierarchical nodality fulfills the objective of neatly subdividing the surface of the earth into "natural regions" without overlaps. Quite the contrary. All of the volumetric dynamic natural regions referred to above are partially open systems. Their surface boundaries are permeable in varying degrees. Energy exchanges and bondings of a very great variety, indeed, are observable in the interacting relationships which have been studied by scientists of many disciplines. This open system aspect is particularly important to living matter. But within the field of study addressed by geographers, the nested system of hierarchical nodality does, indeed, "provide a framework independent of man" (Hartshorne, 1939, 300) against which we can study all earthly things.

In this context people are the dynamic 'atoms' of human geography. At the scale of human society, living on and using the resources of the biosphere, each person is a three dimensional dynamic natural region, bounded by his or her body surface. Before the geographers reading this reject this idea as "off the wall", consider each person's territoriality. We humans are, each of us, open systems, dependent for survival internally upon constant interchange with our external environment. We survive literally as well as metaphorically because we have our feet on the ground; we breath air; and we drink the water of which we are primarily (70-80 percent) composed. As humans we have two distinct advantages over all other animals in the size and cognitive capacity of our forebrain, and its probable consequence in the capacity for linguistic communication. But we must not think of human beings only as three dimensional objects. As interacting parts of the hierarchy of natural regions in the planetary biosphere, humans are active participants. Each person within the self is a neural unit capable of understanding experientially the external environment, especially including other people. From the standpoint of relative location, to each person, self appears to be inside the self. To each of us, all others are outside the self in the external environment as other people. Communication, biological, and social needs, generate cooperative and competitive relationships. Out of such human interactions larger units of territorial and environmental organization have arisen. History records the extension and construction of societal infrastructures, physical, social, administrative, economic, political, and cultural, over the surface of the earth. The impact of humanity upon the world has been immense.

Cognitive and linguistic capacities have immeasurably enhanced mankind's ability to survive by organizing (using and misusing) its external surroundings. They explain how man among all animals has come to both dominate and threaten the quality of life over the entire surface of the earth.

Brains and speech account for both the success story of man's increase in number and impact as well as pose the potential threat to mankind's continued existence. Incidentally,these two facets of human existence underscore the appropriateness of the association of cognitive, linguistic, and geographical scientists.

Within human geography there is, also, the sub-discipline of behavioral geography. When we examine human behavior geographically in the context of this paper, we join forces with cognitive and linguistic science. Each discipline is concerned with what occurs both inside the human individuals' brain/nervous systems and outside the selves of the members of human communities as they interact. In the past the world external to the "mind" was regarded as objective. The "within" was the realm of the subjective. If the human person is treated as a dynamic geographical regional entity, both objectivity and subjectivity have location inside the self as well as outside the self. In place of the "myths of objectivity and subjectivity" we find a new "experientialist alternative" (Lakoff and Johnson, 1980), which unites aspects of all three in providing a role for the creative and active participation of human beings in explaining the patterns of human life as these continue to develop regionally.

3. Conclusion

In conclusion, I would propose that the association of inside/outside studies of human behavior by cognitive, linguistic and geographic scientists fit together so logically as to make possible a hand in glove paradigm for geography. The glove in this metaphor represents the dynamic man-made protective environment derived from the resources of the world environment. The moving hand is symbolic of the body and the mind of the human person, bounded by the skin, on the outside, articulated by the bone, muscle, and tissue and enobled by the blood from the heart, and energy from the mind, on the inside. But we must not forget that the hand of man both designed and fashioned the protective glove. The glove represents the many-natured dynamic three dimensional surfaces with which men and women surround themselves: clothes, personal spaces, buildings, walls, and all of the extensions to the person in terms of language, books, metaphor and art as well as weapons, tools, artifacts, and infrastructures of all sorts. Not only is the hand quicker than the eye, the hand is the great instrument for everything fashioned "by the hand of man". What more apt symbol of man could there be, then, than the extended finger of man's hand from Michelangelo's pictorial metaphor on the Sistine Chapel ceiling reaching toward the finger of God - in the acts of communication, love, music, and of art.

4. References

Hartshorne, Richard (1939) 'The Nature of Geography', Annals of The Association of American Geographers, Lancaster, Penn.

Heaps, Claude W. (1960) 'The Structure of the Universe', Smithsonian Treasury of Science, Webster True (ed.), Simon and Schuster Inc., New York .

Herbertson, Andrew, J. (1905) 'The Major Natural Regions, An Essay in Systematic Geography', Geographical Journal 25, 300-312.

Lakoff, George, and Johnson, Mark (1980) Metaphors We Live By, The University of Chicago Press, Chicago.

Philbrick, Allen K. (1982) 'Hierarchical Nodality in Geographical Time-space', Economic Geography 58(1), 1-19.

Philbrick, Allen K. (1988) 'Nodal Hierarchies and the Concept of Survival', in William J. Coffey (ed.), Geographical Systems and Systems of Geography—Essays in Honour of William Warntz, Department of Geography, The University of Western Ontario, London, Canada, 49-62.

SECTION 2: CULTURAL INFLUENCES ON THE CONCEPTUALIZATION OF GEOGRAPHIC SPACE

Culture represents the collective cognitive models of groups and societies. Anthropologists have long studied human culture, as have cultural geographers. This short section contains three papers, by an anthropologist, a cultural geographer, and a linguist. Each takes a different aspect of the relation between culture, landscape, and geographic space.

In the first paper, **Maria Cátedra**, an anthropologist, explains the socio-cultural perception of space, and specifically addresses the symbolic significance which is associated with specific spatial situations. Space as understood socially is very different from the empty, uniform expanse which mathematicians treat. It is naturally and artificially structured by human activities, and acquires meanings in the process, meanings that are far beyond the immediate spatial situation and which point to very fundamental values in a culture. To exemplify these principles, Cátedra selects the "door" as a special, identified spatial configuration, and sheds light on its meaning, using the anthropological tradition of participative observation. In her paper, she goes beyond the image schemata and their metaphorical usage as discussed by linguist George Lakoff—the door is conceptually part of both the inside/outside image schema and the path schema; indeed, it is at the intersection of those schemata. Cátedra then goes beyond that and evokes the cultural significance of these metaphors, revealing other concepts that are immediately present with these spatial situations. In a deep sense, she is reversing the argument that people often comprehend abstract situations in terms of space. In "through the door", a spatial situation is evoking a rich collection of deep a-spatial concepts, which are typically associated with such a situation. This is not contradicting Lakoff's viewpoint, but complements it, as the metaphorical mapping from space to human situation becomes so ingrained and essential part of a culture that the reverse mapping, from the cultural situation to the spatial becomes natural. An example is when death is thought of as a passing from one world to another. The passage stands for death and death stands for the passage. From a cultural point of view, we cannot differentiate between the two.

Elaine Bjorklund, a cultural geographer, examines a similar theme, stressing the complexity of cultural significance—nothing has just one meaning. Is meaning just found (or lost) in the complex web that relates our perceptions to previous experiences, or to the tenants of our culture? The artefacts that people construct reflect not only the technology used, which imposes constraints on what can be constructed, but to a large degree are embodiments of our cultural values and cognitive abilities. This does not only apply to artefacts such as cars or books, but also to the organization of geographic space: highways and settlement patterns, cities and parks. This is very close to what some geographers have long though of as being the essence of geography.

The section on cultural influences on the conceptualization of geographic space concludes with a short note, in which **Ali Kas**, a linguist, points out a quite different aspect of the relations of geography and culture: Kas examines the ways in which foreign-language "bumper stickers" on automobiles in Turkey are an expression of culture, image, and affiliation with foreign and largely unknown cultures and institutions.

"THROUGH THE DOOR": A VIEW OF SPACE FROM AN ANTHROPOLOGICAL PERSPECTIVE

MARIA CATEDRA
Departamento de Anthropología Social
Universidad Complutense de Madrid
Madrid, Spain

1. Introduction

I have prepared a short paper with some basic ideas about space. I give some intuitive hints on the way anthropologists (or better, an anthropologist *like me*) see the topic. In this paper I suggest a variety of concepts with several examples of what a door may stand for. Why a door? Of course I could have used any other spatial object, but I think the door is a suitable and powerful symbol to express ambiguity, something in which I am very interested. I will try to show how the anthropological perspective may help us to understand symbols, how they are used and how they help people to understand and structure the world. It is one of the ways to observe the ideas in peoples' heads, alternative and hopefully complementary to the methods in other sciences.

I know that for some this paper may be quite obscure. Why this obscurity? Some may think that the translation from my native Spanish to English may be the cause, but there are a number of other translations that went into the making of it (conceptual, personal, experiential, disciplinary, social, and so on); one of the most definitive translations, to me, is the cultural one. This problem of translation bothers me, not only because of the language problem, but because although we immediately recognize that we are translating in reference to language, I do not think we are aware of this fact in other contexts. So it may provoke lack of communication. Anthropologists have long faced this problem because our work is essentially translating cultures. The translation problem is not solved, but there has been an increasing sophistication and awareness from the naive and literal translation of the beginning of the century to now, and we take some pains to be more accurate. What is, really, mapping for the geographers, computer code for GISs, metaphor for linguists, and so on? I think it goes right to the core of the problem of translation. From this paper I think it is worthwhile to point out two things. First, perhaps we need a contrast –other cultures– to find out if what we call natural is just cultural. Second, different realities (whatever this word should mean) may need different ways of expression.

You will wonder about the variety of different doors I use in my text, and from such different places. Normally, my data comes from a single society which I studied, but in this paper I tried to share the feeling for complexity and diversity which is the essence of anthropology. The door as a motive can be seen as a marker for a physical limit, an abstract concept, an image for a social entrance, a metaphor for time, a symbol of spiritual passage or cultural domination, or ideology about women. In some examples the existence or absence of a door is the point of interest. The door becomes a highly-charged, condensed, and thick idea.

I was surprised by the different meanings the participants of the NATO Advanced Study Institute had for the notions of sign, image, symbol and metaphor. Of course, there are many scholars using these heuristic tools and linguistic tropes from different disciplines. Just from the point of view of anthropology they can mean different and sometimes opposite things. So I think it would be useful to distinguish among them.

Within expressive culture we are interested, primarily, in communication. We study symbols because we want to get at what is inside peoples' heads. For instance, I do not focus upon the door itself but on what it expresses about human beings and how different human beings think

D. M. Mark and A. U. Frank (eds.), Cognitive and Linguistic Aspects of Geographic Space, 53–63.

about it. I am sure linguistic and cognitive scholars pretend the same thing but from a different perspective. We primarily focus on diversity and cultural differences. If there is a single modern concept which has been useful to get at people's minds, it is that of symbol. A symbol can be a word like "home" or "daddy", an object like a hand or a flag, or an activity like burning a flag. I said it "can be" because it "can be not" as well. Symbolism is not a thing but a process, a human possibility. A word like "daddy" can be just a reference (the way children call their fathers) but depending on the context it can mean something else not so obvious (for example if Marilyn Monroe says "daddy" to someone). By the same token the context is very important to discover if something is a symbol or a signal. Please consider the different quality of a red piece of cloth in the back of a truck, or on a barricade in Paris in May, 1968. The former context of the very same red piece of cloth is just a plain and concrete notice of "danger". In the last example the symbol encapsulated a set of difficult, abstract, important, contradictory and meaningful philosophical and political ideas about freedom and human beings.

Space has become a quite fresh topic within the new orientations of an anthropology, although there is not much consensus on how to approach it. I will mention just three relevant anthropologists who were working on the matter from different perspectives in the past. Probably Edward Hall (1959, 1966) was one of the first to point out the uses of space to communicate with others and how it was cultural difference that allowed for the perception of differences in the uses of space. He employed a socio-biological position in the continuum between an anthropology in space or anthropology of space (quite a different perspective). Before him, Forde (1934) had taken a position similar to those geographers of the 19th century based on the weight of the environment. Levi-Strauss (1963) has tried to show that spatial configurations are not only the product but the producers of social systems, not only the effect but the cause. But he also said that not only can we not directly experience space and natural conditions but that they do not have existence of their own and they need humans to make them intelligible.

At present one of the most interesting views of space from the cultural point of view is the "qualitative space" described by Quine (1968) and Goodman (1951) and others and applied to anthropology by James W. Fernandez (1972, 1974) who define it as a set of dimensions or continuity axis which provides a fundamental ordering in culture. This is a topographic model of human experience. Mind organizes experience in relation to a prelinguistic qualitative space. Every culture can be defined by the positions of their expressive and material data in several dimensions. Metaphor is the way through which individuals occupy their places and go into the qualitative space. Society can be seen as the movement of individuals within this space.

Zulaika (1987) has applied this theory to the aesthetic-ritual space in the Basque country. The qualitative space is used to express a cultural order evident through architectural shapes, cosmological ideas, bodily forms, norms of behavior and so on. Qualitative units of this logical space are pre-linguistics and abstract entities. They are semantic spaces, systematic expressions of a cultural, imaginary order of aesthetic nature, because they are based on final norms which belong to the deep memory of the group. As a ritual space, they are maps of ordered behavior in time to produce qualitative transactions, and a common semantic structure that condition ways of representation.

So much about theory. Let's open a door.

2. Through The Door

"The solemn geography of human limits..."
Paul Eluard

"The arc de triomphe is a pure door looking at the vacuum, which really is the world itself as well as heavenly space. It is a 'door to heaven' made of stone. The door offers in abbreviated form all the sense of the world: something which expresses and waits at the same time."
Jean Hani

To cross a threshold, to pass through a door, can be seen as something apparently plain and clear, of overwhelming simplicity, which we carry out with the disturbing and absurd unconsciousness we use in every ordinary and daily act. Thus, we should remember what Romano Guardino said once: "Modern man needs to learn the deep value of gestures". To cross the door, to enter into a humble home or into the most magnificent temple is a very serious and solemn thing. The mystery of transition is within the proper dynamics of life and it is the basis of rites of passage as important as birth and death. An inscription at the door of the church of Mozat (France) says: "You who are entering, look to heaven". The door is this nowhere, this cleft that separates a world from another world.

But we will not get into this yet. The category door belongs to a more inclusive class of objects and ideas and it depends on the predicative notion of "closing" or "limiting". The concept of "closing" for example is expressed in an iconic way by walls, fences, doors, and so on. But while the predicative notion is shared by everybody, the expression drastically changes according to cultural norms. A piece of cloth may close more effectively than a door closed with seven keys; for example, the purdah: a thin curtain, sometimes a veil which in India divides the part of the house where woman are secluded. Cultural concepts may form limits as impenetrable as stone walls.

3. Limits and Walls

"...movements of closing and opening are so many, so frequently inverted, so charged with hesitation, that we can conclude with this formula: man is a half-opened being.

The door is a cosmos of the half-opened...the very origin of dreaming where there are wishes and temptations, the temptations to open the being, the wish to conquer every reticent being"
Bachelard

The notion of limit —or discontinuity— is a main element of the symbolic making and representation of spatial systems of human beings. To close or to limit invokes the concept of transcendence, a basic concept for the building of logical or aesthetic relations: to limit a world is to transcend it.

Perhaps one of the first and primary limitations of space is the division between natural space and cultural space. The first space is a chaotic and dangerous place and the second is our secure and ordered habitat. But these very same adjectives are used to speak about public and private space. Limits are not always sharp and take different shapes that make us question our classifications which are important to us despite their imprecision. The division between culture and nature is one of these important classifications. A good example of this division is the uncomfortable closeness (Cátedra, 1989) of men and cattle in huts high up in the mountains among the vaqueiros of Asturias, in Spain. Another example is the difficult dichotomy of inside/outside within the Andalucian patio, probably inherited from the Arab house. Other limits are not visible (to western eyes and hearts), but they are no less effective. Some limits are based on the nature of things and other ones are based on the nature of words, but all of them are symbolic limits.

Enclosures (walls) have traditionally marked the limit between two other spaces: the profane space and the sacred one. Such limits are apparently made to protect the house or the temple (or to be protected *from*). But limits divide ideas as well. The walls of a city are not only military but magical protection as well. They divide the chaotic space and make significant an organized space. Sometimes there is redundant protection. For example: it seems that in Avila there was once a church in front of each of the nine gates of the city walls. A skeptical person may think that there are so many churches in this walled city that the existence of this additional symbolic ring of churches could be random and not so much the necessity of mystic protection. But walls are many and not only made of stones: people from Avila say frequently when speaking about their city "Not only the city has walls, people do as well". A completely different case of limits is found in the Japanese house whose material limits are open (this is not the case of an Arab or Andalucian house) but there is an important boundary constructed from collective social obligations.

The image of the body is another particular way to organize limits. Often the human body, but sometimes animal bodies as well, becomes a model for the territorial unit. The house in some contexts can be seen as an intermediary microcosm between human body and society. A society can be equated to a house divided into several rooms and corridors, with interior walls and doors. The symbolic behavior associated with the limits of a human body —the ways in and out, in this case— (our cultural prescriptions to eat, defecate or copulate) reproduces the behavior of physical or social limits of the group, the rules which determine, how one gets into or out of the group. Symbols based on the human body are used to express different social experiences. Why does the bridegroom take the bride into his arms to get into the house? The everyday experience of going through a door expresses many kinds of entrances; in this case the bridegroom ritually duplicates and so communicates the new state of a woman through a spatial and symbolic gesture.

4. The Threshold, Uncertainty

> "...must a door be open or closed? and will we find in this sentence a truly effective analytical tool for a human passion?"
>
> Bachelard

Every limit, despite its unsurmountable nature, provides an access path. By definition, there is no limit one cannot pass. Physical limits such as walls have doors, magical protection has thresholds and national boundaries have customs; all selectively filter their access. Doors allow access but contain the potential for closing. By definition too, space is a relative concept and intimately connected to social relationships. So an open door can be a metaphor for generosity, hospitality, sexual accessibility, or prosperity but it can be, in certain contexts, a metaphor for death and destruction as well.

The threshold marks and prepares the passing, it is the place where limits open, and it is the subject of a great number of physical and symbolic rituals. Some thresholds mark property limits and others the most intimate places in a home. There are specific cultural rules about who may cross a threshold and under what circumstances. The stage may be the same but it is not the way the threshold is to be signified, perceived or interpreted. Positions are always relative; the very threshold is itself a mediator, linking spaces. Earlier, I alluded to the nature/culture dichotomy associated with the natural environment and human space and now I speak about private and public space. We can observe both these dichotomies within the house, from the spaces for social interaction (hall, office, dining room) to the most intimate rooms related with nakedness and sexual relations (like bedrooms and bathrooms). Ambivalence, and danger itself, are inside the house as well.

5. The Passage and Danger

"Why do we not feel that in each door a small god of threshold is incarnated? Maybe it is the same being who opens and closes the door?...There are two 'beings' at the door, the door invokes in us two directions of dreaming. It is symbolic twice."
Bachelard

The house is not only a physical space but a moral territory. The door is its boundary and this is the reason for both the danger and the strong ritualization of the threshold. Every margin is dangerous, just as every structure of ideas is vulnerable at its borders. This vulnerability is expressed by notions of uncleanliness and contamination, just like orifices in the body or the polluted limits of cities (for instance, the sewers), which are also ill-defined and dirty. This is why ablutions are done so frequently at entrances. This very same lack of definition can be found not only for the passage but also for the passenger. The person who crosses a boundary is physically and symbolically in a special situation: he is between two worlds. He is not here, but he is not yet there. Sometimes he is invisible or blurred or characterized by symbols of death and life, male and female. The symbolism of androgyny is in accordance with the ambiguity of the door –and this is why it is associated with "two beings" or twins. Indeed the door is associated with Jano, the divinity with two faces, the double face of time looking both toward the past and the future. Jano (*Ianus*) for the Romans is the porter who opens the doors of the yearly cycle. For this reason he is usually drawn with a key and thus the door in Latin is *janua* and the first month of the year is *januarius*. But it seems that the true face of Jano is neither one. The true face of Jano is invisible because it is present time, an inaccessible and unattainable instant, between the past which is not any more and the future that is not yet. There are also analogies with the serpent, the animal which appears to die every time it looses its skin, another symbol for the double character of time. Some of these associations appear in an ethnographic example. In several parts of China, according to Van Gennep (1960), a ritual is performed periodically, especially for sick children. The sick child, a Taoist priest, and the father of the child pass through a door specifically made for this occasion and situated in the middle of a room. The door is then moved to the four corners of the room before it is burned. The meaning of this procedure seems to be quite clear: the sick child passes from a dangerous world to another more favorable one, from sickness to health. Behind the door remains the disorder, danger and contamination of the sickness. This transitional quality is found in every single ceremony which is associated with the passing from any social and mental position to any other. So the passenger carries with himself a basic ambiguity: he is dusty from the journey.

It is for this reason that rites of purity and incorporation or preparation for union (for instance by an offering) are frequently performed in front of a door. For the same reason people display an image of a divinity or of a monster to protect the entrance, to which people sometimes offer a coin, a prayer or a sacrifice. So the inscription "Ave Maria" which is written at many doors is not a superfluous gesture. A spatial rite of passage is followed by a spiritual passage. The guardians of the threshold –archers, dragons, lions, sphinx or simple crosses in chapels, houses and cathedrals– still remind us of the step we are taking and that we are about to leave something behind.

6. Initiation, Change and Transformation

"Where a door closes, a door opens"[1]
"The world calls at the other side of the door"
Pierre Albert-Birot

"By door we express the harmony between the sensible and spiritual world"
Jean Hani

The person going through a door, or any other transition, crosses to a place with few, if any, of the attributes of the previous space (the bride crossing the threshold becomes spouse, the woman coming out of the delivery room is a mother). The transition is a process, a transformation. We cross many doors of every kind from our mother's womb to the tomb. There are some critical landmarks in such a continuous passing which every society ritualizes and publicly marks. A literal crossing of the door is the ceremonial pattern of changing residence. A new house is taboo until some rites are carried out which identify its inhabitants, like visiting or consuming food. The rites of entrance have as a counterpart rites of exit, which are either the same or opposite depending on the context. One "inaugurates" new identities as well as houses.

So the door is a symbolic object meaning initiation, change and transformation. Our life is a whole set of doors we open and close. We are born and reborn several times to a new state of being, converted to different social persons and, like the serpent, leave our skins along the way. After death we are still crossing doors, travelling as pilgrims in the last transit. In Book IV of Eneida, Eneas, at the door of the Sybila, sees the reproduction of the labyrinth of Crete. The door here is a symbol for access to another existence, to the cavern (which is both an initiation and a funeral place) and the labyrinth is the way to it. The cavern is the image of the world, the entrance door is the symbol of life and the way out is the symbol of passing on, of death. In different traditions (for example, some European Christian temples) these two doors correspond to the two solstices: the entrance is the door of humankind, the door of summer, that of the shadows and death; the exit door is on the side of light and life, the door of the gods and winter.

The association of the door with initiation and death appears in many other contexts. In the ritual of Osiris from old Egypt, there is a Book of Doors. Hades is like an enormous temple divided in a certain number of rooms separated by doors. Doors are defended by guardians and spirits and are opened by means of rituals. At the beginning of the night when Ra, the sun, is dead, he receives in his boat the pure souls of the deceased and the doors open for them. The dead ones who do not get into the boat wander through the halls forever. According to the Book of Doors, funeral rites revived the dead through a nocturnal rebirth and made them gods through the process of mummification. Hence the eagerness for immortality metaphorically passes through the door.

7. The Door, She. ("La puerta")

"Major monuments are nests of devilish riddles"
Emilio Garcia Gomez

Now let us come back to life and to a beautiful study made by a sensitive old professor of Arabian studies, discussing the architecture of the Alhambra. It is like a mystery novel, a detective story about a door –or perhaps two doors: one which is troublesome because it is superfluous and the other which is surprisingly missing. Garcia Gomez, the author, makes a symbolic archaeology of space.

The so called "very strange" Comares facade (Figure 1) today occupies an absurd place within the monumental whole (see Figure 2, circle 51) of the Alhambra. It was originally the main door (circle 5 on Figure 2), and it was moved to this place by the emperor Charles V so he could build a corner of his own palace into the facade of the Alhambra, as you can see in the more detailed Figure 3. There was enough space to build his palace parallel to the Alhambra, but the emperor allowed an angle of his new palace to intrude into the space where the door was originally sited.

The European building makes an uncivilized imposition onto its Arabian fellow and destroys the wise unity of the whole. The change in the location of the door is a symbol of Christian domination through an aggressive attack on the symbolic door of Islam. The emperor fortunately did not want to demolish the Alhambra but he tried "to conserve it, but reduced, neutralized and effeminated". In this reduction the two most important symbols of the Moslem Alhambra were lost: the throne and the door. This is, in a few words, a manifestation of the pro-European and anti-Arab struggle. Translating thc door to an impossiblc and illogical placc is a kind of humiliation. Garcia Gomez (1988) says: "The door should have been there where we miss it, because any palace needs a solemn exterior door. The door paid for its symbolic strength by being retired ... it became a theater decoration".

The architecture of the Alhambra was written with Arab characters and suffered a European reading, and so the imposed cultural meaning was colored with a strong ethnocentrism. There was, for example, a Christian obsession with finding the harem and to identify the palace as a lustful kind of residence. Garcia Gomez says: "We name parts of old buildings that were not given by their builders nor descriptive of the use by their inhabitants... This is most clear in the case of the Alhambra where the building did not evolve with its environment but changed drastically from one world to another, as seen from a sociological, cultural or historical perspective". Hence another passage, once more a difficult transition between two worlds.

Figure 1. Comares Facade

This example evokes an outrage, or more properly a symbolic cultural "rape". We heard from Garcia Gomez that the palace was "effeminated" by this process. The metaphor is not casual. The sexual symbolism of doors and its association to women is quite common and obvious (although it is relative to the context; for example within the dichotomy of outside door/inside door, where men can be outside and women inside). Frequently the woman is a door and the man is a key. There is an old exclusively male proverb which includes these meanings: "No me aplace puerta que a muchas llaves hace" (I do not like the door which opens with many keys).

Why women? I mentioned before that wild nature is another world, an unknown territory only explored by brave travellers facing danger and risk. But this very danger comes from inside the system. The idea of woman is thoroughly ambivalent at several levels: she is at the periphery of the social field, a potential agent of disorder and she is too close to the natural world, within the limits of nature and culture. Women have been traditionally associated with witchcraft and with the obscure arts of healing. Although she has the power of fertility, she has also the opposite possibility of destruction. So she is the agent for sickness among humans and animals and she provokes desire and disorder in human relations. Her menstrual blood makes her appear as dirty, contaminated and anomalous. Female education is an intent to domesticate her nature, take control of her possibilities and dangers, convert sexuality into reproduction, pass from resistance to rest.

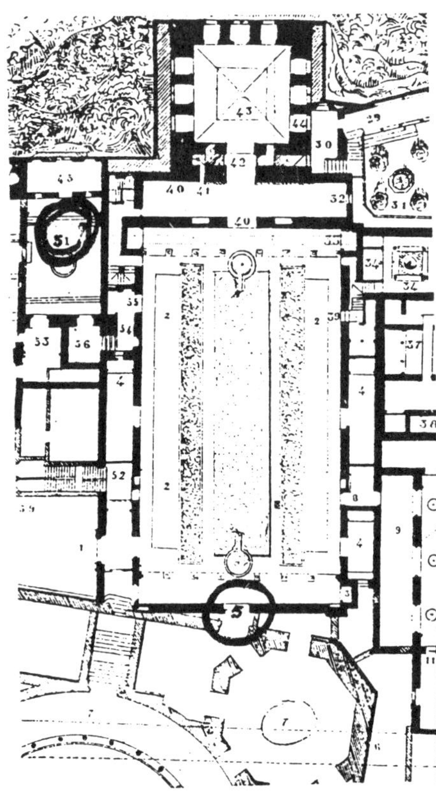

Figure 2. Alhambra (XVIII Century Plan)
51 - Comares Facade
5 - Original Site

Women so considered in the imaginary topology of different cultures are at the limit, within the threshold, peeping through the door, just where the transition takes place from social to natural, from civilized to wild, from historic times to mythical times, from people to things. In the woman seen as a border, social and natural forces converge. Through the image of woman, fertility and blood, life and death come into conjunction. Woman is, like the Alhambra, a riddle; like the door she is a restless uncertainty.

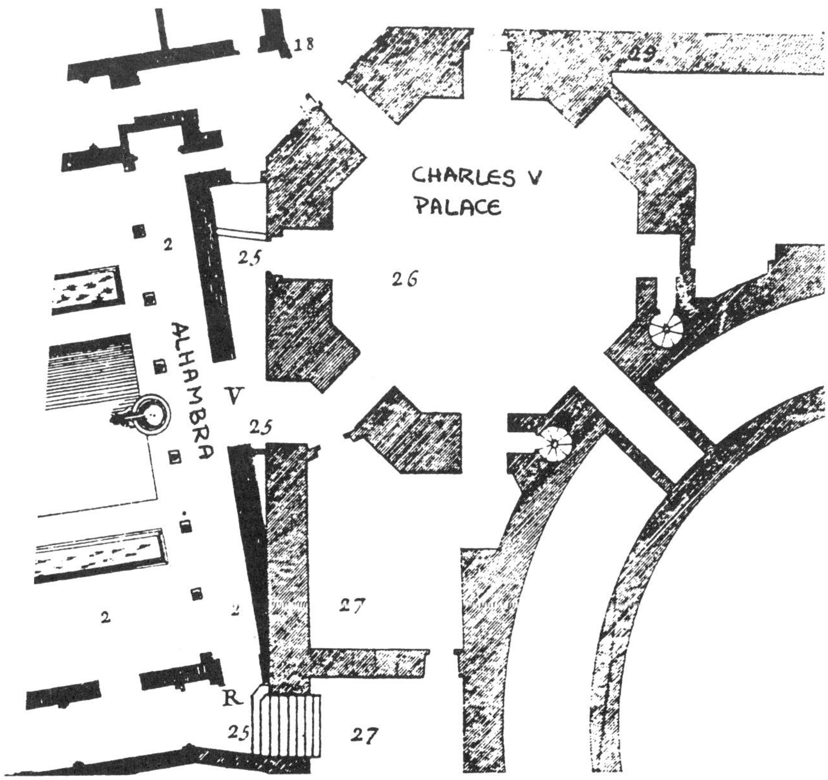

Figure 3. Detail (from E. Garcia Gomez, 1988, 184-201)

8. Notes

1 Inscription of the year 1541 which appeared at the facade of the palace of Davila in Avila, Spain. Although it has an apparently optimistic meaning its original sense is quite different. It is due to a naughty gesture of the owner of the palace. The municipality forbade him to open a door in the city walls so he constructed a second door near the main entrance of the palace and wrote this inscription into its base.

General. In this discussion I have included some basic ideas of space from the anthropological perspective inspired by the classic work of Van Gennep (1960), the more modern analysis of Victor Turner on margins (1967), James W. Fernandez and his "qualitative space" (1974), and Mary Douglas on bodily symbolism (1966, 1970). I have been consulting the work of Joseba Zulaika (1987) as well as the monograph of Francisco Sanchez Perez (1990) where he analyzes the relativity of positions within the house, an idea I have utilized here. I have also been influenced by Gaston Bachelard (1957), Jean Hani (1983), F. Paul-Levy and Marion Segaud (1983), and R. Guenon (1988). On the ambivalence of woman and their place between nature and culture I have previously published (1989: 129-165).

9. Acknowledgements

I thank Andrew Frank, Tony Oliver-Smith, and Geoffrey Edwards for their hard work translating this very metaphorical paper. Also thanks to George Lakoff and Charles Frake, among other scholars from the ASI, for their comments.

10. References

Bachelard, Gaston (1957) La poétique de l'espace, Presses Universitaires de France, Paris. Translated into Spanish : (1965) La poética del espacio, FCE.

Cátedra, Maria (1989) La vida y el mundo del vaqueiro de alzada, CIS-Siglo XXI, Madrid.

Douglas, Mary (1966) Purity and Danger, Routledge and Kegan Paul .

Douglas, Mary (1970) Natural Symbols, Pantheon, New York.

Fernandez, James W. (1972) 'Persuasions and Performances: Of the Beast in Everybody... and the Metaphors of Every Man', Daedalus, Winter 1972: 39-60.

Fernandez, James W. (1974) 'The Mission of Metaphor in Expressive Culture', Current Anthropology 15(2), June 1974: 119-145.

Forde, C. D. (1934) Habitat, Economy and Society, Dutton, New York.

Garcia Gomez, Emilio (1988) Foco de Antigua lus Solre la Alhambra, Publicaciones del Instituto Egipicio de Estudios Islamicos, Madrid.

Goodman, N. (1951) The Structure of Appearance, Harvard university Press, Cambridge, MA.

Guenon, R. (1988) Símbolos fundamentales de la ciencia sagrada, Buenos Aires: Eudeba.

Hall, Edward, T. (1959) The Silent Language, Doubleday and Co.

Hall, Edward, T. (1966) The Hidden Dimension, Doubleday and Co.

Hani, Jean (1983) El simbolismo del templo cristiano, J.J. de Olaveta, Palma de Mallorca (Originally in French, 1978, Le Symbolisme du Temle Chretienne, Guy Tredaniel, Ed.

Levi-Strauss, C. (1962) La Pensée Sauvage, Plon, Paris.

Paul-Levy, F. and Segaud, M. (1983) Anthropologie de l'espace, Centre Georges Pompidou, Paris.

Quine, W. V. O. (1963) Word and Object, MIT Press, Cambridge, MA.

Sanchez Perez, Francisco (1990) La liturgia del espacio, Nerea, Madrid.

Turner, Victor (1967) The Forest of Symbols, Cornell University Press, Ithaca and London.

Van Gennep (1960) The Rites of Passage, The University of Chicago Press, Chicago.

Zulaika, Joseba (1987) Tratado estetico-ritual vasco, Baroja, San Sebastian.

CULTURE AS INPUT AND OUTPUT OF THE COGNITIVE-LINGUISTIC PROCESSES

E. M. BJORKLUND
Department of Geography
University of Western Ontario
London, Ontario, N6A 5C2 Canada

ABSTRACT. There are dynamic input and output functions of culture in the cognitive-linguistic process. Culture, the ongoing outcome of organized experience, is characterized by inputs of differentiation and generalization. The outputs are individual and collective behavior. The design challenge for geographical information systems (GIS) is the accommodation of dynamic interaction of human behavior within and among cultures. The variety of meanings attached to people, places, and things, beliefs, attitudes, and values as timespatial phenomena are organizable using the combinatorial principle.

1. Introduction

In keeping with the theme of this Advanced Study Institute topic —cognitive and linguistic aspects of geographic space— it is appropriate to make explicit the role that culture plays as combined input and output of the cognitive-linguistic processes. The paper by Professor Maria Cátedra (1991), "Through the Door: A View of Space from an Anthropological Perspective", provides an example of the complex ways in which human experience is culturally interpreted and acted upon.

Professor Cátedra's discussion of the door as metaphor provides an entry for ideas of my own related to the consideration of cognitive-linguistic aspects of culture. Her skillful examination of a critical spatial feature common to so many cultures over time and space, expresses many powerful needs (security, privacy), interests (utility, aesthetics), and relationships (inclusion, exclusion). She reveals the complex interactive meanings making use of metaphors attached to this seemingly simple phenomenon within any culture.

2. Culture as Experience

In this context, I want to introduce a statement on culture which, I believe, relates to our discussions.

> "Culture is the ongoing outcome of organized experience. Let 'experience' stand for each individual's perception-cognition acquired from others and from self exploration inside and outside. Let 'organized' stand for the ordered and repeated meanings (and metaphors) assigned to things and their relationships by the group. Let 'outcome' represent the timespatial expression of meanings in form, use, or intention. Let 'ongoing' refer to the

D. M. Mark and A. U. Frank (eds.), Cognitive and Linguistic Aspects of Geographic Space, 65–69.

dynamic continuity of relevant selections of the past with the present with expectations toward the future." Bjorklund (1978).

This statement I find useful because it is anchored explicitly on our sensory and cognitive capacities to seek out, be attentive to, and order information we acquire into meaningful categories. Eleanor Rosch (1977, 1978), George Lakoff (1980, 1987), Mark Johnson (1980, 1987) and others, have developed the experientialist thesis. Their position is that thought is embodied and imaginative; that it has gestalt properties and an ecological structure, and it utilizes cognitive models in structuring information. This position stands as an alternative to objectivist theory. However controversial it may be, it provides a unified theory to account for an integration of mind and body, and posits reasoning as being founded on imagination which functions as a bridge to link sensory data with image-schematic transformations. (Johnson, 1987, 18-40.).

3. Culture Input

I believe work done earlier by the French psychologist Paul Fraisse provides a succinct statement of important imaginative operations involved in cognitive activity. In *The Psychology of Time* , he states that "a basic law of perception is to minimize small differences (tendency toward assimilation) and exaggerate appreciable differences (tendency toward contrast)" (1964, 77). I interpret these apparently oppositional cognitive activities as being related to the basic level categorization described by Rosch (1977) and Lakoff and Johnson (1980). Fraisse suggests that this formulation fits spatial and temporal structures equally well. This statement, in general terms, identifies two fundamental operations of cognitive processes which enable individuals, operating in group contexts, to provide a cultural input function: the cognitive acts of generalization and differentiation. While these input activities are undertaken and acted upon by individuals, they are, nonetheless, in a group context and hence not wholly independent. Sensory intake and transformation by individuals are fed by contact with other people, their earlier acquired experience, and categorization of it. From infancy we are schooled in the world of others until it becomes part of our own reality. Accordingly, the mental and physical worlds constructed by others, cast in cognitive and linguistic structures already developed, become the substance of both individual and collective learning and experience.

One of the universal characteristics of human beings is the necessity to share experience as a basis for learning and communicating. Sharing experience is clearly one of the most complex tasks of our social relations with one another. It involves the mutual focussing of attention on the same or similar phenomena. Selected aspects are articulated to others through body language, speech and its audible, written, and pictographic forms.

Structuring individual and collective experience takes many forms. Entities and identities created in the mind are reified externally. Forms, such as styles of housing and places for livelihood activities, are constructed and, when assembled in some proximal arrangement, are formalized as community infrastructures. Temporal and spatial relationships are developed to regulate and coordinate the social, economic, and political ties of people. Meanings are attached to things. Intentions are expressed. Goals and objectives are set. Among the many forms this process takes, over time some or all of these may be changed as individual and collective experience proceeds and requires alteration.

Linguistically, one of the most powerful ways of ordering experience and communicating it with others is through the use of metaphors and metonomy. Mastery of certain of the prevailing metaphors and metonomies constitutes a necessary requirement for a person to become a full-fledged member of a social group as part of a culture.

The evidence of this input process consists of the collective behaviors of people bound in the social and physical infrastructures created by them to organize and regulate their activities. Daily routines, styles and ways of doing things, the arrays of semiotic displays used to communicate what is expected, what has happened and what is happening, form other evidence of the aggregate effects of cognitive-linguistic processes in operation. As learning occurs or is ignored, information is ordered or reordered, applied or misapplied, judged or misjudged by individuals, and other people in the environment, the outcome becomes part of a culture.

Cultures are nurtured by the inputs of individuals' cognitive-linguistic processes combined with cognitively directed activity patterns of people on a continuing basis. Many kinds of behavior are simply reaffirmed by repetition. Other kinds of behavior are altered by necessity. Sometimes behavior even changes whimsically, although this is subject to the veto prerogatives of the culture.

4. Output as Culture Space

The output of the process is our individual and collective behavior, the artifacts and infrastructures within which it functions, and the communication and other forms of interaction that occur. The interactive output from these behavioral processes constitutes culture as a dynamic cognitive-linguistic process which transforms environment into *culture space*. Culture space is the ongoing outcome of people sharing a territory and infrastructures, faced with the problems of maintaining themselves in that place in relation to their own dynamic beliefs and intentions, and their dynamic relations with other people beyond. The timespatial configurations resulting from social-cultural interactions of individuals and groups becomes the social-cultural environment (output) within a physical environment.

The cognitive-linguistic activities that form the patterns of peoples' lives, identified as input and output respectively, represents only part of the interactive process of culture. As inputs are transformed and displayed as behavior patterns and artifacts, they become outputs. The resulting outputs are the characteristic enduring behaviors of people, their social and physical infrastructures that form the social-cultural environment from which they draw their ongoing experience. Output is selectively used as input by way of a multitude of cognitively organized feedback operations. These feedback operations are ones that both sustain culture and transform it timespatially. Professor Cátedra's example of the Alhambra provides a powerful example of the timespatial transformation of that place as the Islamic design and meanings were 'reduced, neutralized, and effeminated' by Charles V modifications of it. Understanding and explanation of culture anywhere involves consideration of combined effects (input-output) of cognitive-linguistic processes in dynamics of timespace.

5. GIS and Dynamics of Interaction

Geographers and other social scientists have the task of learning what prevailing knowledge and interpretive meanings have been produced from these interactions. These constitute the social-cultural phenomena (units) of our enquiry. There are two major problems confronting us in this task: the cognitive problem to construe accurately the meanings of peoples being investigated, and to find ways to represent their interactive natures in their social-cultural contexts. These confront the age-old problem of the paradox of the observer and the observation so aptly expressed by Margeneau (1950). No phenomena of culture is singular in meaning, nor are these meanings static. The interactive nature of human beings, especially when considered in their ongoing timespatial relations with others and their environment, necessarily produces transformations as generations of people succeed one another, migrations of exogenous people occur, and resources and technologies change. While these are not new ideas in any way, they are ones we have not fully put into practice in our efforts to understand and predict outcomes of human beings on any scale, either scientifically or humanistically. We have tended to treat the phenomena of culture statically and non-interactively. One of the tasks before us is to conceptualize and construct geographical information systems adequately to represent interacting complex human behavior and its social and cultural frameworks. As repeatedly pointed out in the conference, the construction of geographical information systems has the challenging task of finding ways to accommodate dynamic interactions of human beings operating timespatially within and among cultures by building *interacting* information systems. The task before the community of scholars concerned with the design, development and applications of geographical information systems is to be sensitive to the dynamics of human behavior, the variety of meanings that are attached to people, places and things, the variant beliefs, attitudes and values which coexist in cultures, sometimes peaceably, sometimes in conflict.

6. Culture as Ongoing Outcome

In conclusion, three main points may be summarized. First, there are singular *and* aggregate effects of human behavior. Second, arising out of these circumstances multiple meanings are attached to phenomena depending on the roles people play and the activities these produce. Third, places become not static in their meaning or significance but dynamic timespaces.

Confronted with the task of designing systems to describe and analyze human behavior, the systems experts are confronted with three challenging problems. Reality is only partly categorizable based on the principle of mutual exclusion: 'this *or* that'. Reality is also formed on the basis of recognizing the combinatorial principle: 'this *and* that'. Finally, the interaction of these two apparently contradictory tendencies must be built combinatorially into the information systems if they are to fulfill the important functions intended. Furthermore, given the dynamics of human behavior culturally and linguistically, the systems need to be designed to accommodate timespatial changes that result from the combined input-output process.

Culture as an 'ongoing outcome' is, therefore a combinatorially organized experience. Like a door, the dynamic of culture is both its input and its output separately considered, and its input-output combined timespatially.

7. References

Bjorklund, E. M. (1978) 'Culture Space: An Integration of Phenomenological and Empirical Viewpoints', Paper presented at The Free University, Amsterdam and The University of Utrecht, Utrecht, The Netherlands.

Cátedra, M. (1991) Through the Door: A View of Space from an Anthropological Perspective, in D. M. Mark and A. U. Frank (eds.), Cognitive and Linguistic Aspects of Geographic Space, Kluwer, Dordrecht.

Fraisse, Paul (1964) The Psychology of Time, Eyre and Spottiswoode, London.

Johnson, Mark (1987) The Body in the Mind: The Bodily Basis of Imagination and Reason, The University of Chicago Press, Chicago.

Lakoff, George (1987) Women, Fire, and Dangerous Things: What Categories Reveal About the Mind, The University of Chicago Press, Chicago.

Lakoff, George and Johnson, Mark (1980) Metaphors We Live By, The University of Chicago Press, Chicago.

Margeneau, Henry (1950) The Nature of Physical Reality, McGraw-Hill, New York.

Rosch, Eleanor (1977) 'Principles of Categorization', in N. Warrenn, (ed.), Studies in Cross-Cultural Psychology, Academic Press, London.

Rosch, Eleanor (1978) 'Principles of Categorization', in E. Rosch and Lloyd, B.B. (eds.), Cognition and Categorization, Lawrence Erlbaum Associates, Hillsdale, NJ.

DIALOGIC AND ARGUMENTATIVE STRUCTURES OF BUMPER STICKERS

D.A. KAS
Middle East Technical University
Faculty of Education
Department of Foreign Language Education
06531-ODTU-ANKARA
Turkey

ABSTRACT. Bumper stickers are again in fashion in Turkey, and many contain English-language text. The aim of this work is to analyze the dialogic and argumentative structures of these stickers. Typically, the people who use these English-language stickers are young, come from wealthy families and study at English-speaking universities. However, the content of these discourses are unfamiliar to most of the Turkish population, since the subject matter, the language, and the messages of discourses are foreign to the population. The analysis assumes that the person using the bumper sticker wants to establish a dialogue with others. The people who read these stickers start to participate in the dialogue. It is important to investigate the argumentation of these discourses in syntagmatic and paradigmatic levels. The bumper stickers show that members of the young generation want to show that they belong to a privileged social class.

1. Introduction

Bumper stickers have been well known for many years in Middle Eastern countries including Turkey. With these stickers, drivers want to show others what they think, or present their personal problems. The main themes of these kinds of bumper stickers are god, poverty, love, and economic and social problems. Most of these bumper stickers are written in Turkish. But this paper is concerned with bumper stickers written in English, which have only been in fashion since 1989. They are very different from bumper stickers which have been known in Turkey for many years. For one thing, these English-language stickers are never seen on trucks and buses; they are seen only on private cars, especially those of young people. Also, the main themes of these bumper stickers are very different from the themes on traditional ones.

The bumper sticker texts which form the corpus of this research were collected from private cars at the Middle East Technical University campus between December 1989 and May 1990. A selection was made among the bumper stickers collected. In forming the corpus of this work, there are certain reasons for the selection of this particular university campus. Education at this university is in English, and thus every one on the campus is familiar with English. At other universities, the number of English speaking students is very limited. For this reason, the non-Turkish bumper sticker fashion is not seen much among the students of other universities.

D. M. Mark and A. U. Frank (eds.), Cognitive and Linguistic Aspects of Geographic Space, 71–75.

The students who have English-language bumper stickers on their cars have many things in common. All the students, because of the educational policy of their university, can speak and understand English, and also they come from wealthy families. Since they are students, none of them have jobs, and they are the consuming youth. In other words, these students belong to a privileged social class.

1.1 WHY BUMPER STICKERS?

In Turkey, the number of people who can speak and understand English is very low. Thus the use of the English language on bumper stickers is a sign of a social class. To learn a foreign language brings with it certain advantages and disadvantages. These young people who are trying to prove that they are a member of a privileged social class are also in search of an identity. Almost all of the young people who use these bumper stickers are "lonely among a crowd". They have been kept away from politics for nearly 10 years, and they became students who only study their lessons and enter examinations. For these reasons, they want to say something to others: they want to talk to them and to communicate with them. There is satire in almost all of these bumper stickers along with the humor; however, some topics are criticized directly. The owners want to show that they have certain thoughts on those subjects. When using these stickers, they try to be different from others, and to form a social sub-group.

Before continuing, some examples of bumper sticker texts found on the campus will be of interest:

I lover my car, but more the girls

Not too close, I hardly know you

If you read this, you're too damn close

Don't honk, I'm going as fast as I can

This car may be old, but it's in front of you

2. Subject of Discourse

The purpose of every linguistic activity is to communicate with others. We consider bumper stickers as a linguistic activity because a natural language is used and messages are given. For these reasons, each bumper sticker could be considered as a "discourse". We would like to show a correlation between the subjects of these discourses and the young person who displays them. In other words, we would like to point out an opposition between the enunciator and the speaker.

An enunciator is a person who uses the language for him- or herself and who produces the language (Benveniste,1966; Ducrot, 1972, 1980, 1984; Anscombre and Ducrot, 1983). However, a speaker is not always an enunciator (Ducrot, 1980; Berrondonner, 1981; Recanati, 1979). In the bumper stickers, there is an enunciator rather than a speaker, because the subject of the linguistic activity in the bumper stickers, and the person who buys and sticks them on their cars are not the same. The ones who want to affix these stickers to their cars have already made a selection from the discourses of the others that are available on sale. At this point, the "speakers" engage in the activity of selecting from the discourses of others. The speaker wants to identify with the enunciator, but the enunciation of the speaker is loaned to someone else. However, the speaker assumes the responsibility of the discourse, because the subject has made a selection from among the many bumper stickers available in the market. This selection is important because of two points:

a. First, this is an attitude towards the discourses of others. The subject cannot produce what he or she wants, but must buy what others say. This is a part of the process of identification with the activities of others. To select from what others have already said, and to repeat these statements by affixing the bumper stickers to cars, is a passive activity.
b. The second point is that the use of English-language bumper-stickers in Turkey reflects an attitude towards the native language. This is more important than the first attitude, because the subject does not want to use his or her own native language to express himself/herself. Without producing anything in a foreign language, they want to prove themselves by adopting (and accepting) what others have already said. They do this because they need to define their social status by borrowing the discourses of others. This attitude makes the young people who try to form a such social sub-groups strangers to their own culture.

3. Argumentative Structures

The syntactic structures of the bumper stickers are constructed according to the objective of the enunciator. Every discourse has a discursive and enunciated structure that can express the message in an effective and successful way.

One of the common properties of the bumper stickers is to build a dialogue with the addressee. The linguistic structure of each discourse must form the dialogue easily. The addressee in turn shows a reaction after reading the bumper stickers. He or she may make a comment, which is usually accompanied by a smile. Thus the addressee forms a virtual discourse after reading the message. The addressee either refuses or accepts the message, but in either case participates in the discourse.

Examples:

Are you crazy? You follow too closely

Cars, like woman, don't show their performance

The enunciator aims at making the addressee repeat what is on the stickers, whether the addressee refuses or accepts the message of the discourse. The addressee is now in the hands of the enunciator, because the enunciator controls the linguistic reactions of the addressee by deictics and syntagmatic markers:

I'm not deaf, <u>but</u> I'm ignoring you

I always kill for money <u>but</u> if you are my friend I'll kill you for nothing

Flash your lights <u>if</u> you think you are sexy

The linguistic structures of discourse are important in this argumentative process. The addressee always comes to a conclusion by reading the first utterance. If the addressee accepts the first utterance as the correct one, he or she has to accept the ones that follow.

Examples:

We'll get along fine, as soon as you realize I'm God

Insured by the mafia, you hit me...we hit you!

There is dialogic structure in the bumper stickers, and that structure facilitates the formation of a dialogue by using personal pronouns, because personal pronouns designate the partners in the dialogue. In other words, the linguistic structure of the discourse is based on the opposition "you versus I" or "I versus you".

Examples:

Don't follow me...I'm lost too.

Don't talk about yourself. We'll do that after you leave.

If you are rich, I'm single.

You toucha my car, I smasha your face.

Flash your lights, if you think I'm sexy.

The marks of these dialogues are found in the syntactic structure. This dialogic structure forces the partner to continue the dialogue.

Bumper stickers have another common property: they have two or more layers of discourse. In these cases, every discourse refers to another discourse. Each discourse can be read several ways and, as a result, leads to another discourse.

Examples:

This guy knows how to party

The driver has license to kill

Pass quietly, driver is an auto-maniac

Not too close, I hardly know you

This car may be old, but it's in front of you

I love my car, but more the girls

Private bumper stickers, please do not read!

I hate bumper stickers!

The structure of the discourse makes it easier to construct a dialogue, and helps the addressee to participate in the dialogue. In this way the speakers' discourse becomes a collective subject. This is the result of being in a community. Therefore, the students who display bumper stickers in English prove that they are members of this community.

4. Summary

The study of bumper stickers may reveal interesting things about the attitudes of people who use them. The bumper sticker creates a dialog between the "speaker" (who uses the bumper sticker) and the addressee (the one who reads it). In non-English-speaking countries, English-language bumper stickers may be especially revealing of attitudes. In Turkey, they appear to reflect attitudes of young members of privileged social classes

5. References

Anscombre, J.C., Ducrot, O. (1983): L'argumentation dans la langue, Pierre Mardaga, Editeur, Bruxelles.

Benveniste, E. (1966): Problèmes de linguistique générale, Gallimard, Paris.

Berrondonner, A. (1981): 'Eléments de pragmatique linguistique,' Editions de Minuit, Paris.

Ducrot. O. (1972): Dire et ne pas dire' Hermann, Paris.

Ducrot, O. (1980): 'Analyses pragmatiques'. Communications, (32), pp. 11-60.

Ducrot, O. (1984): 'Le dire et le dit,' Editions de Minuit, Paris.

Recanati, F. (1979): La transparence et l'énonciation, pour introduire á la pragmatique, Editions du Scuil, Paris.

SECTION 3: WAYFINDING AND SPATIAL COGNITION

This section brings together six papers which bridge the gap between conceptual modeling and empirical research. Whereas theory and conceptual overviews are important, it is insufficient simply to think in the abstract about what could be or what should be. Only under exceptional circumstances can one deduce properties by considering all alternatives and showing that one solution must be correct. The papers in this section deal with spatial cognition, and each selects an important special case or situation, for which considerable literature exists or for which the authors develop and conduct specific experiments to observe human behavior. The results of these experiments in turn lead to further insight—and then to more questions and more experiments.

In the first paper, **Mark Blades** surveys current theories that have been advanced to explain the development of wayfinding. The seminal work of Jean Piaget has led to studies of the development of human cognition in children, and has produced a theoretical model involving stages of increasing competence, based on increasingly-complex cognitive situations, processes, and capabilities. This theory has influenced most discussions of human spatial cognition, and is picked up again in the experimental paper by Blades, which follows later in the section. A problem with studying spatial cognition is to observe it. A number of methods to elicit information about the spatial concepts that people use have been devised, but these methods do not all lead to the same conclusion. Observing the same aspect of cognition by using two (or more) different methods does not always yield the same result. Blades gives a potential explanation by pointing out that different methods for eliciting information may require different spatial thought processes, and thus lead to different results. He then discusses the special problem of wayfinding, and experiments that addressed specific issues such as memory and task strategies, encoding, metacognition, knowledge, and social mediation. From this discussion, it becomes clear that a wealth of facts follow from the experiments, but no coherent picture emerges. There is a lack of a comprehensive theory, since the theory that underpinned Piaget's initial work appears inadequate, yet no general model has replaced it. This conference seems to confirm that psychological experiments, linguistic analyses, and computer modeling and simulation, will all be needed to develop such a model.

In the paper which follows, **Myke Gluck** discusses wayfinding from a different perspective, starting with a discussion of the work of a planner interested in understanding the urban environment and how people move in it. He then discusses computational models, especially Kuipers' TOUR program. If a successful computer model of human spatial behavior can be built, then its content may embody a (partial) theory of spatial cognition. Simulation of wayfinding and learning about geographic space is one such area. (Spatial language use would be another.) One usually does not assume that the mechanisms used in the programs are duplications of the methods that people use, but the methods defined provide a framework that gives at least a potential explanation, especially when the program's design is based on introspection or informal surveys. It is then possible to compare the behavior of people with the behavior of the algorithm and observe areas where they differ. This would reveal potential for further improvement of the computer model, or provide designs for experiments with human subjects. Examination of erroneous human behavior can also be revealing, since a correct computer model of mind should reproduce such error. If programs produce "correct" results in situations where human subjects often produce "incorrect" results, this reveals a difference between the algorithmic solution and the solution people use. Such an algorithm embodies a formal theory that can potentially explain human behavior. Outcomes from experiments can be compared with the prediction of the theory to confirm or falsify the theory (or aspects of it).

Gluck's paper also points to the very important social and affective component of wayfinding and spatial behavior. For example, models often optimize solution paths so as to minimize route distance, travel time, or cost, assuming "economic behavior" by the people being modeled. However, people often select other paths—are they optimizing, but with erroneous knowledge of costs, or are they correctly optimizing but including other factors in their cost calculation? Or are they "satisficing", that is, finding some satisfactory solution and using it, rather than working on finding a better one? Formalized algorithmic models usually consist of different modules, separating sub-goals and limiting interaction between the modules. They make optimization goals explicit and formalize the situation precisely. Thus issues that are not explained by current theory become evident. For example, current computational models of wayfinding do not include the search for multi-location, generic objects (such as a seafood restaurant or a source for unleaded gasoline). Gluck ends by pointing to the surprising lack of knowledge about what information people need to find their way, and what factors affect these information needs. Clearly the information needs are influenced both by affective and logistic concerns and depend on the environment. Gluck concludes that the method of "sense making", which has been used to understand the information needs in other domains, should be applied to wayfinding.

In his second paper in this section, **Mark Blades** concentrates on the use of maps for finding one's way. Most cartographic research in the past has observed expert users and compared alternative designs. Recently, however, attempts have been made to understand the cognitive processes involved in navigation with maps. Using a map for wayfinding depends on three components: identification of the correspondence between the map and the actual environment; selection of the information from the map which is relevant for the task; and identification of one's own position on the map. Blades discusses these tasks and describes a number of experiments he carried out to observe actual behavior of young children. The analysis of the observation is complicated because it requires assumptions about the strategies used to solve the task and Blades goes on to change the tasks so that they help indicate which strategy was used and contain sub-tasks that require more complex strategies for their solution. In detail however, it is unclear what factors limit the performance of younger children to strategies that use a single object, yet fail if relationships *between* objects are required. Blades then turns his attention to the different strategies one can use in wayfinding and the related specific abilities. Different abilities play together to allow people to understand maps as representation of space, which can be used to direct their behavior towards achieving specific wayfinding goals.

Scott Freundschuh picks up one specific topic in map use, spatial cognition, and wayfinding. He proposes to investigate the influences of street pattern on wayfinding and map-recall competence for populations from two neighborhoods with very different geometries. Specifically, he describes experiments to test what effect regularity in the spatial environment has on the acquisition of spatial knowledge. He selected two areas in the suburbs of Buffalo (New York), one with a regular grid of streets and the other with an irregular street pattern. He asked subjects to perform tasks related to estimation of directions, distances, and locations of landmarks. For each area, half of the subjects knew the test area from daily experience and the other half learned the area only from a map. Subjects were expected to provide more accurate results in the gridded environment. For the irregular environment, map-learners were expected to perform better than residents. These predictions are likely to be true if people can successfully use grid schemas for spatial inference in gridded environments and if in the absence of grids, people cannot easily acquire geometrical (map-like, Euclidean) knowledge from experience.

In a different approach, not based on wayfinding or map use, **Gentry** and **Wakefield** concentrate on measurable aspects of more general spatial cognition. After an extensive review of existing test for spatial cognition, and general spatial abilities, they conclude that most of the known tests measure multiple aspects, often mixing spatial and non-spatial cognitive aspects. In their own experimental work, they ask the subjects to point towards objects (real or imaginary), and they then measure the pointing gestures. Then, they use techniques related to multidimensional scaling to deduce a spatial arrangement from ordinal distance measures between points in the conceptual space, fitting a space with a variable metric to the observed distances. Finally, they apply fractal analysis to determine the dimensionality of the cognitive spaces they have constructed.

In the final paper in this section, **Ian Pratt** examines methods for supporting wayfinding by robots or autonomous vehicles. The mathematical methods needed to support pathfinding involve transformations of the geometry of obstacles and their arrangements. Whereas it seems unlikely that people use these same transformations when planning routes, the paper does draw attention to the possible importance of geometric transformations in mental representations of space.

THE DEVELOPMENT OF THE ABILITIES REQUIRED TO UNDERSTAND SPATIAL REPRESENTATIONS

MARK BLADES
Department of Psychology
University of Sheffield
Sheffield S10 2TN United Kingdom

ABSTRACT. This paper considers the cognitive abilities which are required for using a spatial representation for wayfinding. These abilities include an awareness of correspondence (i.e. that there are specific spatial relationships in common between a map and the environment which it represents); the application of strategies (i.e. the use of spatial strategies to extract appropriate information from the representation which can then be applied in the environment); and the ability to find one's own position on a representation. These abilities are discussed with reference to recent experiments which have investigated the earliest evidence for the development of such abilities in young children. Children from the age of 3 years have some limited appreciation of the correspondence between a representation and the environment; from 4 years of age children can start to use simple maps, and they can understand that their own position can be indicated on a map. However, children of 4 years are generally limited in the spatial strategies which they can apply when using a representation. For example, 4 year olds tend to focus only on a single item of information from the map and though this can be adequate for some map using tasks, in other tasks one item of information may be insufficient to carry out the task successfully. With increasing age, children are able to take into account more information from a representation and they develop progressively more appropriate spatial strategies when using representations to find places or follow routes through the environment.

1. Introduction

Maps and other representations of the world are important sources of information about space, but despite their obvious importance there has been very little research into the cognitive processes which are involved in understanding and using spatial representations. Section 2 of this paper will summarize the different ways that representations can be used, and it will be pointed out that one of the most important uses for a map is for wayfinding. Unlike the use of maps for other purposes the use of a map for wayfinding involves certain component abilities which are specific to that task. More specifically, the task of wayfinding with a map requires an appreciation of the relationship between the map and the environment which it represents, and an awareness that information can be selected from the map in such a way as to aid travel through that environment.

This paper will consider the earliest development of the spatial abilities which are crucial to an understanding of the relationship between a representation and the environment (section 3) and describe how those abilities develop in young children (sections 4-6).

D. M. Mark and A. U. Frank (eds.), Cognitive and Linguistic Aspects of Geographic Space, 81–115.

2. The Use of Spatial Representations

The term spatial representation can be used in two different ways - to refer to the 'cognitive map', or mental image, of an area which an individual constructs from direct experience in the environment (see Golledge, 1987; Heft and Wohlwill, 1987; Mandler, 1983) or to refer to any physical representation of the environment (such as a map or a model). This paper will be concerned with the latter meaning of the term and will concentrate on how children develop the ability to understand physical representations.

A representation such as a map can be employed in a number of ways and these can be broadly summarized under three headings (see Board, 1978). First, as map reading, which is the direct use of the information provided by a map (e.g. to discover whether a particular city is located in a certain country). Secondly, map interpretation which involves going beyond the immediate information provided by the map to draw a particular conclusion (e.g. to estimate the comparative proportion of land given over to urban and agricultural use). Thirdly, maps can be used for navigation, and there are two ways in which a map can be an aid to navigation. One of these can be referred to as indirect (e.g. when a map is used to gain an overall, survey, understanding of an area which can supplement an individual's direct experience of that area, see Presson, 1987) and the other, more direct way when a map is specifically employed in the environment to find a place or a route.

Using a map to find a route requires the ability to relate the map to the environment which it represents - in other words the ability to extract information from the representation in such a way that that information can lead to appropriate behavior in the actual environment. It is this aspect of using a physical representation which will be the focus of the following discussion.

The distinctions made above are not, of course, mutually exclusive because most map tasks require a combination of abilities - the individual who employs a map to follow a route has to be able to 'read' the map and understand the symbols used on the map in order to find the destination before planning the route. Nonetheless, particular aspects of map using require specific skills. For example, Board (1978) has suggested that successful wayfinding with a map includes searching the map (to identify and locate one's own position and the intended destination); orienting the map (either to reference points in the environment or to compass directions); planning an appropriate route and selecting landmarks along the route, from the symbols on the map; and then identifying the landmarks as one travels the actual route until the destination is recognized.

Board's (1978) analysis, though a reasonable description of what is involved in using a representation, is an intuitive one and is not based on any empirical research of the processes required for wayfinding. It can also be noted that depending on the context of the task not all the steps suggested by Board are always necessary for successful map use. For example, some routes might be planned in terms of distances and directions ('go north for 100 meters then west for 50 meters'), or simply directions based on a sequence of junctions ('take the third road on the left, then the first right') rather than by reference to landmarks. Depending on the environment, the representation, and the preferred route there will be many different ways to plan a route to a particular destination.

Despite the acknowledgement by cartographers such as Board (1978) and Keates (1982) that one of the primary uses of spatial representations is for navigation there has been little research into this aspect of their use. The research which has been carried out with adults has concentrated on the abilities of expect users with specialized maps. Much of this research has been conducted by cartographers to consider the comparative effectiveness of different types of maps (rather than the processes involved in understanding a representation). Early examples of this approach include Taylor and Hopkin's (1975) examination of pilots' ability to use aeronautical charts and Hill's (1974) study which compared soldiers' performance with different types of orthophotomaps.

A few cartographic studies have also looked at the design of non-specialized maps (such as street guides and road maps) - for instance, Garland, Haynes and Grubb (1979) asked college students to plan bus journeys with several differently designed transit maps; Carpenter (1979) asked British drivers to plan a route using either an AA or RAC road map, and other experiments have considered the legibility of differently designed tourist maps (e.g. Clarke, 1989; Forrest and Castner, 1985). With only rare exceptions (e.g. Kovach, Surrette and Aamodt, 1988; Yarnal and Coulson, 1982) these studies have been laboratory based and have not usually tested the design of navigational maps in real environments (for a full review see Blades and Spencer, 1987a).

Other research into the use of spatial representations for navigation has considered how individuals actually use maps to find their way through the environment, but again, this research has concentrated on specific user groups. There have been a number of experiments which have examined how visually impaired children and adults use tactile maps for wayfinding (e.g. Dodds, 1989; Morsley and Spencer, 1988; Yngstrom, 1989; Spencer, Blades and Morsley, 1989) and the emphasis of these studies has been on designing effective wayfinding aids for the visually impaired.

It is only recently that detailed attention has been paid to the cognitive processes which may be involved in using spatial representations for navigation - this has been the outcome of a growing literature on the use of maps for orienteering (e.g. Barrell and Cooper, 1986; Crampton, 1988; Murakoshi, 1988; Ottosson, 1986, 1987; Seiler, 1985, 1989) and most of these studies have gone beyond laboratory based experiments to consider how orienteering participants perform in actual navigational tasks.

In summary, the current interest in the use of maps for wayfinding has led to a variety of contemporary research in different fields, but the emphasis of this research has been on the design of maps (i.e. the cartographic studies and much of the research with the visually impaired) and those experiments which have examined the information processing aspects of map using have concentrated on highly experienced and expert map users (i.e. the research with orienteering participants). It is still the case that little is known about how less experienced individuals use spatial representations for practical navigation, and in particular, how the ability to understand such representations develops. The rest of this paper will consider the abilities involved in using a representation and the early development of these abilities.

3. The Development of Understanding Spatial Representations

Until recently, there has only been limited research into any aspect of children's ability to understand spatial representations (for reviews see Boardman, 1989; Spencer *et al.*, 1989). This lack of research has undoubtedly been a consequence of the influence of Piaget's theory of spatial development (Piaget and Inhelder, 1956; Piaget, Inhelder and Szeminska, 1960). Piaget himself did not discuss children's ability to understand representations such as maps, but later researchers have drawn particular assumptions from his theory which have led to the conclusion that young children will have little ability to understand representations in the early stages of cognitive development (for a discussion of these assumptions see Blades, 1989).

In general, most discussions of Piaget's theory have ignored the important distinction which he made between perceptual spatial thought and conceptual (or representational) spatial thought. The former is what a child can understand from direct activity in space — in other words, from the direct perception of the immediate environment. Conceptual spatial thought is the ability to consider spatial features and their relationships when those relationships cannot be perceived directly. In particular, conceptual thought refers to the ability to form a mental representation of space which can be manipulated in spatial problem solving tasks (for example, the ability to mentally rotate a three dimensional object, or the ability to describe the sequence of turns along a remembered route).

When Piaget suggested that children in the earlier stages of cognitive development had limited spatial thought he was referring to limitations of conceptual thought and it is these limitations which he argued the child could not overcome until after about the age of 7 years. Piaget made the point that perceptual spatial abilities develop well in advance of conceptual abilities (Piaget and Inhelder, 1956).

Tasks involving spatial representations are not uniform, and it is important to bear in mind the particular aspects of spatial thought which may be required to perform any specific task. For example, planning a simple route might mean being able to trace a route along the paths or roads marked on a map and this is predominantly a perceptual spatial task, because the map, and therefore all the map information, is constantly in front of the user and there is no need to represent the spatial information in memory. On the other hand, selecting a line of sight from a contour map requires the visualization of the contour information as a three dimensional landscape and such a task may depend primarily on conceptual spatial thought.

Piaget's theory of spatial development should not lead to the assumption that young children's abilities with representations are inevitably limited, because much will depend on the nature of the representation and on the task which is to be performed. If the task is one that relies mainly or exclusively on aspects of perceptual spatial thought there is no reason to conclude (from Piaget's theory) that it will be beyond the abilities of a young child. Nonetheless the failure to distinguish between perceptual and conceptual spatial thought and the overemphasis on Piaget's suggestion that the latter is not achieved until after about the age of 7 years has led to frequent assumptions that younger children will not be able to understand spatial representations. These assumptions have, until recently, led researchers to ignore young children's ability to use representations and the early development of this ability (see Spencer *et al.*, 1989). However, contemporary research has demonstrated that young untrained children do have some ability to understand simple spatial representations and there are now suggestions that under certain conditions children who are just 3 year olds can appreciate a model as a spatial representation (this will be discussed below).

Piaget's theory of spatial development and the results from his research can point to important aspects of spatial ability which are related to the abilities required for using spatial representations. But most of Piaget's studies focused on aspects of conceptual spatial thought with a consequent emphasis on children's mental representation and manipulation of space (see Piaget and Inhelder, 1956) — as pointed out above, he did not directly consider the understanding of physical representations (such as maps and models). In other words, Piaget's theory does have some relevance to children's performance with physical representations, but given the general nature of the theory and its greater relevance to older children's spatial abilities, it does not provide a sufficiently detailed framework to explain very young children's understanding of representations.

The approach taken in this paper will be to discuss children's performance in different tasks with spatial representations in terms of the spatial abilities and strategies which are required for those tasks. In this way it is possible to make comparisons between different experimental tasks on the basis of the common strategies which are required for those tasks. Clearly, certain abilities will be common to all tasks (e.g. the recognition that a map is a symbolic representation is crucial before that representation can be used at all), but specific tasks may also require specific strategies (e.g. when a representation is not aligned, the user may need to take this into account before using it).

From Board's (1978) description of map using it can be argued that several component abilities are necessary when applying the information gained from a spatial representation. These include such abilities as understanding one's own position, orienting the representation, identifying landmarks and so on. In addition it is important to know that the map actually represents the real world.

It is convenient to examine the development of the component abilities separately and the following discussion of children's performance with spatial representations will concentrate on three of these components. Firstly, the ability to understand that a map or model is a representation of an actual environment, and this ability will be called the recognition of the *correspondence* between the representation and the environment. For adults in practical wayfinding tasks this can be taken as 'given' knowledge, because an adult will be able to select a map which is appropriate for the task - when travelling through London one refers to a map of London and not a map of Paris. On the other hand, young children may have an initial difficulty in realizing that a model or map is a representation of an actual part of the world.

Secondly, having understood the correspondence, it is necessary to select information from the representation which can be used to aid wayfinding. The selection of effective information which can be used in the environment will be referred to as the ability to apply an appropriate *strategy*. Thirdly, in many navigational tasks it is necessary to be aware of one's own position 'on' the map and this can be described as the ability of *self location*. These different components of using a representation will be discussed in turn.

4. Correspondence

The ability to recognize that one set of connected information (e.g. a map) has a relationship to a second set of connected information (e.g. an environment) is one aspect of analogical reasoning (DeLoache, 1989a), and Gentner (1983, 1989) has proposed a theory of analogical reasoning which considers such reasoning in the following steps. Firstly an individual has to be aware of the relationship between the two sets of information, and secondly, appropriate correspondences have to be noted between the two sets. Such correspondences can be in terms of what Gentner refers to as 'object' attributes (e.g. if two items share common properties the analogy between them can be based on those properties — 'the sun is like a yellow ball'), or 'relational' attributes (e.g. when the emphasis is on the relationship between the objects in the analogy - 'the hydrogen atom is like the solar system').

Spatial representations, such as maps and scale models, are analogies of the real environments which they portray, and a representation incorporates both object attributes (e.g. a model of a room will include a model of each item of furniture from the room) and relational attributes (the spatial relationships between the items in the model will correspond to the relationships between the furniture in the actual room). If an analogy is recognized it can be at either of the levels suggested by Gentner (1983). For instance, if an individual is shown a model of a room and notices that there is a relationship between the model and the room, this could be because the individual has noted that a particular item of furniture in the model (e.g. a chair) corresponds to an actual chair in the room. This would be a recognition based on an object attribute (and would not require a full appreciation of the relationship between the model and the environment).

Alternatively the individual might note that the chair is between a table and a bookcase in the model and that the chair in the actual room is in a equivalent position between the real table and bookcase. This recognition is based on both object attributes (the furniture) and relational attributes (the relative positions of that furniture) — and this would indicate a better recognition of the correspondence between the two sets of information. I will argue that a recognition of both object and relational attributes between a representation and an environment indicates an awareness that the representation stands for the environment, but a recognition based on object attributes alone cannot be considered a true understanding of the representation as such.

Research in domains other than the spatial domain has shown that young children can recognize analogies in problem solving and story tasks (e.g. Brown, Kane and Long, 1989; Crisafi and Brown, 1986; Gentner and Toupin, 1986), and DeLoache (e.g.1989a) has argued that 3 year old children can recognize spatial analogies (specifically the relationship between a model and a room). DeLoache's claim is important because it implies that very young children have some ability to understand a spatial representation, and if so, this could be seen as the foundation for the development of abilities related to using representations such as maps.

DeLoache (1989a,b, in press) carried out an extensive series of experiments with children aged 2.5 and just 3 years. The typical procedure in these studies was as follows: children were shown a scale model of an actual room both of which contained six unique items of furniture (chair, table, floor pillow etc) and it was explained to the children that the model corresponded to the real room. Then the children saw a miniature toy (original object) hidden behind one of the items of furniture in the model and were asked to go into the actual room and find an equivalent toy (analogous object) which had been hidden under the corresponding piece of furniture in the real room. Finally, the children also retrieved the original toy from the model (to check that any difficulties in finding the analogous toy in the room were not just due to forgetting its location).

DeLoache found that in general the 3 year olds were very successful, but that the 2.5 year olds performed poorly, and this was a very consistent finding across a number of studies which varied the procedure and materials. DeLoache (1989b) noted that although the 3 year olds were generally successful their performance depended to some extent on the explicitness of the instructions which they were given about the correspondence between the model and the room — the less preliminary help they were given about the correspondence the less likely they were to recognize the model-room relationship.

These are important findings because they suggest that the ability to recognize the relationship between a model and the environment which it represents is not present in 2.5 year olds, but by the age of about 3 years that recognition is possible (at least with some help).

These results prompt the question — exactly what are the relationships which the children recognize? DeLoache (1989b) says 'three year old children understand the relation between a model and the larger space it represents' (p. 135) and similar phrases occur in her other papers (e.g. 'the model task requires that the child understands that the model *represents* or *stands for* the room', DeLoache, in press, italics in original). However, it is not always clear whether DeLoache is referring to the recognition simply of object attributes or to some more sophisticated appreciation that the whole model represents the whole room.

In one study, DeLoache (in press), varied the correspondence of the furniture so that the pieces in the model were the same type of furniture but were not identical to the real furniture in the room, and the 3 year olds' performance declined. In another study (DeLoache, in press) the furniture in the room and model were identical, but the items in the model were repositioned so that their position no longer corresponded with the position of the same furniture in the room. For half the children the toy was hidden in the room behind the same piece of furniture that the miniature toy was hidden behind in the model. For the other children the toy was hidden in the same spatial position in the room as its position in the model (irrespective of the furniture). The results showed that the '3 year olds firmly expected that the toys would be hidden with the same items of furniture in the room and the model, even though those items were in different positions within the spaces' (DeLoache, in press). When the toy was hidden in equivalent spatial positions the children had very poor performance and were not able to locate it successfully. These, and similar results reported by DeLoache, imply that the children's performance was based on recognizing the correspondence between individual items of furniture.

From the studies summarized in the last paragraph it would appear that the children relied largely on simply recognizing an equivalent item of furniture in the two spaces (which can be described as a reliance on object attributes) and to imply more than this might overstate young children's understanding of representations. To clarify this point we (Blades and Cooke, in preparation) replicated DeLoache's (1989a,b, in press) experimental procedure, but added an important condition. DeLoache always used *unique* items of furniture as hiding places in the models and rooms, but we included a condition in which two identical pieces of furniture (i.e. hiding places) were placed in the same model.

We argued as follows — if children have a full understanding that a model can represent an environment (i.e. an understanding based on relational attributes) then the presence of two identical hiding places in the model should not effect performance (because children will be able to distinguish those places by relating their position relative to other places in the representation). However, if children's understanding of the analogy between a model and an environment is based simply on object attributes (i.e. individual and separate items) they will have difficulty in identifying the correct item of furniture if that item is one of two identical objects.

4.1. MODEL EXPERIMENT

4.1.1. *Subjects and procedure.* The 35 subjects were all the available children from a preschool playgroup and a preschool nursery class. They were divided into two age groups: 19 children with a mean age of 3 years (range 2 years 7 months to 3 years 3 months) and 16 with a mean age of 4 years 3 months (range 3 years 9 months to 4 years 10 months).

The children were shown two identical model spaces. Two identical models of the same scale were used because DeLoache (1989a) demonstrated that, generally, the greater the similarity between two spaces the more likely that children would recognize the correspondence, and we wanted to maximize the likelihood of the two spaces being treated as equivalent.

Each model was made from a box, 30cm x 30cm x 30cm high, which was open at the top, and included four items of furniture (see Figure 1). In conditions 1 and 2 each model space included four unique items of toy furniture — bed, wardrobe, table, and chair. In conditions 3 and 4 the table was removed and replaced by a second chair which was identical to the first chair, in other words, the models included a bed, wardrobe and two chairs.

A pair of models was placed on the floor in front of a child so that she could see into both models at the same time (to minimize any memory effects). In conditions 1 and 3 the models were aligned, and in conditions 2 and 4 one model was rotated 180° relative to the other.

The children were tested individually in a room which, apart from the models, was empty so as to minimize distractions. They were shown both models during a familiarization period and it was emphasized that both the model spaces were the same. They were also shown two identical miniature toy dogs (A and B) and given four practice trials with these toys. Toy A was placed on an

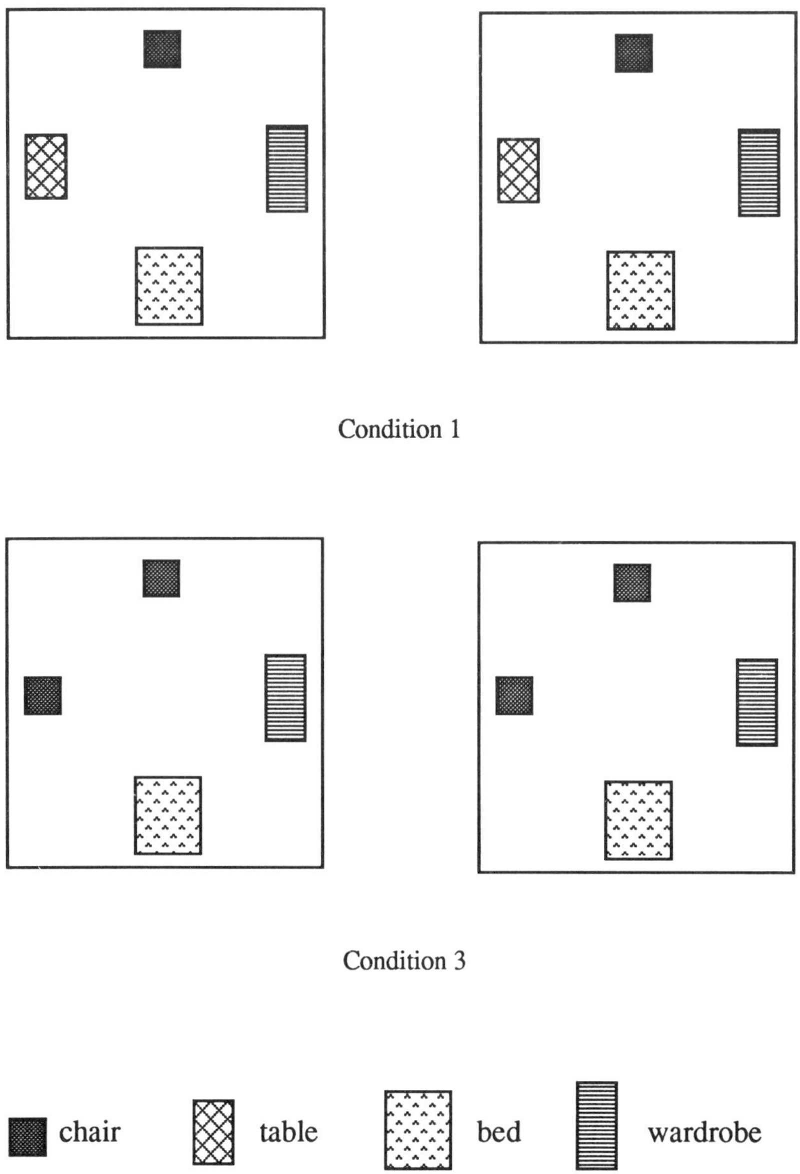

Figure 1. The two layouts used in conditions 1 and 3 of the model experiment. (The pairs of layouts used in conditions 1 and 3 were also used in conditions 2 and 4 respectively, but in the latter conditions one of the pair was rotated 180 degrees relative to the other.)

item of furniture in one model and the child was asked to place toy B on the equivalent piece of furniture in the second model. After the familiarization period the child was given the experimental trials.

In each trial the child watched as toy A was hidden in or under an item of furniture in one of the models. Then, while the child was not watching toy B was hidden in the equivalent place in the second model. The child was allowed to look again, and was told that toy B was hidden in the same place in the second model as toy A which she had seen hidden in the first model. She was asked to find toy B in the second model (the analogous retrieval) and then find the toy which she had seen hidden - toy A in the first model (the original retrieval). Locating toy A served as a check that the child had actually remembered the position of the toy.

Each child carried out 12 trials (the toy was hidden three times under each item of furniture, in random order). There were 12 trials in each condition and the four conditions were administered in the same order (1 to 4), but on separate days.

4.1.2. *Results.* The mean scores for each condition for both analogous and original retrievals are shown in Table 1. In all conditions and for both age groups the scores were better for the original retrieval and this indicates that any difficulty the children had on the analogous trials was not simply due to forgetting the position of toy A or any lack of motivation.

Condition	3 yr olds		4 yr olds	
	Analogous	Original	Analogous	Original
1	9.53	10.11	10.94	12.00
2	9.11	9.53	10.31	11.56
3	8.06	8.84	9.06	10.81
4	6.79	8.21	8.25	9.63

Table 1. Mean scores for each condition and age group for both analogous and original retrievals in the model experiment (maximum score was 12).

Overall the children performed well in the analogous trials (if they had not understood the task and resorted to guessing there was only a 0.25 probability of being correct and they would have achieved only 3 successful trials out of the 12). A 2(age) x 4 (condition) analysis of variance showed that there was no effect for age ($F=2.01$, d.f. 1,33, $p>0.05$), but there was a significant effect for condition ($F= 32.91$, d.f. 3,99, $p<0.01$) and post hoc multiple comparisons (Tukey HSD, $p<0.01$) showed that there was no difference between the mean scores for conditions 1 and 2, (10.17, 9.66) but these scores were better than the mean score in condition 3 (8.51) which was better than the mean score in condition 4 (7.46). In other words, the children were more successful when there were four unique hiding places in the models, and performance declined when the models included identical hiding places.

Of particular importance is how the children performed for each hiding place in the analogous trials. In conditions 1 and 2 (for both ages combined) the mean scores, out of a maximum of 3, for each hiding place did not differ (bed, 2.52; wardrobe, 2.55; table, 2.58; and chair, 2.43). In conditions 3 and 4 the mean scores for the unique hiding places were similar (bed, 2.60; wardrobe,

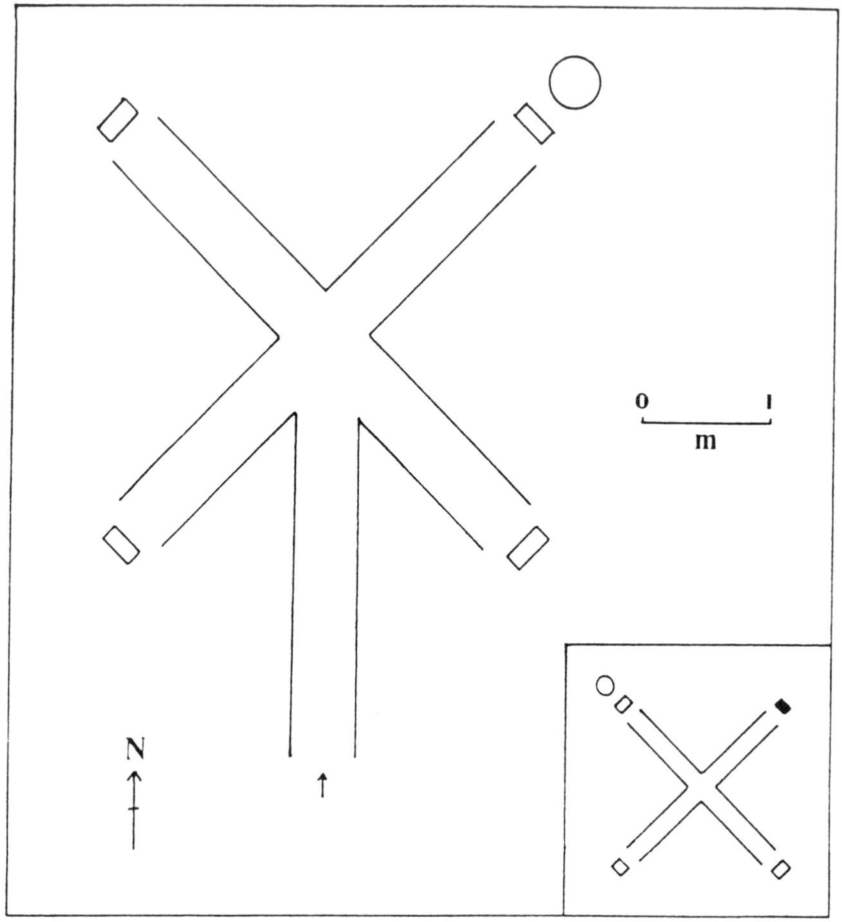

Figure 2. Plan of layout for the crossroads experiment with, inset, an example of a map indicating the position of a hidden toy (the shaded box). The landmark was a red bucket.

2.64), but there was a marked decline for the other two, identical, hiding places (chair, opposite wardrobe, 1.35; chair, opposite bed, 1.54).

From these scores it can be seen that the children were equally successful whenever the analogous toy was hidden in a unique place, but performance was much poorer when the toy was hidden in one of the two identical places (in conditions 3 and 4). When the toy was hidden under one of the chairs in the first model the children had difficulty distinguishing between the chairs in the second model, and more often than not were simply guessing which of the two chairs was the correct hiding place.

Of course, the correct chair could have been successfully identified if the children had noted its relationship to other items in the model (e.g. by encoding the correct location as the 'chair opposite the bed'). The children's reliance on unique hiding places (object attributes), and their inability to distinguish between identical hiding places (which required some appreciation of relational attributes) suggests that the children only had a limited appreciation of the correspondence between the two spaces. They could understand that an item in one space represented an item in the other space, but not the relational correspondence between the two models. In other words, in this experiment, the children's ability was based on the identification of unique objects (or landmarks) which were common to both spaces, and not on a complete understanding of the overall correspondence between those spaces.

4.2. CROSSROADS EXPERIMENT (Blades and Spencer, 1990).

In another study we asked 3 to 6 year olds to use a map (rather than a model) to find a hiding place in a simple layout. The layout consisted of five paths, one of the paths led to the center of a 'crossroads' and at the end of each path of the crossroads there was an identical wooden box in which a small toy could be hidden (see Figure 2). A large red plastic bucket (which could be used as a reference point or landmark) was placed to the north-east of the layout. The children were given maps which were attached to a clipboard so that they could carry the maps with them (to reduce any memory component in the task). The maps showed the crossroads and a red circle to represent the landmark, and the box in which the toy was hidden was shaded.

Each child was given 16 experimental trials with different maps. The children were given the maps (in random order) aligned with the layout or rotated 90°, 180° or 270° clockwise relative to the layout (with four maps of each alignment), and all four possible hiding places were used for each alignment. In each trial the child's task was to walk through the layout and locate the hidden toy in the correct box.

The results from this study complement the findings from the model experiment described in section 4.1. Firstly, even the youngest children had some ability to understand that a spatial representation (in this case a two dimensional map) corresponded to the layout. The 3 year olds in this experiment performed better than would have been expected by chance with aligned maps (irrespective of the toy's hiding place) and with rotated maps (*if* the toy was hidden in the north-east box, next to the landmark). But their ability was limited to those two conditions - when the maps were rotated and the toy was not hidden in the box next to the landmark their performance declined.

Secondly, all the age groups performed best with the aligned maps, and this corresponds to the result from conditions 3 and 4 of the model experiment, that when identical hiding places are included in a task young children's performance will be superior when the representation and the environment are in alignment. It was inferred from the errors of the 4, 5, and 6 year olds with the 180° rotated maps that they sometimes made a false assumption that the map and layout were in alignment (though this effect was not found for the maps of other rotations). This type of error was also found in a map study with young children by Bluestein and Acredolo (1979) which is discussed below.

Thirdly, although all the older age groups performed better with the rotated maps than the 3 year olds, their performance also reflected the effect of the hiding position. They were nearly always correct when the toy was hidden in the north-east box. They were less successful when the toy was in the south-west box, but more likely to find the toy in this box than when it was in either the north-west or south-east boxes (and there was no difference in the children's performance when the toy was hidden in the latter two boxes). The north-east box (next to the landmark) and the south-west box (opposite the landmark) had unique relationships to the landmark and could be located in terms of whether the boxes were near to or far from it. On the other hand, the north-west and south-east boxes were equally distant from the landmark and the children could only distinguish these boxes by considering whether they were to the left or right of the landmark and this may have been difficult in the conditions when the maps were not aligned.

In summary, the map experiment demonstrated that 3 year old children could use a simple two-dimensional representation, but their performance was particularly dependent on being able to identify one box (the box associated with the landmark). Like the model experiment this would suggest that such young children do not have a full understanding of the correspondence between the representation and the environment. The older children showed an age related improvement in the ability to differentiate the correct hiding place and by the age of 6 years the mean score of the children was 11 correct out of 16 trials indicating a comparatively complete understanding of the relational correspondence between the map and the layout.

4.3. SUMMARY

The children's performance in the experiments described in this section and equally young children's success in some conditions of several other map using experiments (e.g. Blades and Spencer 1986a, 1987b; Bluestein and Acredolo, 1979; Freundschuh, 1990; Uttal and Wellman, 1989) demonstrate that in tasks which involve simple maps and layouts, 3 and 4 year children have some understanding of the correspondence between representations and environments. But this understanding may be limited to identifying individual items of information (such as a piece of furniture in a model, or a landmark on a map).

When children can rely on a single item of information to perform a particular task they will be successful. However when children cannot rely on a single item because it is ambiguous (e.g. one of two identical landmarks on a map) then they need to apply an effective strategy to distinguish that item. Such strategies will be discussed in section 5.

Studies which have considered 3 and 4 year olds use of representations in conditions which manipulate the alignment of the representation (e.g. in both experiments described above, and in Bluestein and Acredolo, 1979) have shown that such age groups usually perform better when the model or map is oriented with the environment. [This is also the case for older children (e.g. Liben and Downs, 1989; Presson 1982) and for adults (e.g. Levine, Marchon and Hanley, 1984)]. There are several reasons why children's performance with oriented representations may be better than with rotated representations. Firstly, the correspondence between the representation and the environment may be easier to recognize when the two are in alignment. Secondly, if children falsely assume that a representation is aligned when it is not (as in the 'crossroads' experiment, just described) this will result in errors when using rotated representations.

Thirdly, the spatial strategies which can be applied to tasks involving oriented representations may be less difficult to use. When the representation and environment are in alignment the user may recognize, for example, that anything to right of her position on the representation will also be to her right in the environment, and often comparatively simple, but effective strategies can be based on this realization (see below). But when representations are not oriented the user has to employ a strategy which compensates for the lack of alignment (e.g. physically or mentally rotate the representation, or if the representation is rotated 180° relative to the environment, consider all right turns on the representation as left turns in the environment, and so on). Alternatively, when a representation is not oriented the user can apply a strategy which is not effected by the representation's rotation. For example, identifying landmark B as a place or feature between landmark A and landmark C from a map may be an effective way to locate B in the environment and this form of reasoning is not dependent on the rotation of the map. These aspects of understanding spatial representations will be discussed in the following section.

5. Strategies

In this section it will be pointed out that any individual task which requires the use of a representation may be open to several different strategies and that often more than one of those strategies can lead to a successful solution of the task.

Bluestein and Acredolo (1979) examined 3 to 5 year old children's ability to use a map to locate a hiding place and (like the 'crossroads' study, section 4.2) Bluestein and Acredolo used a number of identical hiding places. Bluestein and Acredolo showed the children a map of a small room which included a table and chairs (in the middle of the room) and four identical green boxes (at the mid-point of each wall) which served as hiding places for a toy. In one corner of the room there was a door, and in each of the other three corners there was a different salient object (e.g. a box painted red) which could be used as a landmark (see Figure 3). The map showed all the furniture and features in the room and it also indicated in which of the green boxes a toy was hidden. The children viewed the map either inside or outside the room, and either correctly oriented with the room or rotated 180° relative to it, and after they had seen where the toy was hidden on the map they had to go and look for the real toy in the actual room.

In the first condition, when the children saw the map inside, and aligned, with the room half the 3 year olds, three-fourths of the 4 year olds and all the 5 year olds were able to locate the toy on at least two trials out of three. Performance was only slightly poorer when children viewed the map outside and aligned with the room (it is possible that some of the difference in the children's performance between conditions 1 and 2 was due to the additional memory factors involved when they saw the map outside the room). When the children saw the maps rotated 180° (either inside or outside the room) the 5 year olds still performed well, but most of the younger children were unable to use the maps successfully. (For a critical summary of some of Bluestein and Acredolo's results, see Liben and Downs, 1989).

How did the children use the map in Bluestein and Acredolo's (1979) experiment? In the first condition of Bluestein and Acredolo's study the children were shown the aligned map exactly in the center of the room and this meant that the actual hiding place in the room was always the green box nearest the box indicated on the map. This could have made the task trivially easy in that condition and required little understanding of the map. Nonetheless, in condition 2 (when the aligned map was shown to the children outside the room) there was no such relationship between the toy's

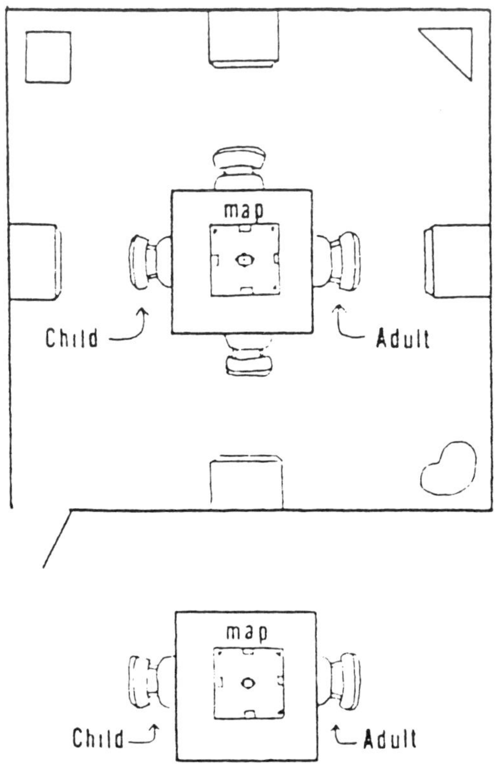

Figure 3. Plan of room used in Bluestein and Acredolo's experiment (copied from their illustration: 1979, 693).

location and the indicated box on the map, and the children were generally successful at finding the toy.

The children may have assumed that the map was always correctly aligned and then, for example, if the toy was hidden in the green box on the left of the map (relative to the child's own viewing position) she could expect the toy to be in the box on her left in the actual room. This may have been the case for the 3 and 4 year olds, because Bluestein and Acredolo noticed that when these age groups made errors with the 180° rotated maps they often chose the box directly opposite the correct one (i.e. the children were using the rotated maps as if they were aligned).

Bluestein and Acredolo found that the 5 year olds were successful irrespective of the map's alignment and they suggested that 'the children had available to them simple relations like "between the door and the red box" to help them cope with the rotation. The 5 year olds. . .took advantage of this information' (1979, 696). Such a strategy would always locate the correct hiding place in the room, but in Bluestein and Acredolo's experiment the children could have employed a less precise strategy and still have been scored as successful. They may simply have encoded from the map that the toy was in the hiding place 'near the door'. (i.e. relied on a single salient landmark as a reference point). There were two hiding places which were each equally distant from the door and if a child guessed between those two hiding places she would have had a 0.5 chance of finding the toy. In Bluestein and Acredolo's experiment children only had three trials and they were rated as successful if they located the hidden toy in just two of those trials. Therefore, some of the children who relied on a 'near to' strategy would sometimes have been scored as successful.

Presson (1982) suggested that children using maps which were not aligned could compensate for the lack of alignment by some form of mental rotation, and an experiment carried out by Presson (1982) found that 6 and 8 year olds performed better with maps which were shown to them rotated 90° compared to maps which were rotated 180°. That the degree of rotation had some effect on ability could be interpreted in terms of the children using a mental rotation strategy, and being able to carry it out better when less rotation (to achieve alignment) was involved. On the other hand, there were no differences in children's performance with any of the three rotated maps included in the crossroads experiment described in section 4.2 and therefore the evidence for mental rotation in map using tasks is only tentative.

Apart from the strategies discussed above, there are many other ways that children could have approached the map task used by Bluestein and Acredolo (1979). For example, they could have attempted to locate the correct green box by counting from a known point on the map (e.g. the door) in a given direction (e.g. clockwise) round the perimeter of the room and noting the number of the box containing the toy ('third box clockwise from door'), and then applied this information in the room itself. Alternatively they could have imagined themselves at a point 'on' the map (e.g. the center) facing a specific direction (e.g. the red box in the corner) and then the hiding place with the toy could have been identified by reference to their own position ('directly left when facing the red box from the center'), and then this information could have been used in the room. Describing these strategies is not meant to imply that young children employed (or would have been capable of employing) such strategies in Bluestein and Acredolo's (1979) experiment. But they are given here as examples of feasible strategies to emphasize that there are several ways that successful performance could have been achieved in Bluestein and Acredolo's task.

The important point is that tasks which involve using spatial representations to find a place may be solvable in different ways. Knowing that a child of a certain age can (or cannot) perform the task does not usually indicate much about how that child solved (or failed) the map problem. Similarly, without knowing what abilities or strategies are required for a particular task it is difficult to say much about the development of those abilities. For example, if it is found that 5 and 8 year olds are equally successful in a map using task this could mean that there are no developmental changes in map using skills on the that task. Alternatively, it could be the case that the different age groups perform the task in different ways. Therefore it is important to construct map using tasks in such a way that the subjects' strategies can be identified, and the studies described in sections 5.1 and 5.2 are examples of map experiments which have considered children's performance in terms of the spatial strategies which they used. The first example is a summary of an experiment in which young children were asked to use a map to locate a particular place (a path junction) in a large scale layout.

5.1. STRATEGIES IN A LARGE SCALE LAYOUT (Blades and Spencer, 1987c)

5.1.1. *Subjects and procedure.* Forty-eight children in three age groups (4.5, 6 and 7.5 years took part in the experiment). A large octagonal layout (7m across) was designed and consisted of an outer path and eight paths leading to the center of the layout (see Figure 4). At the outside end of each of the eight paths there was a box which was open at the top so that children could see inside it, and on the inside base of the each box there was a colored circle. These could be used as landmarks, but because of the depth of the boxes a child could only see into a box when she was standing next to it. In other words, she could only see one landmark at a time.

The maps showed the layout and the colors of the landmarks, and they also showed that all but one of the paths had a roadblock (see Figure 5). There were no roadblocks in the actual layout, but the children were told that they were not supposed to walk down a path in the layout which was marked with a roadblock on the map - their task was to find the path which did not have a roadblock on the map and walk down that path to the center of the maze (the children were not told whether they were correct in each trial). Children walked clockwise from the south of the layout and carried the maps with them so that there were no memory component in the task. The maps were always given to the children so that they were rotated relative to the layout.

5.1.2. *Strategies.* There were three conditions and these can be described by reference to Figures 4 and 5. In condition 1 the path which was shown without a roadblock on the map was the one marked by the yellow landmark. In condition 2 the correct path was one of the two paths marked by red landmarks. In condition 3 (see Figure 5) the correct path was one of the three paths with black landmarks.

In condition 1 it was sufficient to note the color of the landmark at the end of the correct path on the map (yellow, in the above example) and then look for the path with the yellow landmark in the layout. In condition 2, having noted that the correct path is marked by a red landmark) it is necessary to distinguish between the two red landmarks in the layout. There are several strategies which can be used to do this (e.g. if the correct red landmark is the second red, clockwise in Figure 4, this could be encoded as the red after the green landmark). In condition 3, the correct black landmark has to be distinguished from the other two black ones. When (as in Figure 5) the correct black landmark is the third black one it is not sufficient to just encode it as the black after the red, because the second black landmark also follows a red landmark, and if this strategy is used there will be no effective way of discriminating between the paths marked by the second and third black landmarks. These paths can only be identified in the layout if some additional information is encoded from the map (e.g. by considering the correct black landmark as the black after the red after the green).

Figure 4. Example of the environment for the layout experiment. The landmark at the end of each path was a painted circle on the inside base of a cardboard box. In this example the colors were, from south clockwise, green, black, yellow, red, black, green, red, black.

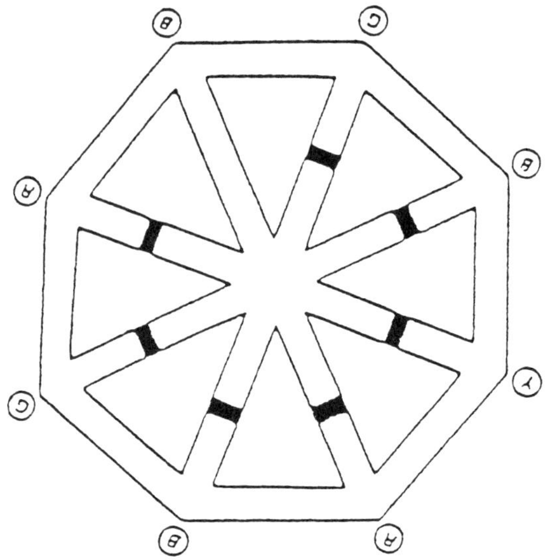

Figure 5. Example of a map used in condition 3 of the layout experiment. The maps were always given to children rotated relative to the layout. For landmark colors see Figure 4.

Several different strategies can be employed in this experiment:

1. 'Note the color of the landmark at the end of the correct path on the map and look for that color landmark in the layout.'

2. 'If the landmark at the end of the correct path on the map is unique, then look for that color landmark in the layout. If it is not unique, then also note the previous landmark on the map.'

3. 'If the landmark at the end of the correct path on the map is unique, then look for that color landmark in the layout. If it is not unique, then also note the following landmark.'

4. 'If the landmark at the end of the correct path on the map is unique, then look for that color landmark in the layout. If it is not unique, then also note the the colors of the preceding and following landmarks'

5. 'If the landmark at the end of the correct path on the map is unique, then look for that color landmark in the layout. If it is not unique, then also note the colors of the two preceding landmarks'

6. 'Locate own position on map and then update own position on the map while walking round the layout, until the map indicates the correct path.' There are at least two ways in which a child can locate herself:-

6a. with reference to Figure 4, a child starting at the south (green) landmark can note that the next landmark as she walks is black. She can then look for the black landmark which is preceded by a green landmark on the map and hence locate her position on the map.

6b. she can search the layout for a place which is uniquely shown on the map. For example, in Figure 4 this would be the yellow landmark. Having found the yellow landmark in the layout she can update her position as she walks round until arriving at the correct path.

7. 'Locate own position on the map, then count the number of paths from own position to correct path on the map. Then count the same number of paths, from present position, in the layout.'

8. 'Locate own position on the map, then orient the map so that it is aligned with the layout. Then the direction of the correct path in the layout will correspond to the direction of the correct path on the map.'

Each strategy will result in a specific pattern of performance across the different conditions, and such patterns can be identified by considering a child's performance. For example, a child who relied on strategy 1 (looking for the color of the landmark at the end of the correct path) would be successful with maps in condition 1, but would not be able to distinguish between the two, or three, identical landmarks in conditions 2 and 3. In the latter conditions the use of strategy 1 would lead incorrectly to the selection of the first path in the layout with a landmark of the expected color (in Figure 4: the north-west path in condition 2, and the south-west path in condition 3).

As other examples: a child who employed strategy 2 would be successful in conditions 1 and 2, but not in condition 3 when she would incorrectly select the north path. Strategy 4 would produce the same pattern of successful trials and the same errors as strategy 2, but strategy 4 can be identified from a child's behavior because when using the latter strategy the child will need to walk past and then back to the selected path.

With the exception of strategies 5 and 6a, all the strategies result in different and observable patterns of behavior in the layout, and by analyzing the children's successes, type of errors, and behavior it was possible to identify the particular strategy used by most of the children in the study (see Blades and Spencer, 1987c for full details).

5.1.3. *Results*. The 7.5 year olds used several of the strategies, and were successful with some or all of the maps, but the majority of the 6 year olds and nearly all the 4.5 year olds used strategy 1 (see Table 2). None of the children used strategies 3, 7 or 8.

Strategy	Age group (yrs) 4.5	6	7.5
1	14	10	2
2	1	-	1
4	-	2	5
5/6a	-	1	3
6b	-	-	2

Table 2. Number of children in each age group using each strategy in layout experiment

The younger children in this experiment relied predominantly on strategy 1 which only took into account a single landmark, and this was not a very successful strategy when there were two or more landmarks of the same color in the layout because strategy 1 did not effectively distinguish between these identical landmarks.

In a second experiment in a slightly different layout which also included identical landmarks (Blades and Spencer, 1987c, Experiment 2) 6 year olds were able to take two landmarks into account when using maps, but 4 year olds rarely did more than just look for the color of the landmark at the end of the correct path on the map. In different conditions of this experiment two factors which might have effected the children's performance were examined. In one condition the children were given feedback after each trial which indicated whether they had found the correct path in the layout, and in another condition all the landmarks were completely and equally visible in the layout, but these variations had no effect on the 4 year olds performance.

In other words, across a number of layouts, the 4 year olds only referred to a single landmark, and this would not imply a full understanding of the spatial relationship between the maps and the layouts, and it leads to the question: why were the youngest children so limited in their performance in these studies? Given that feedback and landmark visibility did not lead to improved success for this age group, the most likely inference is that the youngest children were simply unable to apply an appropriate strategy to 'disambiguate' the identical landmarks, and to examine this possibility a further experiment was carried out.

5.2. BOARD GAME EXPERIMENT

5.2.1. *Subjects and procedure.* Sixty children were selected randomly from a nursery and a first school. There were three age groups with mean ages of 4 years 6 month (range 4 years 4 months to 4 years 9 months), 5 years 5 months (range 5 years 3 months to 5 years 7 months) and 6 years 5 months (range 6 years 1 month to 6 years 7 months).

Three pairs of circular (25 cm diameter) boards were designed. The boards in each pair were identical. One pair of boards (Figure 6) included 12 small painted circles arranged in pairs round the edge of the board, and each pair of circles included a black circle and a circle of a unique color. On the other two pairs of boards (one is illustrated in Figure 7) the 12 painted circles were equally spaced; a black circle alternated with a circle of a unique color.

Children were tested individually, in an empty classroom, and the child sat on the floor with a pair of boards in front of her. The two boards were never oriented, but always rotated at least 90° relative to each other. A small token was placed on one of the circles on one of the boards and the child was asked to place a token on the same circle on the other board.

The first pair of boards (Figure 6) was used as a pretest. A token was placed on one of the colored circles on one of the boards and the child had to find the same colored circle on her board. This was repeated for each of the colored circles (in random order) and the child's board was rotated at least 60^{f} between each trial. All but one child (a 4 year old who was excluded from the experiment) were always able to perform this part of the task successfully. Then, in six further trials, the token was placed on each of the black circles and the child had to select the same black circle on her board.

The pretest therefore consisted of 12 trials (six with colored circles and then six with the black circles). Children who were correct on at least 4 of the 6 trials with the black circles were then tested with the other pairs of boards (Figure 7), and the procedure was similar; on each pair of boards the token was placed on each of the six black circles (in random order and rotating the child's board between trials). Therefore there were a total of 12 test trials altogether.

5.2.2. *Strategies.* The boards were designed so that the children's strategies would be reflected by their scores and performance.

1. No strategy. If the children did not have a strategy and simply selected one of the black circles on their board by guessing they only had a 1 in 6 chance of being correct in each trial in the second half of the pretest. In other words, in this part of the test the children would score less than 4 and would fail the pretest.

2. 'Next to' strategy. This strategy refers to the appreciation that a target place can be encoded by reference to an immediately adjacent landmark. If children used a 'next to' strategy they would have been able to complete the pretest (because the black circles on the boards used in the pretest were always paired with a circle of a unique color). But children using the strategy on the boards where the circles were equidistant from each other would not be able to locate the correct black circle unambiguously.

For example, if the target black circle was the one between the yellow and the blue circles (in Figure 7) the child could have considered it as 'next to the yellow' or 'next to the blue'. However,

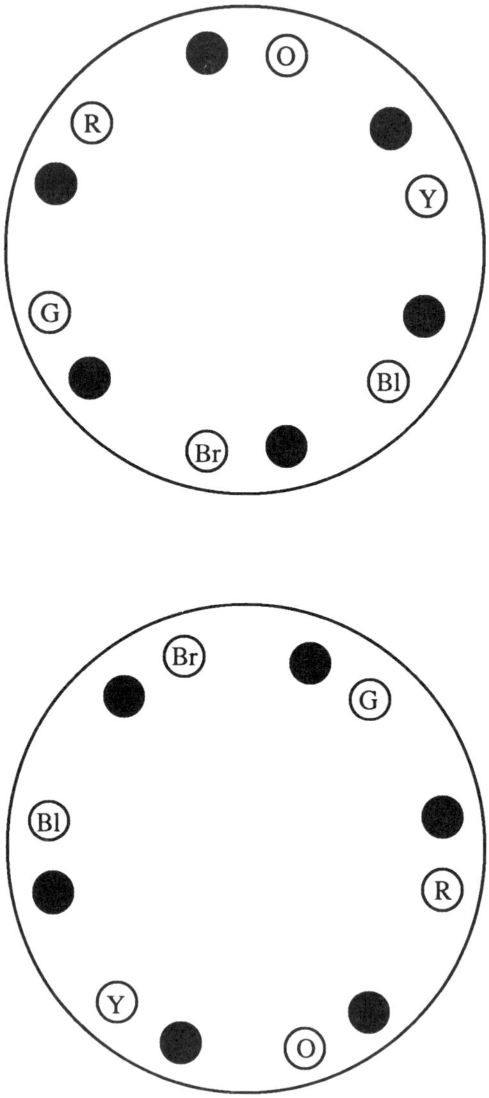

Figure 6. Pair of boards used for the pretest in the board game experiment. There were six black circles and one each of orange, yellow, blue, brown, green and red.

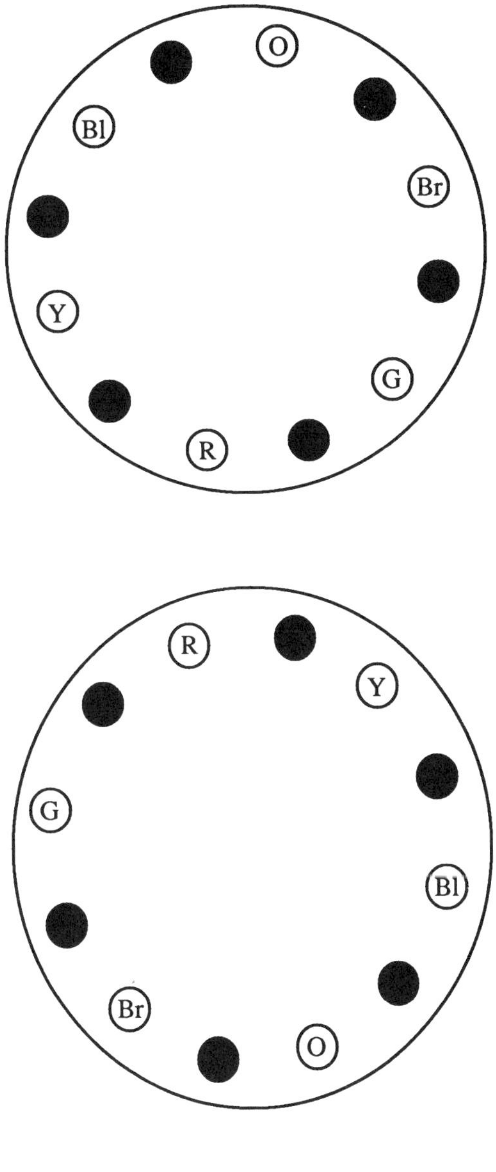

Figure 7. Example of one of the two pairs of boards with equally spaced circles for the board game experiment. There were six black circles and one each of orange, brown, green, red, yellow and blue.

there were two black circles next to the yellow; the correct one and the one between the yellow and the red. Therefore the child would have to guess either of these. Similarly there are two black circles next to the blue; the correct one and the one between the blue and the orange, and the child would have to guess which of these two was correct.

In other words, a 'next to' strategy always leaves the child guessing between two possible black circles on her board. Over a number of trial this strategy will result in a score of 50% correct trials, and all the child's incorrect choices will be to one of the black circles on either side of the correct one.

3. 'Between' strategy. This means identifying the target black circles being between two other colors. Children who used this strategy would pass the pretest and be correct on the other boards because such a strategy always locates the target unambiguously.

5.2.3. *Results*. Children who used the 'next to' strategy on the boards with the equally spaced circles would have been correct on approximately half of their trials and therefore to distinguish between these children and those using a 'between' strategy the following criteria were used. The probability of a child using the 'next to' strategy being correct on a single trial was 0.5, but to make correct choices on 9 or more trials out of 12 with this strategy is only 0.073 (binomial distribution), and therefore children who achieved scores of 9 or better were considered to be using 'between'.

Children who used 'next to' might have had a range of scores lower than 9, but all their incorrect choices would be to a black circle on one or other side of the correct one (which will be referred to as an adjacent error), and children who made 11 or more choices of the correct or adjacent black circle were considered to be using 'next to'.

The results of all the children who could be classified by the above criteria (58 out of the 60 subjects) are shown in Table 3. These results showed marked age differences in performance. More than half the 4 year olds were unable to carry out the pretest and they nearly always responded as if the two boards were aligned. It is possible that some of the children interpreted the word 'same' to mean the same relative position on their board, rather than the same black circle. But the children saw their board being rotated between trials, and in the first six trials of the pretest (with the colored circles) the children understood that the aim of the task was to place their token on the same circle on their board, rather than place it in a corresponding spatial position. It is more likely that the children who failed the pretest had no adequate strategy for identifying the correct black circle.

Age (years)	Failed pretest	Strategy 'next to'	Strategy 'between'
4	13	5	2
5	3	12	3
6	-	6	14

Table 3. Number of children using the different strategies in the board game experiment (for criteria see text).

Most of the 4 year olds who passed the pretest and were then tested on the boards with equally spaced circles used a 'next to' strategy. The majority of the 5 year olds also used a 'next to' strategy. By the age of 6 years most of the children were able to use a 'between' strategy. After the trials were completed each child was asked how they had carried out the task and their answers corresponded to their performance — there was no evidence from these answers that the children attempted any other strategies.

5.3. SUMMARY

The experiments described in this section have considered the spatial strategies which young children employ in tasks which involve transferring information from a representation to an environment or between representations. The layout experiment showed an age related improvement in 4 to 7 year olds' ability to apply appropriate strategies to a map using task, and the results from that study indicated that the youngest children (4 year olds) nearly always used on a comparatively limited strategy which relied on the identification of a single landmark from the map. In the board game experiment the majority of the 4 year olds had no strategy, and a few relied on a 'next to' strategy. In the latter study it was not until the age of 6 that children consistently used a 'between' strategy.

These results correspond to Piaget and Inhelder's (1956) suggestion that it is not until after about the age of 5 years that children can understand the concept of 'between' (though Piaget and Inhelder do not present any direct evidence for this claim). Johnston (1988) has reviewed the studies of children's verbal use of spatial terms. Comprehending and using the term 'next to' (at 2 to 3.5 years in different studies) precedes the comprehension and use of the term 'between '(at 3 to 4 years in different studies). Therefore it would seem likely that children can employ these terms in some contexts before they can incorporate such concepts in effective strategies with spatial representations.

Bluestein and Acredolo (1979) suggested that the successful children in their map using experiment located a toy by considering its hiding place as 'between' two of the landmarks in the room, but as emphasized above, the children in their experiment could have used a variety of strategies. In fact all the strategies described for the layout experiment in section 5.1 (except strategies 1 and 8) could be applied *mutatis mutandis* to the task used by Bluestein and Acredolo (as well as several more which were discussed in section 5). There is no way of determining from Bluestein and Acredolo's results exactly how the successful children in their experiment performed. Because of the generous criterion for success used in that study (2 out of 3 correct trials with only 4 hiding places) at least a few of the children who were classified as successful could have employed a 'next to' strategy.

Nonetheless, the board experiment described in section 5.2 indicates that by the age of about 6 years children are capable of a 'between' strategy and it may have been the case that the successful children (mainly 5 year olds) in Bluestein and Acredolo's experiment also relied on the concept of 'between'.

Some of the strategies (6, 7 and 8) described with reference to the layout experiment necessitated the child being aware of her own position. Only a few of the older children employed these strategies and the 4 year olds did not use these strategies at all (even in other conditions - not discussed above — when these were the only strategies which could be used for successful performance, see Blades and Spencer, 1987c). It may be that young children do not realize that their own position can be identified on a representation and to consider this possibility the experiment described in section 6 was carried out.

6. Self Location

In pilot studies with preschool children (Blades and Spencer, 1986a) we found that 4 year olds could place a doll (to represent themselves) in a model of a room to indicate where they were standing. In another study (Blades and Spencer, 1987b) children were asked to use a map to walk along a simple route through a large scale maze. The route included choice points at three T junctions. In order to make a correct turn in the maze the children had to know their position on the map (so that they knew which junction they had reached) and the 4 year olds in this study were proficient at the task which would suggest that children of this age already have some self-location ability. However, in the maze study the children's movement was deliberately restricted by the paths in the layout and therefore each choice point (the three T junctions in the maze) could only have been one of three possible places on the map. Therefore the following experiment was carried out to discover how children would perform when they had to decide where they were on a map and the choice of places was not so limited.

6.1 SUBJECTS AND PROCEDURE

A total of 120 children were randomly selected from a first school. There were four age groups of 30 children with mean ages of 4 years 11 months (range 4 years 9 months to 5 years), 5 years 8 months (range 5 years 4 months to 5 years 11 months), 6 years 8 months (range 6 years 4 months to 6 years 11 months) and 7 years 8 months (range 7 years 3 months to 8 years).

A grid was drawn in chalk on the floor of a school playground. The total area was 5.7 m x 2.7m and was divided into 19 x 9 squares; each square was 30 cm x 30 cm (see Figure 8). Two buckets (one green and one red) which could be used as landmarks were placed in the area, in the grid squares shown in Figure 8.

A map (scale 1:16) was drawn on white cardboard and showed the grid lines and the position of the two buckets, symbolized by green and red circles. The map (Figure 8) was oval so that the edges and corners of the grid area in the playground were not indicated as such on the map (and therefore children could only relate the map and the playground area by reference to the two landmarks). A small doll (5cm high) was provided for the children and they used the doll to indicate a place on the map.

Each child was tested individually. The child stood in one of the squares in the playground, facing towards the landmarks. The map was held in front of the child (in condition 1 it was aligned with the playground, in condition 2 it was rotated 90°, and in condition 3 it was rotated 180°). The child was asked to pretend that the doll was herself and place it on the map to indicate the square in which she was standing.

Each child took part in one condition and there were 10 positions (presented in random order) in each condition. The positions are shown in Figure 9; positions A-E were adjacent to the landmarks, and the positions V-Z were three squares (linearly or diagonally) from the landmarks. The two landmarks were placed on the same line of grid squares and positions A, B, V and W were also on this line.

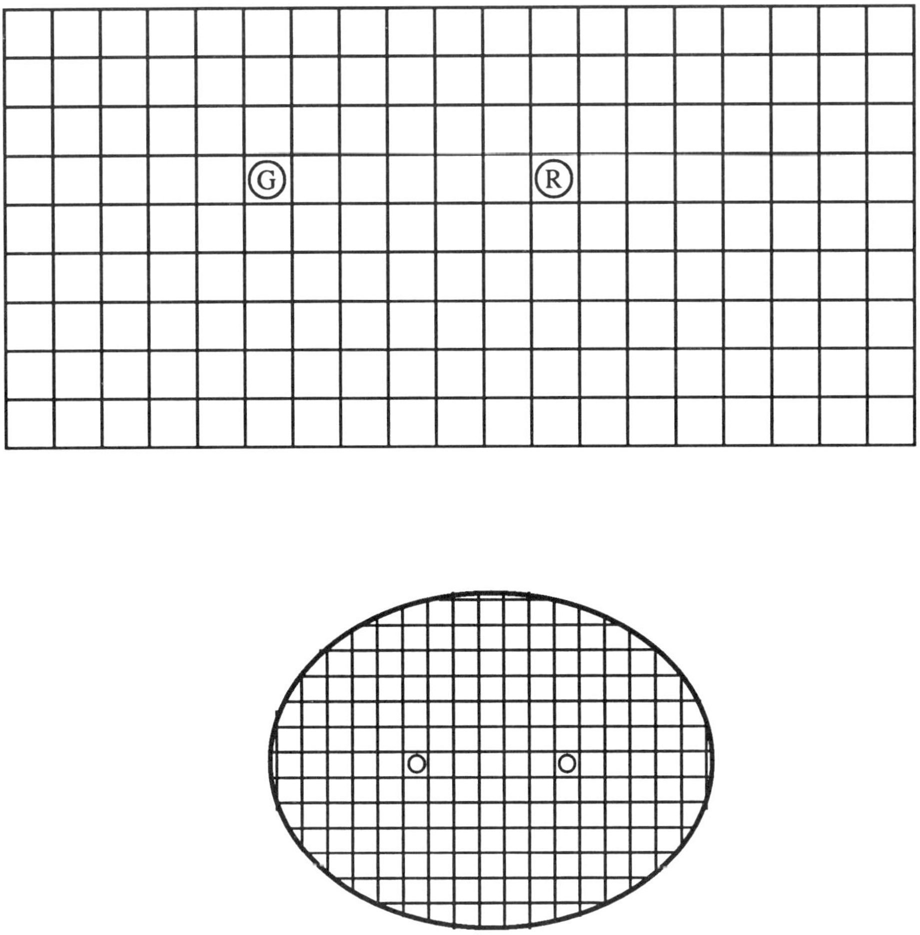

Figure 8. Plan of layout, and map, used for the self location experiment. The layout was chalked on a playground floor; the circles on the plan represent the green and red buckets which were used as landmarks. The map was drawn at scale 1:16 and the landmarks were indicated by coloured circles

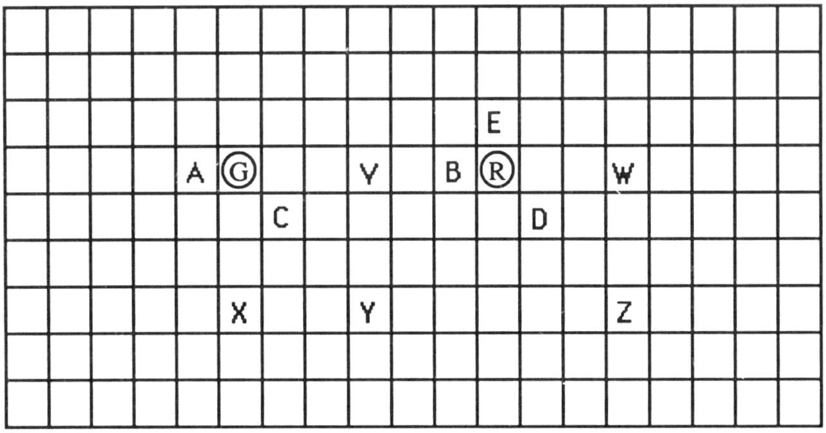

Figure 9. Plan of layout used in the self location experiment showing the positions where the children stood.

6.1.1. *Results*. A child scored one point for placing the doll in the correct square on the map and therefore the maximum score in each condition was 10. There were approximately 170 squares on the map and therefore if a child simply guessed she only had a 1 in 170 chance of being correct in each trial. Table 4 shows, for each age group and condition, the number of children who were correct on at least 4 trials ($p<0.001$, binomial distribution).

Condition	Age (years)			
	4	5	6	7
0°	6	8	8	10
90°	2	2	5	6
180°	1	1	5	7

Table 4. Number of children in each age group and condition in self location experiment who scored 4 or more ($p<0.001$), $n = 10$.

Table 5 shows the mean scores for each age group and condition. A 4 (age) x 3 (condition) x 10 (position) analysis of variance was carried out and this showed a significant effect for age ($F = 16.02$, d.f. = 3, 108, $p<0.001$) and post hoc Tukey HSD ($p<0.05$) comparisons showed that the mean scores of the 6 year olds (4.8) and the 7 year olds (5.6) were both greater than the mean scores of the 4 year olds (2.1) and the 5 year olds (3.3). There was also a significant effect for condition ($F= 34.43$, d.f. = 2, 108, $p<0.01$) with the mean scores for the 0° map (6.2) being greater than the mean scores for the 90° rotated map (3.1) and the 180° map (2.5). There was no difference between the two rotated maps.

Condition	Age (years)			
	4	5	6	7
0°	3.6	6.0	7.0	8.3
90°	1.7	2.3	3.7	4.5
180°	0.9	1.5	3.7	4.0

Table 5. Mean scores for each age group and condition in self location experiment. Maximum score was 10, $n = 10$.

There was also an effect for position ($F= 20.01$, d.f. = 9,972, $p<0.01$). Overall the order of difficulty of the positions was A,B,V,E,W,C,X,D,Y,Z (with best performance being achieved at A and poorest at Z) and this order of difficulty was similar for each of the three conditions.

The children's errors were analyzed. If a child placed the doll in a square on the map which was next to the correct square this was labelled an 'adjacent' error. One type of error on the rotated maps will be referred to as 'orientation' errors — these occurred when a child assumed that a rotated map was aligned. Overall 48% of the errors with the 0° map were adjacent errors. For the 90° map, 28% of the errors were adjacent ones and 43% were orientation errors: for the 180° map, 9% of the errors were adjacent and 59% were orientation errors.

In condition 1 of this experiment (with the aligned map) the majority of children in each age group were able to place the doll correctly. In this condition the children were successful in nearly two-thirds of the trials and in half the incorrect trials they selected a square on the map which was adjacent to the correct square. In particular the 5, 6 and 7 year olds achieved high mean scores in this condition (see Table 4). These results suggest that the concept of self location is within the ability of some 4 year olds and most older children.

There were large variations in the children's success at different positions - in general, positions near a landmark, or in line with the two landmarks were easier to locate than positions further away, and one factor which may have contributed to this effect was inaccurately counting the squares (in the playground or on the map).

Although the children had the concept of self location their performance was poorer when the maps were not aligned and on these maps the most common error was what was labelled an 'orientation' error. One explanation for these errors is that the children only related their own position to one landmark in the area, and ignored the need to take into account both landmarks when the map was rotated. When the map was aligned a child could utilize the grid and rely on only one landmark to find her position correctly — and this may explain the higher scores for the children in condition 1. If the map was not aligned and a child only relied on one landmark she would make an 'orientation' error. For example, when standing at X (see Figure 9) the green landmark was to the north of the child and the red landmark was to the east of the child in the playground. But if, with a 180° rotated map, a child only considered the nearer (green) landmark and counted the number of grid squares to her own position from that landmark in the playground and on the map she would (incorrectly) place the doll to the north of the green landmark on the map. In other words, produce an orientation error.

It may be that although young children do understand that their own position can be indicated on a representation they do not always realize that to identify their position accurately on representations which are not aligned they need to consider two landmarks rather than just one.

7. Conclusions

In this paper some of the component abilities required to understand and use spatial representations have been discussed as separate abilities. These are the abilities to recognize the correspondence between two identical spatial models or between a map and the environment which it represents; the ability to extract information from a representation in such a way that the information can be applied to a task in the environment; and the ability to locate one's own position in a model or on a map.

To some extent these are separate abilities. It is possible to apply a spatial strategy to a simple map when locating a place (e.g. relative to other places shown on the map) without any awareness of one's own position on the map, and this may well be the case in some of the experiments carried out with young children (e.g. in the crossroads experiment in section 4.2; or in Bluestein and Acredolo's study, 1979). Similarly, it may be possible to have some understanding of the correspondence between two spaces (e.g. the younger children's recognition that the same items of furniture were equivalent in the model experiment in section 4.1) without being able to apply any effective strategy when the spaces include identical and therefore ambiguous landmarks.

Clearly, the recognition of the correspondence between a model or map and the environment is a crucial first step in using a spatial representation, but having recognized the correspondence the next step is to transfer information between the representation and the environment. The transfer of information can be from the representation to the environment (e.g. when a hiding place on a map is located as next to a landmark included on the map and that landmark can be identified in the environment) or from the environment to the map (e.g. when locating one's own position relative to places in the environment and then finding those places on a map). The difficulty of the task will depend on the complexity of the strategy which is required to transfer the information.

The results described above suggest that at the age of 2.5 years children have little or no ability to understand the correspondence between two spaces (DeLoache 1989a). By the age of about 3 years children can recognize the correspondence between identical features in two spaces (e.g. model experiment, section 4.1; DeLoache, 1989a). However, 3 year olds only have a limited ability to transfer information from a representation. Their performance is limited to tasks which involve using a representation to locate a particular place, such as an object's hiding place. When locating a particular place 3 year olds can use models or maps which are aligned (e.g. model experiment; crossroads experiment; Blades and Spencer, 1986a; Bluestein and Acredolo, 1979), though they may have some difficulty when there is more than one of the same type of place (as in condition 3 of the model experiment). When the map or model is not aligned with the environment 3 year olds can only use the representation if the target place is a unique feature (in condition 4 of the model experiment) or immediately adjacent to a salient feature (in the crossroads experiment).

By the age of 4 years children can use aligned representations to follow simple routes (Blades and Spencer, 1986a, 1987b) and can locate target places with rotated maps even when that place is not immediately associated with a salient feature, as long as there is a clear relationship between the target place and a unique feature (e.g. in the crossroads experiment). However, 4 year olds usually have difficulty with rotated maps when a target place cannot be related to a unique feature and some additional information has to be taken into consideration. There is a tendency for 4 and 5 year olds to focus on only one item of information from the representation (e.g. only the landmark at the end of the correct path in the layout experiment, section 5.1; or only one of the landmarks in the self location experiment, section 6). It is not until after the age of about 6 years that children spontaneously employ strategies which require the consideration of two or more items of information from the representation and the spatial relationships of those items (e.g. when using the 'between' strategy in the board experiment, section 5.2).

It is possible to see the development of strategies in terms of the number of items of information which are required to carry out the strategy - with younger children being limited to spatial strategies which involve no more than a single item of information, and older children being able to combine progressively more items from the representation to generate more appropriate spatial strategies.

However it is not clear why this should be the case - it is unlikely that the younger children's performance reflects any cognitive limitations such as a more constrained working memory capacity. For example, a strategy such as 'next to' (in the board experiment) requires the consideration of a black circle, a colored circle, and their relationship ('next to'). It might be expected that retaining this information would be within the memory capacity of most 4 year olds.

Yet few children of that age spontaneously employed such a strategy. Therefore the question of why young children only demonstrate limited spatial strategies is not immediately answerable. One tentative suggestion, put forward by Ottosson (1988), is that children may need to understand spatial relationships in the world before they can appreciate those relationships on a representation. However, at present we know little about the way that young children encode relational information in everyday contexts, and that may be an important focus for research in the future.

In summary, the experiments described in this paper have demonstrated that young children have some understanding of spatial representations, and these findings contradict the assumptions drawn from Piaget's theory that children before the age of about 7 years will be unable to use representations (see section 3). The experiments have shown that there is an age related relationship between untrained children's accurate performance in map tasks and the complexity of the required strategy (i.e. the amount of information and spatial relationships which have to be encoded from the representation for successful performance); and it is now possible to sketch the early development of map performance in terms of the spatial strategies which children apply to representations, but as yet it is not known how children first learn these strategies.

8. References

Barrell, G. V. and Cooper, P. J. (1986) 'Cognitive processes in orienteering: the interpretation of contours and responses to the map as a whole', Scientific Journal of Orienteering 2, 25-46.

Blades, M. (1989) 'Children's ability to learn about the environment from direct experience and from spatial representations', Children's Environments Quarterly 6, 4-14.

Blades, M. and Spencer, C. (1986a) 'Map use by young children', Geography 71, 47-52.

Blades, M. and Spencer, C. (1987a) 'How do people use maps to navigate through the world?', Cartographica 24, 64-75.

Blades, M. and Spencer, C. (1987b) 'The use of maps by 4-6 year old children in a large-scale maze', British Journal of Developmental Psychology 5, 19-24.

Blades, M. and Spencer, C. (1987c) 'Young children's strategies when using maps with landmarks', Journal of Environmental Psychology 7, 201-217.

Blades, M. and Spencer, C. (1990) 'The development of 3-6 year olds' map using ability: the relative importance of landmarks and map alignment', J. Genetic Psychology 151, 181-194.

Bluestein, M. and Acredolo, L. (1979) 'Developmental changes in map reading skill', Child Development 50, 691-697.

Board, C. (1978) 'Map reading tasks appropriate in experimental studies in cartographic communication', Canadian Cartographer 15, 1-12.

Boardman, D. (1989) 'The development of graphicacy: children's understanding of maps', Geography 74, 321-331.

Brown, A. L., Kane, M. J. and Long, C. (1989) 'Analogical transfer in young children; analogies as tools for communication and exposition', Applied Cognitive Psychology 3, 275-293.

Carpenter, S. M. (1979) Drivers' route choice project, Transport Studies Unit, Oxford University.

Clarke, L. M. (1989) 'An experimental investigation of the communicative efficiency of point symbols on tourist maps', Cartographic Journal 26, 105-110.

Crampton, J. W. (1988) 'The cognitive processes of being lost', Scientific Journal of Orienteering 4, 34-46.

Crisafi, M. A. and Brown, A. L. (1986) 'Analogical transfer in very young children', Child Development 57, 953-968

DeLoache, J. S. (1989a) 'The development of representation in young children', in H.W. Reese (ed.), Advances in Child Development 22, 1-39.

DeLoache, J. S. (1989b) 'Young children's understanding of the correspondence between a scale model and a larger space', Cognitive Development 4, 121-139.

DeLoache, J. S. (in press) 'Young children's understanding of models', in R. Fivush and J. Hudson (eds.), Knowing and Remembering in Young Children, Cambridge University Press, New York.

Dodds, A. G. (1989) 'Tactile maps: a psychologist's perspective', Cartographic Journal 26, 3-6.

Forrest, D. and Castner, H. W. (1985) 'The design and perception of point symbols for tourist maps', Cartographic Journal 2, 11-19.

Freundschuh, S. (1990) 'Can young children use maps to navigate?', Cartographica 27(1), 54-66.

Garland, H. C., Haynes, J. J. and Grubb, G. C. (1979) 'Transit map color coding and street detail: effects on trip planning performance', Environment and Behavior 11, 162-184.

Gentner, D. (1983) 'Structure-mapping: a theoretical framework for analogy', Cognitive Science 7, 155-170.

Gentner, D. (1989) 'The mechanisms of analogical reasoning', in S. Vosniadou and A. Ortony (eds.), Similarity and Analogical Reasoning, Cambridge University Press, Cambridge.

Gentner, D. and Toupin, C. (1986) 'Systematicity and similarity in the development of analogy', Cognitive Science 10, 277-300.

Golledge, R. G. (1987) 'Environmental cognition', in D. Stokols and I. Altman (eds.), Handbook of Environmental Psychology, Vol. 1, Wiley, New York.

Heft, H. and Wohlwill, J. F. (1987) 'Environmental cognition in children', in D. Stokols and I. Altman (eds.), Handbook of Environmental Psychology, Vol. 1, Wiley, New York.

Hill, A. R. (1974) Cartographic performance: an evaluation of orthophotomaps, Experimental Cartographic Unit, Royal College of Art, London. Final Technical Report, AD 783519.

Johnston, J. R. (1988) 'Children's verbal representation of spatial location', in J. Stiles-Davis, M. Kritchevsky and U. Bellugi (eds.), Spatial Cognition. Brain Bases and Development, Erlbaum, Hillsdale, New Jersey.

Keates, J. S. (1982) Understanding Maps, Longman, London.

Kovach, R. C., Surrette, M. A. and Aamodt, M. G. (1988) 'Following informal street maps. Effects of map design', Environment and Behavior 20, 683-699.

Levine, M., Marchon, I. and Hanley, G. (1984) 'The placement and misplacement of you-are-here maps', Environment and Behavior 16, 139-157.

Liben, L. S. and Downs, R. M. (1989) 'Understanding maps as symbols: the development of map concepts in children', in H.W. Reese (ed.), Advances in Child Development and Behavior 22, Academic Press, San Diego, California.

Mandler, J. M. (1983) 'Representation', in P.H. Mussen (ed.), Handbook of Child Psychology 3, Wiley, New York.

Morsley, K and Spencer, C. (1988) 'The use of maps and models by visually impaired children between the ages of 5 and 11: how useful are they?', Proceedings, 2nd International Symposium on Maps and Graphics for Visually Handicapped People, (London, April 1988), Nottingham,University of Nottingham Press, 54-64.

Murakoshi, S. (1988) 'Information processing in photo-orienteering: how do we relocate ourselves?', Scientific Journal of Orienteering 4, 14-33.

Ottosson, T. (1986) 'Cognitive processes in orienteering: an outline of a theoretical frame of reference and some preliminary data', Scientific Journal of Orienteering 2, 75-101.

Ottosson, T. (1987) 'Map-reading and wayfinding', Goteborg Studies in Educational Sciences 65, Acta Universitatis Gothoburgensis, Goteborg.

Ottosson, T. (1988) 'What does it take to read a map?', Scientific Journal of Orienteering 4, 97-106.

Piaget, J. and Inhelder, B. (1956) The Child's Conception of Space, Routledge and Kegan Paul, London.

Piaget, J., Inhelder, B. and Szeminska, A. (1960) The Child's Conception of Geometry, Routledge and Kegan Paul, London.

Presson, C. C. (1982) 'The development of map reading skills', Child Development 53, 196-199.

Presson, C. C. (1987) 'The development of spatial cognition: secondary sources of spatial information', in N. Eisenberg (ed.), Contemporary Topics in Developmental Psychology, Wiley, New York.

Seiler, R. (1985) 'The psychological structure of information-seeking and decision making in route-choice situations in orienteering', Scientific Journal of Orienteering 1, 24-34.

Seiler, R. (1989) 'Route planning and route choice. An empirical investigation into information processing and decision making in orienteering', Scientific Journal of Orienteering 5, 74-84.

Spencer, C., Blades, M. and Morsley, K. (1989) The Child in the Physical Environment: the Development of Spatial Knowledge and Cognition, Wiley, Chichester.

Taylor, R. M. and Hopkin, V. D. (1975) 'Ergonomic principles and map design', Applied Ergonomics 6, 196-204.

Uttal, D. H. and Wellman, H. M. (1989) 'Young children's representation of spatial information acquired from maps', Developmental Psychology 25, 128-138.

Yarnal, C. M. M. and Coulson, M. R. C. (1982) 'Recreational map design and map use: an experiment', Cartographic Journal 19, 16-27.

Yngstrom, A. (1989) 'Map reading techniques for visually impaired', Scientific Journal of Orienteering 5, 12-20.

MAKING SENSE OF HUMAN WAYFINDING: REVIEW OF COGNITIVE AND LINGUISTIC KNOWLEDGE FOR PERSONAL NAVIGATION WITH A NEW RESEARCH DIRECTION

MYKE GLUCK
School of Information Studies
Syracuse University
Syracuse, New York 13244 USA

ABSTRACT. This article provides a summary of the basic knowledge of human wayfinding found in the literature. It explores the historical roots of the theories (competence literature) and some of the more inductive empirical results (performance literature) of wayfinding. It will point out the limited scope of these investigations compared to the richness of actual human wayfinding, especially in describing the information needs for wayfinding. To address this lack of knowledge the use of the"sense-making" investigation method is suggested. The method is described and several empirical studies are proposed.

1. Introduction

The question that motivated this investigation was:

1) What is the state of knowledge of the cognitive processes that shape the way humans navigate and communicate about navigating on foot or by vehicle?

Or more simply, what do we know about :

1a) How we find our way in the physical world?
1b How we communicate directions to each other for finding our way?
1c) How our verbal and visual abilities contribute to finding our way?
1d) What we need to find our way?

Human wayfinding describes the process humans use to orient and navigate on foot or by vehicle. The overall goal of wayfinding is to accurately relocate from one place to another in large-scale space. Large-scale space is defined in this literature as space whose structure is revealed by integrating local observations over time, rather than being perceived from a single vantage point (Kuipers, 1983, 346). [Note this is different from the usual cartographic use of the term "large-scale space."]

Wayfinding encompasses many different situations humans frequently find themselves negotiating. These include travelling on foot about a building, a city, or a trail in the woods as well as finding the way while driving about a city or across country. A distinction is made between the

D. M. Mark and A. U. Frank (eds.), Cognitive and Linguistic Aspects of Geographic Space, 117–135.

kinesthetic sense of orientation and wayfinding. The kinesthetic sense is seen as a low level reflex for maintaining equilibrium while wayfinding is seen as the use of several higher level cognitive processes for navigating. These various cognitive skills include reading and following a map, navigating to a landmark that can be seen, giving or receiving directions, and retracing a path.

Wayfinding is a common, natural human experience. Primitive societies of wanderers and seafarers have developed sophisticated techniques for wayfinding and a rich language for communicating directions (Mark and Frank, 1989; Mark, Svorou and Zubin, 1988; Talmy, 1983). These cognitive tasks all have the common goal of getting from one place to another, even though some tasks involve actual movement while others do not.

The literature on wayfinding can be divided into two overlapping categories: competence and performance. The literature of competence discusses cognitive models of human wayfinding. These are sometimes thought experiments (e.g., Kuipers, 1978) similar to the construction of grammars by linguists. Some models are based on cognitive experiments or meta-level analysis of the literature (e.g., Garling, Book and Lindberg, 1984). The competence literature comes from the fields of cognitive science, artificial intelligence, geography, and urban planning.

The literature of performance discusses the ability of humans to find their way. This includes map reading and following (e.g., Kovach, Surrette and Aamodt, 1988; Streeter and Vitello, 1986; Simutis and Barsam, 1983), wayfinding while driving (e.g., Streeter, Vitello, and Wonsiewicz, 1985), production and comprehension of directions (e.g., Vanetti and Allen, 1988; Mark, 1990), and ability to judge distances and angles (e.g., Sadalla and Montello, 1989; Nasar *et al.*, 1985). The performance literature describes wayfinding in a manner that is similar to psycholinguists' approach to language: It generally investigates the accuracy of human cognitive maps (instantiated by sketch maps) and compares it with the reality of the environment and assesses how effective these maps are for wayfinding. The performance literature comes from the fields of cognitive science, psychophysics and psychometrics, human factors, and urban planning.

This article assumes humans do have a mental map in long-term memory that corresponds to their perception of the real world. However, it will only explore a few of the proposed representations (1). Also, there is no discussion presented of how children develop wayfinding skills (2) or the analysis of wayfinding by other special populations such as the aged or blind.

2. The Literature of Wayfinding

2.1 SUMMARY OF THE WAYFINDING LITERATURE

The investigations into the construction of mental representations of large-scale spatial environments have resulted in several generally accepted principles. (Review in Leiser and Zilbershatz, 1989; Sadalla and Montello, 1989; Golledge *et al.*, 1985; Saisa and Garling, 1987; Wickens, 1984; Evans, 1980; Thorndyke, 1980; Thorndyke and Hayes-Roth, 1982; Siasa *et al.*, 1986; McNamara, 1986; McNamara, 1984; Moar and Carlton, 1982; Hirtle and Jonides, 1985; Siegel and White, 1975; Golledge and Zannaras, 1975; Hart and Moore, 1973; Marble and Bowlby, 1968.)

Individuals learn about their environment quite incrementally. First, the knowledge is of particular locations which are not spatially connected one to the other. However, all points are not equal: People tend to give landmarks and more "important" places in the network a special status. Increasing familiarity with the environment leads to learning routes which connect the locations. In the early stages of spatial understanding routes tend to have a directional bias. This is demonstrated by recall and evaluation being more readily achieved in the presented direction than in the reverse. Also, these routes tend to be integrated into a whole network and not in isolation from each other. Further familiarity leads to survey or map knowledge in which locations are linked efficiently and spatial relations such as distance and orientation are abstracted. The organization is primarily topological since the number of nodes seems to be a better predictor of distance than the actual distance. Distance estimates also seem to have a directional bias: people underestimate distances from a regular point to a landmark but not from the landmark to the regular point or between two regular points. Also, judgments concerning landmarks tend to be faster. Regions are readily differentiated and stronger associations of points within than across regions are formed.

Patterns for travel are developed early and are not willingly modified even when information providing more efficient routes becomes available. Heavily traveled areas yield well developed images while areas less well travelled only provide enough information to arrive at destinations from well-known locations. In multi-stop trips humans tend to minimize total trip distance and not local minimum distances. Memory for turns seems to be linked to the body's orthogonal axes since the least disorienting turns are those near 0, 90, 180 degrees from the direction of forward motions. (Vanetti and Allen, 1988, Evans, 1980).

A study of Parisian cab drivers (Pailhous, 1970) concluded that knowledge was organized into two complementary structures: a basic network and a secondary network. The basic network is a skeleton that contains approximately 10% of the streets and is well-known and highly integrated. The remaining 90% of locations are in the secondary map. Information in the secondary network is related to the basic network but not to other points in the secondary network. Thus, typical travel plans call for pickup at a secondary point with travel to a basic point, travel from this basic point to a basic point near the destination, and travel to the destination from the nearby basic point. However, this study also showed that experts such as cab drivers, in contrast to the general public, do tend to improve the efficiency of their routes rapidly and continually.

In summary four levels of knowledge emerge from these studies. The lowest level is knowledge of a collection of landmarks. The next level is a knowledge of a collection of unordered production rules used to choose the correct direction to achieve a goal when at a node (intersection). The next level is rote knowledge of sequences of production rules reproducible when not even in the environment. The highest level is survey knowledge. (There is some controversy over the representation of survey knowledge but it tends to be like that of a map.)

2.2 COMPETENCE LITERATURE

2.2.1 *Roots in Urban Planning*. The first well documented attempt to understand how people find their way was a byproduct of a study by Lynch (Lynch, 1960). Lynch did a comparative study of the central district of three U.S. cities (Jersey City, NJ, Boston, MA, and Los Angeles, CA). His goal was to describe how people viewed each city. The study included extensive interviews that had subjects make sketch maps and wayfind, and also professional observations of the cities. He tried to describe the imaging (sic) of the city.

He discussed such observations as: Where the Bunker Hill section of Boston actually ends versus where people think it ends, blind spots and other distortions on sketch maps, what people liked and disliked about their city, and how people found their way about each city district. From the interviews and observations he created a five class taxonomy of a city's geographic features:

- landmarks
- nodes (intersection points)
- paths
- edges (terminators or boundaries)
- regions

These were a distillation of the common features in the representations his subjects presented of their cities. His goal was not an analysis of wayfinding but of discovering what gave a city an appealing image and how that might affect the quality of city living; the taxonomy and wayfinding information were merely a byproduct. Interestingly, this taxonomy has been the basis for all wayfinding discussions since its presentation (e.g., Appleyard, 1970; Stea and Downs, 1970; Kuipers, 1982; Garling, Book and Lindberg, 1984). Subsequently, Lynch and Appleyard (Appleyard, Lynch and Myer, 1963) described how to apply the principles of design discussed by Lynch (Lynch, 1960) to the design of highway systems.

Appleyard was a student of Lynch (3) and he was especially interested in studying the layperson's or user's perceptions of the environment: Conclusions such as residents on the most heavily traveled streets were the least satisfied and knew the fewest neighbors (Appleyard, 1981) are re-occurring themes in his work. After analyzing sketch maps Appleyard expanded Lynch's model to include loosely connected highly structured areas. Lynch and Appleyard were the first to articulate the fact that errors in cognitive maps are generally metrical (distance) and rarely topological (relational).

2.2.2 *Computational Models*. Kuipers developed the first computational theory of wayfinding found in the literature. The resulting LISP program is called TOUR (Kuipers, 1977, 1978, 1982, 1983). TOUR copes with incomplete spatial knowledge of the environment and so learns about the environment as more information is received. Thus, TOUR forms a basis for studying the communication of spatial information for wayfinding. TOUR incorporates three types of spatial information: sensorimotor procedures, topological relations and metrical relations. The sensorimotor procedures constitute a set of sequential actions required to travel from one place to another; the topological relations describe the non-metrical attributes of the environment such as containment, connectivity, and order; and the metrical relations describe knowledge of the measurable attributes of the environment such as distance and direction as well as the means to compute them.

The procedural knowledge is based on knowledge of two sorts of objects: VIEWs (V) and ACTIONs (A) (Kuipers, 1983; Mark, 1986). A VIEW (V) is what see seen at a particular place with a particular orientation. The attributes of a VIEW allow discriminating between Vs. An ACTION (A) is a movement that changes the V. This might be a turn or translation relative to either a rectangular or polar coordinate reference frame. The linking of ACTION A to VIEW V constitutes procedural learning (A -> V) when such a pair is relocated from short-term memory to long-term memory. Thus, replay of a particular movement will be executed upon verification of a particular VIEW. A ROUTE is then a sequence of such (V -> A) pairs. If the pairs are ordered:

$$[(V1\text{->}A1) \text{->} (V2\text{->}A2)]$$

then the whole ROUTE may be replayed without traveling. If they are unordered, then replay results in: "I can't tell you how to get there, but I can show you." (Mark, 1986; Kuipers, 1983).

Topological knowledge grows from the procedural knowledge in this model by developing the concept of PLACE from a cycle of Vs from the same point, PLACEs then may be common to more than one ROUTE. A PATH is then defined as a sequence of PLACEs and may correspond to a ROUTE. PATHs connect PLACES dividing two dimensional space into REGIONS. Scaling also enters this model since REGIONs at one scale may be PLACEs at a lower resolution. With only this knowledge sketch maps may be very distorted but effective for navigation.

Metrical information is acquired by linking turn information or translation measurements (time or distance) to a polar or rectangular coordinate reference frame. This knowledge then becomes the basis for estimating distances and indicating directions.

Note these three stages of knowledge development (procedural, topological, and metrical) correspond to the difficulty of wayfinding tasks. For example, giving or following directions requires procedural knowledge while making an accurate map requires topological or metrical level knowledge.

McCalla built a computerized planning system that was to simulate the actions of a taxicab driver making his or her way about a city (McCalla and Schneider, 1979, 1980; McCalla, Schneider and Reid, 1982). The name of the simulated driver was Elmer. The architecture of ELMER consisted of a MAP, PLANNER and EXECUTOR modules. A plan to go from point A to point B would be devised by the planner using route information from the MAP, and sent to the EXECUTOR to be performed. The EXECUTOR would modify the plan based upon pragmatic knowledge of the world such as how to stop at lights and stop signs, how to avoid running over pedestrians, and how to maintain the vehicle's speed. Once a plan had been successfully executed it would be sent to the MAP as a ROUTE which included the obstacles encountered along the way. A crucial part of a ROUTE is its associations to other ROUTEs. This resulted in the development of a complete planner capable of operating in a "dynamic microworld" and able to learn from past experiences.

A window is presented to ELMER showing "him" what "he" can "see" at any moment of travel and ELMER provides speed and direction information. ELMER does not control what the window will present to him as he "drives" across the microworld but, on the other hand, the window must obey his every command. (The system as described had an ad hoc mechanism for window information, not a real microworld set of images.).

A Plan would include two sets of information: A core consisting of directional information and secondary information of prioritized controllers for ELMER. Such controllers included a red light handler and speed maintainer. When conflicts of control occur actions are executed based on priorities and exceptions. For example, "maintain speed but don't hit the pedestrian" might result in a minor series of turns if the adjacent lane were available.

The module of interest is the PLANNER. The origin and destination served as inputs and subroutes that link them are produced. Often the trip requires more than the linking of known ROUTEs in sequence. When this occurs higher level non-primitive ROUTEs called SUPERROUTEs that contain potential linkages are investigated; when these fail, splicing of ROUTEs is attempted to complete the trip plan from A to B. The PLANNER is not an optimal planner nor can it handle catastrophic execution errors (for example, a previously successful ROUTE that is no longer available because of road repairs).

The TRAVELLER is a detailed computational model that simulates learning of spatial networks (Leiser and Zilbershatz, 1989).

When the system is first started it has no network information and as paths are built it does not know if they are optimal. The system simply tracks what nodes were reached from what other nodes and the actions needed to do so. Inputs to the TRAVELLER consist of links between neighboring locations and a sequence of randomly defined planning assignments. For each assignment TRAVELLER plans a route. If a route can be constructed it is stored with all other successful plans. The output of the system is a cognitive map: The collection of all successful routes.

The basic components or production rules of the model are typical artificial intelligence condition-action pair constructs of the form:

(present location = S1) and (goal = S2) --> (action = A)

with S1 and S2 adjacent nodes. The system searches for new routes in a forward, breadth first search. All nodes accessible from the origin node are scanned for the goal node. If the goal node is found the route is completed; else, the system scans, one by one, the nodes accessible to the origin node in a succession of concentric circles of increasing radius. When the system finds a route (which it will if there is sufficient memory), it saves the new route as a sequence of production rules. The model does not know ahead of time which nodes will be encountered (much like humans who can only read the sign when an intersection is reached).

The routes have a directional bias in this model. When both a route and return are known there may not be any specific spatial relationship. However, only one route between two points in a given direction will be constructed. TRAVELLER develops similar routes for nearby points and exploits this in an extended version using regional centroids similar to the Parisian cab drivers' basic points (Pailhous, 1970). The learning in this system is developmental and not incremental since different starting points and assignments may result in quite different cognitive maps.

TRAVELLER imitates human empirical data in several ways: its route knowledge is oriented, integrated with early pattern development, and it is able to use regional centroids to generate basic and secondary networks. The authors feel a conceptual leap to organize the empirical knowledge is necessary since the empiricist approach that aims to document facts about cognitive maps is unlikely to produce an integrated theory. Consequently, programming this model forced specification of not-well-understood features which the authors realize must be validated. They do not claim the model is complete since people do have alternative ways of processing spatial knowledge. The model addresses only one of these mental structures but the fundamental one in the view of the authors.

TRAVELLER, TOUR, and ELMER differ in their learning and final products. TOUR learns the whole network eventually and the sequence of tasks has no influence on the final map. TOUR may take time but will always result in a route. (Unfortunately searching is of the order n-squared with n nodes in the system.).

TRAVELLER develops similar routes for nearby points and uses them to construct regional centroids in its extended version. TOUR may know how to get from A to C but not from B to C even when B is on the route from A. TRAVELLER does not have this limitation but may not have concurrent knowledge to go from C back to A. ELMER will usually find a path but the difficulty in doing so is highly dependent upon the state of the system.

McMap is a natural language parser for a very limited domain (Riesbeck, 1980). It takes as input natural language text of wayfinding directions and verifies its logical consistency -- not its real world validity. The program has two basic loops, one to decide upon the class for an input and a second to process the inputs. Inputs are of three classes: MOTIONs that tell what to do, DESCRIPTIONs that tell what a place looks like, and COMMENTs that describe the direction-giving process itself. For example, a MOTION: I-5 South; a DESCRIPTION: on the left is a restaurant; and a COMMENT: you can't miss it. The program includes inference that clarifies comments, expands fragments, and establishes links of subsequent instructions to previous open statements. The inference uses expansion rules, classifier rules, clarity rules, and 'cruciality' rules. Many of the rules are very heuristic and may easily be wrong.

Riesbeck's major contribution, besides a rather nice parser, is his finding that most instruction following and direction clarity requires little or no spatial reasoning. People often just verify the clarity of directions and assume spatial specifics will take care of themselves during the trip. However, anecdotal information indicates that some people routinely make a mental map of directions as they are received to expedite their subsequent wayfinding.

2.2.3 *An Information Processing Model.* An information processing model for wayfinding is presented by Garling (Garling, Book and Lindberg, 1984). This information processing theory uses action plans to link internal information processing to actual behavior (based on Russell and Ward, 1982). A travel plan, a type of action plan, is a predetermined course of action to reach a desired destination with minimal investment of effort. Travel plans, a subset of action plans, are then seen as the motivation for developing cognitive maps. Three interrelated elements constitute a cognitive map under the theory: places, spatial relations between places, and travel plans.

Although the theory acknowledges the validity of Lynch's taxonomy it claims that the city planner's elements are just a subset of the psychologically meaningful cognitive map elements. Garling collapses Lynch's taxonomy by describing PLACE as a collection of cognitive map elements that have function and spatial scale. This theory includes doorways, mailboxes, streets, parks, crossings, buildings, landmarks, towns, and districts as examples of PLACE. SPATIAL RELATION is another category of cognitive mapping units introduced in this theory. These are defined as a property of pairs of PLACEs. Three such relations are listed: inclusion, metric spatial relation, and proximity relations.

The theory claims humans usually use local reference frames since humans do not have a magnetic sense and often align their cognitive maps with body axes. Also, it incorporates the fact that humans have difficulty in distinguishing ordinal from metric relations especially when the metric relations are stored at low resolution.

The four assumptions of this theory are derived from the cognitive mapping literature and, thus, provide a summary of that literature from an information processing point of view:

(1) Action plans determine behavior, and when the action plan requires movement, a travel plan is made part of the action plan.
(2) The cognitive map is a semantic long-term memory representation of properties of the environment. The content is directed by function, and when that function is movement, information about places is stored.
(3) Travel plans are formed by a hierarchical process that includes information from and about the environment. Travel plans often lead to inferences that are stored in the cognitive map (i.e. learning occurs).
(4) As travel plans are executed revisions are often necessary. These revisions also provide learning that alters the cognitive map. The type of information acquired upon revision may be significantly different from that needed in travel plan formation.

The authors acknowledge that the theory of travel plans is no more than a skeleton; however, some of their recent work has begun to put flesh on the skeleton by looking at the selection and sequencing of travel tasks (Saisa and Garling, 1987; Lindberg and Garling, 1987; Garling, Book and Lindberg, 1986).

Also, Garling (Garling, Book and Lindberg, 1986) does list three variables for classifying environments that might be used for forecasting problems with wayfinding: degree of differentiation of the environment, degree of visual access, and complexity of layout. These three variables may also be seen as a starting point for a general description of human needs for wayfinding.

2.3 PERFORMANCE LITERATURE

2.3.1 *Psychometric, Psychophysic Investigations.* Psychometric investigations into wayfinding and navigation are based on collecting individuals' perceptions of distances, angles, or locations. A typical experiment obtains pairwise distance judgments between points (often buildings on a university campus), constructs a matrix of the averaged data values, and uses the matrix as input to multidimensional scaling (MDS) or psychophysical curve fitting algorithms When MDS is used a composite map is made and hierarchical groupings are determined. When curve fitting is done perceived distances are plotted as a function of actual distance, and Steven's law is assumed: The perceived distance is a constant multiple of the actual distance raised to some power. The intent in either of these procedures is to mirror the significant features of the cognitive map.

In one experiment, subjects' accuracy in distance estimates was tested in a laboratory setting using two different placement tasks (Baird and Wagner, 1983). The chaining treatment had subjects place one object on a map board, select and place the object that belonged closest to the previously placed object, and continue until all objects were placed. The landmark treatment had subjects select 2-3 landmark objects and then proceed to place all the remaining objects in relation to the landmarks. The landmark treatment produced more accurate distance estimates than the chaining treatment.

The significance of these results in light of Kuipers work and the lack of attention to direction in assessing relationships among points are questioned by Attneave (Attneave, 1983). He points out that humans talk of a sense of direction, not a sense of distance. Also the work of Lynch and Appleyard in pointing out that most errors in cognitive maps are metrical, not topological, is important in evaluating the significance of any psychometric distance estimating experiment.

Psychometric distance estimate accuracy was compared with methods used for imaging in another study (Foley and Cohen, 1984). Distance estimates were compared with familiarity with the environment and found to be consistent with other work claiming the estimates are better the more familiar subjects are with the environment (no surprise here). However, they also found that experience with the environment showed a significant interaction with imaging types: more familiarity with the environment seemed to generate more cartographic-like imagery.

Another form of the psychometric investigation was performed to identify the boundaries of the piedmont, midlands, and lowcountry regions of South Carolina by residents of that state (Lloyd and Steinke, 1986). Subjects were placed in front of a CRT displaying a state map of South Carolina. Previously determined points were highlighted on the map and subjects indicated the region they believed contained the point. Reaction times were also recorded under the assumption that longer reaction times would indicate points in the transition zones between regions. The researchers also conducted interviews with subjects and had them draw sketch maps. A high degree of agreement was found among the respondents since the transition zones were very narrow and the sketch maps were very similar to the computerized point selection maps.

2.3.2 *Human Factors and Individual Differences Investigations.* Typical of the human factors experiments in wayfinding is a study by Streeter (Streeter, Vitello and Wonsiewicz, 1985). Automobile drivers were divided into four groups to compare the effectiveness of different navigational aids in providing assistance in following routes in unfamiliar environments. Members of the control group were given standard state maps and the address of the destination. The three other treatment groups consisted of subjects with a customized route map, subjects with a audio taped set of instructions, and subjects with both audio tape and customized map. Subjects were from various professions and different types of routes were used that were randomly assigned to subjects.

The audio-tape-only group took significantly less time and less distance, and made significantly fewer errors in accomplishing the task. This should not be surprising to anyone who has tried to follow a map alone in a car versus having a passenger give directions. (Other results were not counter-intuitive, either.) However, the implications for car guidance systems are clear: don't distract the driver from the driving task. Attention to the driving task is maintained with an audio tape in this experiment but aircraft-like heads-up displays may also be of value.

A review of experiments of individual differences in map reading and spatial learning is provided by Thorndyke (Thorndyke and Goldin, 1983). Assumptions similar to those of Kuipers are discussed: spatial knowledge consists of landmark knowledge (scene data), procedural knowledge (action sequences), and survey knowledge (metric and topological data). Major conclusions derived from the review are:

1) Spatial knowledge sources should be selected on the basis of task requirements (information processing similar to Garling's travel plans),
2) Simulated navigation should be used as a knowledge source when environmental access is restricted (e.g., military strike forces),
3) Map use and spatial knowledge acquisition require different and independent cognitive skills (only the latter is correlated with cognitive mapping),
4) Psychometric components of spatial ability can predict success on complex tasks such as learning a new environment by navigation, and
5) Different knowledge acquisition strategies are appropriate for different individuals and tasks. However, no typology of the strategies used by good mappers is presented.

3. The Introductory Questions Revisited

Reviewing the initial questions that motivated this article will provide a further summary of the state of knowledge about wayfinding.

1a) How do we find our way in the physical world?

Lynch, Garling, and the computational models provide insights that appear to provide a first approximation to the processes humans use to wayfind. These analyses describe basic travel from a known static point to another static point.

However, human wayfinding is much richer than these models present. People adjust to traffic congestion, road repairs, weather, and lighting conditions. The potential optimizing functions are not restricted to minimal distance or even minimal effort. We may try to minimize time, avoid busy intersections, avoid less travelled paths, or travel at night rather than during the day or vice-versa for many logistical and affective reasons. These situations are occasionally mentioned in the literature (e.g., Streeter, Vitello and Wonsiewicz, 1985) but no model accounts for them. Also, wayfinding may well entail following a moving object (police in hot pursuit, tornado spotters, etc.), avoiding a moving object such as a snowstorm, or trying to find needed resources such as food or gas that do not have unique destination solutions. More importantly, the literature does not discuss being lost and how we get "unlost" explicitly.

1b) How do we communicate directions to each other for finding our way?

Riesbeck's McMap is able to parse natural language directions and verify their clarity. Streeter's work on taped instructions seems to reinforce Riesbeck's thinking about the lack of spatial input needed to follow directions.

Unfortunately, the situations in which language is being analyzed are limited. Many of the situations listed in summarizing question (1a) may need additional categories of cognitive map objects, not just more words in the same categories. Garling's theorizing about the urban planners' taxonomy being a subset of the vocabulary humans need to describe the world in more psychologically meaningful ways seems quite apt.

1c) How do our verbal and visual abilities contribute to finding our way?

Vanetti and Foley separately describe humans as quite robust in their ability to adapt to the environment based on their individual skill levels. When human spatial skills are weak verbal skills may compensate and vice-versa. The work of Thorndyke, Hayes-Roth and others on individual differences indicates there are differences but again humans find alternative (though perhaps suboptimal) solutions. Unfortunately, no catalog of abilities and how best to use them has been presented.

1d) What do we need to find our way?

Garling's work, the TRAVELLER model, and Streeter's work begin to describe some needs for human wayfinding. Travel plans and the information processing model may well present a framework for the human mechanisms for wayfinding. Garling (Garling, Book and Lindberg, 1986) as mentioned does list three variables for classifying environments that generally describe human needs for wayfinding. However, this classification is too general and appears to confound scene visualization with needs for navigation. No model describes how a particular individual makes his or her way or what information needs arise in arriving at a precise point in time and space. Unfortunately, the models that are presented seem to be simplistic representations of a complex process. They seem to be simply convenient 'black box' algorithms producing outputs similar to those humans would produce under very constrained situations.

People may use different networks in parallel to wayfind; perhaps topological and metrical information are really on different networks. These possibilities suggest parallelism may be rampant in human wayfinding but none of the models account for a set of multiple cognitive maps. For example, no model includes multiple paths between locations nor a decision model for when a particular path might be chosen. Perhaps the cognitive map needs to be seen as a cognitive atlas with many thematic maps overlaying the same space

The psychometric and human factor investigations are creative and also point the way toward understanding the processes of wayfinding. They provide interesting snapshots at points in time but do not seem to provide (at least not yet) much insight into the needs for wayfinding. Unfortunately, these experiments look at pieces of the puzzle and seem to be prone to reductionist error. They imply that if all these pieces were put together then we would understand the process.

I am not so sure: Wayfinding is a process and picking apart the pieces may not result in an accurate reconstruction without having a better sense of the correct sampling rate.

4. Does Research on Wayfinding Matter, Really?

Except as an intellectual exercise, does understanding human wayfinding have any real significance or importance? There are several areas in which a more complete understanding of wayfinding will prove quite useful.

Automobile drivers, hikers, and cyclists probably constitute a reasonable market for navigational aids for wayfinding (McGranaghan, Mark and Gould, 1987; Mark and McGranaghan, 1986; Judge, 1990). If the price were right the rental car market would certainly make use of such devices. Also, the large number of trip planning packets that the automobile clubs provide point toward a market of vacationers and business people. The large number of topographic maps sold to hikers and campers also indicates a potential market for such devices. A standard direction provider fashioned after McMap and using audio tapes might be sufficient for some market segments. However, if a system were developed that could provide mid-course corrections it might also prove valuable to police, ambulance, and delivery personnel. Note, mixed navigational technologies combining satellite systems such as GPS (Global Positioning System) and inertial navigation equipment can provide the accuracy needed for such activities (Judge, 1990). However, an ideal system might be expected to cope with traffic congestion, road repair, accidents, inclement weather, fallen trees, washed out bridges, as well as personal preferences such as scenic routes, fastest routes, shortest routes, routes with good restaurants, and so forth.

Many futurists predict that personal aircraft will begin to replace personal automobiles early in the next century. The highways of today are expected to become the skyways of tomorrow. There will be a need for new sorts of navigational aids to support such transport. The cognitive map may need to become a cognitive globe and everyday trip planning may have to incorporate spherical geometry. Such a change may result in radically altering human awareness of the spatial environment. Thus, understanding current methods of wayfinding and knowledge of human limitations may assist in the transition to a new transportation infrastructure.

The knowledge of how people wayfind may also be useful in system designs and interfaces for geographic information systems (GIS), and more general Hypertext/Hypermedia systems. Realizing that humans rarely make topological errors, frequently make metric errors, and generally have a basic and secondary network for navigation suggests possible enhancements to such systems. Also, noting individual differences in spatial and verbal abilities may suggest user customization paradigms. Such modifications might improve the recall, precision, time of retrieval, and user satisfaction of such systems.

5. A Research Direction *Not* Indicated in the Literature: Information Needs

The literature on wayfinding sees either competence or performance as the stimulus for study. Both of these approaches strive to see the final cognitive map or take snapshots at almost random events in the process of developing the cognitive map. Unfortunately, both the affective and logistical concerns in wayfinding and their interaction have not been fully explored with either of these approaches. Affective concerns involve the full range of human emotions: including fear, hate, love, need for novelty, need for connecting with others, and so forth. The logistic concerns are more task-oriented and include Garling's three variables (degree of differentiation of the environment, degree of visual access, and complexity of layout), but extend to navigational aids and well beyond.

Consequently, I would like to suggest an alternative focus: To systematically explore the information needs and not the products of wayfinding. That is, to seek answers to these questions that are derived from the initial set but are unanswered by the literature:

1) What is the range of situations that elicit wayfinding behaviors? More specifically: How, and why, do people re-route themselves? How may police in hot pursuit be supported by navigational aids and when should they not attempt hot pursuit? What sorts of multi-solution situations are involved in wayfinding (gas stations, restrooms, restaurants, etc.), and how is navigation altered under these differing goals? Resolution of these sorts of questions would result in the development of a typology of wayfinding scenarios. and decision support systems.
2) What are the affective and logistical information needs for wayfinding in different situations? That is: What are the uncertainties, questions and gaps experienced by humans as they explore the environment in wayfinding through a series of points in space and time associated with different situations? More specifically: Are the techniques of wayfinding different under different environmental conditions? Do people have the same needs and navigate with the same assumptions at night or during a storm as they do during a sunny day? Are humans really lost in a fog?
3) How are these needs resolved? That is: How do creative, intelligent creatures overcome these uncertainties, resolve these questions and bridge gaps in finding their way within a particular situation to arrive at a desired place in space and time?

In most experiments in the wayfinding literature when subjects get lost they are informed of the error and guided back to the "straight and narrow" (e.g., Passini and Proulx, 1988) (4). That is, false starts while navigating and the means of their resolution have not been explicitly investigated. Also, past studies have generally been biased by a preselected route or by preselection of the origin and destination by researchers. Also, the few studies that have looked at trip planning seem to only be concerned with limited optimization functions (e.g., Saisa and Garling, 1987). Consequently, past studies have not looked at the complete time line of events associated with wayfinding.

Since the competence and performance foci using sketch maps and studying speed and accuracy in wayfinding do not explore the affective, and only partially investigate the logistical concerns of wayfinding, another approach seems warranted.

6. A Methodology for Investigating Information Needs: Sense-making

The use of the "sense-making" approach (Dervin *et al.*, 1982, 1976; Dervin and Martin, 1983; Nilan *et al.*, 1989; Dervin and Nilan, 1986) in wayfinding research designs may provide data and analysis that past methods have not been able to extract adequately —the affective as well as logistical needs for wayfinding.

6.1 SENSE-MAKING'S CONCEPTUAL FRAMEWORK

Dervin (Dervin, 1983; Dervin and Nilan, 1986) pictures humans as intelligent, creative creatures capable of making sense of the world by incorporating knowledge from within themselves and from the external world. She metaphorically represents the cognitive states of humans as movement continually along a road in time and space. The perception of the current position along this road is a function of where we have been (experiences, environment, etc.) as well as where we are (present) and where we will be going (future). Thus, "sense-making" allows forward movement along the cognitive road. That is, if a person has no concerns or doubts or anomalous states of knowledge (Belkin, Oddy and Brooks, 1982) then forward movement is made along this road; however, gaps often appear that represent a need for making sense of the world before forward movement can be resumed. These gaps are a direct consequence of Dervin's perspective of a human's view of reality as intersubjective and constantly changing.

Previous studies of wayfinding as described have usually viewed people in traditional terms by assessing them demographically (age, sex, etc.) and by analyzing their cognitive map as a reflection of current understanding of spatial relationships. These previous studies attempted to find those aspects of humans that are constant in time and space so that stable generalizations about them can be made. Often such analyses see human diversity as too complex and revert to a simplified "experts' view" as the basis for model design (e.g., TOUR, ELMER, and TRAVELLER).

Dervin's approach allows each person to represent his or her own reality; however, Nilan posits (Dervin and Nilan, 1986) that this does not lead to chaos in analyzing human behavior since humans are connected by shared socio-cultural experiences. This overlap of experiences limits the range of diversity present at a particular point in time. This can be summed up in the cliche that "we went to different high schools together." Thus "sense-making" is a creative human process of understanding the world at a particular point in time and space limited by our physiological capacities, our present, past and future. Dervin (Dervin 1983, 5) consequently posits that:

> 'sense-making behavior can be predicted more successfully within the framework of a model which focuses on changing situations as predictors rather than such constant across time-space attributes as personality or demography [sex, race, age, etc.].'

Thus, scientific prediction about human behavior changes from trying to make "If...Then..." implications: If you are male, Then you act this way; to predicting a sequence of events in the form "Then...Then...": first one does this and may need this, Then one does that and may need that. This results in our particular case in the potential to develop a more comprehensive understanding of wayfinding based on a time line of events during wayfinding. It will also naturally result in developing navigational aids that give people what they need when they need it to find their way. Previous "sense-making" studies have lead to clearer pictures of information needs and their resolution in other domains (e.g., Dervin *et al.*, 1982, 1976; Dervin and Martin, 1983, Nilan *et al.*, 1989). These have included improved support for cancer patients, wider circulation for a newspaper, a radically different context-sensitive help system for desktop publishing, improved decision rules for public administrators, modified design specifications and training for word processing, development of a multivector weather information system, and so forth. Consequently, there is no reason to believe that such an approach applied to wayfinding would fail to deepen understanding of the needs of this very basic human cognitive process.

The "sense-making" instruments (structured neutral questioning time line interviews) would allow examination of the affective and logistical uncertainties encountered during wayfinding, potential solutions and their expected assistance, actual solutions and their actual assistance, and the aids used in resolving the uncertainties and their degree of success. The focus would be on the information needs for wayfinding as opposed to competence or performance. Such studies would not look at the product of the process alone (the cognitive map) but at the process itself.

7.2 PROTOTYPICAL INVESTIGATIONS USING "SENSE-MAKING"

Possible studies of wayfinding using "sense-making" might include:

- Reviewing a recent actual wayfinding experience of subjects using the structured instruments of "sense-making". This would be the most natural study for applying the method.
- Placing people in an unfamiliar, large city and having them select a landmark as a destination for wayfinding, and, after task completion or failure, interviewing them using the "sense-making" method.
- A laboratory version of the above process might be accomplished with a computer simulation and full motion video, again followed by a structured interview.
- Exploration of wayfinding and trip planning experiences with drivers, hikers, orienteerers, or auto rally participants would be quite amenable to use with this method.
- Individuals might be asked to select a goal and to develop and execute a travel plan, much like a scavenger hunt, followed by a structured "sense-making" interview.
- Situations in which subjects start lost or find themselves lost and get "unlost" might be devised, followed by a "sense-making" interview.

Such experiments would explore the cognitive uncertainties and how they were managed and resolved. The effects of unexpected situations, moving destinations, execution at different times of day or night might all be investigated. The creativity of humans in wayfinding seems never to have been fully studied, even though possible alternative means of wayfinding have been discussed.

7. Closing

Previous research has lead to many significant findings in understanding the cognitive aspects of human wayfinding. However, the "sense-making" method applied to wayfinding would explore and analyze the information needs of the process of wayfinding that have not been emphasized in the research literature because of its focus on assessment of performance and competence. The results from such investigations may readily lead to (but are certainly not limited to):

- Increased sensitivity and attention to human emotional needs in constructing travel routes and roadways, in the design of vehicles, and in the design of both traditional and "high-tech" navigational aids,
- Improved road signs, roadway markings, and lighting,
- Customized navigational aids including novel thematic maps for tourists and travelers, and
- Improved user interfaces to Geographic Information Systems (GIS) and other information retrieval systems including Hypertext and Hypermedia systems.

8. Notes

1. A central assumption often only implicit in much of the literature is the concept of a mental map or a map in the head (Evans, 1980 for review; Trowbridge, 1913 for initial discussion). The existence of such mental maps is based on people's ability to sketch maps. However, the actual representation in long-term memory is not clear at all (Vanetti and Allen, 1988; Foley and Cohen, 1984). The picture is further clouded by the work of Vanetti and Allen who demonstrated that high verbal ability may compensate for low spatial ability.
2. For review of the early literature on the development of human cognitive maps see Piaget and Inhelder (1967) or Hart and Moore (1973).
3. For a review of the majority of Appleyard's work, see Anthony, 1983.
4. Streeter (Streeter, Vitello, and Wonsiewicz, 1985) did not help subjects but did indirectly monitor the process of getting "unlost" by counting the number of wrong turns and measuring the travel times.

9. References

Anthony, K. H. (1983) 'Major Themes in the Work of Donald Appleyard', Environment and Behavior 15(4), 411-418.

Appleyard, D. (1981) Livable Streets, The MIT Press, Cambridge, MA.

Appleyard, D. (1970) 'Styles and Methods of Structuring a City', Environment and Behavior 2, 100-117.

Appleyard, D., Lynch, K. and Myer, J. R. (1963) The View from the Road, The MIT Press, Cambridge, MA.

Attneave, F. (1983) 'Concerning Cognitive Maps: Discussion of Baird and Kuipers', In Pick, H.L. and Acredolo, L.P. (Eds.), Spatial Orientation: Research, Theory, and Application, Plenum Press, New York, 361-365.

Baird, J. C. and Wagner, M. (1983) 'Modelling the Creation of Cognitive Maps', In Pick, H.L. and Acredolo, L.P. (Eds.), Spatial Orientation: Research, Theory, and Application, Plenum Press, New York, 321-344.

Belkin, N. J. Oddy, R. N. and Brooks, H. M. (1982) 'ASK for Information Retrieval: Part I: Background and Theory', Journal of Documentation 38(2), 61-71.

Dervin, B. (1983) 'An Overview of Sense-Making Research: Concepts, Methods, and Results to Date',Paper presented at the annual meeting of the International Communication Association, Dallas, Texas.

Dervin, B., Jacobson, T., and Nilan, M. (1982) ' When Cancer Strikes: How Cancer Patients Make Sense out of their Health Situations', Report to the Office of Cancer Communications, National Cancer Institute, Procurement order #263-MD-102094-3.

Dervin, B., Zweizig, D., Banister, M., Gabriel, M., Hall, E., and Kwan, C. (1976) 'The Development of Strategies for Dealing with the Information Needs of Urban Residents. Phase I: the Citizen Study', Final Report on Project L0035JA to the Office of Libraries and Learning Resources, U.S. Office of Education, U.S. Department of Health, Education, and Welfare, (ERIC:ED136791).

Dervin, B. and Martin, M. (1983) 'Sense-making profiles of Message Queing on Seattle Times Leisure Time Coverage', Report to the Seattle Times.

Dervin, B. and Nilan, M. (1986) 'Information Needs and Uses', in Williams, Martha (Ed.), Annual Review of Information Science and Technology (ARISTA) 21, Knowledge Industry Publications, 3-33.

Evans, G. W. (1980) 'Environmental Cognition.', Psychological Bulletin. 88(2), 259-287.

Evans, G. W. and Pezdek (1980) 'Cognitive Mapping: Knowledge of Real World Distance and Location Information', Journal of Experimental Psychology: Human Learning and Memory 6, 13-24.

Foley, J. E. and Cohen, A. J. (1984) 'Working Mental Representations of the Environment', Environment and Behavior 16(6), 713-729.

Garling, T., Book, A., and Lindberg, E. (1986) 'Spatial Orientation and Wayfinding in the Designed Environment: A Conceptual Analysis and Some Suggestions for Postoccupancy Evaluation', Journal of Arch. Plan. Res. 3, 55-64.

Garling, T., Book, A., and Lindberg, E. (1984) 'Cognitive Mapping of Large-Scale Environments: The Interrelationship of Action Plans, Acquisition, and Orientation', Environment and Behavior 16(1), 3-34.

Golledge, R.G., Smith, T. R., Pelligrino, J., Doherty, S., and Marshall, S. P. (1985) 'A Conceptual Model and Empirical Analysis of Children's acquisition of Spatial Knowledge', Journal of Environmental Psychology 5, 125-152.

Golledge, R. G. and Zannaras, G. (1975) 'Cognitive Approaches to the Analysis of Human Spatial Behavior', In Ittelson, W. (ed.), Environment and Cognition, Seminar Press, New York.

Hart, R. A. and Moore, G. T. (1973) 'The Development of Spatial Cognition: A Review', in Downs, R.R. and Stea, D. (Eds.), Image and Environment -- Cognitive Mapping and Spatial Behavior, Aldine, Chicago, Ill., 246-288.

Hirtle, S. C. and Jonides, J. (1985) 'Evidence of Hierarchies in Cognitive Maps', Memory and Cognition 13, 208-217.

Judge, P. C. (1990) 'Computers to Help Drivers Find Their Way', New York Times, Apr 24, 1990, D12.

Kovach, R. C., Surrette, M. A. and Aamodt, M. G. (1988) 'Following Informal Street Maps: Effects of Map Design', Environment and Behavior 20(6), 683-699.

Kuipers, B. (1983) 'The Cognitive Map: Could It Have Been Any Other Way?', In Pick, H.L. and Acredolo, L.P. (Eds.), Spatial Orientation: Research, Theory, and Application, Plenum Press, New York, 345-359.

Kuipers, B. (1982) 'The Map in the Head Metaphor', Environment and Behavior 14, 202-220.

Kuipers, B. (1978) 'Modeling Spatial Knowledge', Cognitive Science 2, 129-153.

Kuipers, B. (1977) Representing Knowledge of Large-Scale Space, The MIT AI Lab, TR-418, (Doctoral Thesis, MIT Mathematics Dept.), Cambridge, MA.

Leiser, D. and Zilbershatz, A. (1989) 'The Traveller: A Computational Model of Spatial Network Learning', Environment and Behavior 21(4), 435-463.

Lindberg, E. and Garling, T. (1987) 'Memory for Spatial Location in Two-Dimensional Arrays', Acta Psychologica 64, 151-166.

Lloyd, R. and Steinke, T. (1986) 'The Identification of Regional Boundaries on Cognitive Maps', Professional Geographer 38(2), 149-159.

Lynch, K. (1960) Image of the City, The MIT Press, Cambridge, MA.

Marble, D. F. and Bowlby, S. R. (1968) 'Shopping Alternatives and Recurrent Travel Patterns', in Horton, F. (ed.), Geographic Studies of Urban Transportation and Network Analysis, Northwestern Univ.Studies in Geography 16, Northwestern Univ. Press, Evanston, Ill.

Mark, D. M. (1990) 'The House on the Lake: Toward a Cross-Linguistic Computational Model of Geographic Relations', Draft manuscript, NCGIA, SUNY Buffalo, Dept. of Geography.

Mark, D. M. and Frank, A. U. (1989) 'Concepts of Space and Spatial Language", Proceedings, International Symposium on Computer-Assisted Cartography (Auto-Carto 9), Baltimore, MD., 538-556.

Mark, D. M. and McGranaghan, M. (1986) 'Effective Provision of Navigational Assistance to Drivers: A Cognitive Science Approach', in Blakemore, Michael (ed.), Proceedings, Auto Carto: International Conference on the Acquisition, Management, and Presentation of Spatial Data, London, International Cartographic Association (ISBN: 0854063129).

Mark, D. M., Svorou, S., and Zubin, D. (1988) 'Spatial Terms and Spatial Concepts: Geographic, Cognitive, and Linguistic Perspectives', Proceedings, International Geographic Information Systems (IGIS) Symposium: The Research Agenda 2: Technical Issues and the Research Agenda, NASA, Arlington, Virginia, 101-112.

Marr, D. (1982) Vision: A Computational Investigation into the Human Representation and Processing of Visual Information, W. H. Freeman, San Francisco.

McCalla, G. I. and Schneider, P. F. (1979) 'The Execution of Plans in an Independent Dynamic Microworld', Proceedings, Sixth International Joint Conference of Artificial Intelligence, Tokyo, Japan.

McCalla, G. I. and Schneider, P. F. (1980) 'Planning in a Dynamic Microworld', in Proceedings, Third Biennial Conference of the Canadian Society for the Computational Studies of Intelligence, Victoria, British Columbia, 248-255.

McCalla, G. I., Reid, L. and Schneider, P. F. (1982) 'Plan Creation, Plan Execution, and Knowledge Acquisition in a Dynamic Microworld', International Journal of Man-Machine Studies 16, 89-112.

McGranaghan, M., Mark D. M., and Gould, M. D. (1987) 'Automated Provision of Navigational Assistance to Drivers', American Cartographer 14(2), 121-138.

McNamara, T. P. (1984) 'Mental Representations of Knowledge Acquired From Maps', Journal of Experimental Psychology, Learning Memory and Cognition 10, 723-732.

McNamara, T. P., (1986) 'The Mental Representations of Spatial Relations', Cognitive Psychology 18, 87-121.

Moar, I and Carlton, I. R. (1982) 'Memory for Routes', Quarterly Journal of Experimental Psychology 34a, 381-394.

Nasar, Jack L., Valencia, H., Omar, Z., Chueh, S., Hwang, J. (1985) 'Out of Sight Further From Mind: Destination Visibility and Distance Perception', Environment and Behavior 17(5), 627-639.

Nilan, M. S., Newby, G. B., Paik, W., and Lopatin, K. (1989) 'User-oriented Interfaces for Computer Systems: A User-defined Online Help System for Desktop Publishing', Proceedings, American Association for Information Science.

Pailhous, J. (1970) La Representation de l'Espace Urbain -- L'exemple du Chauffeur de Taxi, Paris: Presses Universitaires de France.

Piaget, J. and Inhelder, B. (1967) The Child's Conception of Space, (orig in French, 1948), Norton, New York.

Passini, R. and Proulx, G. (1988) 'Wayfinding Without Vision: An Experiment with Congenitally Totally Blind People', Environment and Behavior 20(2), 227-252.

Riesbeck, C. (1980) '"You Can't Miss It!": Judging Clarity of Directions', Cognitive Science 4, 285-303.

Russell, J. A. and Ward, L. M. (1982) 'Environmental Psychology', Annual Review of Psychology 32, 651-688.

Sadalla, E. K. and Montello, D. R. (1989) 'Remembering Changes in Direction', Environment and Behavior 21(3), 346-363.

Siasa, J. and Garling, T. (1987) 'Sequential Spatial Choices in the Large-Scale Environment', Environment and Behavior 19(5), 614-635.

Siasa, J., Svensson-Garling, A., Garling, T., and Lindberg, E. (1986) 'Interurban Cognitive Distance: The Relationship between Judgements of Straight-line Distances, Travel Distances, and Travel Time', Geographical Analysis 18, 167-174.

Siegel, A. W. and White, S. H. (1975) 'The Development of Spatial Representations of Large-Scale Environments', in Reese, H.W. (ed.), Advance in Child Development and Behavior 10, Academic Press, New York.

Simutis, Z. M. and Barsam, H. F. (1983) 'Terrain Visualization and Map Reading', In Pick, H.L. and Acredolo, L.P. (Eds.), Spatial Orientation: Research, Theory, and Application, Plenum Press, New York, 161-193.

Stea, D. and Downs, R. M. (1970) 'From the Outside Looking In At the Inside Looking Out', Environment and Behavior 2, 3-12.

Streeter, L. A. and Vitello, D. (1986) 'A Profile of Drivers' Map-Reading Abilities', Human Factors 28(2), 223-239.

Streeter, L. A., Vitello, D. and Wonsiewicz, S. A. (1985) 'How to Tell People Where to Go: Comparing Navigational Aids', International Journal of Man-Machine Studies 22, 549-562.

Talmy, L. (1983) 'How Language Structures Space', In Pick, H.L. and Acredolo, L.P. (Eds.), Spatial Orientation: Research, Theory, and Application, Plenum Press, New York.

Thorndyke, P. (1980) 'Performance Models for Spatial and Locational Cognition', Technical Report R-2676-ONR. Rand Corp., Santa Monica, CA.

Thorndyke, P. M. and Goldin, S. E. (1983) 'Spatial Learning and Reasoning Skills', In Pick, H.L. and Acredolo, L.P. (Eds.), Spatial Orientation: Research, Theory, and Application, Plenum Press, New York, 195-217.

Thorndyke, P. and Hayes-Roth, B. (1982) 'Differences in Spatial Knowledge acquired from Maps and Navigation', Cognitive Psychology 14, 560-589

Trowbridge, C. C. (1913) 'On Fundamental Methods of Orientation and Imaginary maps', Science 38, 888-897.

Vanetti, E. J. and Allen, G. L. (1988) 'Communicating Environmental Knowledge: The Spatial Impact of Verbal and Spatial Abilities on the Production and Comprehension of Route Directions', Environment and Behavior 20(6), 667-682.

Wickens, C. D. (1984) Engineering Psychology and Human Performance, Merrill, Columbus, Ohio.

WAYFINDING THEORY AND RESEARCH: THE NEED FOR A NEW APPROACH

MARK BLADES
Department of Psychology
University of Sheffield
Sheffield S10 2TN United Kingdom

ABSTRACT. Wayfinding is the ability to learn a route through the environment. This paper considers the theories (Piaget, Inhelder and Szeminska, 1960; Siegel and White, 1975) which have been put forward to explain the development of wayfinding and which have provided a framework for much of the related empirical research. It is argued that the evidence does not clearly support those theories, and it is suggested that recent research in other areas of cognitive development can offer alternative approaches to the consideration of how wayfinding abilities develop.

1. Introduction

For the purposes of this paper, wayfinding will be defined as the ability to learn and remember a route through the environment. Route knowledge is, of course, not an isolated component of environmental awareness and wayfinding in familiar environments will depend on a person's overall understanding of the area and that person's plans and intentions when deciding how to travel between several different places (see Garling, 1989; Garling, Book, and Lindberg, 1984; Hayes-Roth and Hayes-Roth, 1979; Smith, Pellegrino, and Golledge, 1982). But to keep this review within manageable proportions I will concentrate on the research which has considered how a single route is learnt, and the emphasis will be on how children learn a novel route - partly because the majority of relevant studies have involved young subjects, and partly because of my own interests as a developmental psychologist.

The first part of the paper (section 2) will consider in some detail the theories of environmental development put forward by Piaget (Piaget and Inhelder, 1956; Piaget, Inhelder and Szeminska, 1960) and by Siegel and White (1975). These theories have been influential and have provided the starting point and framework for much of the research which will be discussed. It will be pointed out that that the theories are not as clearly expressed as is sometimes implied in the literature, but they do nonetheless, offer particular hypotheses about the development of wayfinding, and in section 5 the hypotheses will be considered in the light of the empirical evidence.

There are already a number of recent and extensive reviews of the literature on environmental cognition and wayfinding (see Cohen, 1985; Golledge, 1987; Heft and Wohlwill, 1985; Spencer, Blades and Morsley, 1989), and rather than attempt to duplicate those reviews, this paper will concentrate on those aspects of the environmental research which can be considered in the context of contemporary approaches and themes within developmental psychology (see sections 6).

D. M. Mark and A. U. Frank (eds.), Cognitive and Linguistic Aspects of Geographic Space, 137–165.

2. Theories of Environmental Cognition

A large proportion of the research into children and adults wayfinding has been carried out explicitly or implicitly within the theoretical frameworks of environmental cognition put forward by Piaget and by Siegel and White (1975). The latter theory is derived directly from Piaget's original theory, but because it differs from Piaget's in several important respects I will give separate summaries of the two theories. The underlying assumption behind both is that from repeated direct experience in the environment an individual progresses through a series of stages of environmental knowledge culminating in an overall understanding of the spatial relationships between all the places within that environment. Such an understanding has been referred to by a variety of terms 'full configurational' (Piaget), survey knowledge (Shemyakin, 1961; Siegel and White, 1975), vector knowledge (Byrne, 1979), or more generally as a cognitive map.

2.1. PIAGET'S THEORY

Piaget tested children's knowledge of their own home area by asking children of different ages to construct a model of the area (from material which was provided, such as toy houses and 'pieces of wood' to represent parks, public squares and bridges) and to draw a plan of a route from school to a familiar landmark (in a sand-tray or on paper). The children were asked about their models or plans as they produced them.

Piaget divided children's performance into three stages. In stage I children have little understanding beyond a general sense of direction and the ability to recognize individual places in the environment. In stage II Piaget makes the distinction between performance based on direct activity in the environment and performance which is dependent on a mental representation (i.e. memory) of the environment. As an example of performance based on direct activity in space Piaget points out that children in stage II are often capable of finding their own way along a familiar route and as an example of performance based on a mental representation Piaget refers to children describing a route from memory.

Piaget suggests that in stage II children base their descriptions of a route on their memory not of the environment as such, but their memory of their movement through the environment. For example, he quotes one child (aged 6 years) describing a familiar route 'I go straight along, I turn there, I go straight along again, I turn there, then I keep going straight and I turn once more' (Piaget *et al*, 1960, 11). This particular child did not refer to a single landmark along the route.

Other children drew plans of a known route, but when asked to place specific landmarks along the route they were unable to do so accurately. These results led Piaget to argue that 'in the process of remembering their own actions, these subjects are led to mention landmarks. But these are simply tacked on to their recollections...Accordingly, astonishing as it may seem, subjects fail to respect the true positions of landmarks and do not even preserve the order in which they come to them' (Piaget *et al*, 1960, 12).

In stage IIIA children are able to describe more about a route and will often include landmarks at choice points along the route, but Piaget noted that children in this stage tended to place the landmarks in sub-groups. The landmarks in each sub-group along a route might be accurately positioned relative to each other, but the different sub-groups were not accurately related to each other. It is not until stage IIIB that children produce a route as 'a coherent whole' with all the landmarks placed in relationship to each other. Children's specific ability to describe routes in the last stage corresponds with their performance in drawing plans. Piaget refers to IIIA as the stage of partial coordination of landmarks - children can portray groups of landmarks relatively accurately, but the groups are not interconnected. Stage IIIB is called the stage of full coordination - when all features are described as a single comprehensive group.

Piaget's description of children's environmental representation is inseparable from his theory of spatial development, and in turn his more general theory of cognitive development (for a summary of his theory of spatial development see Robinson and Petchenik, 1976; for a summary of the general theory see Ginsburg and Opper, 1969; Flavell, 1985). By referring to his overall theory Piaget is able to offer explanations for the children's limitations in each stage of environmental knowledge. Just to give a couple of examples: Piaget explains young children's early reliance on memory for actual movement through the environment as a consequence of their dependency on physical motor schema. When discussing children's inaccurate placing of landmarks along a route in stage II Piaget refers to children in that stage being limited to topological spatial relationships (i.e. relationships relying on simple notions of proximity) and he emphasizes that such relationships are influenced by subjective criteria (e.g. home and school may be inaccurately shown as close together because of the importance of those two places for the child). These limitations are a reflection of children's general cognitive abilities in the preoperational stage of development.

It is not possible here to go into details of Piaget's spatial and cognitive theories (see Brainerd, 1978a,b; Cox, 1986; Light, 1987; Morss, 1987; Presson and Somerville, 1985, for relevant commentaries). Those theories are mentioned here to emphasize that when Piaget discussed children's ability to represent routes he was not only putting forward a description of the progression of this ability, he was also attempting to explain the nature of that progression by reference to children's underlying cognitive abilities.

Piaget's theory of the development of environmental knowledge has been extremely influential both directly, and indirectly through secondary sources (e.g. Hart and Moore, 1973), and Piaget's work is also the foundation for the theory put forward by Siegel and White (1975)

2.2 SIEGEL AND WHITE'S THEORY

Siegel and White (1975) propose a sequence in the development of environmental knowledge, which includes a knowledge of landmarks, routes, 'minimaps' and survey knowledge. According to Siegel and White the first stage of children's environmental learning is noticing and remembering landmarks. In the second stage children can use the landmarks as reference points for their behavior in the environment: in other words, the decisions and actions which a child carries out at a particular landmark become associated with that landmark and the sequence of landmarks provides a basis for an associated sequence of decisions so that the child is able to follow a complete route.

In the third stage children can form clusters of landmarks which Siegel and White label 'minimaps' (and this term is directly equivalent to Piaget's phrase 'partial coordination' used to describe children's knowledge in his stage IIIA). In Siegel and White's fourth stage children achieve a survey representation of the environment (this is the same as Piaget's stage IIIB of 'full coordination') and the Siegel and White's description of the transition from their stage three to stage four is taken from Piaget's description of the progression from his stage IIIA to IIIB.

Siegel and White's theory is very similar to Piaget's, but there are several points of contrast. Piaget was solely concerned with the development of children's knowledge, but Siegel and White emphasized the similarity between children's stages of development and the stages of development experienced by an adult learning a new area. It needs to be stressed that Piaget considered that children passed through stages (from the most limited to least limited understanding) of environmental knowledge because of their underlying cognitive limitations (see above). Adults, of course, do not have such cognitive limitations and therefore if adults progress through a series of stages which appear similar to the children's stages described by Piaget it will not necessarily be for the same reasons.

Although Piaget's and Siegel and White's theories are similar (and essentially the same for the later stages of development) they do differ in one important respect. Piaget did not clearly distinguish between route and landmark learning - though the implication of his discussion is that very young children (in the preoperational stage of cognitive development) remember familiar routes on the basis of their physical movements along the route and only later 'attach' landmark information to their knowledge of the route. Older children (in the operational periods of development) will integrate route and landmark knowledge in their memory of a familiar route. In contrast, Siegel and White argue that landmark knowledge precedes route knowledge.

Both Piaget and Siegel and White provide descriptions of environmental development, but they differ in terms of the degree of explanation put forward. As mentioned above, Piaget explained the development of children's environmental knowledge as a consequence of the development of their cognitive abilities and therefore he can suggest why children might be limited in a particular stage. Piaget is less clear how children progress from one stage of environmental knowledge to the next, though reference can be made to his general theory of development for some potential reasons (see Brainerd, 1978a). For example, he discusses development in terms of maturation and experience and refers to the concept of 'cognitive conflict'. Cognitive conflict means that a child will have expectations about the outcome of a task or procedure and that if from direct experience of the task the child finds that her expectations are incorrect this will engender a state of conflict. When the child is at an appropriate stage of development she can benefit from this conflict to progress to a better understanding of the task. Two points stem from this; firstly, a young child will not necessarily gain from experience because she will not have the mental processes to benefit from the experience, and secondly, learning is the consequence of the child's own direct experience.

Except when they rely on Piaget's arguments (for the later stages of environmental knowledge) Siegel and White offer less in terms of explanation and do not suggest how individuals progress through the different stages of environmental knowledge. Their emphasis on the necessity of locomotion and practical activity implies that such progression is dependent on repeated direct experience in the environment, but it is not clear how repeated experience can lead to more detailed and accurate encoding of environmental knowledge.

2.3. SUMMARY

To summarize the preceding discussion. Both Piaget's and Siegel and White's theories are stage theories of the development of environmental knowledge. Both offer a similar description of children's development (and Siegel and White apply the same description to adults learning novel environments) but whereas Piaget did not distinguish between landmark and route knowledge, Siegel and White suggest that learning landmarks is a preliminary stage to learning routes.

Siegel and White's theory is primarily descriptive, but Piaget provides explanations of children's performance in terms of their cognitive development. Although both theories stress the need for direct experience of the environment, neither specifies either the amount of experience or the type of experience which is necessary to achieve a particular stage of development (although Piaget argues that the achievement of each stage is constrained by the child's cognitive level).

Both theories appear to offer several predictions about the development of environmental knowledge and in particular the development of wayfinding ability, which is the primary concern of this paper, and these will be discussed together with recent empirical evidence in section 5, but first it is necessary to point out the difficulty of defining the key terms which are typically used in descriptions of wayfinding ability.

3. Problems of Definition

The term landmark has been used in a variety of ways. Piaget used the term very generally to refer to any of the toy features included in the model layout which he asked children to construct; to refer to any place which was mentioned by children in their verbal descriptions; and sometimes to refer to named roads on the plans drawn by children. Siegel and White (1975) give several descriptions of landmarks: as the beginning or end points of routes (e.g. home or school); as unique 'configurations of perceptual events' at a particular location (such as the intersection of two roads); and as 'intermediate course maintaining devices' (i.e. any recognizable place along the route which confirms that the individual is on the intended route).

It is useful to distinguish between two different meanings of the word landmark (Anooshian, 1988; Blades, 1989). Both meanings are used, but not distinguished by Siegel and White. The term landmark can be used, in a weak sense, to refer to any feature which is salient for an individual. I consider that when referring to a feature which has been noticed or remembered in isolation it is usually preferable to define such a feature simply as a 'place'.

However, a particular feature has little value unless an individual knows what action to take after recognizing it. To be a landmark, in the precise sense of that term, a feature needs to be more than just an isolated place and has to be linked in memory to information which indicates how the individual should act when approaching the landmark. This information necessitates knowing such information as: whether one is facing towards or away from the landmark; one's own position in the environment relative to the landmark (and this requires locating oneself with reference to at least one other feature as well as the landmark itself); a knowledge of where the landmark is (i.e. on which route or routes it is to be expected); and the action which is necessary at the landmark in order to reach the next point along the route (turn left, go straight ahead, etc.). The term 'landmark' is best restricted to features which are specifically used to locate oneself in the environment and act as a guide to wayfinding.

The point which needs emphasis is that for a landmark to be effective it must be related to a great deal of other environmental information, but once this additional information is considered it becomes difficult to separate the terms 'landmark' and 'route'. Consider a simple example. The layout illustrated in Figure 1 consists of six differently colored, but otherwise identical, cylinders (landmarks) in an unbounded space. Assume that you can only view one landmark at a time and that from your present position you can see landmark C and want to reach D. Knowing that you are approaching C is not very helpful because you may be approaching C from any direction. You need to know, for example, that you are between B and C which identifies your direction of travel. If you want to reach D you need to know the relationship between C and D so that you are aware of the need to turn right at C. In this series of decision making processes no single landmark is useful on it own - it is necessary to understand the relationship between three landmarks. That relationship is effectively a route; C is can only be employed as a landmark because it is known to be on the route B, C, D. (For a more detailed discussion of the complexity of the relationship between landmarks and routes see Kuipers, 1978; and for recent discussions of the use of the term landmark see Newcombe, 1988; Pick, Montello, and Somerville, 1988; Presson and Montello, 1988).

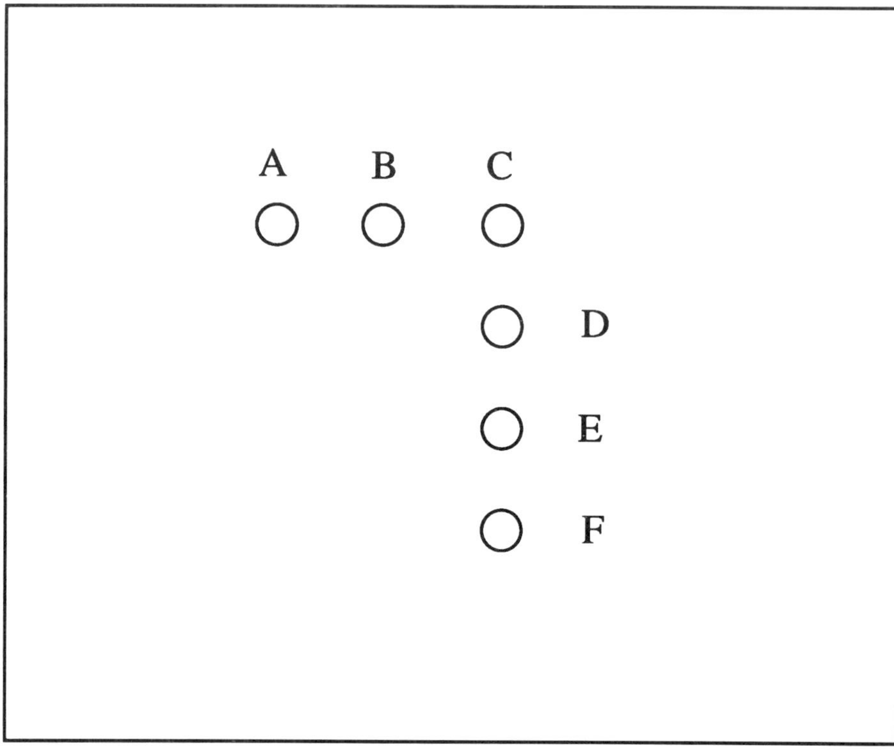

Figure 1. Layout of six differently-colored cylinders (see text).

In other words, if an environmental feature is to function as an effective landmark it must already be fully or partially integrated into the route sequence of which it is part. Therefore, it is difficult to consider landmark knowledge and route knowledge as independent stages of environmental cognition.

An additional problem is to define the term route. If we know that there is a path between B and C does this constitute a route? Perhaps such a suggestion would be considered too narrow a use of the term route, which is usually employed in reference to a series of landmarks and decisions, and possibly the path between B and C would be better considered as a partial route, but this prompts the question of how long does a route have to be. This is rather like asking how long is a piece of string. Several experimenters have used routes consisting of only a starting point, two landmarks and an end point (e.g. Hazen, Lockman and Pick, 1978; Waller, 1985) and this would be the equivalent, in the above example, of considering B,C,D,E as a route, but these features are only part of the longer route A to F and therefore an argument could be made that B to E is only a partial route. Of course, in the real world any route is only part of a longer route. (For a useful discussion of routes see Leiser, 1987; Leiser and Zilbershatz, 1989). In considering and comparing the wayfinding research it is important to bear in mind the different operational definitions of landmark and route which have been used in different experiments.

4. Note on Methodologies

A large number of methods have been used to discover an individual's knowledge about the environment, these have been well reviewed elsewhere (e.g. Golledge, 1987) and therefore only a few comments about the different methods will be made here. Subjects have been asked to give verbal or written descriptions, draw sketch maps, construct models, identify drawings, photographs and films, estimate distances and directions, or actually travel through experimental layouts or real environments.

The particular method employed will depend on the purpose of the study; for example some methodologies (such as estimating distances and direction) have been used to elicit an individual's configurational knowledge of space, and others (such as drawing route plans, or walking through an environment) are more specific to the consideration of an individual's wayfinding performance. Also, some consideration must be given to the abilities of the subject (for instance, it would be inappropriate to ask very young children to estimate metric distances or draw maps).

The question of methodology involves the issues of convergent validity, and reliability. Convergent validity means that if the environmental knowledge of a subject is tested with two, or more, methods, then the subject's performance should be equivalent for each method, so for example, if a person is asked to estimate distances and draw a scale map between places in two separate tasks, then that person's estimate of the distance from place A to place B and the distance shown on the map should be the same. Newcombe (1985) has reviewed a number of studies of environmental cognition which have used more than a single method and has concluded that studies which have been carried out 'in an effort to establish convergent validity have had mixed outcomes. Although some have been successful, the overall picture is that different techniques frequently reveal different results' (1985, 296).

The lack of convergent validity is a major problem for studies of environmental cognition, but it can be considered in the context of Liben's (1981) discussion of spatial representation. Liben describes three types of spatial representation: spatial products, which refers to any external representation of space (models, maps, written descriptions, etc.); spatial thought which refers to the mental manipulation of spatial information (e.g. mentally 'travelling' a remembered route or counting the number of windows in your home); and spatial storage which refers to the store of spatial information which an individual has in memory.

The techniques employed in studies of environmental cognition inevitably tap the interaction between spatial thought and spatial storage. For example, asking a person to express their knowledge (spatial storage) by drawing a sketch map (a spatial product) requires that person to imagine the environment from an aerial perspective (spatial thought). Therefore, when inferring an individual's environmental knowledge it is important to take into account how different methods employed to elicit that knowledge might require different spatial thought processes. For example, if a child is unable to produce a sketch map this could be the result of either the child's lack of environmental knowledge, or the child's inability to draw. Although such a point is often made in the literature, it is still the case that many studies of environmental cognition and wayfinding confound spatial thought and spatial storage and this leads to difficulties in interpreting individuals' performance.

The issue of reliability has not often been addressed. The reliability of a method means that the same method should produce the same result when used with the same subject. For example, if a person is asked to judge the distance from place A to place B on two separate occasions that judgment should be the same each time. Most studies simply assume that the methods used will be reliable, and this is in marked contrast to other areas of psychology (e.g. the assessment of intelligence) where the reliability of the methods used to test an individual's performance would be examined and reported.

Blades (in press) reviewed the studies which have used sketch maps as a technique to assess environmental knowledge and pointed out that all of these studies had only collected sketch maps at one point in time. In other words, the experimenters had assumed that the sketch drawn by a subject reflected a consistent spatial product, but if this is not the case and the same subject produces a different map each time she is asked to draw the same area then sketch maps would have little value as representations of that person's environmental knowledge. Blades (in press) carried out an experiment in which the same subjects were asked to draw the same, familiar, route on two occasions separated by a week (so that any effects for increased knowledge of the route would be minimal) and found that subjects produced very consistent sketches - in other words, sketch maps do provide a reliable method of data collection. It would be reassuring to know that many of the other methods used to test environmental cognition are equally reliable.

The environments used to examine spatial knowledge include different types of model layouts (e.g. Herman, Roth, Miranda and Getz, 1982); individual rooms, such as classrooms (e.g. Liben, Moore and Golbeck, 1982); specially constructed experimental layouts which subjects can travel through (e.g. Hazen *et al,* 1978); individual buildings (e.g. Garling, Lindberg and Mantyla, 1983); and several real world settings such as a shopping precinct, Passini (1984), a school complex, Cousins, Siegel and Maxwell (1983), a university campus, Evans, Marrero and Butler (1981) and various urban and natural environments, e.g. Okabe, Aoki and Hamamoto (1986). For a review and discussion of such environments, see Weatherford (1985). Given the range of different settings which have been used there is inevitably a difficulty in making comparisons between different studies.

In most experiments the environment is a constant - in other words, all subjects experience the same environment and the variables which are examined are 'external' to the environment (e.g. the experimenter may be interested in examining the effects of varying the degree of experience in the same environment, varying the type of experience in the same environment, or considering age related differences in learning the same environment). Some studies have manipulated the environment itself and and have tested different groups of subjects in variations of the same environment (e.g. Evans, Skorpanich, Garling, Bryant and Bresolin, 1984; Rowen and Hardwick, 1983), but very few studies have tested the same subjects in manipulated environments so that direct comparisons can be made between subjects' performance in different settings.

One further point - some experiments have carried out case studies of an individual's environmental learning (e.g. Canter 1977; Golledge, Smith, Pellegrino, Doherty and Marshall, 1985; Wood and Beck, 1989), or reported analyses of individual performance (e.g. Golledge, Rivizzigno and Spector, 1976; Lloyd, 1989). But the majority of the research employs experimental designs which make comparisons within or between groups. This is so typical of psychological research that it usually goes without comment, but it needs to be pointed out that studies involving group comparisons can, of course, say little or nothing about individual differences.

5. The Empirical Research

As the emphasis of this paper is on the development of wayfinding I will concentrate on those experiments which have indicated something about children's and adults' knowledge of routes and relate the findings from those experiments to the principle theories which were described above. Many studies have considered not only route knowledge, but also aspects of configurational understanding and I will refer to these aspects of the research when the experimental results are indicative of wayfinding ability.

5.1. WAYFINDING IN FAMILIAR ENVIRONMENTS

Cousins et al.(1983) tested 7, 10 and 13 year old children's knowledge of their own school environment. The children carried out a series of tasks. First they were asked to lead the experimenter along the most direct route between two designated points in the school area, and they did this for three different routes. The children were then shown eight photographs of landmarks along each route (five of these were scenes on the route and three were not directly visible from the route) and had to select the correct five pictures, and then order the photographs in the sequence they occurred along the route. They were also asked to carry out a 'route scaling' task (i.e. place the photographs in a line so that the distance between the photographs in the line represented the scaled distance between the features along the actual route). The children were then retested in the environment at specific 'sighting locations' (usually off the route) and asked to point and to estimate the distance to out-of-sight target places.

All three age groups were very proficient at walking the routes in the actual environment. There were also high levels of performance on the landmark selection and ordering tasks. These measures can be taken as indications of the children's landmark and route knowledge. The other tasks (which required an understanding of the configuration of the environment) showed effects for both age and for different routes.

Individual analysis of the children's results demonstrated that the pattern of the children's success showed a sequence of performance - children were only successful on the configurational tasks if they were successful on the route ordering and landmark identification tasks, and this can be taken as an indication of the developmental progression put forward by Siegel and White (1975) - i.e. that landmark knowledge precedes route knowledge which precedes a configurational understanding of the environment. The study by Cousins *et al* (1983) is important because it is one of the very few experiments which have examined children's performance in a naturalistic and familiar area and it demonstrates that environmental knowledge may be stage like in development.

Nonetheless, detailed consideration of the results of Cousin *et al* (1983) results does not entirely support the theory put forward by Siegel and White (1975). All the children were very proficient at walking the test routes in the actual environment and this raises the question of how wayfinding ability should be assessed. If practical performance is taken as the criterion of successful ability then there is no doubt that even the youngest children in the study by Cousin *et al.* were competent wayfinders. They were also equally successful on the landmark selection and landmark ordering tasks. If landmark knowledge precedes route knowledge it might have been expected that there would be differences in performance on these two tasks, but this did not appear to be the case. In fact, the experiment can say little about the stages of landmark and route learning because the children were already very familiar with the environment and this experience might account for the lack of difference between landmark and route knowledge.

It can be assumed that the majority of the children included in the study by Cousin's *et al* were in the concrete operational or formal operational period of development and given Piaget's description of environmental development it would be expected that all the children would be successful on the wayfinding tasks.

Other studies have also considered children's abilities in familiar environments, but have used artificial layouts in which the children have received specific and controlled experience. Cohen and Schuepfer (1980) showed 7, 11 and 23 year olds six photographic slides of a route through a series of connected halls. Each slide portrayed a particular hall and its intersections with three other halls, and the intersections were marked by different toy animals which served as landmarks. The subjects learnt a route through the halls by viewing the slides in the correct sequence along the route. Having learnt the route they were then shown the same sequence of halls, but without the landmarks present and for each of the six slides they were asked to indicate the correct turn to the next hall, and the landmarks associated with all the intersections.

There was no difference between the age groups for the number of times which they had to view the slide sequence before learning the whole route successfully - all the groups learnt the route after an average of 7.5 viewings. However, when the landmarks were removed there was an age related effect, the 7 year olds made more errors than the 11 year olds who made more errors than the 23 year olds, and this implies that the younger the children the more dependent they were on the presence of the landmarks for successful navigation. Overall, the older the subjects the more landmarks they remembered, but all three age groups recalled more landmarks which marked correct turns than landmarks which marked incorrect turns, and this would suggest that all the subjects were sensitive to the importance of the landmarks at correct junctions.

In a second experiment, Cohen and Schuepfer (1980) also had similar aged subjects learn the route through the slide sequence and then asked them to draw or construct a plan of the hall. The youngest subjects tended to produce a linear arrangement of the route, the 11 year olds showed some partial coordination of the route, and the oldest subjects produced accurate plans. Cohen and Schuepfer argued that the better performance of the older subjects (in the first experiment) when the landmarks were not present in the slides was due to their better understanding of the overall configuration of the halls.

Hazen *et al.*(1978) also taught subjects the route through an artificial environment; their subjects were aged 3, 4, 5 and 6 years of age and the test layout consisted of four or six small rooms arranged in two rows of two or three rooms. Each room contained a single toy animal (landmark) in the center of the room. The children were repeatedly walked along either a U shaped route or a zig-zag (Z) route through all the rooms. They were trained until they could both correctly follow the route and, as they walked the route, correctly predict which landmark they would see in the next room they would enter (before they actually entered the room). The number of trials needed to learn the route depended on the condition, and varied from six trials (for the 5 year olds learning the Z route through the 6 room layout) to three trials (for the 5 year olds learning the U route through the 4 room layout).

After learning the route, the children were asked to walk the route in the reverse direction and as in the learning trials, they had to predict the landmark in the next room along the route before they entered it. For all age groups the ability to walk the reverse route was almost perfect and the children had high scores for predicting the landmarks.

An experiment by Uttal and Wellman (1989) included a condition with a similar procedure to the study by Hazen *et al* (1979). Four and five year olds learnt a U shaped route through six rooms, each of which had a single landmark. The results were similar to those found by Hazen et al; the children required five trips through the rooms to learn the route and were able to perform the route reversal test successfully. Allen, Kirasic and Beard (1989) had 6, 9 and 11 year olds learn a route through a maze with four choice points each of which was marked by a distinct landmark. The youngest children required seven trials to learn the route and the other two age groups required four trials, but there was no difference between the groups on a route reversal task.

At the end of their study Hazen *et al.*(1979) gave the children small boxes to represent the rooms and miniature animals to represent the landmarks and asked them to construct a model of the environment through which they had walked. To do this successfully the children needed to understand the configuration of the rooms. The children made frequent errors, either because they did not place the boxes in the correct pattern, or because they did not include the landmarks in the correct rooms. In other words, although the children were capable of finding the route through the rooms in the wayfinding task this did not necessarily result in a full appreciation of the rooms' spatial configuration.

Although the studies of Cohen and Schuepfer's (1980) and Hazen *et al.* (1978) provide much evidence about the learning of routes, neither offers a direct test of Siegel and White's (1975) hypothesis that landmark learning precedes route learning. As the subjects were only tested after having learnt a route it is not possible to say whether the subjects passed through a stage of learning information about the landmarks prior to learning the full route.

Hazen *et al.*(1978) suggested that the young children's very good route performance in their experiment could have been a consequence of the children having learnt a motor sequence through the rooms. This would correspond to Piaget's suggestion that earliest route learning is based on a memory for physical movement through an environment. But this does not seem to be a likely explanation, because in the study by Hazen *et al.* the children were asked to walk the route in the reverse direction, and the physical movements they made on the reverse trip would not correspond to the movements they had learnt during the training trials. If anything, the finding that the children were competent at the route reversal task would contradict Piaget's argument that route learning is based on motor schemes.

5.1. WAYFINDING IN UNFAMILIAR ENVIRONMENTS

In contrast to the above experiments which examined subjects' wayfinding abilities in familiar environments, several studies have considered the performance of subjects in unfamiliar areas. All these studies have tested individuals in real world settings, but have usually only included young subjects. For examples of research with adults see Garling, Book, Lindberg and Nilsson (1981) who found that adults could recall, almost perfectly, the order of landmarks along a walked route after a single experience of the route; and Heft (1979) who found that after one guided experience, adults could retrace a route which included 22 intersections with an average of only three mistakes (see section 6.2).

Gale, Golledge, Pellegrino and Doherty (1990) asked 16 children aged between 9 and 12 year old children to learn routes through a uniform urban environment. The children were taken over the route by the experiment in the first trial and were then asked to retrace the same route on four subsequent trials - if the children made an error in following the route on the later trials they were corrected by the experimenter, who pointed out that they had made a mistake and this allowed the children the opportunity to recognize and correct the error (a 'realized mistake'). If the child did not correct the error she was asked to return to the location where the error had occurred and make another route decision (an 'unrealized' mistake).

In discussing the results Gale *et al.*state 'With respect to the ... field navigation task, there was strong evidence of relatively rapid and systematic learning across all measures. By the end of the third trial, learning was virtually complete. Two subjects, both 12 year old girls, actually experienced one-trial learning, not once making an unrealized mistake and only one subject made an unrealized mistake on the final day' (1990, 11). These results clearly indicate that children of this age can learn a route very quickly, and sometimes after only a single walk along the route.

Torell (1990) considered four 6 year olds and four 10 year olds in a large and comparatively complex environment. She initially guided the children (two at a time) through the area and then on two or three following occasions the children found their own way. After each occasion the children were asked to construct a sketch map of the area (by directing the experimenter's drawing), to select photographs of features in the area, and arrange them in the correct route order. Torell did not specifically analyze the children's performance as they walked the route after the first, guided, trial, but she does say that the children had 'remarkable way-finding skill' and 'although the stimulus area was large and topographically very complex with several choice points and forks along the route, only a few signs of hesitations were shown by the children during the second walk. On only one occasion, the boys in the older group, made a mistake in the choice of route.' (1.97). Even though this must, in the absence of any reported data, remain a slightly subjective statement it does imply that even young children in an unfamiliar environment can learn a route effectively after very little experience.

The report by Torell (1990) is confirmed by other studies which have examined young children's wayfinding ability. Cornell and Hay (1984) studied 84 five year olds and 84 eight year olds. Equal numbers of each age group were tested in one of three conditions. In one condition the children experienced a guided walk along a route (through a university campus) which included seven choice points. In the second condition the children viewed a videotape of the same route, and in the third condition they saw a slide sequence of 56 views along the route.

Although the children were not explicitly told to remember the route, at each choice point on the walked route the child was advised to look around and look back at the way from which she had come. In the videotape condition the film included a 360° camera pan at each choice point, and in the slide condition four slides illustrated the view in four directions from the choice point.

The children only experienced the route once and were then tested by 'travelling' the route in either the same or reverse direction, and at each choice point they were asked to indicate the way to proceed. The 8 year olds made fewer errors than the 5 year olds in all conditions. The mean number of errors were 2.4 and 2.0 in the slide and video conditions, and in these conditions the children were poorer when tested on the reverse route compared to following the route in the original direction of travel. In the walked condition children only made 0.7 errors and the direction of travel had no effect on performance.

The experiment by Cornell and Hay (1984) demonstrated that children as young as 5 years can remember a route after a single experience, and the significant differences between children's performance in the different conditions also points to the difficulty of extrapolating from laboratory simulations to real world behavior.

One further study has considered very young children: Darvizeh and Spencer (1984) took twenty 3 and 4 year olds along two short routes through an urban area. For each route, half the children (in the 'unaided' condition) simply walked the route, but for the other children (in the 'aided' condition) the experimenter referred to features related to choice points on the route (e.g. a cobbled section of road, a tree, or a poster). The features were mentioned in the course of conversation and were not specifically indicated as potential landmarks. The children only walked the routes once, and then three days later they were asked to retrace the route. All the children in the aided condition were able to find the route with some prompts from the experimenter to remind the children to make a route choice, but with no advice about which direction to take. A third of the children in the unaided condition were also able to retrace the route.

Taken together these experiments of wayfinding in novel environments, despite the different ages of the children and the different environments, demonstrate that children can often learn a route from a single experience, and there are clear indications that the ability to do so develops. The unaided 3 and 4 year olds in Darvizeh and Spencer's (1984) study were not always successful, and in Cornell and Hay's (1984) experiment the 8 year olds were markedly better than the 5 year olds. However, there is no evidence that this development consists of a stage of learning about landmarks and then routes. The fact that many children could retrace a route after a single journey suggests that route knowledge can be effectively learnt without a preliminary stage of landmark knowledge and this does not support Siegel and White's (1975) hypothesis.

Piaget, did not distinguish between landmark and route learning, but he did emphasize young, preoperational, children's dependency on motor schemes for remembering routes. Without independent tests of children's cognitive abilities it is not possible to be certain of their developmental stage, but given the age of the younger children in Cornell and Hay's (1984) experiment it may be assumed that many of these children were in the preoperational stage of development. Nonetheless, these children were proficient (in the walked condition) at retracing the route in the reverse direction and this required a different and novel pattern of physical movement. (The same point was made with reference to the findings of Hazen *et al.*discussed in section 5.1). Therefore a remembered motor scheme does not appear to be essential for successful wayfinding.

5.3. GENERAL CONCLUSIONS

The experiments described above were selected because they included tests of wayfinding ability. As noted before, several of these studies (e.g. Cousins *et al*, 1983; Cohen and Schuepfer, 1980; Hazen *et al.*1978) also examined children's abilities to appreciate the configuration of the spaces which they had experienced. From these and other studies with both children and adults (e.g. Evans *et al*, 1981) it is clear that an understanding of routes is achieved before a full appreciation of the configuration of the environment. But the focus of this paper is on the earliest stages of environmental knowledge and the development of landmark and route knowledge. The evidence, particularly from experiments which have tested subjects in unfamiliar environments, demonstrates that routes can be successfully remembered after minimal experience (at all ages which have been examined) and this evidence does not support either of the sequences of early environmental learning proposed by Siegel and White (1975) or by Piaget.

6. The Development of Environmental Understanding

Until recently, the influence of Piaget's and Siegel and White's theories has led to many experiments which have considered the development of environmental learning and wayfinding in terms of stages. The dependent variables in these experiments have frequently been measures of landmark and route knowledge, so that conclusions are based on the performance of different age groups' abilities in terms of such environmental features (e.g. number of landmarks recalled, accuracy of distance and direction estimates between landmarks, or success in retracing a route).

Undoubtedly these measures are important for an understanding of an individual's environmental representation at any one period of time. They are more limited for answering questions related to the processes which underlie abilities such as wayfinding. At this point it is worth noting the divergence (with some notable exceptions, e.g. Allen, 1985; Garling, Book and Lindberg, 1985; Golledge *et al*, 1985) between much of the environmental literature on wayfinding (which is after all a matter of memory and problem solving) and the developmental literature concerned with learning and memory processes. To make a rather simplistic comparison: the bibliography of a standard text on memory development (Kail, 1990) includes very few references to environmental memory and the reviews of the environmental literature (see introduction) have some, but not many references, to the current research and themes in developmental psychology. The latter (see Siegler, 1983) focuses on specific issues such as task strategies, attention and memory processes, metacognition and knowledge, and more generally on the influence of theorists like Vygotsky (in contradistinction to Piaget). Considering children's wayfinding in terms of these topics highlights some of the issues involved in the development of wayfinding abilities.

6.1. MEMORY STRATEGIES

The ability to accurately remember information depends on having effective memory strategies at the time of processing that information. For example, rehearsal (repeating information), elaboration (attaching novel information to previously establish knowledge) and grouping (organizing information into meaningful 'chunks'). These processes may play a part in learning information which is relevant for wayfinding. Little can be said for the role of rehearsal in the absence of any direct empirical studies, but elaboration has been shown to be a beneficial strategy. Cohen and Cohen (1982) took 6 and 10 year olds along paths between five landmarks (chair, wastebasket, desk, TV table and a box, in an otherwise empty room). The children experienced these paths simply as a walk; or they performed a different activity at each landmark; or they carried out a linked activity which involved picking up a pencil at the first landmark, paper at the second, signing the paper at the third and so on, until they had completed a 'letter-writing and mailing' task. The children were then asked (in another room) to estimate distances between the landmarks. For both age groups children in the linked activity condition were best at judging the distances.

In a similar experiment, Herman and Roth (1984) had 6 and 9 year olds experience a layout of items and models (stick, fence, man, house etc.). Half the children simply walked through the layout; the other half were told a story which linked all the landmarks in the order of the walk and then they were asked to reconstruct the layout. Both ages performed better when had been given the story connecting the landmarks. The Cohen and Cohen (1982) and Herman and Roth (1984) studies tested the children's configurational understanding rather than wayfinding per se, but they indicate that spatial information which is linked in memory (i.e. elaborated) is more accurately remembered.

Grouping is a phenomenon which is frequently reported in spatial memory tasks for both adults (e.g. Hirtle and Mascolo, 1986; McNamara and LeSueur, 1989; McNamara, Hardy and Hirtle, 1989; Stevens and Coupe, 1978) and children (e.g. Acredolo and Boulter, 1984; DeLoache and Todd, 1988; Liben, 1982). For example Hirtle and Mascolo (1986) asked subjects to learn a 'map' with 10 named landmarks. The map included the names of two groups of landmarks; a recreation cluster (playground, golf course, dock, beach and pond) and a government cluster (Town Hall, Police Station, Court House, Post Office and Bank). After learning the map, subjects estimated the distances between the landmarks, and Hirtle and Mascolo found that subjects judged the distance between landmarks within the same cluster as less than the distance between equally spaced landmarks which were in different clusters. In other words, the subjects tended to generate spatial groupings of the landmarks in terms of the relatedness implied by their names.

As well as in the configurational tasks referred to above, grouping or chunking can also be found in subjects' memory for routes. Allen (1981) had 7, 10 and 20 year olds view a route pictured on a sequence of sixty photographic slides. They were asked to divide the route into as many divisions as they thought appropriate and there was general agreement, by all three age groups, that the route could be divided into six segments (a park area, two different sections of a college campus, and three distinct residential areas). In a second experiment, Allen (1981), confirmed the validity of this segmentation by demonstrating that the route segmentation affected how subjects made judgments of the comparative distance between a reference scene and two target scenes along the route. Subjects' judgements about which target scene was further from the reference scene were influenced by whether or not the target scenes were in the same or different segments of the route. Allen's experiments suggest that route knowledge may be encoded in memory in terms of distinct chunks of information reflecting environmental changes along the route.

6.2. TASK STRATEGIES

At the risk of some simplification, memory processes can be referred to as non-specific strategies in the sense that the same or similar memory strategies can be applied across a variety of different tasks. There are also task-specific strategies (i.e. those strategies which are necessary to complete a particular problem or task). Of course, the same problem may be solvable by a number of different strategies.

The influence of Piaget's and Siegel and White's theories has led to an assumption that wayfinding is particularly dependent on landmark knowledge. However, it is possible to envisage successful wayfinding in the absence of any specific landmark information. For example, a route can be encoded simply by reference to turns at junctions such as 'first left, then second right, then first right' and so on (see Allen *et al*, 1989). Recognition of scenes (see Heft, 1983), route segments (Allen, 1981), names of places or roads (Blades and Spencer, 1986) and a sense of orientation based on an understanding of the configuration of the environment (Cohen and Schuepfer, 1980), may also be important. The individual's choice and combination of such information will result in various wayfinding strategies.

Very few studies have considered actual wayfinding in terms of the different strategies used by subjects. One important study, with adults, was carried out by Heft (1979) and his study is exceptional for three reasons. Firstly, it examined adults' actual wayfinding in a novel area; secondly, it manipulated the environment, and thirdly, it considered the strategies which individual's used. The subjects were taken over a route, through a Biological Reserve, which included 22 intersections. There were two conditions; in the first no changes were made to the natural setting, but in the second condition distinctive artificial landmarks (flags, road markers, lawn ornaments, etc.) were placed at the intersections. After a single walk along the route the subjects were asked to retrace it and at each intersection they explained the basis for their route decision. They also completed a questionnaire about their previous experience of wayfinding in natural environments.

There was no difference in wayfinding performance in the two conditions, though there were differences in the strategies used. Not surprisingly, the subjects in the second condition often relied on the artificial landmarks as a guide to wayfinding (at more than half the intersections). Subjects in the first condition reported more reliance on remembering specific turns and the characteristics of the trial as well as natural landmarks and the general topography of the area. Both groups relied equally on geographic orientation (i.e. made decisions based on the location of an out-of-sight feature). Overall, successful performance correlated with the use of natural landmarks as guides to wayfinding, and the use of natural landmarks correlated with previous experience in natural environments.

Heft's (1979) study showed that subjects were more likely to use distinctive landmarks when these were present, but in the absence of such landmarks various other strategies were used (and these did not result in any difference in successful performance). It also indicated that previous experience in similar areas influenced the subjects' choice of wayfinding strategy. Heft's experiment is important for demonstrating that subjects do have a range of different wayfinding strategies, which may have been built up from past experience, and that these strategies can be combined and adapted in response to the environmental information which is available.

In the experiment, referred to in section 5.1, by Allen *et al.*(1989) the experimenters considered the strategies which children used after they had learnt the route through the maze. The children retraced the learnt route and at each choice point they were asked how they knew which way to go: 24 children referred to a general sense of knowing; 34 mentioned the landmark at a choice point and 2 children used a directional term (such as left or right). Allen et al.found that the children were very consistent in the type of responses which they gave, and that there was an age related effect in the type of response. The younger children (6 year olds) were more likely to refer to a general sense of knowing while 9 and 11 years made reference to the landmarks, and the two children who used directions were from the oldest age group. If the children were accurately reporting how they remembered the turns through the maze, these results would indicate that they relied on different wayfinding strategies.

6.3. ENCODING

Siegler (1978) has suggested that to develop any problem solving strategy, or to progress from simple to more sophisticated problem solutions, an individual has to select and encode information from the problem domain so that the information can be incorporated into an effective solution. According to Siegler, the development of problem solving strategies can be seen as age (or experience) related improvements in the ability to identify appropriate information from the problem domain.

The results described in section 5.2 indicated that both children and adults can remember a route from minimal experience imply that environmental learning can be a very rapid and comparatively effortless process. It has been argued that spatial knowledge can be encoded automatically (e.g. Hasher and Zacks, 1979), but recent evidence does not support this argument (see Naveh-Benjamin, 1987, 1988) and there is evidence from several studies of environmental cognition that attention is an important factor in learning. For example, Lindberg and Garling (1981) found that adults' ability to learn the configurational layout of a series of paths was poorer when the subjects were required to perform a counting task while they walked through the layout, and experiments such as Darvizeh and Spencer's (1984) described above, found that young children's memory for a route was improved if their attention was drawn to potential landmarks at choice points.

Although children are often capable of recalling a great deal of information about special or interesting places along a route (e.g. Spencer and Darvizeh, 1983) it has been shown that young children (compared to older children or adults) are less proficient at selecting information which will be beneficial for successful wayfinding. Allen, Kirasic, Siegel and Herman (1979) showed 7 year olds, 10 year olds, and college students a sequence of 52 slides portraying a route through an urban environment. The subjects were asked to select the nine slides from the sequence which they thought would be most useful for remembering where they were along the route. The adults selected scenes which were associated with choice points or intersections along the route. Although both groups of children picked more scenes near choice points than would have been expected by chance, only half the scenes selected by the 10 year olds and a fourth of the scenes chosen by the 7 year olds included choice points - in contrast to the adult subjects, the children were more likely to select salient or attractive views from the middle of blocks.

In a second experiment Allen *et al.*(1979) tested the 10 year olds' ability to make distance judgements between scenes along the route (an aspect of configurational knowledge rather than wayfinding per se). When these children were asked to make distance judgements between scenes selected by other 10 year olds they were less accurate than when making judgements between scenes chosen by adults. Considered together, the experiments by Allen indicate that 10 year olds were less able to spontaneously select helpful features as reference points, but nonetheless they could benefit from the more useful features selected by the adults. In a similar study, with only adult subjects, Hardwick, Woolridge and Rinalducci (1983) found individual differences in performance - subjects' ability to produce a sketch map of an equally familiar college campus correlated with their ability to select effective choice point scenes from a slide presentation of an unfamiliar route.

These experiments (together with Heft's 1979 study) indicate that there may be age and individual differences in the ability to encode appropriate information from the environment. Clearly, the ability to select appropriate wayfinding features will be an important component in the development of effective wayfinding strategies.

It should be noted that the studies described above focused on particular scenes or environmental features, though other aspects of the environment may also offer potential cues for wayfinding. For example, Gale *et al.*(1990) noted that some of the children in their route learning experiment through an urban area (see section 5.2) remembered (and presumably used) street names as they learnt the route. Research with the visually impaired has demonstrated that visually impaired subjects can rely on a variety of auditory and tactile cues (see Passini and Proulx, 1988; Spencer *et al*, 1989) and, of course, such information is equally available to sighted individuals. As yet, the relative importance of scenes, landmarks, street names and other environmental cues in the formation of effective wayfinding strategies has not been considered.

6.4. METACOGNITION

There are two uses of the term 'metacognition'. In one sense it can refer to an individual's ability to transfer and adapt successful problem solving strategies from a known task to a novel task which is recognized as being similar in some way to the known task (see Campione, Brown and Ferrara, 1982). One can speculate that the subjects in Heft's (1979) experiment carried out the wayfinding task so successfully because they were able to recognize the similarity between that task and previous experiences in similar environments. But in the absence of any research into wayfinding strategies it is not possible to say if or how individuals might adapt strategies learnt in one environment to another or whether learning one route will have a beneficial or negative effect on learning a similar route (see Elmes, 1988).

In an alternative sense the term metacognition can mean an individual's self awareness of her own abilities and performance. It can be expected that adults have a well developed understanding of their own wayfinding abilities, both in general (e.g. adults can provide self reports about how competent they feel at finding their way around, see Kozlowski and Bryant, 1977), and more specifically (e.g. Garling *et al*, 1981, asked subjects to indicate how confident they felt about their judgements of the position of particular target features).

Self-awareness of performance allows the adult to adjust her wayfinding strategies to the demands of the task (e.g. if you know that you are lost, then ask for directions). Little is known about this aspect of environmental ability - given that children only gradually develop a self-awareness of their own abilities in other domains (see Flavell, 1985) it might be assumed that at least children, and perhaps sometimes adults, have inaccurate perceptions of their own abilities. For example, Herman (1980) allowed 5 and 8 year olds free exploration of a model 'town' layout. They were allowed as long as they liked to walk round the layout, until they said that they could remember the locations of all the model buildings and then they were asked to reconstruct the town from memory. Although the children had claimed that they knew all the locations, their reconstruction performance was poor. It would seem to be the case, at least for the configurational knowledge in Herman's (1980) experiment, that children often have an inaccurate understanding of their own abilities.

6.5. KNOWLEDGE

The latter aspect of metacognition relies on knowledge, and the term knowledge is used here in the same sense that it is usually employed in the literature on cognitive development to mean not just specific knowledge of a task from previously having carried out that task (which is better referred to as experience), but the general knowledge which an individual can bring to bear on any task. Few adults are ever in the situation where they cannot make assumptions about a new area from their general knowledge of other environments. For instance, an adult who is confronted with finding a route through a previously unvisited city already knows a lot about cities and this general information can contribute to successful wayfinding. Devlin (1976) studied the way that wives of navel personnel (who frequently changed location) learnt a new town, and suggested that the subjects very rapid learning of the area was a reflection of their general understanding of urban layouts. This suggestion would seem appropriate in the light of research which has highlighted the importance of knowledge based schemas (see Nelson, 1986; Nelson, Fivush, Hudson, and Lucariello, 1983) and the evidence for well established spatial schemas (e.g. Brewer and Treyens, 1981; Friedman and Brudos, 1988; Salmaso, Baroni, Job, and Peron, 1983).

Less is known about the development of spatial schemas, though some studies have indicated the role of such schemas in memory tasks (e.g. Axia and Caravaggi, 1987; Hock, Romanski, Galie and Williams, 1978) and Blades and Banham (in preparation) have demonstrated the effect of spatial schema in preschool children. Four year old children were shown various models of a kitchen layout with several appropriate items of model furniture (but crucially excluding an item such as a cooker). The children were asked to reconstruct the models accurately from memory by selecting from a large variety of model furniture. Their success depended on how closely the original model reflected a typical kitchen layout, but irrespective of the 'typicality' of the layout, the majority of children incorrectly included key items (such as the cooker) which had not been present.

As yet little is known about the development of spatial schemas of large-scale areas like a city (Dijkink and Elbers, 1981) but it can be expected that an individual's general understanding of the layout and pattern of typical urban areas will contribute to effective wayfinding in such environments. More specifically, with reference to routes, Allen (1988); Allen, Siegel and Rosinski (1978) and Pittenger and Jenkins (1979) have shown that adults can learn a route (from a slide presentation) even when the scenes from the route are presented out of sequence and that subjects have difficulty distinguishing between scenes which they have actually viewed and scenes which were not viewed but were part of the route. Discussing adults' ability to extract route information from randomized scenes, Allen et al.say 'no doubt they were aided by a general knowledge of how to go about the task of cognitively representing large-scale environments, for example, a schema for spatial events or route 'metaknowledge'. This general knowledge that spatial information can be cognitively structured in route form apparently provided the frame with which the mosaic of [scenes] were organized' (1978, 628).

6.6. SOCIAL MEDIATION

The experiments concerned with wayfinding have considered individual children or adults in tasks where they are required to recall or retrace a route or layout, and this rather gives the impression of the lone individual learning from direct interaction with the environment. What is rarely, if ever, remarked upon is that wayfinding takes place in highly structured social contexts. It is uncommon for an adult to explore a novel area on her own without what I will refer to as social mediation. This may be direct, when for example a person travels through a new environment with another person (who may or may not already be familiar with the area) or it may be indirect, when the person relies on culturally shared representations of the environment (for example, maps). It is certainly the case for young children that unfamiliar environments are always experienced in the company of either adults or other children.

In this regard it is worth noting Beck and Wood's (1976) study of a group of American teenagers as they visited different cities in Europe. They identified several 'personalities' including 'rangers' (students who were happy to explore on their own), 'mixers' (who liked to explore, but only as members of a group) and 'fixers' (who were hesitant about exploring except as passive members of a group). As Beck and Wood suggest, the 'mixer' is probably the most typical type of traveller.

Clearly, the presence of other people will influence the environmental information which is noted and, for instance, when a child is guided along a route by an adult who already knows the environment, the child is effectively provided with a preexisting schema. Vygotsky (1987) has stressed the importance of social interaction in cognitive development and learning, and how a child or adult may benefit from the support of more experienced individuals in performing a task. (see Gauvain and Rogoff, 1989a; Wood, 1988; Brown and Palincsar, 1989)

As route learning usually takes place in a social context it is to be expected that the mediation of other individuals will have a significant effect on a person's learning and knowledge. The effect can be in one of two directions - either the presence of a more experienced guide will benefit the person who is learning the route by structuring the experience of the route, or alternatively, the presence of another person who can be depended on to find the correct way will reduce the need of the individual to pay attention to the environment.

There are several studies which have indicated the importance of the role of an adult in influencing what a child learns about an environment. Herman (1980) found that both 5 and 8 year olds learnt more configurational knowledge about a model town when their movement through the town was directed by an adult. Golbeck, Rand and Soundy (1986) also found that 4 year olds' configurational knowledge was enhanced by an adult who directed the children's attention in a task which required the children to reconstruct their classroom.

On the other hand, Feldman and Acredolo (1979) gave 4 and 10 year olds the experience of a short walk through a series of halls, and during the walk the children found a hidden object. At the end of the walk the children were asked to return to the place where the object had been found.

Children took part in two conditions - either a 'passive' condition in which they were led round the halls with the experiment holding the child's hand, or an 'active' condition when the children were allowed to explore for themselves. The older children located the place equally accurately in both conditions, but the younger children were more accurate in the active condition. One contributory factor in the children's performance may have been, as Acredolo and Feldman suggest, that 'the younger children are more likely than the older children to abrogate responsibility for the route to the guiding adult, thus turning their attention to nonspatial matters' (1979, 703).

Poag, Cohen and Weatherford (1983) tested 5 and 8 year olds' memory for distances between features in a room, which they had experienced in several different conditions. The younger children were more accurate when they had explored the room themselves, rather than under an adult's guidance. Additionally, Poag et al.found an effect for the way in which the children were guided. If, during the experience of the room with the adult, the children were specifically told to notice the features' relationships, the younger children were more accurate than if they had simply been told to 'look around'.

The findings from these studies are contradictory. The results of Herman's (1980) and Golbeck *et al* (1986) suggest that configurational knowledge is better learnt with the help of an adult, but Poag *et al.*(1983) found that, the younger children in their experiment gained more from self-directed activity, and this latter result corresponds to Feldman and Acredolo's (1979) findings. No doubt, several factors (which probably interact) account for these different results; for example, the complexity of the task, the age of the children and the specific guidance provided (as Poag et al.demonstrated). Nonetheless, these studies clearly demonstrate that the presence of another individual will have an effect on what is learnt.

The experiments discussed above considered children's configurational knowledge rather than wayfinding, but it can be expected that adult mediation will also effect wayfinding performance. For example, Darvizeh and Spencer (1984), (section 5.2) found that preschool children were better able to recall a route when potential landmarks were mentioned as the children walked the route. If information from a more experienced individual can influence environmental knowledge this raises the question of what sort of information is communicated and how that information is communicated.

In the Darvizeh and Spencer (1984) experiment the landmarks were simply referred to in the course of conversation - they were not specifically identified as landmarks. Nonetheless, the children benefited from such references. It would seem likely that the effectiveness of communicated information will often depend on how that information is structured, and the abilities of the receiver. For example, Waller and Harris (1988) asked children to give directions along a simple route to other children. They found that 8 year olds spontaneously made adjustments in the way that they gave route directions to a 5 year old and were aware that a younger child might have difficulty remembering the directions if they were not phrased clearly. Several studies have shown that there are age-related and cultural differences in the ability to give and understand directions or describe layouts (e.g. Brewster and Blades, 1989; Flavell, Green, and Flavell, 1985; Gauvain and Rogoff, 1989b; Spencer and Darvizeh, 1983; Waller, 1985, 1986), but these have not specifically considered how the communicated knowledge might be an aid to actual wayfinding.

As well as stressing the importance of communicated knowledge Vygotsky also suggested that a crucial component in learning is a child's ability to gain control over her own actions via language. For example, as language ability develops a child can 'talk herself' through the stages of a problem in order to plan the necessary actions to solve that problem, or put another way, an individual can learn from 'self-communication'.

There is little environmental research related to this point, but Allen *et al.*(1989) (section 6.2) have argued that route learning may depend on verbal mediation. They found a relationship between how quickly children learnt a route through a maze and how well the children were able to give a verbal description of the correct route, and suggested 'that verbal mediation facilitates the rate of maze learning. It seems reasonable to posit, for example, that rehearsal of appropriate cues and associated movements would contribute to rapid learning. No doubt, verbal mediation is of central importance in providing a description of a route.' (Allen *et al,* 1989, 127). If this is the case, it would correspond with Vygotsky's claim that learning is often mediated through language.

One further aspect of communication refers to the use of spatial representations. Representations (whether a freehand sketch of a route, or a professionally produced cartographic map) can be an important aid to successful wayfinding. Several experiments have shown that information learnt from cartographic maps or building plans can influence an individual's configurational knowledge of an area (e.g. Moeser, 1988; Lloyd, 1989; Presson, 1987; Rossano and Warren, 1989; Thorndyke and Hayes-Roth, 1982), and some recent studies have examined adults' ability to use a map for wayfinding, but such studies have concentrated on the performance of expert map users and orienteers (e.g. Ottosson, 1987; Seiler, 1989).

Although it can be assumed that most adults have some understanding of street guides and road maps, there has been virtually no research into how people actually use such maps for 'everyday' wayfinding (see review by Blades and Spencer, 1987a), and only a little is known about the development of map using abilities. Uttal and Wellman (1989) demonstrated that very young children can benefit from viewing a map before walking through a novel area. They asked two groups of 4 and 5 year olds to learn a route through a layout of six rooms; one group simply learnt the route from direct experience of walking through the rooms, but the other group viewed a map (which showed the rooms and the landmarks in each) prior to the actual experience. The children who had seen the map required significantly fewer trips through the rooms to learn the route correctly. Blades and Spencer (1987b) have suggested that map using depends on the development of appropriate spatial strategies. They asked 4 to 8 year olds to use maps to find a route through a simple maze and found that there was an age-related improvement in the children's ability to select relevant information from the maps.

7. Summary and Conclusions

Until recently, much of the research into the development of wayfinding has been conducted within the framework provided by the theories put forward by Piaget and by Siegel and White (1975) and these have undoubtedly stimulated many important studies. The crux of Piaget's theory is that an individual's wayfinding abilities are dependent on the individual's cognitive abilities and experience of the environment. However, several studies (e.g. Darvizeh and Spencer, 1984) have shown that very young children, at an early stage of cognitive ability, can learn a route after a single experience, even when they are required to retrace the route in reverse (e.g. Cornell and Hay, 1984). These studies have been carried out in natural environments when all wayfinding cues are potentially available. In other experiments, in artificial environments with very limited wayfinding cues (such as a single landmark in an otherwise undifferentiated room) young children can learn a route after only a small number of journeys along the route (see section 5.1).

Siegel and White (1975) placed less emphasis on the individual's level of cognitive ability, but stressed the need for experience, suggesting that experience leads to a knowledge of landmarks and consequently a knowledge of how landmarks are linked as a route. But the evidence that both children and adults can perform wayfinding tasks successfully with little experience implies that landmarks and routes (if indeed these represent separate components of wayfinding, see section 3) are learnt conjointly and very quickly.

The influence of the two theories has led to a concentration on examining the relative importance of environmental features such as landmarks and routes, and on assessing the amount and type of experience which is required for wayfinding. This has had the tendency to bias the research away from the cognitive processes which are in involved in learning a route, and it is only comparatively recently that such processes have been considered (in the experiments summarized in section 6).

There is something of a paradox in that the original description of environmental learning (Piaget's) was built directly on the foundation of a comprehensive theory of cognitive development, while later theories and empirical work related to wayfinding have tended to diverge from the contemporaneous research in cognitive development and have rarely addressed the issues raised by that research. For example, questions about wayfinding strategies, metacognitive abilities, knowledge, and the role of social mediation have rarely been asked and have received little attention.

Possibly the most important of the above questions is the matter of individual strategies. Both Heft (1979) with adults, and Allen *et al.*(1989) with children have demonstrated that different individuals, in the same route learning task may solve the problem of finding the way with very different strategies. Similarly, Blades and Spencer (1987b) showed that children using a map to find a route performed that task with a variety of strategies. These findings suggest that those experiments which have relied on comparisons between groups of subjects may have subsumed many individual differences in wayfinding performance. In particular, it is difficult to be certain how wayfinding abilities 'develop'. Obviously, any single route can be encoded in several different ways - as a series of landmarks, as a list of directions at choice points, as a series of street names, and so on, and it is not clear from the present wayfinding research exactly what role these different encodings my play in learning a new route, or their comparative importance once a route is well known. It is to be hoped that future wayfinding studies will address some of the issues raised by the research into the development of other cognitive abilities and consider in more detail the processes which contribute to successful route learning.

8. References

Acredolo, L. P. and Boulter, L. T. (1984) 'Effects of hierarchical organization on children's judgments of distance and direction', Journal of Experimental Child Psychology 37, 409-425.

Allen, G. L. (1981) 'A developmental perspective on the effects of 'subdividing' macrospatial experience', Journal of Experimental Psychology: Human Learning and Memory 7, 120-132.

Allen, G.L. (1985) 'Strengthening weak links in the study of the development of macrospatial cognition', in R. Cohen (ed.), The Development of Spatial Cognition, Erlbaum, Hillsdale, New Jersey.

Allen, G.L. (1988) 'The acquisition of spatial knowledge under conditions of temporospatial discontinuity', Psychological Research 50, 183-190.

Allen, G.L., Kirasic, K. C. and Beard, R. L. (1989) 'Children's expressions of spatial knowledge', Journal of Experimental Child Psychology 48, 114-130.

Allen, G. L., Siegel, A. W. and Rosinski, R. R. (1978) 'The role of perceptual context in structuring spatial knowledge', Journal of Experimental Psychology: Human Learning and Memory 6, 617-630.

Allen, G. L., Kirasic, K. C., Siegel, A. W. and Herman, J. F. (1979) 'Developmental issues in cognitive mapping: the selection and utilization of environmental landmarks', Child Development 50, 1062-1070.

Anooshian, L. J. (1988) 'Places versus procedures in spatial cognition: alternative approaches to defining and remembering landmarks', British Journal of Developmental Psychology 6, 389-390.

Axia, G. and Caravaggi, D. (1987) 'Effects of spatial arrangement on 4 and 6 year old children's memory', Perceptual and Motor Skills 65, 283-293.

Beck, R. J. and Wood, D. (1976) 'Cognitive transformation of information from urban geographic fields to mental maps', Environment and Behavior 8, 199-238.

Blades, M. (1989) 'Children's ability to learn about the environment from direct experience and from spatial representations', Children's Environments Quarterly 6, 4-14.

Blades, M. (in press) 'The reliability of data collected from sketch maps', Journal of Environmental Psychology.

Blades, M. and Spencer, C. (1986) 'Map use in the environment and educating children to use maps', Journal of Environmental Education and Information 5, 187-204.

Blades, M. and Spencer, C. (1987a) 'How do people use maps to navigate through the world?', Cartographica 24, 64-75.

Blades, M. and Spencer, C. (1987b) 'Young children's strategies when using maps with landmarks', Journal of Environmental Psychology 7, 201-217.

Brainerd, C. J. (1978a) Piaget's Theory of Intelligence, Prentice-Hall, Englewood Cliffs, New Jersey.

Brainerd, C. J. (1978b) 'The stage question in cognitive-developmental theory', Behavioral and Brain Sciences 2, 173-213.

Brewer, W. F. and Treyens, J. C. (1981) 'Role of schemata in memory for places', Cognitive Psychology 13, 207-230.

Brewster, S. and Blades, M. (1989) 'Which way to go? Giving directions in the environment and from maps', Journal of Environmental Education and Information 8, 141-156.

Brown, A. L. and Palincsar, A. S. (1989) 'Guided, cooperative learning and individual knowledge acquisition', in L.B. Resnick (ed.), Knowing, Learning and Instruction, Erlbaum, Hillsdale, New Jersey.

Bryne, R. W. (1979) 'Memory for urban geography', Quarterly Journal of Experimental Psychology 31, 1-8.

Campione, J. C., Brown, A. L. and Ferrara, R. A. (1982) 'Mental retardation and intelligence', in R.J. Sternberg (ed.), Handbook of Human Intelligence. Cambridge University Press, Cambridge.

Canter, D. (1977) The Psychology of Place, Architectural Press, London.

Cohen, R. (ed.) (1985) The Development of Spatial Cognition, Erlbaum, Hillsdale, New Jersey.

Cohen, R. and Schuepfer, T. (1980) 'The representation of landmarks and routes', Child Development 51, 1065-1071.

Cohen, S. and Cohen, R. (1982) 'Distance estimates of children as a function of type of activity in the environment', Child Development 53, 834-837.

Cornell, E. H. and Hay, D. H. (1984) 'Children's acquisition of a route via different media', Environment and Behaviour 16, 627-641.

Cousins, J. H., Siegel, A. W. and Maxwell, S. E. (1983) 'Wayfinding and cognitive mapping in large-scale environments: a test of a developmental model', Journal of Experimental Child Psychology 35, 1-20.

Cox, M. V. (1986) The Child's Point of View: the Development of Cognition and Language, Harvester Press, Brighton, Sussex.

Darvizeh, Z. and Spencer, C. P. (1984) 'How do young children learn novel routes? The importance of landmarks in the child's retracing routes through the large scale environment', Journal of Environmental Education and Information 3, 97-105.

DeLoache, J.S. and Todd, C.M. (1988) 'Young children's use of spatial categorization as a mnemonic strategy', Journal of Experimental Child Psychology 46, 1-20.

Devlin, A.S. (1976) 'The small town cognitive map: adjusting to a new environment', in G. T. Moore and R. G. Golledge (eds.,) Environmental Knowing, Dowden, Hutchinson and Ross, Stroudsburg, PA.

Dijkink, G. and Elbers, E. (1981) 'The development of geographic representation in children', Tijdschrift voor Economische en Sociale Geografie 72, 2-16.

Elmes, D.G. (1988) 'Interference in spatial memory', Journal of Experimental Psychology: Learning, Memory and Cognition 14, 668-675.

Evans, G.W., Marrero, D.G. and Butler, P.A. (1981) 'Environmental learning and cognitive mapping', Environment and Behavior 13, 83-104.

Evans, G. W., Skorpanich, A., Garling, T., Bryant, K. J. and Bresolin, B. (1984) 'The effects of pathway configuration, landmarks and stress on environmental cognition', Journal of Environmental Psychology 4, 323-335.

Feldman, A and Acredolo, L. (1979) 'The effect of active versus passive exploration on memory for spatial location in children', Child Development 50, 698-704.

Flavell, J. H. (1985) Cognitive Development (2nd edition), Prentice Hall, Englewood Cliffs, New Jersey.

Flavell, J. H., Green, F. L. and Flavell, E. R. (1985) 'The road not taken: understanding the implications of initial uncertainty in evaluating spatial directions', Developmental Psychology 21, 207-216.

Friedman, W. J. and Brudos, S. L. (1988) 'On routes and routines: the early development of spatial and temporal representations', Cognitive Development 3, 167-182.

Gale, N., Golledge, R. G., Pellegrino, J. W. and Doherty, S. (1990) 'The acquisition and integration of route knowledge in an unfamiliar neighborhood', Journal of Environmental Psychology 10, 3-25.

Garling, T. (1989) 'The role of cognitive maps in spatial decisions', Journal of Environmental Psychology 9, 269-278.

Garling, T., Book, A. and Lindberg, E. (1984) 'Cognitive mapping of large-scale environments: the interrelationship of action plans, acquisition, and orientation', Environment and Behavior 16, 3-34.

Garling, T., Book, A. and Lindberg, E. (1985) 'Adults memory representations of the spatial properties of their everyday physical environment', in R. Cohen (ed.), The Development of Spatial Cognition, Erlbaum, Hillsdale, New Jersey.

Garling, T., Lindberg, E. and Mantyla, T. (1983) 'Orientation in buildings: effects of familiarity, visual access, and orientation aids', Journal of Applied Psychology 68, 177-186.

Garling, T., Book, A., Lindberg, E. and Nilsson, T. (1981) 'Memory for the spatial layout of the everyday physical environment: factors affecting rate of acquisition', Journal of Environmental Psychology 1, 263-277.

Gauvain, M. and Rogoff, B. (1989a) 'Collaborative problem solving and children's planning skills', Developmental Psychology 25, 139-151.

Gauvain, M. and Rogoff, B. (1989b) 'Ways of speaking about space: the development of children's skill in communicating spatial knowledge', Cognitive Development 4, 295-307.

Ginsburg, H. and Opper, S. (1969) Piaget's Theory of Intellectual Development, Prentice Hall, Englewood Cliffs, New Jersey.

Golbeck, S. L., Rand, M. and Soundy, C. (1986) 'Constructing a model of a large-scale space with the space in view: effects on preschoolers of guidance and cognitive restructuring', Merrill-Palmer Quarterly 32, 187-203.

Golledge, R. G. (1987) 'Environmental cognition', in D. Stokols and I. Altman (eds.), Handbook of Environmental Psychology 1, New York: Wiley.

Golledge, R. G., Rivizzigno, V. L. and Spector, A. (1976) 'Learning about a city: Analysis by multidimensional scaling', in R. G. Golledge and G. Rushton (eds.), Spatial Choice and Spatial Behavior, Ohio State University Press, Columbus.

Golledge, R. G., Smith, T. R., Pellegrino, J. W., Doherty, S., and Marshall, S. P. (1985) 'A conceptual model and empirical analysis of children's acquisition of spatial knowledge', Journal of Environmental Psychology 5, 125-152.

Hardwick, D. A., Woolridge, S. C. and Rinalducci, E. J. (1983) 'Selection of landmarks as a correlate of cognitive map organization', Psychological Reports 53, 807-813.

Hart, R. A. and Moore, G. T. (1973) 'The development of spatial cognition: a review', in R. M. Downs and D. Stea (eds.), Image and Environment: Cognitive Mapping and Spatial Behaviour, Aldine, Chicago.

Hasher, L. and Zacks, R. T. (1979) 'Automatic and effortful processes in memory', Journal of Experimental Psychology: General 108, 356-388.

Hayes-Roth, B. and Hayes-Roth, F. (1979) 'A cognitive model of planning', Cognitive Science 3, 275-310.

Hazen, N. L., Lockman, J. L. and Pick, H. L. (1978) 'The development of children's representations of large-scale environments', Child Development 49, 623-636.

Heft, H. (1979) 'The role of environmental features in route-learning: two exploratory studies of wayfinding', Environmental Psychology and Nonverbal Behaviour 3, 172-185.

Heft, H. (1983) 'Wayfinding as the perception of information over time', Population and Environment: Behavioural and Social Issues 6, 133-150.

Heft, H. and Wohlwill, J. F. (1987) 'Environmental cognition in children', in D. Stokols and I. Altman (eds.), Handbook of Environmental Psychology 1, New York: Wiley.

Herman, J. F. (1980) 'Children's cognitive maps of large-scale spaces: effects of exploration, direction, and repeated experience', Journal of Child Psychology 29, 126-143.

Herman, J. F. and Roth, S. F. (1984) 'Children's incidental memory for spatial locations in a large-scale environment: taking a tour down memory lane', Merrill-Palmer Quarterly 30, 87-102.

Herman, J. F., Roth, S. F., Miranda, C. and Getz, M. (1982) 'Children's memory for spatial locations: the influence of recall perspective and type of environment', Journal of Experimental Child Psychology 34, 257-273.

Hirtle, S. C. and Mascolo, M. F. (1986) 'Effect of clustering on the memory of spatial locations', Journal of Experimental Psychology: Learning, Memory and Cognition 12, 182-189.

Hock, H. S., Romanski, L., Galie, A. and Williams, C. S. (1978) 'Real world schemata and scene recognition in adults and children', Memory and Cognition 6, 423-431.

Kail, R. (1990) The Development of Memory in Children (3rd edition), Freeman, New York.

Kozlowski, L. and Bryant, K. J. (1977) 'Sense of direction, spatial orientation, and cognitive maps', Journal of Experimental Psychology: Human Perception and Performance 3, 590-598.

Kuipers, B. (1978) 'Modeling spatial knowledge', Cognitive Science 2, 129-153.

Leiser, D. (1987) 'The changing relations of representation and cognitive structure during the development of a cognitive map', New Ideas in Psychology 5, 95-110.

Leiser, D. and Zilbershatz, A. (1989) 'The Traveller. A computational model of spatial network learning', Environment and Behavior 21, 435-463.

Liben, L. S. (1981) 'Spatial representation and behavior: multiple perspectives', in L. S. Liben, A. H. Patterson and N. Newcombe (eds.), Spatial Representation and Behavior Across the Life Span, Academic Press, New York.

Liben, L. S. (1982) 'Children's large-scale spatial cognition: is the measure the message?', in R. Cohen (ed.), New Directions for Child Development, no.15. Children's Conceptions of Spatial Relationships, Jossey-Bass, San Francisco.

Liben, L. S., Moore, M. L. and Golbeck, S. L. (1982) 'Preschoolers' knowledge of their classroom environment; evidence from small-scale and life-size spatial tasks', Child Development 53, 1275-1284.

Light, P. (1987) 'Taking roles', in J. Bruner and H. Haste (eds.), Making Sense, The Child's Construction of the World, Methuen, London.

Lindberg, E. and Garling, T. (1981) 'Acquisition of locational information about reference points during locomotion with and without a concurrent task: effects of number of reference points', Scandinavian Journal of Psychology 22, 109-115.

Lloyd, R. (1989) 'Cognitive maps: encoding and decoding information', Annals of the Association of American Geographers 79, 101-124.

McNamara, T. P. and LeSueur, L. L. (1989) 'Mental representations of spatial and nonspatial relations', The Quarterly Journal of Experimental Psychology 41A, 215-233.

McNamara, T. P., Hardy, J. K. and Hirtle, S. C. (1989) 'Subjective hierarchies in spatial memory', Journal of Experimental Psychology: Learning, Memory and Cognition 15, 211-227.

Moeser, S. D. (1988) 'Cognitive mapping in a complex building', Environment and Behavior 20, 21-49.

Morss, J. R. (1987) 'The construction of perspectives: Piaget's alternative to spatial egocentrism', International Journal of Behavioral Development 10, 263-279.

Naveh-Benjamin, M. (1987) 'Coding of spatial location information: an automatic process?', Journal of Experimental Psychology: Learning Memory and Cognition 13, 595-605.

Naveh-Benjamin, M. (1988) 'Recognition memory of spatial location information: another failure to support automaticity', Memory and Cognition 16, 437-445.

Nelson, K. (ed). (1986) Event Knowledge. Structure and Function in Development, Erlbaum, Hillsdale, New Jersey.

Nelson, K., Fivush, R., Hudson, J. and Lucariello, J. (1983) 'Scripts and the development of memory', in M. T. H. Chi (ed.), Contributions to Human Development, vol 9, Trends in Memory Development Research, Karger, New York.

Newcombe, N. (1985) 'Methods for the study of spatial cognition', in R. Cohen (ed.), The Development of Spatial Cognition, Erlbaum, Hillsdale, New Jersey.

Newcombe, N. (1988) 'The paradox of proximity in early spatial representation', British Journal of Developmental Psychology 6, 376-378.

Okabe, A, Aoki, K and Hamamoto, W. (1986) 'Distance and direction judgment in a large-scale natural environment. Effects of a slope and winding trail', Environment and Behavior 18, 755-772.

Ottosson, T. (1987) 'Map reading and wayfinding', Goteborg Studies in Educational Sciences, vol. 65.

Passini, R. (1984) Wayfinding in Architecture, Van Nostrand Reinhold Company, New York.

Passini, R. and Proulx, G. (1988) 'Wayfinding without vision', Environment and Behavior 20, 227-252.

Piaget, J. and Inhelder, B. (1956) The Child's Conception of Space, Routledge and Kegan Paul, London.

Piaget, J., Inhelder, B. and Szeminska, A. (1960) The Child's Conception of Geometry, Routledge and Kegan Paul, London.

Pick, H. L., Montello, D. R. and Somerville, S. C. (1988. 'Landmarks and the coordination and integration of spatial information', British Journal of Developmental Psychology 6, 372-375.

Pittenger, J. B. and Jenkins, J. J. (1979) 'Apprehension of pictorial events: the case of a moving observer in a static environment', Bulletin of the Psychonomic Society 13, 117-120.

Poag, C. K., Cohen, R. and Weatherford, D. L. (1983) 'Spatial representations of young children: the role of self- versus adult-directed movement and viewing', Journal of Experimental Child Psychology 35, 172-179.

Presson, C. C. (1987) 'The development of spatial cognition: secondary uses of spatial information', in N. Eisenberg (ed.), Contemporary Topics in Developmental Psychology, Wiley and Sons, New York.

Presson, C. C. and Montello, D. R. (1988) 'Points of reference in spatial cognition: stalking the elusive landmark', British Journal of Developmental Psychology **6**, 378-381.

Presson, C. C. and Somerville, S. C. (1985) 'Beyond egocentrism: A new look at the beginnings of spatial representation', in H. M. Wellman (ed.), Children's Searching: The Development of Search Skill and Spatial Representation, Erlbaum, Hillsdale, New Jersey.

Robinson, A. H. and Petchenik, B. B. (1976) The Nature of Maps, University of Chicago Press, Chicago.

Rossano, M. J. and Warren, D. H. (1989) 'Misaligned maps lead to predictable errors', Perception 18, 215-229.

Rowen, R.B. and Hardwick, D.A. (1983) 'Effects of landmark salience and direction of travel on young children's memory for spatial location', Journal of Psychology 113, 271-276.

Salmaso, P., Baroni, M. R., Job, R. and Peron, E. M. (1983) 'Schematic information, attention, and memory for places', Journal of Experimental Psychology: Learning, Memory and Cognition 9, 263-268.

Seiler, R. (1989) 'Route planning and route choice', Scientific Journal of Orienteering 5, 74-84.

Shemyakin, F. N. (1961) 'Orientation in space', in Psychological Science in the U.S.S.R., vol 1. Office of Technical Services, U.S. Department of Commerce, Washington.

Siegel, A. W. and White, S. H. (1975) 'The development of spatial representations of large-scale environments', in H. W. Reese (ed.), Advances in Child Development and Behaviour 10, Academic Press, New York.

Siegler, R. S. (1978) 'The origins of scientific reasoning', in R .S. Siegler (ed.), Children's Thinking: What Develops?, Wiley and Sons, New York.

Siegler, R. S. (1983) 'Information processing approaches to development', in P. H. Mussen (ed.), Handbook of Child Psychology, Wiley and Sons, New York.

Smith, T. R., Pellegrino, J. W. and Golledge, R. G. (1982) 'Computational process modelling of spatial cognition and behavior', Geographical Analysis 14, 305-325.

Spencer, C. and Darvizeh, Z. (1983) 'Young children's place-descriptions, maps and route-finding: a comparison of nursery school children in Iran and Britain', International Journal of Early Childhood 15, 26-31.

Spencer, C., Blades, M. and Morsley, K. (1989) The Child in the Physical Environment: The Development of Spatial Knowledge and Cognition, Wiley, Chichester.

Stevens, A and Coupe, P. (1978) 'Distortions in judged spatial relations', Cognitive Psychology 10, 422-437.

Thorndyke, P. W. and Hayes-Roth, B. (1982) 'Differences in spatial knowledge acquired from maps and navigation', Cognitive Psychology 14, 560-589.

Torell, G. (1990) 'The acquisition and development of environmental cognition in children', Goteborg Psychological Reports 20, No.1.

Uttal, D. H. and Wellman, H. M. (1989) 'Young children's representation of spatial information acquired from maps', Developmental Psychology 25, 128-138.

Vygotsky, L. S. (1987) Collected Works, vol.1. (R. W. Rieber and A. S. Carton, eds.), Plenum, New York.

Waller, G. (1985) 'Linear organization of spatial instructions: development of comprehension and production', First Language 6, 53-67.

Waller, G. (1986) 'The development of route knowledge: multiple dimensions?', Journal of Environmental Psychology 6, 109-119.

Waller, G. and Harris, P. L. (1988) 'Who's going where?: Children's route descriptions for peers and younger children', British Journal of Developmental Psychology 6, 137-143.

Weatherford, D. L. (1985) 'Representing and manipulating spatial information from different environments: models to neighbourhoods', in R. Cohen (ed.), The Development of Spatial Cognition, Erlbaum, Hillsdale, New Jersey.

Wood, D. and Beck, R. (1989) 'Janine Eber maps London: individual dimensions of cognitive imagery', Journal of Environmental Psychology 9, 1-26.

Wood, D. (1988) How Children Think and Learn, Blackwell, Oxford.

THE EFFECT OF THE PATTERN OF THE ENVIRONMENT ON SPATIAL KNOWLEDGE ACQUISITION

SCOTT M. FREUNDSCHUH
National Center for Geographic Information and Analysis
Department of Geography
State University of New York at Buffalo
Buffalo, New York 14261 USA

ABSTRACT In the literature, there is a question as to whether or not survey (map-view) knowledge can be acquired from procedural (route) knowledge. Specifically, Thorndyke and Hayes-Roth (1982) contend that in a new environment, procedural knowledge is the basis for navigation, but that survey knowledge is acquired with increased familiarity of that environment. In contrast, Lloyd (1989) suggests that there is no transition from procedural to survey knowledge. This particular study (in progress) addresses the geographical behavior of wayfinding, and more specifically, explores spatial knowledge acquisition and representation. It proposes a model for spatial knowledge based upon previous studies, the pattern of the environment (gridded vs winding or serpentine), and the source of spatial knowledge (direct navigation experience vs maps) and it attempts to discover, through experimental research, the relative accuracy of peoples' configurational knowledge by investigating the roles that arrangement of the environment and source of spatial knowledge play in the acquisition of spatial knowledge.

1. Introduction

Geographers study spatial relationships. Though geographers seem to understand spatial relationships intuitively, they have proceeded in the absence of "a general theory of spatial relationships" (Abler, 1987, 304). Abler stated that:

> "Fundamental spatial concepts have not been formalized mathematically and elegantly. Cardinal directions are relative concepts, as are ideas basic to geography such as near, far, touching, adjacent, left of, right of, inside, outside, above, below, upon, and beneath" (1987, 306).

Research Initiative 2, *Languages of Spatial Relations*, of the National Center for Geographic Information and Analysis (see NCGIA, 1988) has set as its primary goal the development of a theory of spatial relations. In this initiative, two components of spatial relations were identified. The first is the cognitive component, which encompasses the theoretical issues of "how people think about geographic space [and] the influence of cognitive models of geographic space upon decision making and behavior" (Mark *et al.*, 1989, 4). The second is the formal component which encompasses practical issues such as "user interfaces, e.g., natural language, selection of best

D. M. Mark and A. U. Frank (eds.), Cognitive and Linguistic Aspects of Geographic Space, 167–183.

words in menus, design and structure of possible queries; and fundamental database designs" (Mark *et al.*, 1989, 4).

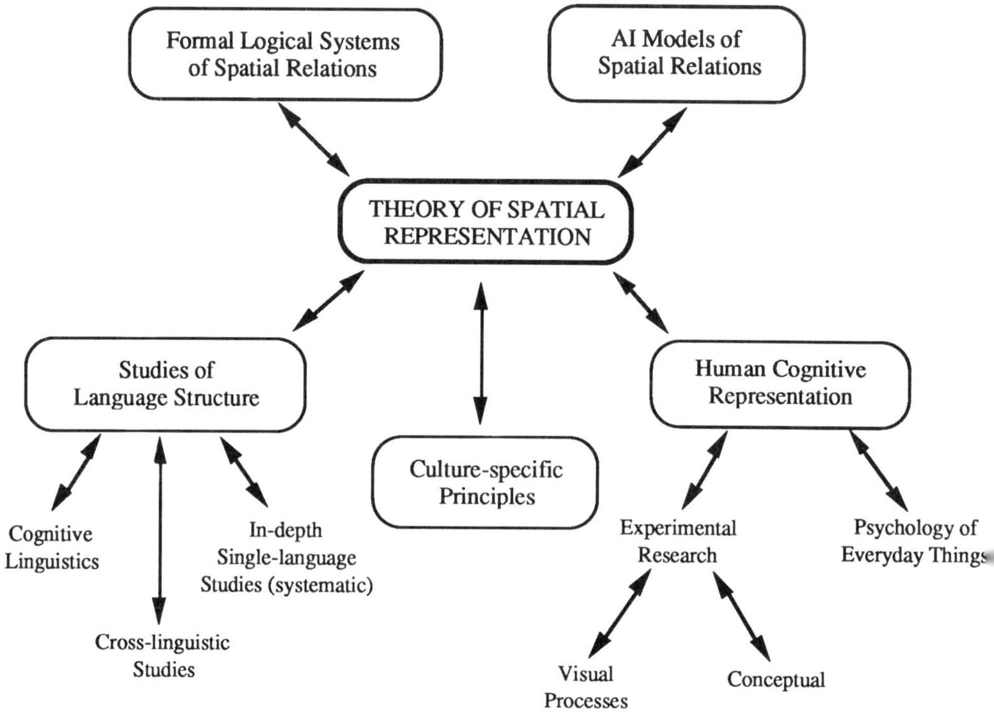

Figure 1. Interrelationships among the many aspects of a theory of spatial representation (Adapted from Mark *et al.*, 1989).

Zubin (in Mark *et al.*, 1989) states that there are several approaches to a theory of spatial representation (see figure 1) that include: *Formal/logical Systems of Spatial Relations*; *Studies of Language Structure* including cross-linguistic studies, cognitive linguistics, and in-depth single-language studies; *Human Cognitive Representation* including experimental research in both visual processes and conceptual models, and the psychology of everyday experience; *Culture-specific Principles*; and *AI Models of Spatial Relations*. His model threads together the conceptual aspects that Research Initiative 2 is most concerned with.

In the framework of Zubin's model, this research is concerned with aspects of *Human Cognitive Representation.* The study of human cognition, which in the past was considered only within the domain of psychology, is having a profound impact on numerous disciplines, including geography. Abler (1988) discussed the current and future emphasis of the awards program in Geography at the National Science Foundation (NSF). He listed five categories within which supported research falls, one of them being "Locational Behavior and Decision Making". Abler states:

"Statistical explanations regarding resource use, regional change, and settlement systems provide more penetrating insights when they are complemented by knowledge of how individuals or groups actually make the decisions that produce observed patterns, processes, and relationships...... Geographic behavior and decision making will continue to attract research on many facets of human locational behavior. Wayfinding research will become especially important in the next several years; it is of great theoretical interest, and it is a kind of geographic behavior more amenable to experimental research than most others. Applications such as computer-based, onboard navigation systems in vehicles and vessels will also provide incentives for research" (1988, 136).

Clearly, there is a strong focus on human behavior in the discipline of geography, indicating the importance of this topic.

This particular study addresses the geographical behavior of wayfinding, and more specifically explores spatial knowledge acquisition/representation. It will include a brief discussion about existing spatial knowledge models/theories; a model for spatial knowledge based upon previous studies, the pattern of the environment (gridded vs winding or serpentine), and the source of spatial knowledge (direct navigation experience vs maps); a research design proposal to explore the relative accuracy of peoples configurational knowledge by investigating the roles that arrangement of the environment and source of spatial knowledge play in the acquisition of spatial knowledge.

2. Literature Review

For the most part, models of spatial knowledge suggest that humans possess at least two kinds of spatial knowledge: procedural or route knowledge, and survey or map-view knowledge. There are several schools of thought as to how many other kinds of spatial knowledge exist. Some suggest only two kinds, while others purport three or more: Trowbridge (1913) refers to domi-centric and ego-centric knowledge; Tolman (1948) refers to strip map and comprehensive map knowledge; Shemyakin (1962) refers to route map and survey map knowledge; Piaget and Inhelder (1968) refer to topological, projective, and Euclidean knowledge; Gibson (1969) refers to perspective structure and invariant structure; Hart and Moore (1973) refer to egocentric, fixed, and coordinated orientation; Siegel and White (1975) refer to landmark, route learning, and configurational knowledge; Kuipers (1978) refers to sensorimotor, topological, and metrical knowledge; Thorndyke and Hayes-Roth (1982) refer to procedural and survey knowledge; Golledge (1988) refers to declarative, procedural, and configurational knowledge; Stern and Leiser (1988) refer to landmark, route, and survey knowledge. To further complicate this domain, the terminology used to label these different kinds of spatial knowledge vary from model to model. For the sake of brevity, a comprehensive literature review concerning spatial knowledge models will not be included here. (For a discussion of spatial knowledge models, see Hart and Moore, 1973; Heft and Wohlwill, 1987; and Golledge, 1987). Instead, two significant points from these previous theories are brought to light.

The first point, which is a reiteration of what has already been said, is that there is general agreement that humans possess at least two kinds of spatial knowledge. The first has been referred to as strip-map, route map, projective, topological, and procedural knowledge. Ignoring terminology for the moment, this kind of spatial knowledge can be described as a linear representation of space. The second, which has been referred to as ego-centric, comprehensive map, survey, Euclidean, configurational, and metrical knowledge, can be described as a two dimensional representation of space which includes metric measures such as angles and distances. In other words, this second type of spatial knowledge is akin to the geographic information shown on maps.

The second point to be raised is that for the most part, these models describe or define kinds of spatial knowledge in discrete "packages". Types of spatial knowledge have been defined within the framework of each model, and presented as individual entities seemingly separate from one another. Though it would seem sensible to consider how these kinds of spatial knowledge are related and how they build upon one another, none of the models (with the possible exception of Kuiper's model) explore this. There is an attempt here to integrate kinds of spatial knowledge.

3. A Model Of Spatial (Geographical) Knowledge

I would like to put forth a model that suggests three kinds of hierarchical spatial (geographical) knowledge. For the benefit of this research, knowledge is "any information that can be represented as either declarative knowledge or procedural knowledge, e.g., in the form of rules, entries in data memory or another data base, or control strategies. Knowledge may be specific to a task domain or general enough to be independent of all domains" (Brownston *et al.*, 1985, 448). Declarative knowledge has been defined as "knowledge that can be retrieved and stored but cannot be immediately executed; to be effective, it must be interpreted by procedural knowledge" (Brownston *et al.*, 1985, 445). In addition, procedural knowledge has been defined as "knowledge that can be immediately executed using declarative knowledge as data but that may not be examined" (Brownston *et al.*, 1985, 452).

The first kind of spatial knowledge in this model is *Geographic Facts*. Geographic facts are facts about places that one believes to be true, and can be placed within the following typology:

- "**X exists**": One knows that a place exits, but has no knowledge of what it is like, or where it is (e.g., a typical response to X is, "Oh yeah, I've heard of that place.");

- "**X has characteristics**": These characteristics are descriptive in nature (e.g., New York is a large city, Buffalo has a large Polish population);

- "**X is located**": Location of "X" is always in reference to another place or feature (e.g., Ogden is in Utah, Duluth is on Lake Superior, Buffalo is about 60 miles west of Rochester, the Boulevard Mall is at the intersection of Maple and Niagara Falls Boulevard.). This has been referred to in the literature as landmark knowledge.

In support of this typology, Berry (1964) refers to *geographic facts* as "single observation(s) recorded from the spatial point of view. [These] observation(s) refer to a single characteristic at a single place or location" (1964, 5). Berry further states that a geographic fact can be a single characteristic at several places, or a group of characteristics at one place. Sources for geographical facts are direct real-world experiences, maps, television and movies, books and newspapers, and other people.

The second kind of geographical knowledge is *Route Knowledge*. This knowledge is the "linking together" of geographical facts therefore, enabling travel between two places. At a minimum, route knowledge consists of a set of choice points [landmarks], the paths that link these landmarks [roads, etc.], and the action executed at these landmarks [go left/right, go straight]. Route knowledge is usually acquired through direct navigation experience, but it is possible that route knowledge can be gained from other sources such as map reading (i.e., route planning), directions from other people, and from text.

The third kind of geographical knowledge is *Configurational Knowledge*. This type of knowledge threads together route knowledge to form a "network" between several or more places, and is acquired most easily from maps and other sources of vertical perspective. Depending on a number of factors (such as source of knowledge and arrangement of the environment), configurational knowledge of an environment may be a simple network of routes where distances may or may not be known (i.e., the topology). On the other hand, it may be a network of routes where distance and direction is known, as well as the pattern the routes form (i.e., topology plus metrics).

This model, instead of treating kinds of spatial knowledge as separate components, integrates or "brings together" kinds of spatial knowledge. Here, spatial knowledge is considered not simply a static representation of space, but instead is a dynamic representation that becomes refined through time (see figure 2). This diagram shows that within the framework of spatial knowledge, or if you will, a cognitive map, there exists geographic facts, route knowledge, and configurational knowledge. Geographic facts are a necessary component of both route knowledge and configurational knowledge. In addition, geographic information that is lower in this hierarchy is retrievable from higher levels. In other words, one could extract geographic facts from both route knowledge and configurational knowledge. It is also possible that some higher level geographic information is retrievable from lower levels, but not all. For instance, it is possible to integrate route knowledge to "build" configurational knowledge, but it seems unlikely that configurational knowledge could be accessed from geographic facts.

It is of interest not only to define kinds of spatial knowledge, but to explore how these types of spatial knowledge are interconnected. Golledge (in Mark *et al.*, 1989) has identified three kinds of spatial knowledge: declarative, procedural, and configurational. He states that configurational knowledge is more than linking procedural rules and the declarative base, but that the nature of these links is not clearly understood. In addition, he states the importance of studying how configurational knowledge is acquired or assembled from the other kinds of spatial knowledge. Though this question is intriguing, it is not new. Tolman (1973) raised the same question, though it was stated in a different manner.

Tolman and his students performed a number of maze running experiments with rats, and formulated some interesting hypotheses. The experiments that were most interesting (see Ritchie, 1948) examined the spatial abilities of these rodents. As controversial as the experiments are, I do not intend to dwell on their credibility (for a discussion of Tolman's work, see Amundson, 1983), but instead elaborate on a question that Tolman felt was "humanly significant" and an "exciting problem": "What are the conditions which favor [the development of] narrow strip maps [i.e., route knowledge] and what are those that tend to favor [the development of] broad comprehensive maps [i.e., topologic plus metrical configurational knowledge] not only in rats but also in men? (Tolman, 1973, 47)".

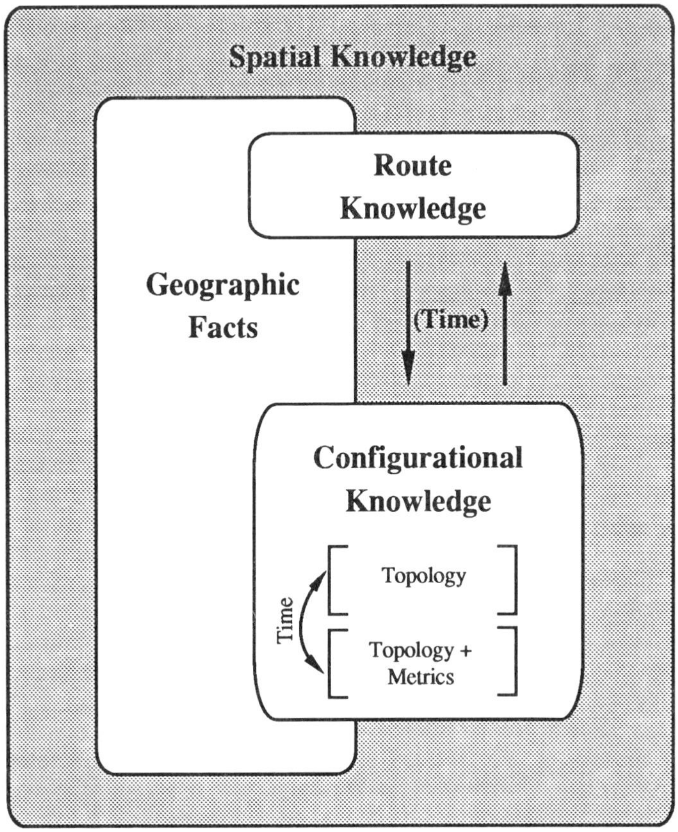

Figure 2. Spatial knowledge, considered not simply as a static representation of space, but instead as a dynamic representation that becomes refined through time.

4. The Research Question

In the literature, there is a question as to whether or not topographic plus metrical configurational knowledge can be acquired from route knowledge. Specifically, Thorndyke and Hayes-Roth (1982) contend that in a new environment, route knowledge is the basis for navigation, but that topologic plus metrical knowledge is acquired with increased familiarity of that environment. In contrast, Lloyd (1989) suggests that there is no transition from route to topologic plus metrical configurational knowledge.

It is possible that certain factors have a significant effect on the acquisition of spatial knowledge. One such factor is the arrangement of the environment. Arrangement of the environment involves the patterns formed by the paths or routes required to travel between places. One arrangement is a gridded pattern where paths are straight lines, and intersect at right angles (e.g., a gridded road pattern). Such gridded patterns are here considered regular environments. A different arrangement is one in which the paths are not necessarily straight, and the intersections between paths may or may not be right angles (e.g., winding or serpentine road pattern). This pattern is here considered an irregular environment.

Another factor that may affect the acquisition of spatial knowledge is that of "density of geographic information." Density can be defined as a function of scope (the size of the window on the environment) and the amount of geographic information, such as landmarks and roads, that one utilizes for navigation. Density is difficult to quantify for several reasons. First of all, how much information does one require to be able to navigate? Most likely, the amount of required information is not the same for all people. Secondly, exactly what geographic information is used? Is one person's fleeting perception another person's landmark?

A third factor that may affect the acquisition of spatial knowledge is variation in individual spatial cognitive ability. Gould (1989) discusses individual differences in spatial cognition as a function of familiarity or "experience" within an environment. Acredolo (1982, as reported in Gould, 1989) stresses that experience navigating within an environment contributes to the ability to find one's way through an unfamiliar environment. Like "density of geographic information", variation in individual spatial cognitive ability is difficult to measure, though some studies (Kozlowski and Bryant, 1977; Streeter et al., 1985) have shown that people who rate themselves as having a good sense of direction proved so in field testing; likewise, those who rate themselves as having a poor sense of direction performed poorly in field testing (Gould, 1989).

Though it is recognized that these three factors may affect the acquisition of spatial knowledge, this paper will explore only the effect of arrangement of the environment on spatial knowledge acquisition.

5. Previous Research On Learning Large-Scale Environments

Thorndyke and Hayes-Roth (1982) have explored knowledge representations for areas of small scope (first floor of The Rand Corporation building in Santa Monica). In their experiment, subjects learned the first floor of the Rand Building by one of two methods: by memorizing a map (this group consisted of college students), or by navigating in the environment (this group consisted of employees who worked in the Rand Building). The premise for these two methods is that navigating within an environment results in the acquisition of route knowledge, and that map learning results in the acquisition of map-like configurational knowledge (Thorndyke and Hayes-Roth, 1982). After learning this environment, the subjects performed a number of spatial judgements based on their knowledge of the environment, and the results were compared. The spatial judgements "included placing the locations [of landmarks] in the proper spatial configurations on an external map [i.e., a location task], pointing in the direction of unseen [landmarks] from [other landmarks] in the environment [i.e., an orientation task], and estimating locations along routes [i.e., route distance estimations] and along a straight line [i.e., Euclidean distance estimations]" (1982, 562).

For Euclidean distance estimates, Thorndyke and Hayes-Roth found that "map-learning subjects were more accurate than the least experienced navigation subjects but no more accurate than more experienced [navigation] subjects" (1982, 577). For route distance estimates, navigation subjects were significantly more accurate than map-learning subjects. For orientation tasks, "navigation subjects were far more accurate than map-learning subjects" (1982, 580). For location tasks, they found that "overall, map-learning subjects were more accurate than navigation subjects. However, the accuracy of the most experienced navigation subjects did not differ reliably from that of the map-learning subjects" (1982, 581). These results support Thorndyke and Hayes- Roth's assertion that "as navigation subjects acquire more [navigational experience within an environment], they induce [map-like configurational] knowledge of the environment" (1982, 572). This assertion is supported by other studies which report similar results (Allen, Siegel and Rosinski, 1978; Appleyard, 1970; Golledge and Zannaras, 1973; Herman and Siegel, 1977; Kozlowski and Bryant, 1977; Ladd, 1970).

In contrast to Thorndyke and Hayes-Roth, Lloyd (1989) has explored knowledge acquisition for areas of larger scope (i.e., the city of Columbia, South Carolina). In Lloyd's experiment, as in Thorndyke and Hayes-Roth's, subjects learned an environment either via navigation (this group consisted of residents of Columbia), or a map (this group consisted of college students). Subjects then performed a series of location tasks based upon their knowledge of the environment. These tasks consisted of locating "a set of landmarks with respect to three reference points [on a map]" (12). Lloyd found that map-learning subjects were significantly more accurate in location estimates, as well as significantly faster in task performance, than were navigation subjects. The length of time within the environment had no effect on the accuracy of the navigation subjects estimates. Lloyd's navigation subjects had lived in the test area for a minimum of three years, and an average of eleven years. These results appear to conflict with Thorndyke's and Hayes-Roth's results, and do not support their assertion that map-like configurational knowledge is acquired from route knowledge with increased navigation experience. There are several reasons for these opposing results.

First , the test conditions for the two experiments were very different. In Lloyd's experiment, the subjects performed the spatial judgement tasks removed from the environment upon which the estimates were made. This is in contrast to Thorndyke's and Hayes-Roth's experiment where subjects performed the spatial judgement tasks within the environment upon which the estimates were made. Second, in the literature, survey or configurational knowledge has been described as the ability to manipulate magnitudes such as distance, direction, and relative position. Subjects in Lloyd's experiment only performed one spatial judgement, relative position. Therefore, Lloyd's results should be considered only a partial demonstration of configurational knowledge. A third reason is that the testing environments for the two experiments were markedly different. The environment utilized in Thorndyke's and Hayes-Roth's experiment was highly regular and was of relatively small scope (approximately 18,200 sq. ft.), whereas, the environment used by Lloyd was highly irregular and was of a larger scope (approximately 25 sq. mi.).

It is interesting to note that in light of work by Lynch (1960), these two studies fail to consider the effect of the regularity of the environment on the acquisition of spatial knowledge. In *The Image of the City*, Lynch (1960) discussed the role that environmental images play in social development, emotional security, [and] human experience, as well as *movement within the environment.* Lynch's discussion of the impact of an ordered environment on further acquisition of environmental information needs to be brought forward. He contends that the strategic link for way-finding is the image one has of the environment. In an excerpt from his book, Lynch states:

> Obviously a clear image enables one to move about easily and quickly: to find a friend's house or a policeman or a button store. But an ordered environment can do more than this; it may serve as a broad frame of reference, an organizer of activity or belief or knowledge. On the basis of a structural understanding of Manhattan, for example, one can order a substantial quantity of facts and fancies about the nature of the world we live in. Like any good framework, such a

structure gives the individual a possibility of choice and a starting point for the acquisition of further information. A clear image of the surroundings is thus a useful basis for individual growth (1960, 4).

Clearly, Lynch believes that the arrangement of the environment (i.e., the paths) has a significant impact on how humans perceive and learn an environment. To support his contentions, Lynch performed several analyses of three contrasting city centers: Boston, Massachusetts; Jersey City, New Jersey; and Los Angeles, California. I will focus on Boston (an irregular environment) and Los Angeles (a regular environment).

Lynch described Boston as "vivid in form and full of locational difficulties", whereas Los Angeles was characterized as a "new city... with a gridiron plan in its central area" (1960, 14-15). In his interviews with Bostonians, Lynch found Boston to be "a city of very different districts and of crooked, confusing paths" (1960, 17). In contrast, residents of Los Angeles, though they sometimes confused one downtown street for another, had little "difficulty in maintaining direction on the paths" (1960, 38). Residents of Boston were easily disoriented and had, in some cases, highly distorted cognitive maps of the city. Many of the residents believed major roads in Boston were parallel, when in fact they were not. On the other hand, for the most part, residents of Los Angeles had little difficulty maintaining orientation and realized that roads in the central part of the city formed a grid. These examples raise the question: how does the arrangement of the environment affect the acquisition of spatial knowledge?

Given Lynch's results, it would seem that both Thorndyke's and Hayes-Roth's, and Lloyd's results are reasonable, and one could certainly expect similar results under similar circumstances. The test environment utilized by Thorndyke and Hayes-Roth, like Los Angeles, was highly regular. The navigation learners were able to learn the pattern of this environment and make spatial judgements more accurately than map-learning subjects. On the other hand, the test environment in Lloyd's experiment, like Boston, was quite irregular. The navigation learners found it difficult to learn the pattern of this environment and subsequently, the map-learners were able to make more accurate spatial judgements than the navigation learners. Conceptually, these results can be interwoven into the spatial knowledge model proposed earlier (figure 2) resulting in the following diagram (see figure 3).

This diagram is complex and rich in information. It is a conceptual model that shows a dynamic relationship between route knowledge, configurational knowledge, arrangement and scope of the environment, and the dimension of time when learning an environment via navigation. Essentially, the model indicates that initially one acquires route knowledge of an environment, and with increased navigation experience within this environment, configurational (topology only) knowledge is acquired. After the topology is learned, the arrangement and scope of the environment begin to play a significant role in determining, in a sense, the "accuracy" and "extent" of one's knowledge about the metrical properties of this environment. In a regular environment of small scope, it is possible that both topology and metrical properties of an environment can be learned, whereas in an irregular environment of fairly large scope, due to the more complex nature of the geographic information, it is possible that only topology is learned.

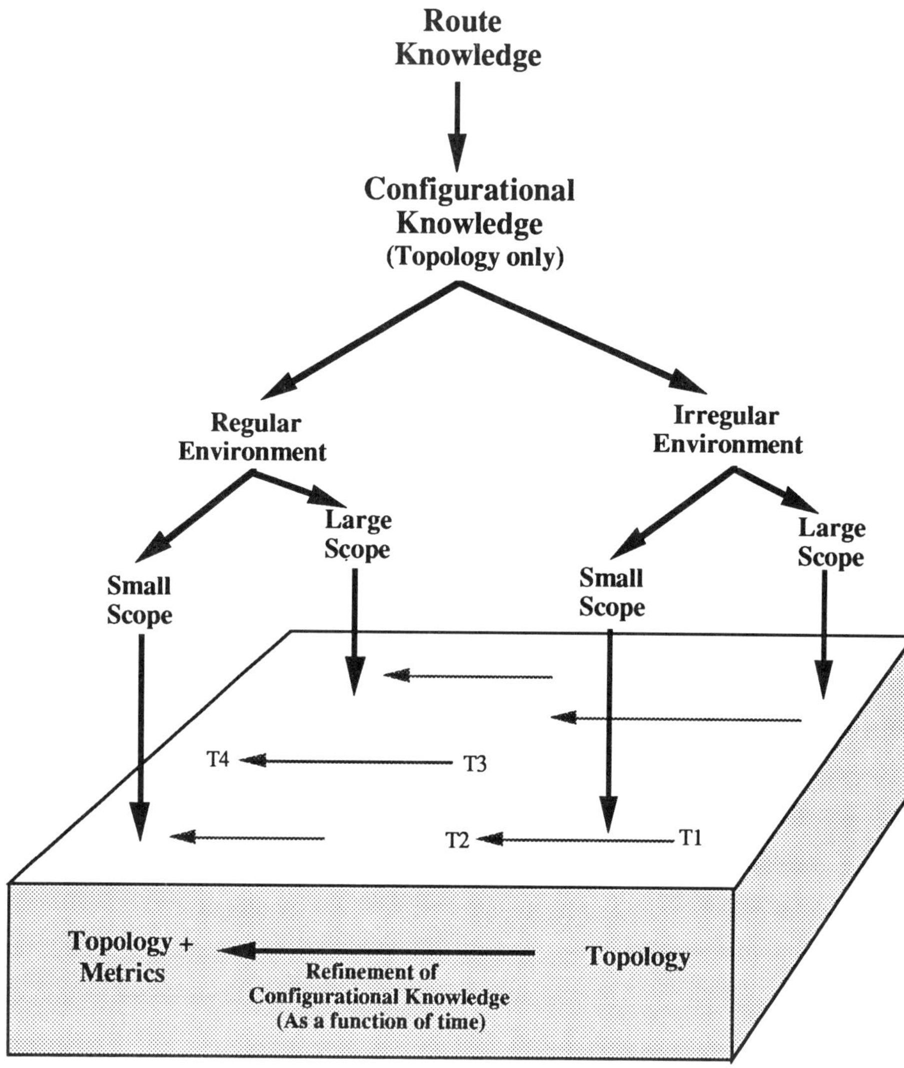

Figure 3. The relationship of route and configurational knowledge, and the arrangement and scope of an environment to the amount of navigation experience (length of time) in an environment.

6. A New Approach To The Problem

This paper attempts to discover what effect regularity of the environment has on the acquisition of spatial knowledge. Specifically, does regularity of the environment promote the acquisition of topologic and metrical properties [configurational knowledge] of an environment from route knowledge in subjects who learn an environment through navigation only.

Figures 4 and 5 represent the hypotheses that this experiment is to explore. The y axis in these two diagrams shows a lineage of spatial learning from route knowledge to topologic plus metrical configurational knowledge of an environment. In comparison, Tobler (1989) stated that one first learns the configuration of an environment, and then learns scale (proportions), distance, orientation, and location, resulting in a map-like representation of that environment. Unlike Tobler's progression, this lineage differs in several ways. Tobler's progression begins with acquisition of topological (configurational) knowledge, followed by the learning of scale (proportions) and distance. In contrast, this lineage begins with the learning of routes, followed by the acquisition of proportions and distances, and then the topology; learning the configuration of an environment is not a necessary prerequisite to learning proportions and distances.

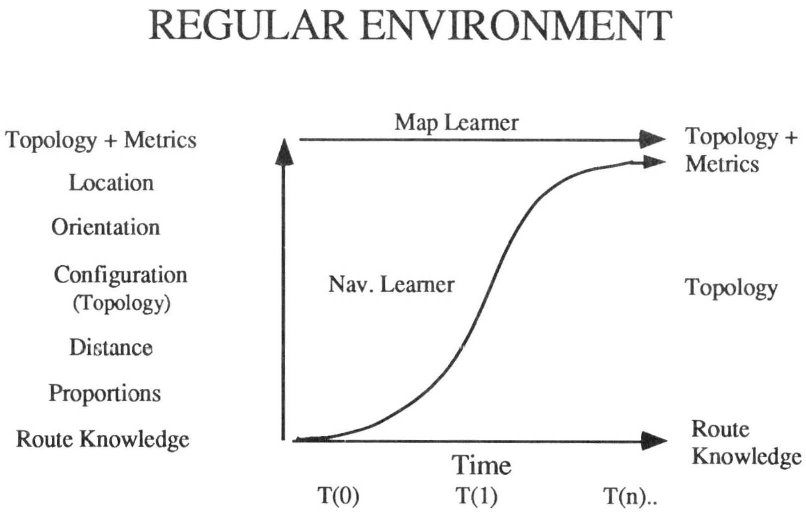

Figure 4. In a regular environment, a map-learner would acquire topologic plus metrical configurational knowledge at the outset, whereas a navigation learner would initially acquire route knowledge, and with increased experience in this environment, would acquire topologic plus metrical configurational knowledge.

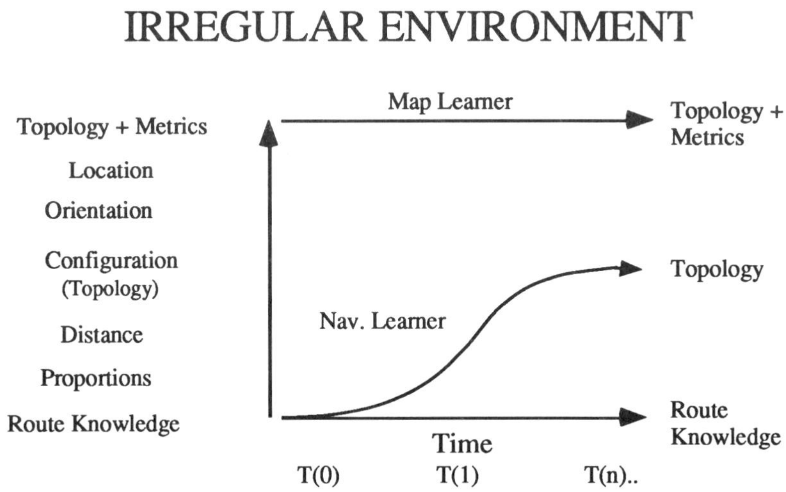

Figure 5. In an irregular environment, a map-learner would acquire topologic plus metrical configurational knowledge at the outset, whereas a navigation learner would initially acquire route knowledge of this environment, but would not acquire topologic plus metrical configurational knowledge.

An interpretation of figure 4 is that in a regular environment, a map-learner would acquire topologic plus metrical configurational knowledge whereas, a navigation learner would initially acquire route knowledge, and through time within this environment, would acquire topologic plus metrical configurational knowledge. Conversely, figure 5 indicates that in an irregular environment, a map-learner would again acquire topologic plus metrical configurational knowledge, whereas a navigation learner would initially acquire route knowledge of this environment, but would not acquire topologic plus metrical configurational knowledge. At best, only proportions, distances and topology may be learned.

7. Research Design Proposal

7.1 THE TESTING ENVIRONMENTS

The two testing environments chosen for this study are approximately the same size, 1.5 square miles, and are both "outside of building" spaces. Testing environment 1 is Fire District 2 in the Town of Tonawanda, New York. This is a regular environment, an environmental situation similar to that of Thorndyke's & Hayes-Roth's. Six landmarks have been selected in the area, and they are:

- Kenilworth Fire Hall;
- Lincoln Park Water Tower;
- Corner of Englewood and Kenmore Ave;

- Corner of Parker Blvd and Sheridan Drive;
- Corner of Decatur Ave and Niagara Falls Blvd;
- Curtis Park.

These landmarks were chosen from a list of 12 possible landmarks. Ten residents in the area were asked to rank the 12 possible landmarks from most familiar to least familiar, and from this ranking, the six landmarks that were ranked most often as top six were chosen. Testing environment 2 is the Walton Woods Area in the Town of Amherst, New York. This is an irregular environment, an environmental situation similar to that of Lloyd's. Using the same ranking procedure as described above, six landmarks have also been selected in this area, and they are:

- Getzville Fire Hall;
- The pool in Walton Woods Park;
- Corner of Sylvan Pkwy and Little Robin Road;
- Amherst Police Station;
- Corner of Little Robin Road and Robin Ridge Road;
- Corner of Millersport Highway and North Forest Road.

8. The Experiment

8.1 SUBJECTS

The sample will consist of 40 people (20 female, 20 male). Of these 40 people, 20 of them will be high-school students who reside in testing environment 1, and the other 20 will be high school students who reside in testing environment 2. Each group of 20 subjects are navigation learners (i.e., they have only navigation experience) in the testing environment in which they reside, and are map learners (i.e., have only map knowledge) in the other testing environment. The subjects will be recruited from local high-schools. Each subject will be offered $25.00 as reimbursement for participating in the experiment.

8.2 PROCEDURE

The map learning subjects will be divided into two groups based upon the amount of time spent studying maps of the testing environments. The first group will study the map until they can accurately label the six landmarks on an unlabeled copy of the map, as well as label cardinal directions and the roads that the subjects will be estimating route distances from. The second group will study the map until they reach the same criterion set for the first group, and then study the map for an additional 30 minutes.

The navigation subjects will also be divided into two groups, but in this case, group division is based upon the amount of experience within the testing environment (i.e., length of residence). Because the range of the subjects' length of residence is unknown at this time, division into three groups will be done after the sample is obtained.

All subjects will be told that the purpose of the study is to assess the accuracy of people's spatial knowledge given different types and amounts of learning experience, and that they will be asked to perform a number of distance estimations, orientation estimations, and location estimations. To compensate for a possible learning effect of subjects being tested in one environment, and then in the other, 10 of the 20 subjects who are residents of testing environment 1 will be tested first in environment 1 and then in environment 2. The other 10 will be tested first in environment 2, and then in environment 1. The same procedure will be used with the 20 subjects who are residents of testing environment 2. Subjects will be tested individually, and will be taken to each landmark in turn. At each landmark, a compass wheel with a diameter of twenty-four inches will be placed against the wall of a building (or other permanent structure) closest to the landmark. The compass wheel displays rays numbered from 0 to 350 degrees in 10 degree increments, and will be oriented with the 0 degree ray aligned North-South. The subject will then be informed that a series of locations (i.e., the five other landmarks) within the testing environment will be recited. For each location, the subject is to indicate on the compass, to the nearest degree, the direction to the center of the other five destination points. The responses will be written by the experimenter on a form which the subject will not be able to reference. Upon completing the five estimates at the first landmark, the subject will be taken to each of the remaining five landmarks (in a varied order), and the above procedure will be repeated.

The second spatial task that the subjects will be asked to perform is a series of distance estimations. The six landmarks are paired such that no two landmarks are paired more than once (there were 15 pairs in all). Each subject will be asked to first estimate the distance in tenths of miles on the shortest path between paired landmarks. In order to avoid any ambiguity, the shortest path will be explicitly indicated to the subject in the form of a written list of the roads that comprise each shortest path. The subject will then be asked to estimate the straight line [Euclidean] distance in tenths of miles between paired landmarks. To aid distance estimations, the subjects will be given three reference distances from each testing environment. In testing environment 1, the distances will be from Kenmore Avenue to Decatur Avenue along Niagara Falls Boulevard (0.80 miles), from Parker Avenue to Parkhurst Avenue along Decatur Avenue (0.375 miles), and from Parker Avenue to Niagara Falls Boulevard along Sheridan Drive (0.75 miles). In testing environment 2, the distances will be from John James Audubon to Campbell Boulevard along Dodge Road (0.70 miles), from Dodge Road to Millersport Highway along Campbell Boulevard (0.60 miles), and from John James Audubon to Millersport Highway along Sylvan Parkway (0.40 miles).

The last spatial task that subjects will be asked to perform is a set of twenty location estimations. Each subject will be given a booklet containing twenty 8.5 x 11 inch pages. On each page, two labeled dots designate the location of two of the six landmarks. The upper left corner of the page contains the names of these two labeled dots. The upper right corner of each page has the name of one of the other four landmarks. The subject will be asked to place a dot on the page indicating the location of this landmark with respect to the two labeled landmarks on the page. To avoid the effect of subjects remembering the locations of dots on a page, the two landmarks displayed on each page will be rotated "x" degrees from their "cardinal positions" in the real world. Random numbers will be used to select rotation angles. There are two constraints on these map displays: (1) the scale of the map is the same as the original map studied by the map-learning subjects; (2) the correct location of the destination point is within the boundaries of the page. Unlimited time will be given to complete this task.

8.3 RESULTS AND ANALYSIS

Of course, because this is only a research design proposal, there are currently no tangible results to report here. At this time, data collection has recently been completed, and analysis of the data will soon begin. It is hoped that analysis of the data support the hypotheses described by figures 4 and

5. This wealth of data will provide an excellent opportunity to explore differences in spatial knowledge based upon the arrangement of the environment; the source of spatial knowledge; the length of time for knowledge acquisition; and gender.

9. Acknowledgements

This research represents part of Research Initiative 2, "*Languages of Spatial Relations*", of the National Center for Geographic Information and Analysis, supported by a grant from the U. S. National Science Foundation (SES-88-10917). Support by the NSF and the NCGIA is gratefully acknowledged. In addition, support from David M. Mark, for attendance at the NATO Advanced Study Institute, is gratefully acknowledged.

The author's current address is: Department of Geography, Memorial University of Newfoundland, St. John's, Newfoundland, Canada A1B 3X9.

10. References

Abler, R. F. (1987) 'What Shall We Say? To whom shall we speak?', Annals of the Association of American Geographers 77(4), 511-524.

Abler, R. F. (1988) 'Awards, Rewards, and Excellence: Keeping geography alive and well', The Professional Geographer 40(2), 135-140.

Acredolo, L. (1982) 'The Role of Familiarity', in R. Cohen (Ed.), New Directions for Child Development: Children's Conceptions of Spatial Relationships 15, Jossey-Bass, San Francisco.

Amundson, R. (1983) 'Psychology and Epistemology: The place versus response controversy', Cognition 20, 127-153.

Allen, G. W., Siegel, A. W. and Rosinski, R. R. (1978) 'The role of Perceptual Context in Structuring Spatial Knowledge', Journal of Experimental Psychology 4, 617-630.

Appleyard, D. (1970) 'Styles and Methods of Structuring a City', Environment and Behavior 2, 100-118.

Berry, B. J. L. (1964) 'Approaches to Regional Analysis: A synthesis', Annals of the Association of American Geographers 54(1), 2-11.

Brownston, L., Farrell, R., Kant, E., and Martin, N. (1985) Programming Expert Systems in OPS 5: An introduction to rule-based programing, Addison-Wesley, Reading, Mass.

Gibson, E. J. (1969) Principles of Perceptual Learning and Development, Appleton-Century-Crofts, New York.

Golledge, R. and Zannaras, G. (1973) 'Cognitive Approaches to the Analysis of Human Spatial Behavior', in W. Ittleson (Ed.), Environment and Cognition, Academic Press, New York

Golledge, R. (1987) 'Environmental Cognition', in Stokols, D., and Altman, I. (Eds.), Handbook of Environmental Psychology, John Wiley and Sons, New York.

Golledge, R. (1988) 'Integrating Spatial Knowledge', Paper presented at the International Geographical Union Conference, Sydney, Australia.

Gould, M. D. (1989) 'The Provision of Usable Navigation Assistance: Considering individual cognitive ability', Proceedings, First Vehicle Navigation and Information System Conference (VNIS '89), Toronto, Canada, 443-447.

Hart, R. A. and Moore, G. T. (1973) 'The Development of Spatial Cognition: A review', in Downs, R. M. and Stea, D. (Eds.), Image and Environment, Aldine, Chicago.

Heft, H. and Wohlwill, J. F. (1987) 'Environmental Cognition in Children', in Stokols, D., and Altman, I. (Eds.), Handbook of Environmental Psychology, John Wiley and Sons, New York.

Herman, J. F. and Siegel, A. W. (1977) 'The Development of Spatial Representations of Large-scale Environments', University of Pittsburgh: Learning Research and Development Center.

Kozlowski, L. and Bryant, K. (1977) 'Sense of Direction, Spatial Orientation, and Cognitive Maps', Journal of Experimental Psychology 3, 590-598.

Kuipers, B. (1978) 'Modeling Spatial Knowledge', Cognitive Science 2, 129-153.

Kuipers, B. (1983b) 'Modeling Human Knowledge of Routes: Partial knowledge and individual variation', Proceedings, National Conference on Artificial Intelligence, AAAI 1983 Conference.

Ladd, F. (1970) 'Black Youths' View of their Environment: Neighborhood Maps', Environment and Behavior 2, 64-79.

Lloyd, R. (1989) 'Cognitive Maps: Encoding and Decoding Information', Annals of the Association of American Geographers 79, 101-124.

Lynch, K. (1960) The Image of the City, The MIT Press, Cambridge, Mass.

Mark, D. M., and Frank, A. U. (1989) 'Concepts of Space and Spatial Language', Proceedings, International Symposium on Computer-Assisted Cartography (Auto-Carto 9), Baltimore, 538-556.

Mark, D. M., Frank, A. U., Egenhofer, M. J., Freundschuh, S. M., McGranaghan, M. and White, R. M. (1989) 'Languages of Spatial Relations: Initiative Two Specialist Meeting Report', Technical Paper 89-2, National Center for Geographic Information and Analysis, University of California, Santa Barbara.

NCGIA (1988) 'Proposal to the Geography and Regional Science Program at the National Science Foundation', Technical Paper 88-1, National Center for Geographic Information and Analysis, University of California ,Santa Barbara.

Piaget, J. and Inhelder, B. (1967) The Child's Conception of Space, Norton, New York

Ritchie, B. F. (1948) 'Spatial Learning in Rats', Ph.D. Thesis, Library of the University of California, Berkeley, CA.

Siegel, A. W., and White, S. H. (1975) 'The Development of Spatial Representations of Large-scale Environments', in H. W. Reese (Ed.), Advances in Child Development and Behavior 10, Academic Press, New York

Shemyakin, F. N. (1962) 'Orientation in Space', in B. G. Ananyev *et al.* (Eds.), Psychological Sciences of the USSR (Vol. 1, Report No. 62-11083) Office of Technical Services, Washington, DC.

Stern, E. and Leiser, D. (1988) 'Levels of Spatial Knowledge and Urban Travel Modeling', Geographic Analysis 20(2), 140-155.

Streeter, L. A., Vitello, D. and Wonsiewicz, S. A. (1985) 'How to Tell People Where to go: Comparing navigation aids', International Journal of Man/Machine Systems 22(5), 549-562.

Thorndyke, W. and Hayes-Roth, B. (1982) 'Differences in Spatial Knowledge Acquired from Maps and Navigation', Cognitive Psychology 12, 137-175.

Tobler, Waldo (1989) Personal communication, June.

Tolman, E. (1973) 'Cognitive Maps in Mice and Men', Reprinted in Downs, Roger M. and Stea, David (Eds.), Image and Environment, Aldine Publishing Company, Chicago.

Trowbridge, C. C. (1913) 'On Fundamental Methods of Orientation and Imaginary Maps', Science 38, 888-897.

METHODS FOR MEASURING SPATIAL COGNITION

THOMAS A. GENTRY and JAMES A. WAKEFIELD, JR.
Department of Philosophy and Cognitive Studies
California State University, Stanislaus
Turlock, California 95380 USA

ABSTRACT. Measures of spatial cognition include a diverse array of methodologies and reflect a major dimension in the study of individual differences. Several spatial abilities assessment methods are reviewed that could be incorporated into the development of geographic informations systems. A new experimental procedure, using pointing behaviors, digital video technology and fractal geometry, to estimate the dimensionality of imagined environments is described. It is suggested that the *fractal dimension* (D) represents a useful measure for the study of spatial cognition.

1. Introduction

Mark and Frank (1990) describe a perspective on the relationships between experiential models of geographic space and geographic information systems (GIS) that provides the focus for our review of some methods used to measure spatial cognition. In particular, we address two issues raised by the Mark and Frank paper; the relevance of individual differences in spatial cognition, and the types of geometry relevant to cognitive maps.

1.1 COGNITIVE CATEGORIES AS A CENTRAL PROBLEM

In their overview chapter Mark and Frank (1990) have emphasized the contributions from linguistics to a central question in the study of human cognition; on what basis do we form mental categories about the world? Their review of the work by Rosch (1973, 1978), Neisser (1976), Johnson (1987), Lakoff (1987), and others, points up the difficulty in developing a general model for the way humans categorize their experiences. Mark and Frank adopt the philosophical perspective of "experiential realism" to organize their attack on the problem of cognitive categories, and provide a rationale for using this approach in the development of GIS projects.

The view that human categorizations of the world are of central importance in any model of the mind is strongly supported by data and theories from many areas in both cognitive science and neuroscience. Edelman (1979, 1985, 1987, 1988, 1989) has developed a detailed account of how brains may create categories and he believes that "a fundamental task of neuroscience is thus to show how, in a particular species, the structure and function of the nervous system permit perceptual categorization to occur as a basis for learning and meaningful adaptive behavior" (Edelman, 1987, 4). The Edelman model is extensive in scope and detail, ranging from the molecular to the cognitive, and provides a perspective that is congruent on many points with the

D. M. Mark and A. U. Frank (eds.), Cognitive and Linguistic Aspects of Geographic Space, 185–217.

views of Mark and Frank (1990). For example, both approaches emphasize the afferent/efferent integration in the formation of mental categories:

> "*Experiential* models of the world are based on sensorimotor and visual experiences with environments, and form in individual minds as the associated bodies and senses experience their worlds." (Mark and Frank, 1990)

> "While sensation and perhaps certain aspects of perception can proceed without a contribution of the motor apparatus, perceptual categorization depends upon the interplay between local cortical sensory maps and local motor maps; these, together with thalamic nuclei, basal ganglia, and cerebellum, interact to form the global mappings that permit the definition of objects as a result of continual motor activity." (Edelman, 1987, 210)

1.2 COGNITIVE CATEGORIES ARE DYNAMIC.

A second major point of agreement between a neuroscience account of categorization, as represented by Edelman, and the perspective from the linguistic view of experiential realism concerns the dynamics of mental categories. The notion that cognitive categories can be defined with the precision of logical or mathematical sets is rejected in favor of approaches that recognize the often very "fuzzy" and changeable fashion in which nervous systems organize sensory inputs from the world. It is this problem that Edelman has used to formulate a central question in his research:

> "According to the theoretical position taken here, the world of real stimuli is, in general, polymorphous. But if the world of stimuli is a world of polymorphous sets, the disjunctive properties of which can change depending on the context, the circumstance, or the adaptive needs of the animal, how can an animal use its brain to categorize and generalize in an adaptive manner?" (Edelman, 1984, 12)

In a similar fashion, Mark and Frank acknowledge the dynamic structure of categories, including those used by cognitive scientists, when they are describing the "image-schemata" model of Johnson (1987):

> "For any particular domain of investigation, one conceptual schema may be more useful than others. It is more likely that the most appropriate schema will change from problem to problem. Also, the schema themselves may change with each use. It is not an issue of whether one particular schema is 'correct' or not, but rather is an issue of how useful some particular schema is for some particular situation." (Mark and Frank, 1990)

1.3 COGNITION DEPENDS ON "SPATIALIZATION"

A third area of general agreement, and the one which most directly relates to our research reviewed in Section 3, concerns the "spatial" nature of cognition. Many paths lead to the conclusion that a *spatialization* of the "real" world is needed for cognitive processes. Mark and Frank review evidence for the spatial character of thinking as exemplified in linguistic analysis. Edelman's account of spatial abilities arises out of the cytoarchitecture and functioning of the nervous system. The work we describe has its origins in psychophysics, psychodiagnostics, and developmental psychology. This convergence on the spatial nature of mind has provided the basis for broad interdisciplinary collaboration, as reflected in the contributions to this volume. Mark and Frank (1990) correctly note that the "spatialization of inherently non-spatial concepts makes results from geography, as the science investigating space and spatial relations, applicable to other domains." In a reciprocal fashion it is expected that research on spatial cognition will benefit the development of geographic information systems.

1.4 INDIVIDUAL DIFFERENCES IN "SPATIALIZATION"

The three points of general agreement outlined above allow for broad collaboration on GIS projects, but future progress will most likely reflect the resolution of some issues that at present represent different perspectives. One area of apparent divergence concerns the general significance of individual variation in cognition and spatial abilities in particular. Mark and Frank would appear to be overly optimistic in minimizing the role individual differences might play in the application of experiential models to GIS.

> "Due to the physiological similarities that exist among individual beings, it appears that most people experience their environments in similar ways. Thus, we can expect that the basic features of individual experiential models of geographic space, while inherently personal, will have much in common across individuals" (Mark and Frank, 1990).

This may be more a matter of emphasis than any disagreement over the relevant facts, as they also acknowledge GIS problems seemingly associated with individual differences:

> "Many of the systems [GISs] constructed were either extremely limited in their capabilities to exploit spatial location, or the methods used were mathematically well-defined but not necessarily 'intuitive', i.e. they did not agree with the spatial concepts used by all their users" (Mark and Frank, 1990).

The issue of whether cognitive variation between individuals represents a problem depends on the objectives of the project. A possible strategy with computer based GIS is to detect the critical aspects associated with user differences, and adjust the system for a best fit between the system and the individual. This general solution is often implemented in software by allowing users to indicate their level of expertise and adjusting the command structure accordingly. If individual differences represent potential problems in the development of GIS then an understanding of the variation likely to be encountered should be helpful. In Section 2 we review some research on individual differences in cognition, especially those associated with spatial abilities.

1.5 WHAT IS THE GEOMETRY OF SPATIAL COGNITION?

The quantitative methods used to characterize human cognition represents an area of continuing research and theory. As in the case of individual differences, the article by Mark and Frank (1990) reflects two perspectives. On the one hand they point out that Euclidean geometry remains the principle measurement system in geography and surveying, and on the other they conclude that "the concepts applied by users of geographic information are not exactly Euclidean."

The selection of a quantitative approach to any problem is, *per se*, a reflection of individual differences. Poincare (1900) asserted that mathematicians used either analytic or geometric methods, but not both. Price (1975) described the genesis of western science as having been shaped in large measure by the combining of Babylonian arithmetical knowledge and Greek geometry, but he believed that we remain divided into "visualists and the verbal thinkers." Gould (1989) provides compelling anecdotal examples of this distinction when describing individual differences in the ability to reconstruct three dimensional organisms from flattened two dimensional fossils (see pages 100-101).

The psychometric literature reviewed below indicates that a verbal/spatial dichotomy is a recurring theme in the assessment of cognitive abilities; however, this is a relativistic distinction analogous (perhaps even homologous) with handedness, where we can use both limbs for many tasks but are usually more skillful with either the left or the right. For example, when Hernandez (1990, 1991) questions the appropriateness of pursuing the development of better geometries for GIS he may be expressing more how his mind works on problems than any inherent advantage of analytic sets. It was certainly the case in the present paper that one author did the algebraic thinking and the other the geometric.

In Section 3 we review a method that may be useful in developing new quantitative measures of spatial cognition, and discuss a geometry we believe best describes the results. We conclude with a proposed synthesis of the problems associated with the individual differences and the selection of a quantitative method by suggesting that fractal geometry offers some unique advantages for the integration of spatial cognition and geographic information systems.

2. Individual Differences In Spatial Cognition

There is no single word that can follow *spatial* in the title of this section and include all of the relevant subject matter. In addition to spatial cognition, a search of the research literature will encounter labels such as spatial *ability*, spatial *behavior*, spatial *memory*, spatial *orientation*, spatial *perception*, spatial *performance*, spatial *reasoning*, and spatial *thinking*.

The concept of a spatial ability that plays a fundamental role in human cognition has a long history. Mach (1886, 1959) used "space sensations" as the bedrock for his analysis of the mind. More recently, Jaynes (1976) has argued that *spatialization* is "The first and most primitive aspect of consciousness... Moreover, things that in the physical-behavioral world do not have a spatial quality are made to have such in consciousness. Otherwise we cannot be conscious of them" (p. 59-60).

The principle objective of this section is to outline some of the methods that are available to measure cognitive spatial processes in order to assess the individual variation in our species. However, it is also important to be aware that the greater proportion of the literature on how organisms cope with the spatial aspects of an environment is derived from other species. Comparative animal research on spatial behavior has important implications for the cognitive analysis of geographic space since it demonstrates that several spatial abilities had evolved in nervous systems long before the appearance of hominids and language (Olton, 1978, 1982; Schone, 1984; Ellen and Thinus-Blanc, 1987; Wehner and Menzel, 1990).

Some pre-human space related brain functions are (1) the ability to recognize places previously visited, (2) memory of topological arrangements so that a goal can be reached from a novel starting point, and (3) route selection between a set of places that minimizes the distance traveled (Olton, 1982). Animals also exhibit a wide range of species specific spatial behaviors based on diverse sensory capacities (Alcock, 1989; McFarland, 1985). The evolution of increasing complexity in fossilized records of spatial behaviors has been demonstrated in successive species from as early as the Cambrian era (Seilacher, 1967), and the differential evolution of visuospatial capacities has been suggested by Lynn (1987) as an account for some of the contemporary human variation in cognitive abilities. A principle message from evolutionary science is that variation is the rule, and, when considering anything as complex as spatial cognition, we should expect individual differences. Edelman (1987) identifies numerous sources for neuronal differences that can give rise to cognitive variability, including variations in cell morphology, connection patterns, cytoarchitectonics, neurotransmitters, metabolic and synaptic dynamics, neuronal transport, and interactions with glia (p. 57). If no two brains are alike to begin with, and if no two brains have the same environmental experiences, then it would seem prudent to make sufficient allowances for individual differences in any account of spatial cognition.

The following review of the methods used to quantify spatial cognition is not immune from the general problems associated with categorization and our list does not attempt to be exhaustive. The primary focus is on techniques that could be incorporated into GIS technology. Thus, we have concentrated on performance measures and psychometric tests that can, in principle, be adopted in computer based systems. Golledge and Timmermans (1990) have developed a more extensive taxonomy of the behavioral research on spatial cognition relevant to geography in which they identify 15 general models or methodologies.

2.1 PSYCHOMETRIC METHODS

Individual differences in separate cognitive abilities have been identified and used extensively in applied psychology and research (Wakefield and Goad, 1982). Among these separate abilities is spatial ability. While Galton's (1874) early work on the measurement of separate abilities was at too fine a level to identify a spatial factor and Binet's (Binet and Simon, 1905) work was at too general a level, a separate spatial factor has been identified at least since the 1930s in Thurstone's work. Using factor analysis, he identified seven primary mental abilities, that is seven sets of items that could be separated by their high correlations with other members of their own factor and lower correlations with items outside their own factor. The seven factors were called Verbal, Space, Reasoning, Number, Rote Memory, Word Fluency, and Perceptual Speed. Toward the end of Thurstone's career, it was shown that these factors, although separable, were not independent and would yield a general factor when they were subjected to further factor analysis (Eysenck, 1939). Thurstone (Thurstone and Thurstone, 1941) agreed that this was correct.

More recently, Vernon (1961, 1979) has suggested that intellectual factors can be reasonably organized into two large factors which he called (1) verbal, educational and (2) mechanical, spatial. These in turn may be broken down into smaller less stable factors. A third factor, attention or freedom from distractibility, has recently been added to the list of large general intellectual factors (Kaufman, 1979).

It should be noted that the visual spatial factor is positively correlated with the other intellectual factors. Dissatisfaction with the psychometric concept of general intelligence (*g*) has encouraged fitful quests for multiple intelligences throughout this century. Notable failures include Thomson's (1941) sampling theory and Guilford's (1967) over elaborate Structure of Intellect model which easily reduces to *g*. At least one large project (Kirk and Kirk, 1971) explicitly attempted to construct measures of visual abilities that were independent of other (verbal) abilities. Although the resulting test, (the *Illinois Test of Psycholinguistic Abilities*), showed lower correlations between visual and verbal tests than were typical, the correlations were still positive, indicating that it is very difficult, if not impossible, to measure visual spatial abilities (a) reliably and (b) independently of other cognitive abilities. Spearman (1904) had suggested that this would be the case very early in the history of testing.

Psychometrically, spatial ability occupies a place of importance as one of the two or three large, but correlated, cognitive factors. In this position, certain disappointments are inevitable. For example, clinical use of spatial ability largely centers on comparing it with verbal ability. The Wechsler scales with their verbal and performance (mostly spatial) scales are widely used (Matarazzo and Herman, 1985). This comparison takes the form of a difference between the verbal and performance scales. Such differences between correlated scales are notoriously unreliable. Nevertheless, behavioral and emotional difficulties as well as neurological distinctions between left and right hemisphere performance have been associated with differences between verbal and spatial ability.

Educational uses for spatial ability tests have been widespread. Overenthusiastic claims, especially for academic therapy, have produced a general disappointment with visual spatial tests such as the Frostig and the Bender-Gestalt. The primary reason for this disappointment is that schools are highly verbal, rather than visual spatial environments, and verbal abilities correlate much higher (.50 to .70) with academic achievement than do spatial abilities (about .20). This holds for all major ethnic groups (Kaufman, 1979). It should be noted that the newest major factor, attention, with academic correlates even lower than those for spatial ability, is currently headed down the same road. A second reason for the disappointment with spatial tests is the mixed results of matching visual treatments with visual learners and verbal treatments with verbal learners (Wakefield and Goad, 1982). Identifying visual and verbal learners depends on the difference between correlated test scores and produces the same difficulties in educational settings as it does in clinical settings.

The existence of lower spatial ability, but normal verbal ability, in Turner syndrome women who have a single X chromosome and no Y and the opposite pattern in men with an extra X has led to attempts to map the verbal versus spatial ability distinction onto the X chromosome. In general, such attempts have not been successful (Loehlin, Willerman, and Horn, 1988), although some large studies (Martin and Eaves, 1977; Martin, Jardine and Eaves, 1984) associate *within* family environmental variance with specific abilities and *between* family environmental variance, as well as hereditary variance, with *g*.

Recent behavior genetic studies of separate cognitive abilities, including spatial abilities, suggest that verbal and spatial abilities show somewhat higher heritabilities than other cognitive abilities (Plomin, 1986). It appears that specific genetic influences on spatial as opposed to verbal abilities have not at present been identified. Multivariate analyses suggest that intercorrelations among cognitive abilities, including spatial ability, are largely due to genetic covariance (Plomin *et al.*, 1990). While there may be different genes that affect different abilities to different degrees, most studies suggest that a set of genes definitely exists that influences all intellectual abilities together.

2.2 TESTS OF SPATIAL ABILITY

Currently, tests of spatial ability are included in most multi-aptitude test batteries. In addition, there are several independent tests that tap this factor that are included in psychological assessments. The selection of tests described below is not an exhaustive list of spatial abilities tests but is meant to represent tests that are widely used in applied psychology.

2.2.1 *Batteries.* As examples of batteries that include spatial abilities tests, the Wechsler series, the Stanford-Binet, and the Differential Aptitudes Tests will be discussed.

The Wechsler (1974, 1981) series is widely used by clinicians and includes measures of all the major intellectual factors. The Wechsler Preschool and Primary Scale of Intelligence—Revised (WPPSI-R) can be used with children as young as three years of age. The Wechsler Intelligence Scale for Children—Revised (WISC-R) is used during the school years from age six through 16. The Wechsler Adult Intelligence Scale—Revised (WAIS-R) is used for persons aged 16 and older. All three Wechsler tests include measures of the "perceptual organization" factor. The block design test, the picture completion test, and the object assembly test are visual-spatial tests that tap the same general ability.

The block design test is generally considered the best of the perceptual tests on the Wechsler. It consists of a series of abstract red and white designs of increasing difficulty that the subject must reproduce with a set of nine red and white cubes. Each of the items has a time limit, and bonus points are given for fast performance. Visual ability as well as motor skill influence scores on this test.

The picture completion test consists of a series of drawings of familiar objects and situations in which one important detail is omitted. The subject is asked to tell or show what is missing in the picture. This test does not require motor coordination which is a factor in many other spatial tests. Lack of familiarity with the objects shown in the drawings can lower scores.

The object assembly test is a series of four puzzles of familiar objects. It is largely redundant with the block design test except that it uses familiar objects rather than abstract designs. Lack of familiarity with the objects and motor skill influence scores on this test.

The fourth edition of the Stanford-Binet (Thorndike *et al.*, 1986) is the most recent revision of this venerable test and is used from age two through 23. Unlike the previous editions, the fourth edition yields more than a single IQ score. In fact, the fourth edition is one of the few tests constructed with a factor analytic model of intelligence in mind. An "abstract/visual reasoning" factor is explicitly scored from four separate tests. These tests are named pattern analysis, copying, matrices, and paper folding and cutting.

The pattern analysis test is really two tests. The easier items are administered with a form board with a circle, triangle, and square cut out of them. After these items, the rest of the test is given with a set of black and white blocks. The subject is asked to reproduce a set of abstract designs with these blocks.

The copying test, used for younger subjects, is also two tests. The earlier items require the subject to reproduce models built by the examiner using a set of solid green blocks. The harder items require the subject to copy geometric designs with a pencil. Fine motor coordination influences scores on this test heavily.

The matrices test is a test of non-verbal (i.e., visual) reasoning. A series of square matrices with designs in all but one space are presented. The subject must choose from a panel of possibilities the design that best completes the matrix. No fine motor skills are required for this test, but successful completion may be accomplished by verbal mediation rendering the test less of a visual test than it appears on the surface.

The paper folding and cutting test is used with older subjects. After the examiner demonstrates folding a piece of paper and cutting it to make a pattern, the subject is presented a series of graphic instructions for folding and cutting paper and is asked to choose the drawing from a panel that shows how the paper would look after the instructions were followed. No fine motor coordination is required for this test.

While the Wechsler and the Stanford-Binet are individually administered and for that reason very expensive, several group test batteries also include spatial tests. An example is the Differential

Aptitude Tests (DAT: Bennett *et al.*, 1984). This battery includes three tests of spatial ability—abstract reasoning, mechanical reasoning, and space relations.

The abstract reasoning test consists of items with a sequence of figures. The subject is asked to choose which item from a panel would come next in the sequence. This test requires the subject to focus on visual sequences rather than two dimensional patterns typical of other visual tests.

The mechanical reasoning test items contain drawings of moving bodies, gears, pulleys, people carrying things, etc. along with a verbal question asking the subject to decide which object will move faster, which person is bearing more weight, which direction gears or pulleys will move, etc. Familiarity with mechanical devices can influence scores on this test.

The space relations test consists of items showing a flat paper pattern followed by a set of three dimensional drawings showing how the paper pattern might look after it is folded together. The subject must choose which three dimensional object will result from the two dimensional pattern. The use of three dimensional drawings on this test differs from the more typical use of two dimensional patterns on spatial tests.

2.2.2 *Independent Visual Spatial Tests.* Three visual spatial tests that stand alone will be discussed—the Bender-Gestalt, Raven's Progressive Matrices, and human figure drawings. Each of these tests is used as a measure of general intelligence and as a measure of visual-spatial ability depending on the purpose of testing. As previously discussed, spatial ability correlates positively with other intellectual factors, justifying the dual use of these tests.

The most widely used test of perceptual-motor abilities is the Bender-Gestalt (Koppitz, 1975). This is a quick procedure that is used with subjects as young as five years. Nine stimulus cards with abstract geometric designs are shown to the subject one at a time. The subject is asked to copy the figure and errors are scored. In addition to spatial ability, other factors such as fine motor coordination, attention, and emotional factors influence this test.

Raven's Progressive Matrices (Raven *et al.*, 1985) was designed as a measure of general intelligence, but its visual, non-verbal form places it in the category of spatial tests under consideration. Each item on this test consists of a continuous abstract pattern, very much like wallpaper, from which a piece has been cut. The subject must choose the correct piece from a set of choices.

Human figure drawings, such as the Goodenough-Harris Drawing Test (Harris, 1963), are widely used as measures of general ability as well as of spatial ability. The subject is asked to draw a picture of a man, a woman, or himself without a specific stimulus to copy and without verbal guidance about what to include in the drawing. The drawing is scored for a specific set of features and characteristics. Differences in fine motor coordination and with representational art may be expected to influence scores on these tests.

2.2.3 *Visual illusion measures.* The phenomenology associated with illusions of space are noted for exhibiting variability and have been employed to measure individual differences. Illusions of motion occur when a moving pattern is stopped after a period of visual fixation. Variation in the duration and magnitude of the reported after effect associated with a "water fall" or rotating spiral pattern have been used in attempts to assess visuospatial neurological impairment. However, the difficulty of developing reliability estimates for this class of illusions, relative to other measures, has discouraged their use.

Illusions of linear extent, area, curvature, or direction can be assessed and have been used to characterize individual differences. Perhaps the most familiar example of a direction illusion is the "rod and frame" test where the observer attempts to set a line to the vertical with respect to gravity in the presence of a surrounding frame that is tilted (Asch and Witkin 1948; Coren and Hoy, 1986). Coren and Porac (1987) studied 26 visual-geometric illusions in combination with 5 measures of spatial cognition and found that individual differences in the magnitude of the illusions were significantly related to the spatial test scores. However, the relationships were complex such that individuals who had high scores on measures of spatial ability showed lower susceptibility to illusions of linear extent but greater susceptibility to illusions of area and direction.

2.2.4 *Wayfinding and map use.* Blades (1991a) has reviewed the literature on the development of wayfinding abilities and stresses the importance in recognizing that individuals can use different strategies. For example, a route can be learned "as a series of landmarks, as a list of directions at choice points, as a series of street names, and so on, and it is not clear from the present wayfinding research exactly what role these different encodings may play in learning a new route, or their comparative importance once a route is well known" (Blades, 1991a).

Blades (1991b), Blades and Spencer (1986a, 1987a,b, 1990) and Spencer, Blades, and Morsley (1989) have devised test environments and associated maps that allow for several different general strategies to be used in solving a wayfinding problem. Their procedures have demonstrated individual differences in children between age groups and within age groups. Blades and Spencer (1986b, 1987c) reviewed the literature on adult wayfinding abilities and identified problems using maps associated with individual differences in cognitive abilities and strategies employed.

2.2.5 *Sketch maps and verbal directions.* Commonly used methods for conveying spatial information about the world are to give verbal directions or draw a sketch map or both. These behaviors have been used for the study of individual differences in spatial abilities and provide an important source of data for the study of cognitive maps in humans.

Miller and Santoni (1986) compared several measures of spatial ability and experience with written directions given after viewing an unfamiliar map. They confirmed the frequently reported sex difference using the spatial relations subtest of the Differential Aptitude Test (Bennett *et al.*, 1959); however, they found that the higher male scores covaried at about the same level of significance with measures of education in mathematics and level of active travel (driving, bicycling, walking). Perhaps most relevant to the development of geographic information systems was the relatively high correlation (.43, $p < 0.01$) between DAT scores and the ability of subjects to use short term memory to give directions to others. (In essence, the DAT spatial relations test reflects an individuals' ability to *imagine* the manipulation of spatial patterns.) Miller and Santoni also found significant correlations between accuracy in map use and math courses (.37, $p < 0.01$) and the measure of active travel (0.27, $p < 0.05$). They also reported a complex matrix of interactions between the characteristics of the subjects and the characteristics of four different maps. Use of landmarks, cardinal directions, distances, and angularity of streets interacted with the gender, age, educational, and experiential variables.

Brewster and Blades (1989) have reported individual differences in the early development of abilities to give oral directions after viewing an unfamiliar map. They found developmental differences not only in the amount of detail but in the type of information used to convey directions. Blades (in press) has reviewed the use of sketch maps to communicate route information and studied the issue of test-retest reliability. Using the criterion of road names, landmarks, and road sections he found high correlations (0.62 - 0.88) between these measures for sketch maps drawn one week apart. The main source of variance was reasonably associated with the subjects familiarity with the locale being sketched. There was also a decrement in the correlations if the retest sketch map was drawn from the reverse starting point on the route specified, but the correlations remained high and significant and Blades concludes that this technique is a reliable method for data collection.

2.2.6 *Imagined environments*. Maps can be regarded as aids for the imagination of spatial arrangements that have not been directly experienced or can not be directly perceived in a single perceptual act. Consequently, the study of imagined environments should contribute to our understanding of the relationships between spatial cognition and geographic information systems.

A recent experiment by Franklin and Tversky (1990) used a paradigm where response time measurements assessed the effects of variables on cognitive tasks in imagined environments. A frequently cited example of this approach is the study by Shepard and Metzler (1971) where a response latency measure was used to support the conclusion that mental rotation of a three dimensional test object was needed to determine if it matched a sample object. Franklin and Tversky (1990) employed narratives about 10 different environments, each containing five different target objects, and recorded the subjects' response times for correctly locating targets in the imagined scenes. In general, their method required the subjects to imagine a particular environment based on the narratives and then imagine themselves in different body orientations in those environments.

Based on the mean response times (RTs) from all five experiments, Franklin and Tversky summarized their results as follows: "...for the upright observer, identification times were fastest for the head/feet axis, slower for the front/back axis, and slowest for the left/right axis, and for the reclining observer, RTs were fastest for the front/back axis, slower for the head/feet axis, and slowest for the left/right axis." This finding supports the "salience ordering of the three dimensions of everyday space (up-down >> foward-back >> left-right)" described in Mark and Frank (1990).

Franklin and Tversky (1990) interpret their results as evidence for a "spatial framework" model in contrast to theories they identified as (1) equiavailability and (2) mental transformation models. As with other experiments that use temporal measures to study spatial abilities, the conclusions of Franklin and Tversky are inherently more inferential than if the dependent variables had directly reflected measures of extent. In our work, with pointing behaviors, we are attempting to study the spatial properties of imagined environments with spatial measurements.

2.2.7 *Pointing*. The ability for humans to point in the direction of real or imagined targets has been employed in several types of research on spatial abilities. Clinical uses of pointing have been reported from the middle of the 19th century (von Graefe, 1854) and are used to test for a range of neurological disorders (Riddoch, 1917; Holmes, 1918; von Noorden, 1971; Bock and Kommerel, 1986). In the treatment of autistic children, pointing behaviors have been used to replace "autistic leading" and as an intermediary step in the acquisition of verbal behavior (Carr and Kemp, 1989).

The clinical studies and use of pointing behaviors to improve autistic dysfunctions are consistent with the view that pointing represents a very early manifestation of an individual's cognitive mapping abilities. As with all gross motor movements the researcher is confronted with the problem of response specification and measurement, and the literature contains a range of methods that vary in complexity, cost, and precision.

Duhamel, Pinek, and Brouchon (1986) measured pointing towards auditory targets in blindfolded subjects by rating whether the index finger of the pointing hand was either to the left or right of a sound source. Using this relatively uncomplicated (and inexpensive) method they found significant differences between left versus right handed subjects (right handers were more accurate) and between which arm/hand was used, irrespective of handedness (left hand/arms were more accurate).

Presson *et al.* (1989) devised a hand-held circular dial with a pointer that could be moved by the subjects to indicate directions in experiments on spatial memory for maps and paths. Their objective was to determine the effects of scale and they found that small scale displays were coded in an orientation specific way, while very large scale displays were remembered "in a more orientation-free manner." They cite their work as support for the "view that there are distinct spatial representations, one more perceptual and episodic and one more integrated and model-like, that have developed to meet different demands faced by mobile organisms."

The advent of the microcomputer has been accompanied by pointing devices that offer new opportunities for the assessment of individual differences in spatial abilities. Ott *et al.* (1987) constructed a head-mounted, computer controlled device to record pointing towards 80 light emitting diodes contained within the head-gear. The subjects can not see their pointing behaviors which is important for clinical testing. Jones (1989) has used mouse, joy stick, and track ball devices to manipulate the computer monitor cursor in studies where subjects are "pointing" to targets displayed on the screen. He found that estimates of processing rates (bits/sec.), based on the method of Fitts (1954, Fitts and Peterson, 1964; Langolf *et al.*, 1976), were substantially slower for the computer pointing tasks than for problems of similar difficulty measured with direct pointing. Sholl (1987) used a modified joy stick in a "point-to-unseen-targets" test of cognitive maps for geographic regions of different size. Her results suggest that "orienting schemata direct orientation with respect to local environments, but that orientation with respect to large geographical regions is supported by a different type of cognitive structure." This finding is similar to the study by Presson *et al.* (1989) cited above.

2.3 IMPLICATIONS OF SPATIAL ABILITIES RESEARCH FOR GIS

The psychometric literature indicates that a spatial ability can not be completely isolated from general intelligence, but is best considered as a major factor in a multidimensional model of cognition. Individual differences on spatial tests have been associated with hereditary, developmental, and experiential sources of variance. It is also reasonable to expect that individuals will differ in the strategy used to acquire spatial knowledge about an environment, and the same individual may adopt different strategies under different conditions (e.g. changes in scale). Consequently, the GIS designer should expect interactions between the system and user differences in spatial abilities and strategies. It is recommended that GIS programs incorporate an assessment component that can adjust the control and display features to the cognitive characteristics of the users.

3. Pointing Behavior And The Geometry Of Spatial Cognition

"They went that-a-way", accompanied by an extended arm and pointing fore-finger, may have been the original geographic information system in hominid evolution. The use of gestures to convey spatial information remains a ubiquitous method for informing others about the immediate environment and actions to be taken. Often this use of arms and hands is unaccompanied by linguistic information, especially in circumstances that require stealth (e.g. military patrols) or where the ambient noise levels preclude auditory communications (e.g. control of traffic). Arm and hand communications have been developed into sophisticated sign languages and it has been argued that gestural behaviors are the origin of language (Hewes, 1973). What ever role gestures may have played in the evolution of language, their primeval roots necessarily arise from the nervous system used for locomotion and manipulation.

The behavior of interest in our research consists of subjects, with eyes closed, pointing simultaneously with both hands to the spatial regions where they are imagining the location of two targets specified by the experimenter. This paired comparison method of having subjects point simultaneously towards two imaginary targets can be recorded on videotape or captured directly by computers equipped with video "framegrabbers". Figure 1 is a digitized image of the pointing behavior used in our research.

Figure 1 Example of an observer pointing simultaneously towards two imagined targets. The ping-pong balls on the finger tips facilitate the subsequent measurements.

The resulting video images of subjects pointing with both hands towards two imagined targets can be measured for the relative (ordinal) inter-target distance as represented by the two ping-pong balls on the index fingers which facilitate the measurements. These distances can then be used as input to programs such as the KYST multidimensional scaling software developed at Bell Telephone Laboratories (Kruskal *et al.*, 1977).

3.1 COMPUTING METRIC SPACES WITH ORDINAL DATA

Psychological data is commonly collected as a series of ordinal judgements, where observers are asked to indicate whether stimulus A is larger, heavier, brighter, etc., than stimulus B. Since Fechner's *Elemente der Psychophysik* was first published in 1860 there has been a continuing quest to extract quantitative models from relativistic judgements, and that enterprise has become one of the central threads in the history of psychology. The hope for improved metric models of mental processes was substantially advanced when Abelson and Tukey (1963) demonstrated that if all the relative ordinal distances between the elements in a set are known it becomes possible to develop quantitative models of the set. That insight, in combination with the rapid iterative operations afforded by computers, has yielded the field of multidimensional scaling (MDS) techniques used for creating metric models based on relativistic human judgements (Davison, 1983; Green *et al.*, 1989; Kruskal and Wish, 1978; Young, 1987). The MDS methods continue to be applied to a diversity of psychological issues, including such labile phenomenology as pain (Clark *et al.*, 1989) and happiness (Thomas and Stock, 1988).

From its early development, MDS analysis programs have been used on problems relevant to the design of geographic information systems (Golledge and Rushton, 1972), and that trend is continuing in research on cognitive maps (Birnbaum *et al.*, 1989), perception of environmental quality (Chokor, 1989; Garling, 1980; Ma, 1987), and human factors (Coury,1987; Pegels and Sekar, 1989; Polzella and Reid, 1989). There is a good rationale for using the multidimensional scaling methods in the study of spatial cognition since the MDS algorithms were originally derived from geometric solutions for sets of relative distance measures. It may be useful here to recall the work of Shepard (1966) in reviewing the foundations for deriving metric structure from ordinal data. If the reader carries out Sheppard's example (1966, 291) of arranging markers according to his ordinal equation, see [1] below, s/he will intuitively appreciate what geometric properties are encoded in relative distance information. And, equally important, what spatial information is not contained in ordinal data.

[1] AB > BD > AC > BE > DE > CE > CD > AE > AD > BC.

While a small number of inter-point distances will set severe constraints on the *dimensionality* and *shape* of the solution that satisfies the ordinal relationships in a set of distance data, the *size*, *orientation*, and *reflection* of the resulting configuration are indeterminate. The heuristic value of this demonstration lies in the possible homology with the brain's functions used for cognitive tasks where size, orientation and reflection are not essential, and, in fact, may have to be ignored in favor of the more ecologically adaptive information of form recognition, categorization, and spatial position.

The ability to move about in either imaginary scenes or the real world requires a spatial ability that can correct for the constantly changing size, rotation, position, and reflection in features of the environment. That is, the spatial mechanisms in the brain need to be sufficiently independent of scaling concerns so that it can categorize the environment despite the constantly changing sensory inputs. This ability has commonly been discussed under the topic of "perceptual constancies" and much of this neural work appears to be accomplished without conscious attention to the task.

3.2 THE GEOMETRY OF POINTING BEHAVIOR

The following procedure was originally reported in Gentry *et al.* (1986), and a more detailed account is available in Goodman (1990). Standing subjects are simultaneously videotaped from two orthogonal perspectives (frontal and lateral). Directions for the subjects and trial by trial target pairs are pre-recorded and played back on an audio tape system to control the experimental sequence. Target words used to elicit pointing behavior can include people, places, and things familiar to the subjects, and in the original study a stimulus tape consisted of two presentations for each of the pairs possible with a set of nine target names, or 72 trials per session. The initial presentation for each pair was randomly arranged with the second occurrence counter balanced according to an ABBA format.

The initial analysis of the videotapes collected using this procedure was accomplished by attaching a transparent overlay, ruled in millimeters along the edges, over the face of a 24-inch TV monitor. Measurements were made with a moveable T-square to determine the vertical and horizontal coordinates for the two ping-pong ball tipped finger positions obtained for each trial (see Figure 1). This laborious method has subsequently been replaced by using direct video input to microcomputers equipped with framegrabbers that digitize the image and the XY coordinates are automatically stored in a spreadsheet.

The initial trials consist of recording the subject's subjective sense of orientation and direction with their eyes closed by having them point out compass and clock-face positions. Also, we ask the subject to indicate, by pointing, where their most central sense of "self" is located. This pointing to one's ego-center has been included because the common assumption that it is in the head, some small distance behind the eyes, is not always what is reported. Some subjects locate their ego-center in the upper chest on the midline in the region of the C-5 dermotome. Traditionally a subject's ego-center is assumed to be within the head (Howard and Templeton, 1966), but, given the suggestion by Jaynes (1976) that the origin of the subjective sense is visceral, it seems prudent to have subjects point directly to their ego-centers, particularly if that concept is to play a role in spatial cognition theory.

Following the preliminary videotape recording of the subject's sense of orientation and ego-center, the target words were presented singly to determine whether or not the subject can perform the pointing task with all stimulus words to be used in that session. After the demonstration that the subject can point towards where they are imagining the people, places, or things elicited by the stimulus words, a random presentation of all possible target pairs is conducted. The resulting videotapes are subsequently analyzed to determine the distances on each trial between the pointing fingers and a computer program combined the target pair replication trials to produce the average distances between the ping-pong ball tipped fingers for the two orthogonal views which were then used to compute the distance values needed for the multidimensional scaling analysis.

It should be noted that this bi-brachial method could be replaced with a single arm procedure if the previously mentioned ego-center or a body reference point is used to derive the relative distance measurements needed for the MDS analysis. Other variations on the method include eliminating the trail by trial procedure and adopting a stream of consciousness format. For example, subjects can be asked to imagine an area such as a familiar room, building, or neighborhood and then begin pointing towards the images as they are remembered, identifying them so that the narrative is recorded on the audio track of the videotape pointing record. The key behavior in our procedure is the human ability to indicate, by pointing, the spatial position of where they are imagining the location of something.

3.3 A MULTIDIMENSIONAL ANALYSIS OF POINTING BEHAVIOR

At a previous NATO Advanced Study Institute entitled *New Advances in the Analysis of Visual Behavior*, Dodwell (1982) presented a paper on the "Geometrical Approaches to Visual Processing" which reviewed evidence that Euclidean space is an unsatisfactory geometry for visual perception. The Dodwell (1982) account began with Luneburg's (1947) geometrical argument that a binocular projection onto a "Cyclopian" visual space requires a hyperbolic space model (see also Blank, 1959). A more radical departure from the Euclidean assumptions was developed by Watson (1978) who used a series of experiments on visual illusions and figural after-effects to argue "...that visual space is non-Euclidean, and that the departure from Euclidean geometry is the result of the interaction between the elements with the visual field. .If the space which the brain constructs from the information received by the eyes, is non-Euclidean and dependent upon context, then it is not surprising that the subjective judgements made under the influence of varying context, are at variance with the physical Euclidean measurements (which do not depend on context)."

Watson's (1978) idea that the geometry of visual perception changes with the "context" of the visual field, suggests that geometric dimensions might also be used as dependent variables in research on cognition, and it was that concept which has provided the rationale for our experiments on the pointing behaviors that reflect a person's spatial localization of imaginary targets. In our initial multidimensional analysis the goodness of fit between the data and the geometry employed is called "stress," and it is a residual sum of squares, a positive dimensionless number that can be expressed as percentages (Kruskal, 1964a). The stress declines as the computer program obtains increasingly better fits to the data through successive iterations which have the goal of obtaining a monotonic relationship between the experimental dissimilarities (or similarities) and the distances in the resulting spatial configuration. If the reader has completed the example in equation [1] and plotted a final configuration, then s/he has directly experienced the iterative task performed by the computer in seeking a solution. The original papers by Kruskal (1964a, b) provide details on the numerical method, and other discussions of the stress measure can be found in Shepard (1974), Kruskal and Wish (1978), and De Leeuw and Stoop (1984).

The stress can be altered in two ways for a set of experimental data points such as those in our example, either through changes in the number of dimensions used in the analysis or changes in the "r" exponent of the Minkowski r-metric used in the KYST-2 program (Kruskal *et al.*, 1977). If the "r" in the Minkowski r-metric is set to 2 then the analysis space is Euclidean. Manipulating "r" indicated that the data in our study did not benefit from using more than three dimensions as the stress in 4D space remained essentially zero under all values of "r" tested. However, when values of "r" other than 2 were used in 3D space the distance data derived from the pointing behaviors was best fit by a non-Euclidean geometry as the stress for r=2 (Euclidean) was greater than at some smaller and larger r values.

Finding lower stress values on *both* sides of r=2 represented a problem of how to interpret the results. This difficulty with the KYST multidimensional analysis is not trivial, either with respect to the practical costs of computing the best "r" value or the theoretical justification for using "r" values other than 2 (Shepard, 1974). The stress values in 3D Euclidean space ranged from 7.4% to 1.0%, and the overall impression was that our pointing behavior data needed generally more than two and less than three dimensions, a rather crude result at best.

The stress reduction as a function of dimension varied between subjects and within subjects as well as across target categories. In general, the cognitive processes represented by the pointing behavior fits in three or less dimensions, but whether the space is Euclidean is doubtful. What was evident in the three subjects used in the original experiment was the large individual differences in their spatial representations of the target people, places, and things. Figures 2a, 2c, and 2e show two dimensional "cognitive maps" derived from the MDS analysis of the observers pointing to geographic targets.[1]

Figures 2a, c and e represent one of the general problems with using the MDS approach for constructing psychological spaces, especially the questions associated with orientation, reflection, and scale of the final configuration. If the original data is derived exclusively from judgements about geographic locations the MDS configuration can be superimposed on an "objective" map of the territory and oriented and anchored to the cartographic representation (e.g., Birnbaum *et al.* 1989). But if the targets have no geographic reality outside of the subjects imagination then the final arrangement of the MDS plots will be necessarily arbitrary.[2]

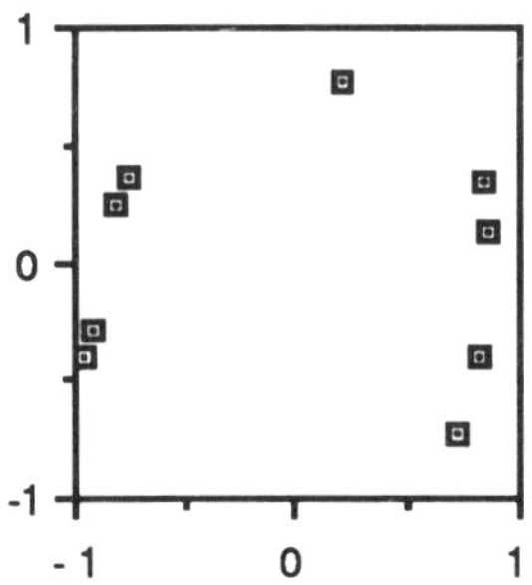

Figure 2a - Observer 1 MDS Places

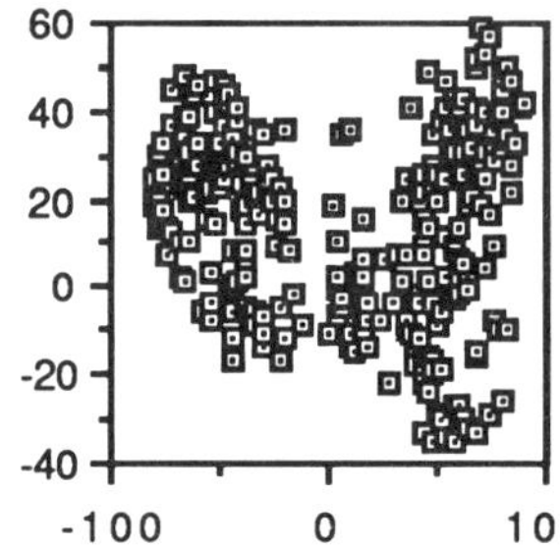

Figure 2b - Observer 1 C-O-P Places

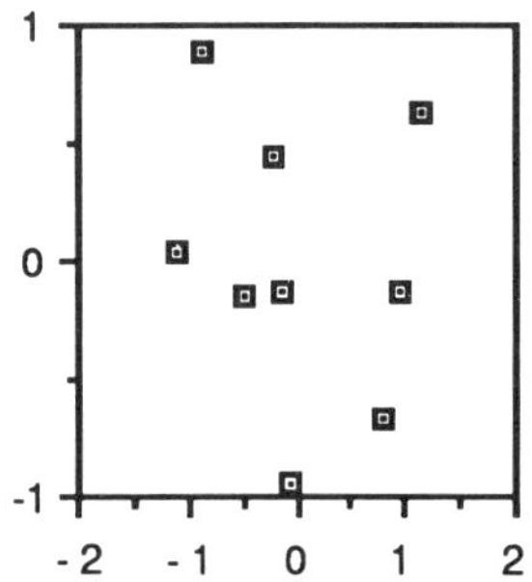

Figure 2c - Observer 2 MDS Places

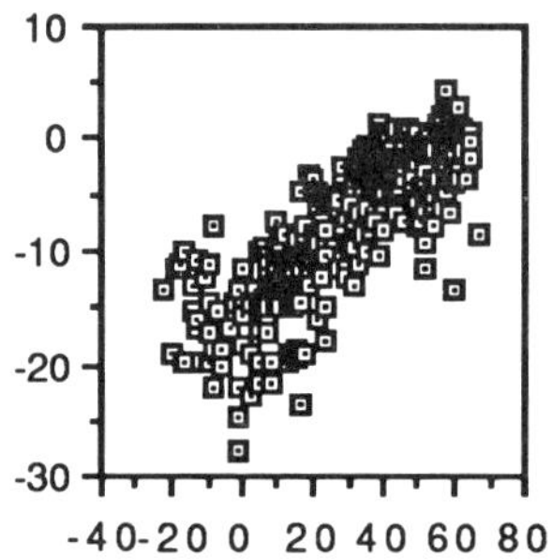

Figure 2d - Observer 2 C-O-P Places

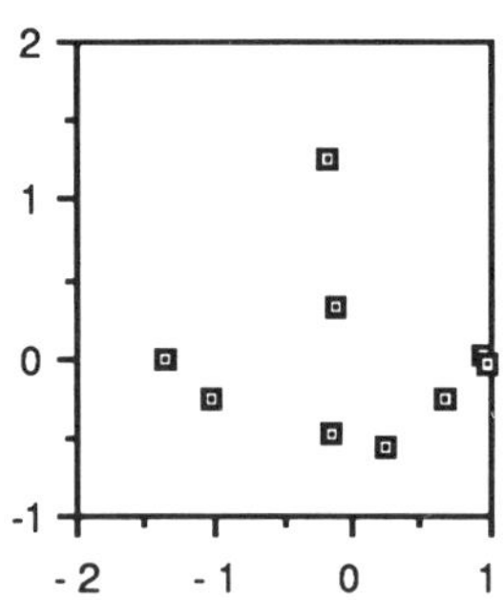

Figure 2e - Observer 3 MDS Places

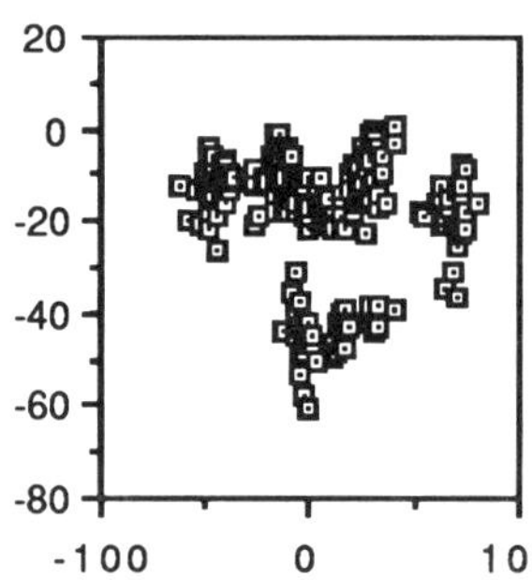

Figure 2f - Observer 3 C-O-P Places

Figure 2. Cognitive maps for 9 geographic places derived from the KYST multidimensional scaling analysis (MDS plots 2a, 2c, and 2e) and the cumulative cloud-of-points (C-O-P) data of the pointing behavior for the same sessions as recorded from the frontal view videotapes (2b, 2d, and 2f).

Figures 2a, c, and e provide two dimensional plots for nine geographic targets derived from the KYST-MDS analysis, and the graphs to the right of each MDS figure are the cumulative pointing positions of the three subjects during the same sessions. While it is possible to use the record of the actual pointing behaviors to orientate the KYST plots to some extent (for example, the MDS plots in Figures 2a and 2e would correspond better to the actual pointing data in Figures 2b and 2f if they were rotated 180 degrees around the horizontal axis), it is also apparent that the dynamic quality of the original behavior is lost in the MDS analysis. It is the dynamic property of the pointing behavior that may provide improved estimates for the "dimensionality" of an individual's spatial models of the world.

3.4 A FRACTAL ANALYSIS OF POINTING BEHAVIOR

The history of the present research is a reflection of contemporary developments in mathematics and how some scientists are beginning to describe the geometric properties of the natural world. The videotapes of pointing behavior used in this report were originally collected to explore the dimensionality of individual cognitive maps using a multidimensional scaling method. The same data tapes have subsequently been re-analyzed using quantitative methods associated with the field of non-linear problems known collectively as Chaos mathematics.

Because of its application in so many areas of research, Chaos mathematics has numerous origins (for historical reviews see Gleick, 1987; Peterson, 1988; Briggs and Peat, 1989), but among its principle progenitors, Mandelbrot's two volumes (1977, 1982) can be credited with introducing the term "fractal"[3] into the general vocabulary of science, and it is towards a fractal view of spatial cognition that this report is directed. The general utility of fractal measurements is being tested in applications ranging from astronomy (Calzetti *et al.*, 1989; Martinez and Jones, 1990; Wen,*et al.*, 1989; Zimbardo *et al.*, 1990)) to particles on the order of a few hundred nanometers (Klingen and Roth, 1989).

Research areas of particular relevance to the study of spatial cognition are those that describe the fractal nature of the environments experienced by humans and those that demonstrate changes in the fractal properties of behavioral or physiological characteristics of organisms in response to the world. The former are significant in that it is a reasonable hypothesis that cognitive maps will reflect the salient properties of the territory mapped. The research on *changes* in the fractal properties of a living system are an indication that this geometry has utility as a dependent variable.

3.4.1 Fractal environments. An initial reason for the enthusiasm towards fractal mathematics was the ability to generate computer graphics that had the appearance of natural geographic features (Mandelbrot, 1977, 1983). A subsequent development has been the rapidly growing literature on extracting the fractal dimension of natural phenomena at all scales of observation (Batty *et al.*, 1989; Brammer, 1989; Huang and Turcotte, 1990; Mann and Wasilewski, 1990; Nelson and Jaggard, 1990; Piech and Piech, 1990; Turcotte, 1989; Voss, 1989; Wang *et al.*, 1990).

3.4.2 Fractal biology. The study of living systems has also begun to utilize the potential of fractal geometry to characterize changes in physiological structures and functions (Barnsley *et al.*, 1987; Goldberger *et al.*, 1990; Gutierrez and Almirall, 1989; Iannaccone, 1990; Nelson *et al.*, 1990; Obert *et al.*, 1990; Takahashi,1989; West, 1990). For example, Fisher (1989) has reported that the shapes of two-dimensional vesicles can vary between fractal and non-fractal states depending on physiological parameters relevant to the tissues. Potential health applications include the use of fractal dimension estimates to characterize diagnostic images such as mammographic patterns (Caldwell *et al.*, 1990).

3.4.3 Fractal brains and cognition. The prospect that nervous systems have fractal properties and that cognition reflects those fractal features is being described at several levels of observation, including the ionic dynamics of nerve conduction (Liebovitch and Sullivan, 1987; Liebovitch and Toth, 1990), the structure of sensory receptor systems (Mainster, 1990; Morigiwa *et al.*, 1989), the patterns of neural spike trains in sensory nerves (Teich, 1989), the dynamics of sensory discriminations (Skarda and Freeman, 1987), and the cognitive processes associated with electroencephalography (Arle and Simon, 1990; Xu and Xu, 1988). The human discrimination and enjoyment of fractal forms may well reflect the nature of the mind's constructions (Cutting and Garvin, 1987; Knill *et al.*, 1990), and our method of studying the pointing towards imagined targets is being developed to examine that possibility.

The "cloud-of-points" (C-O-P) graphs in Figure 2 are two dimensional projections of the original behaviors sampled at nearly regular intervals[4] and can be regarded as a pseudo-phase space (Moon, 1987) where the plane of the plot is a temporal cross-section of the periodic pointing movements. The points also have an appearance similar to a computer monitor during the early stages in the iteration of a difference equation, such as the $Z(n+1) = Zn^2 + C$ quadratic map exploited by Mandelbrot. The general spatial appearance of the cloud-of-points data in Figure 2 suggests a fractal quality and that an estimate of the fractal dimension (D) can be derived by the "covering procedure" (Moon, 1987) or Box Counting Theorem (Barnsley, 1988).

3.5 DETERMINING FRACTAL DIMENSIONS

Whether or not an aspect of nature can be described as fractal depends on differences in the physical processes that produced the structure. For example, terrain formed of shale is reported as exhibiting "fractal behavior", while topographic features formed of granite apparently do not (Piech and Piech, 1990). Similarly, some fault lines associated with tectonic processes are fractal and some are not (Korvin, 1989).

What method should be used to estimate D for a particular set of data presents the experimentalist with practical problems to be solved before this measure can have utility. Gilbert (1989) cautions that "The use of fractal geometry as a quantifier of topography must be qualified by two considerations, (1) scale must be specified in the form of a bandwidth under consideration; and (2) the particulars of the data analysis techniques must be explicitly stated." These concerns are discussed in Moon (1987) and Barnsley (1988).

Moon (1987) provides a taxonomy of some current methods used to estimate D, (1) capacity dimension, (2) pointwise dimension, (3) correlation dimension, and (4) information dimension. The different methods have different requirements for application and in the present study the data are treated as a "capacity dimension". The fractal dimension is not a fixed property to be discovered, but is a class of dependent variables that can be used to characterize geometric data that exhibits self-similarity over a range of scale lengths. "The fractal dimension of a set is a number which tells how densely the set occupies the metric space in which it lies. It is invariant under various stretchings and squeezings of the underlying space. This makes the fractal dimension meaningful as an experimental observable; it possesses a certain robustness, and is independent of the measurement units" (Barnsley, 1988). It is also possible to extract useful information about the fractal dimension of a higher dimensional "cloud-of-points" from its projection on a two dimensional plane (Nelson *et al.*, 1990).

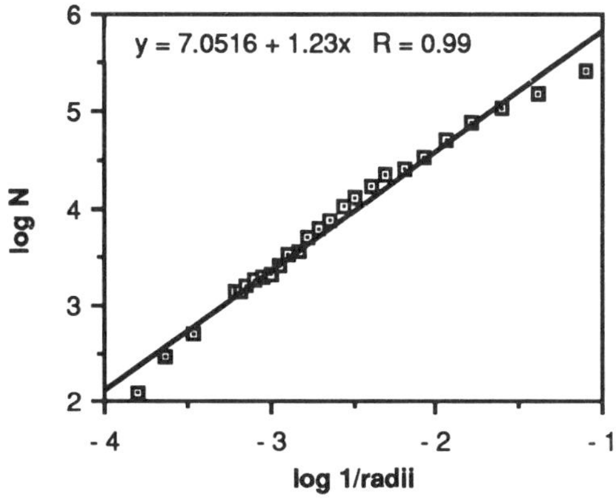

Figure 3A - Observer 1

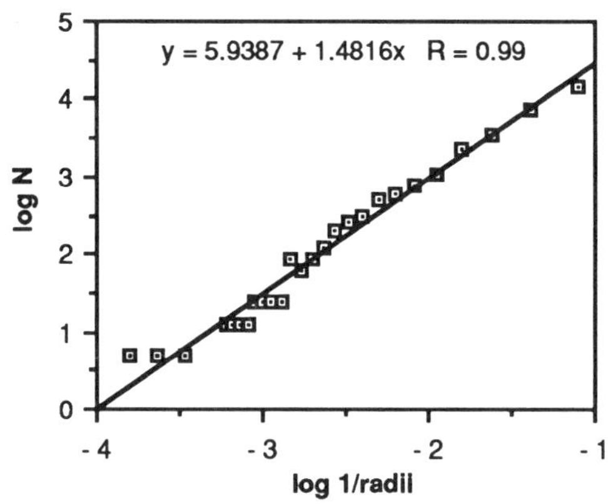

Figure 3B - Observer 2

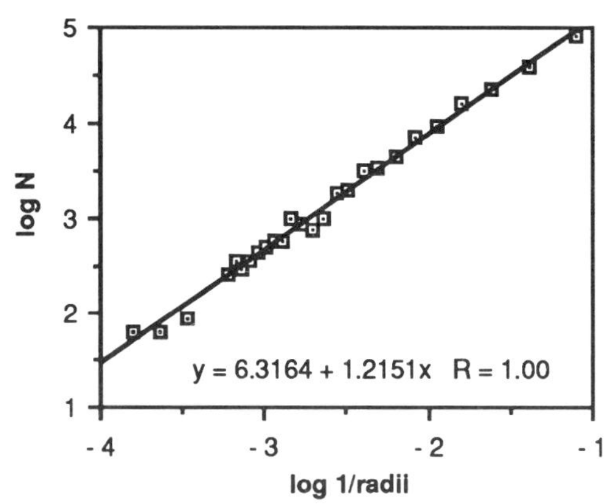

Figure 3C - Observer 3

[Figure 3 caption on next page]

Figures 3A, B, and C. Regression functions for the number of circles needed to cover the cloud-of-points examples from Figure 2 (log N) as a function of the radii of the covering circles (log 1/radii). Slopes of the regression lines are the estimate of the fractal dimension (D). The high "R" values indicate that the clouds-of-points in Figure 2 can be considered fractals.

3.6 THE FRACTAL DIMENSION OF POINTING BEHAVIOR PATTERNS

Figures 3A, B, and C are the log-log regression plots for the number of circles needed to cover the cloud-of-points examples in Figure 2 as a function of the reciprocal of the radii of the circles used, where the radii ranged from 3 to 44.5 mm.

Figures 3A-C exhibit good fits to power functions which suggests that the clouds-of-points in Figure 2 are fractal, with slopes yielding estimates of the fractal dimension (D) of 1.23, 1.48, and 1.22. Table 1 provides the complete set of D values based on the cloud-of-points derived from three observers pointing towards the imagined location of nine targets in six different sessions. In three conditions the targets are all of the same category type (people, places, or things) and in three cases the targets were an equal mixture of the different target types.

TARGET TYPE	OBSERVER 1	OBSERVER 2	OBSERVER 3	MEAN D
PEOPLE	D= 1.314	D = 1.493	D = 1.443	D = 1.4
PLACES	D = 1.230	D = 1.482	D = 1.215	D = 1.3
THINGS	D = 1.208	D = 1.335	D = 1.364	D = 1.2
MIX 1	D = 1.221	D = 1.387	D = 1.277	D = 1.3
MIX 2	D = 1.238	D = 1.458	D = 1.400	D = 1.4
MIX 3	D = 1.265	D = 1.402	D = 1.364	D = 1.3
MEAN D	D = 1.2	D = 1.4	D = 1.3	

Table 1. Estimates of the Fractal Dimension for the Frontal 2D View

Based on Table 1 we can advance the hypothesis that human spatial cognition operates in less than three topological dimensions. A somewhat "flattened" imagination is not unexpected considering that a main source of spatial information is two dimensional (the retina), and that the three dimensional view of the world has to be constructed from two slightly different plane perspectives. When the analysis was performed on the orthogonal view videotapes of the same experimental sessions the mean estimates for D ranged from 1.3 to 1.5. Using the combined results we can tentatively estimate that the fractal dimension for imagined environments can range from 2.5 to 2.9.

However, the potential uses for this method are associated with learning what variables contribute to the inter- and intra-individual differences in D.

3.7 THE INTER-OBSERVER ESTIMATES OF D

Both the cloud-of-points patterns (Figure 2) and the estimated values of D (Table 1) suggest that the pointing behavior method is sensitive to individual differences in how humans spatially organize their representations of people, places, and things. Table 2 shows the correlations among the observers for all estimates of D.

	Obs 1	Obs 2	Obs 3
Obs 1	---		
Obs 2	.62	---	
Obs 3	.60	.06	---

Table 2. Inter-Observer Correlations for D

The values in Table 2 reflect several sources of variation, including the types of target categories, the specific target words, the effects of practice, and individual differences in cognitive functions that influence the pointing behavior and/or interact with the other sources of variance. Since this report is limited to considerations of the method used, it is unrealistic to draw any firm conclusions, but a basis for future hypothesis testing is possible.

Observer 1 showed good agreement with both of the other subjects who did not agree between themselves. Inspection of the data in Table 1 suggests that all three observers had a relatively high dimensionality for people (they were all psychologists!). Perhaps a high dimensionality here indicates a complex system of interrelationships. Observer 1 had relatively low dimensionality for places and things, Observer 2 had high dimensionality for places, but low for things, and Observer 3 had high dimensionality for things but low for places.

An iterative intercolumnar correlational analysis was conducted for the six target types. (This analysis includes computing the correlations among a set of items (i.e. the target types) and then on successive iterations, the columns of correlations are correlated. This usually produces a matrix of +1's and -1's that sort the items into two groups.) Table 3 shows the correlations among the target types across raters above the diagonal and the third iteration of iterative intercolumnar correlation analysis below the diagonal.

	PEOPLE	PLACES	THINGS	MIX 1	MIX 2	MIX 3
PEOPLE	-----------	.68	.90	.91	1.00	1.00
PLACES	-1.0	-----------	.29	.93	.67	.68
THINGS	1.0	-1.0	-----------	.63	.91	.90
MIX 1	-1.0	1.0	-1.0	----------	.90	.91
MIX 2	1.0	-1.0	1.0	-1.0	-----------	1.00
MIX 3	1.0	-1.0	1.0	-1.0	1.0	-----------

Table 3. Iterative intercolumnar Correlational Analysis

The third iteration produced a matrix of plus and minus 1.0's that grouped people, things, mix 2, and mix 3 into one group with dimensionalities that covaried across the observers, and places and mix 1 fit into a second group with dimensionalities that covaried somewhat differently across the raters.

Group 1={people, things, mix 2, mix 3}
Group 2={places, mix 1}

It is possible that the "movability" of the imagined targets might be important. Group 1 with people and things may result from the dimensionality (i.e. complexity) of rating positions of movable items occupying space. Group 2 with places may result from the dimensionality of rating space itself (or at least the immovable anchor points in space.

4. Conclusion

We conclude with a few remarks on how this work fits in with the current efforts to understand how the brain produces cognition. The initial results of our pointing behavior approach to spatial cognition (Gentry, Goodman, Wakefield, and Wright, 1986) used a method (multidimensional scaling) that can produce "maps of the mind" with a reassuring familiarity similar to topographic maps. This approach is still popular and includes attempts to compare the cognitive maps derived from a multidimensional analysis of subjective estimates of geographic distances with two dimensional cartographic versions of the "same" terrain (e.g. Birnbaum, Anderson, and Hynan, 1989). This type of "mind mapping" is consistent with the brain models that localize different memories in different areas and structures. In contrast to the localization model, there is a growing consensus, both in the neurosciences and computer sciences that simulate brain function, that the neural basis for cognition consists of widely distributed and massively parallel architectures, and that memory involves the differential activity or patterns of synaptic events throughout large areas of the brain (Edelman, 1987, 1989; Rosenfield, 1988).

That the cytoarchitecture of the brain includes extensive synaptic interconnections between neurons has been known for a century (Cajal, 1889, 1911), but how that connectivity produces cognition is only now being unraveled. Here, as in so many fields, the computer is playing a central role in the process of discovery and simulation. In addition, the computer simulations of the nervous system are beginning to expand the design of computing strategies, producing the neural network programs that are demonstrating advantages in the solution to some problems. One example, which has relevance to spatial cognition, is the "traveling salesman problem" (TSP), where the task is to find the shortest route between N sites, where each site is visited only once and the path traveled returns to the starting point. Neural network simulators are particularly well suited for solving the TSP (Kirkpatrick *et al.*, 1983; Hopfield and Tank, 1985) and microcomputer programs for demonstrating the neural network solution for the TSP are available (Klimasaukas, 1988; McClelland and Rummelhart, 1988).

The characteristic of neural networks (both organic and silicon based) that is especially relevant to the "fractal" analysis we have applied to spatial cognition is the evidence for non-linearity in these systems (see section 3.4.3). This non-linearity has been demonstrated with artificial neural networks consisting of only a few elements (cells) that exhibit the full range of behaviors characteristic of the bifurcation diagram, including the chaotic regimes (van der Maas *et al.*, 1990).

The same reasoning is being applied to the development of associative memory networks that are explicitly designed to have a fractal structure (Baram, 1989; Boone *et al.*, 1990). We speculate that our pointing behavior results are consistent with the idea that memories (imagination) are stored in fractal neural networks.

Our current operating assumption is that the properties of spatial cognition are determined by the brain's mechanisms that form categories using widely distributed and massively parallel neural networks that precludes any localization of memory or fixed cognitive maps of the world. This cytoarchitecture yields a fractal mapping of all that we know, such that the same neural circuits can encode the spatial information for memories independent of scaling considerations. This hypothesized fractal property of the mind could account for the perceptual constancies of objects both in direct experience and in imagined environments.

5. Acknowledgements

The contributions of the following individuals are gratefully acknowledged: Richard Alter, Thomas Carter, Art Costa, Tim Gable, James Goodwin, Julie Gorman, Sherri Kwiatkowski, Kelly Lautt, Debbie Lillie, Barbara Moose, Phil Moose, and John Sarraille. This research has been supported by grants from the California State University Academic Computing Enhancement Institute, the California State University, Stanislaus Office of the Vice President for Academic Affairs and the College of Arts, Letters, and Sciences.

6. Notes

1 The geographic names used to elicit directional pointing included three local towns (Turlock, Modesto, and Oakdale, California), three U.S. cities (Chicago, Los Angeles, and Washington,D.C.) and three countries (Egypt, Japan, and Vietnam).

2 In the study being described the targets included names such as Snoopy and Marilyn Monroe, and general object categories such as television, carrot, and boat. While observers have no difficulty in giving a location to these targets they reported making-up scenes in order to give the target a context. Post experiment debriefings (which are also recorded on the videotapes) provided examples of these imaginary "locations" which often changed with the context of the other targets used in the same session or trial.

3 Hausdorff or Hausdorff-Besicovitch dimension is sometime used instead of fractal dimension, but for interdisciplinary communications the latter seems preferable. A literature search using the Current Contents data base yielded 482 recent publications for an average of 1.2 articles per day published with "fractal(s)" in the title.

4 The pacing for the trials was one every ten seconds and the observers held the pointing position until the pre-recorded audio tape said "ok". The variance in the "nearly regular intervals" was due to differences in response times and the video frame selected for measurement.

7. References

Abelson, R.P., and Tukey, J.W. (1963) 'Efficient utilization of nonnumerical information in quantitative analysis: General theory and the case of simple order', Annals Mathematical Statistics 34, 1347-1369.

Alcock, J. (1989) Animal behavior: an evolutionary approach, 4th ed., Sunderland, MA: Sinauer Associates.

Arle, J.E. and Simon, R.H. (1990 'An application of fractal dimension to the detection of transients in the electroencephalogram', Electroencephalography and Clinical Neurophysiology 75, 296-305.

Asch, S.E. and Witkin, H.A. (1948) 'Studies in space orientation: II. Perception of the upright with displaced visual fields and with body tilted', Journal of Experimental Psychology 38, 455-477.

Baram, Y. (1989) 'Associative memory in fractal neural networks', IEEE Transactions on Systems, Man, and Cybernetics 19, 1133-1141.

Barnsley, M.F. (1988) Fractals Everywhere, Academic Press:New York

Barnsley, M.F., Massopust, P., Strickland, H., and Sloan, A.D. (1987) 'Fractal modeling of biological structures', Annals of the New York Academy of Sciences 504, 179-94.

Batty, M., Longley, P., and Fotheringham S. (1989) 'Urban growth and form - scaling, fractal geometry, and diffusion-limited aggregation', Environment and Planning A 21, 1447-1472.

Bennett, G. K., Seashore, H. G., and Wesman, A. G. (1984) Differential Aptitude Tests: Technical Supplement, San Antonio, TX: Psychological Corp.

Binet, A., and Simon, T. (1905) 'Methodes nouvelles pour le diagnostic du niveau intellectuel des anormaux', Annee Psychologigue 11, 191-244.

Birnbaum, M.H., Anderson, C.J., and Hynan, L.G. (1989) 'Two operations for "ratios" and "differences" of distances on the mental map', Journal of Experimental Psychology: Human Perception and Performance 15, 785-796.

Blades, M. (in press) 'The reliability of data collected from sketch maps', Journal of Environmental Psychology.

Blades, M. (1991a) 'The development of wayfinding abilities', in D.M. Mark and A.U. Frank (eds.), Cognitive and Linguistic Aspects of Geographic Space, Dordrecht: Kluwer Academic Publishers.

Blades, M. (1991b) 'The development of the abilities required to understand spatial representations', in D.M. Mark and A.U. Frank (eds.), Cognitive and Linguistic Aspects of Geographic Space, Dordrecht: Kluwer Academic Publishers.

Blades, M. and Spencer, C. (1986a) 'Map use by young children', Geography 71, 47-52.

Blades, M. and Spencer, C. (1986b) 'The implications of psychological theory and methodology for cognitive cartography', Cartographica 23, 1-13.

Blades, M. and Spencer, C. (1987a) 'Young children's strategies when using maps with landmarks', Journal of Environmental Psychology 7, 201-217.

Blades, M. and Spencer, C. (1987b) 'The use of maps by 4-6 year-old children in a large scale maze', British Journal of Developmental Psychology 5, 19-24.

Blades, M. and Spencer, C. (1987c) 'How do people use maps to navigatc through the world?', Cartographica 24, 64-75.

Blades, M. and Spencer, C. (1990) 'The development of 3- to 6-year-olds' map using ability: the relative importance of landmarks and map alignment', The Journal of Genetic Psychology 151, 181-194.

Blank, A.A. (1959) 'The Luneberg theory of binocular space perception', in S. Koch (ed.) Psychology: A Study of a Science, Volume 1. Sensory, Perceptual and Physiological Formulations, MacGraw-Hill: New York

Bock, O. and Kommerell, G. (1986) 'Visual localization after strabismus surgery is compatible with the "outflow" theory', Vision Research 26, 1825-1829.

Boone, B.G., Constantikes, K.T., Fry, R.L., Gilbert, A.S. (1990) 'New directions in missile guidance - signal processing based on neural networks and fractal modeling', Johns Hopkins APL Technical Digest 11, 28-36.

Boring, Edwin G. (1950) A History of Experimental Psychology, 2nd Edition, Prentice-Hall: Englewood Cliffs, New Jersey.

Bowers, C.A. and LaBarba, R.C. (1988) 'Sex differences in the lateralization of spatial abilities: a spatial component analysis of extreme group scores', Brain and Cognition 8, 165-177.

Brammer, R.F. (1989) 'Unified image computing based on fractals and chaos model techniques', Optical Engineering 28, 726-734.

Brewster, S. and Blades, M. (1989) 'Which way to go?: Children's ability to give directions in the environment and from maps', Journal of Environmental Education and Information 8, 141-156.

Briggs, J. and Peat, F.D. (1989) Turbulent Mirror: An illustrated guide to Chaos theory and the science of wholeness, Harper and Row: New York.

Cajal, S. Ramon y (1911) Histologie du Systeme Nerveux de l'Homme et des Vertebres, Maloine:Paris. [Note: the 1889 date in the text refers to Boring's (1950) attribution to Cajal for the first description of the synapse].

Caldwell C.B., Stapleton, S.J., Holdsworth, D.W., Jong, R.A., Weiser, W.J., Cooke, G., and Yaffe, M.J. (1990) 'Characterization of mammographic parenchymal pattern by fractal dimension', Physics in Medicine and Biology 35, 235-47.

Calzetti, D., Giavalisco, M. and Ruffini, R. (1989) 'The angular two-point correlation function and the cellular fractal structure of the Universe', Astronomy and Astrophysics 226, 1-8.

Carr, E.G. and Kemp, D.C. (1989) 'Functional equivalence of autistic leading and communicative pointing: analysis and treatment', Journal of Autism and Developmental Disorders 19, 561-578.

Chokor, B.A. (1989) 'The evaluation of environmental quality for planning in the 3rd-world - A multidimensional scaling approach and a case study of Ibadan, Nigeria', Third World Planning Review 11, 189-210.

Clark, W.C., Ferrer-Brechner, T., Janal, M.N., Carroll, J.D., and Yang, J.C. (1989) 'The dimensions of pain: a multidimensional scaling comparison of cancer patients and healthy volunteers', Pain 37, 23-32.

Coren, S and Hoy, V.S. (1986) 'An orientation illusion analog to the rod and frame: relational effects in the magnitude of the distortion', Perception and Psychophysics 39, 159-163.

Coren, S. and Porac, C. (1987) 'Individual differences in visual-geometric illusions: predictions from measures of spatial cognitive abilities', Perception and Psychophysics 41, 211-219.

Coury, B.G. (1987) 'Multidimensional scaling as a method for assessing internal conceptual models of inspection tasks', Ergonomics 30, 959-973.

Cutting, J.E., and Garvin, J.J. (1987) 'Fractal curves and complexity', Perception and Psychophysics 42, 365-70.

Davison, M.L. (1983) Multidimensional scaling, New York: Wiley.

De Leeuw, J. and Stoop, I. (1984) 'Upper bounds for Kruskal's Stress', Psychometrika 49, 391-402.

Dodwell, P.C. (1977) 'Criteria for a neuropsychological theory of perception', Cahiers de Psychologie 20, 175-182.

Dodwell, P.C. (1982) 'Geometrical approaches to visual processing', in D.J. Ingle, M.A. Goodale, and R.J.W. Mansfield (eds.), Analysis of Visual Behavior, The MIT Press, Cambridge, MA.

Duhamel, J-R., Pinek, B., and Brouchon, M. (1986) 'Manual pointing to auditory targets: performances of right versus left handed subjects', Cortex 22, 633-638.

Edelman, G.M. (1979) 'Group selection and phasic reentrant signaling: A theory of higher brain function', in F.O. Schmitt and F.G. Worden (eds.), The Neurosciences Fourth Study Program, Cambridge, MA: MIT Press.

Edelman, G.M. (1985) 'Neural Darwinism: Population thinking and higher brain function', in M. Shafto (ed.), How We Know: Nobel Conference XX, New York: Harper and Row.

Edelman, G.M. (1987) Neural Darwinism, Basic Books: New York.

Edelman, G.M. (1988) Topobiology: An Introduction to Molecular Embryology, New York: Basic Books.

Edelman, G.E. (1989) The Remembered Present, Basic Books: New York

Ellen, P. and Thinus-Blanc, C (1987) 'Cognitive processes and spatial orientation in animal and man', Proceedings of the NATO Advanced Study Institute on Cognitive Processes and Spatial Orientation in Animal and Man (1985: La Baume-les-Aix, France) Dordrecht: Kluwer Academic Publishers.

Eysenck, H. J. (1939) 'Primary mental abilities', British Journal of Educational Psychology 9, 270-279.

Fisher, M.E. (1989) 'Fractal and nonfractal shapes in two-dimensional vesicles', Physica D 38, 112-118.

Fitts, P.M. (1954) 'The information capacity of the human motor system in controlling the amplitude of human movement', Journal of Experimental Psychology 47, 381-391.

Fitts, P.M. and Peterson, J.R. (1964) 'Information capacity of discrete motor responses', Journal of Experimental Psychology 67, 103-112.

Franklin, N. and Tversky, B. (1990) 'Searching imagined environments', Journal of Experimental Psychology: General 119, 63-76.

Galton, F. (1874) English men of science: Their nature and nurture, London: Macmillan.

Garling, T. (1980) 'A comparison of multidimensional scaling with the semantic differential technique as methods for structural analysis of environmental perception and cognition', Umea psychological reports, No.155. Umea, Sweden: University of Umea.

Gentry,T.A., Goodman, N., Wakefield, J.A. Jr., and Wright, R.R. (1986) 'Geometric properties of individual human image spaces', Paper presented to the 10th American Imagery Conference, San Francisco, CA.

Gilbert, L.E. (1989) 'Are topographic data sets fractal?', Pure and Applied Geophysics 131, 241-254.

Gleick, J. (1987) Chaos: Making a New Science, Viking:New York

Goldberger, A.L., Rigney, D.R., and West, B.J. (1990) 'Chaos and fractals in human physiology', Scientific American 262, 43-49.

Golledge, R.G. and Rushton, G. (1972) Multidimensional scaling: review and geographical applications, Washington: Association of American Geographers.

Golledge, R.G. and Timmermans, H. (1990) Applications of behavioural research on spatial problems I: Cognition. Progress in Human Geography 14, 57-99.

Gould, S.J. (1989) Wonderful life: The Burgess shale and the nature of history, New York: Norton.

Goodman, N. (1990) A multidimensional analysis of human spatial cognition, Unpublished Masters Thesis, California State University, Stanislaus.

Graefe, A. von (1854) Beitrage zur Physiologie und Pathologie der schiefen Augenmuskeln, Graefes Archiv Ophthal 1, 1-81.

Green, P.E, Carmone, F.J., Jr., and Smith, S.M (1989) Multidimensional scaling: concepts and applications, North Scituate, MA: Allyn and Bacon.

Guilford, J. P. (1967) The nature of human intelligence, New York: McGraw-Hill.

Gutierrez, E. and Almirall, H. (1989) 'Temporal properties of some biological systems and their fractal attractors', Bulletin of Mathematical Biology 51, 785-800.

Harris, D. B. (1963) Children's drawings as measures of intellectual maturity: A revision and extension of the Goodenough Draw-a-Man Test, San Diego: Harcourt Brace Jovanovich.

Hernandez, D. (1990) 'Relative representation of spatial knowledge: the 2- D case', Technical report FKI-135-90, Institut fur Informatikm, Technische Universitat Munchen.

Hernandez, D. (1991) 'Relative representation of spatial knowledge: the 2- D case', in D.M. Mark and A.U. Frank (eds.), Cognitive and Linguistic Aspects of Geographic Space, Dordrecht: Kluwer Academic Publishers.

Hewes, G.W. (1973) 'Primate communication and the gestural origin of language', Current Anthropology 14, 5-24.

Holmes, G. (1918) 'Disturbances of vision by central lesions', British Journal of Ophthalmology 2, 353-384.

Hopfield, J.J. and Tank, D.W. (1985) '"Neural" computation of decisions in optimization problems', Biological Cybernetics 52, 141-152.

Howard, I.P. and Templeton, W.B. (1966) Human Spatial Orientation, John Wiley Sons: London.

Huang, J. and Turcotte, D.L. (1990) 'Fractal image analysis - application to the topography of Oregon and synthetic images', Journal of the Optical Society of America A-Optics and Image Science 7, 1124-1130.

Iannaccone, P.M. (1990) 'Fractal geometry in mosaic organs: a new interpretation of mosaic pattern', Faseb Journal 4, 1508-12.

Jaynes, J. (1976) The Origin of Consciousness in the Breakdown of the Bicameral Mind, Houghton Mifflin Co., Boston.

Johnson, M. (1987) The Body in the Mind: The Bodily Basis of Meaning, Imagination, and Reason, Chicago: University of Chicago Press.

Jones, T. (1989) 'Psychology of computer use: XVI. Effect of computer-pointing devices on children's processing rate', Perceptual and Motor Skills 69, 1259-1263.

Kaufman, A. S. (1979) Intelligent testing with the WISC-R, New York: Wiley.

Kirk, S. A., and Kirk, W. D. (1971) Psycholinguistic learning disabilities, Urbana: University of Illinois Press.

Kirkpatrick, S., Gelatt, C.D., and Vecchi, M.P. (1983) 'Optimization by simulated annealing', Science 220, 671-680.

Klimasaukas, C.C. (1988) NeuralWorks: An Introduction to Neural Computing, version 2.0. NeuralWare: Pittsburgh PA.

Klingen, H.-J. and Roth, P. (1989) 'Size analysis and fractal dimension of diesel particles based on REM measurements with an automatic imaging system', Journal of Aerosol Science 20, 861-864.

Knill, D.C., Field, D., and Kersten, D. (1990) 'Human discrimination of fractal images', Journal of the Optical Society of America. Part A, Optics and Image Science 7, 1113-1123.

Koppitz, E. M. (1975) The Bender Gestalt Test for young children: Volume II: Research and application, 1963-1973, Orlando, FL: Grune and Stratton.

Korvin, G. (1989) 'Fractured but not fractal: Fragmentation of the Gulf of Suez basement', Pure and Applied Geophysics 131, 289-305.

Kruskal, J.B. (1964a) 'Multidimensional scaling by optimizing goodness of fit to a nonmetric hypothesis', Psychometrika 29(1), 1-27.

Kruskal, J.B. (1964b) 'Nonmetric multidimensional scaling: a numerical method', Psychometrika 29(2), 115-129.

Kruskal, J.B. and Wish, M. (1978) Multidimensional scaling, Beverly Hills: Sage Publications.

Kruskal, J.B., Young, F.W., and Seery, J.B. (1977) How to use KYST-2A: A very flexible program to do multidimensional scaling and unfolding, Bell Laboratories: Murray Hill, NJ.

Lakoff, G. (1987) Women, Fire, and Dangerous Things: What Categories Reveal About the Mind, Chicago: University of Chicago Press.

Langolf, G.D., Chaffin, D.B., and Foulke, J.A. (1976) 'An investigation of Fitts' law using a wide range of movement amplitudes', Journal of Motor Behavior 8, 113-128.

Liebovitch, L.S. and Sullivan, J.M. (1987) 'Fractal analysis of a voltage-dependent potassium channel from cultured mouse hippocampal neurons', Biophysical Journal 52, 979-988.

Liebovitch, L.S. and Toth, T.I. (1990) 'Using fractals to understand the opening and closing of ion channels', Annals of Biomedical Engineering 18, 177-194.

Loehlin, J. C., Willerman, L., and Horn J. M. (1988) 'Human behavior genetics', Annual Review of Psychology 39, 101-133.

Luneberg, G. (1947) Mathematical Analysis of Binocular Vision, Princeton, N.J.: Princeton University Press.

Lynn, R. (1987) 'The intelligence of mongoloids: a psychometric, evolutionary, and neurological theory', Journal of Personality and Individual Differences 8, 813-844.

Ma, Y-C. A. (1987) A perceptual map analysis for ten Pacific Rim countries for vacationing purposes by the use of multi-dimensional scaling technique, Ph.D. Dissertation: University of California, Riverside.

Mach, E. (1886) Die Analyse der Empfindungen und das Verhaltnis des Psychischen zun Physischen, [The Analysis of Sensations and the Relation of the Physical to the Psychical, 5th Edition, 1959, Dover: New York].

Mainster, M.A. (1990) 'The fractal properties of retinal vessels: embryological and clinical implications', Eye 4, 5-41.

Mandelbrot, B. B. (1967) 'How long is the coast of Britain?: Statistical self-similarity and fractional dimension', Science 155, 636-638.

Mandelbrot, B. B. (1977) Fractals: Form, Chance, and Dimension, W.H. Freeman and Co., San Francisco.

Mandelbrot, B. B. (1983) The Fractal Geometry of Nature, W.H. Freeman and Co., New York.

Mann, R. and Wasilewski, M.C. (1990) 'Towards a fractal computer graphic basis for characterization of catalyst pore structure by image reconstruction', Chemical Engineering Research and Design 68, 177-184.

Mark, D.M. and Frank, A.U. (1990) 'Experiential and formal models of geographic space', NCGIA Technical Paper 90-10, Santa Barbara, CA: National Center for Geographic Information and Analysis.

Martin, N.G., and Eaves, L.J. (1977) 'The genetical analysis of covariance structure', Heredity 38, 79-95.

Martin, N.G., Jardine, R., Eaves, L.J. (1984) 'Is there only one set of genes for different abilities?: A reanalysis of the National Merit Scholarship Qualifying Test (NMSQT) data', Behavior Genetics 14, 355-370.

Martinez, V.J. and Jones, B.J.T. (1990) 'Why the Universe is not a fractal', Monthly Notices of The Royal Astronomical Society 242, 517-521.

Matarazzo, J.D., and Herman, D.O. (1985) 'Clinical uses of the WAIS-R: Base rates of differences between VIQ and PIQ in the WAIS-R standardization sample', in B.B. Wolman (ed.), Handbook of intelligence: Theories, measurements, and applications, New York: Wiley.

McClelland, J.L. and Rummelhart, D.E. (1988) Explorations in Parallel Distributed Processing, MIT Press: Cambridge, Mass.

McFarland, D. (1985) Animal behavior: psychobiology, ethology, and evolution, Menlo Park, Benjamin/Cummings Pub. Co.

Miller, L.K. and Santoni, V. (1986) 'Sex differences in spatial abilities: Strategic and experiential correlates', Acta Psychologia 62, 225-235.

Moon, F.C. (1987) Chaotic Vibrations, Wiley:New York

Morigiwa, K., Tauchi, M., and Fukuda, Y. (1989) 'Fractal analysis of ganglion cell dendritic branching patterns of the rat and cat retinae', Neuroscience Research Supplement 10, S131-139.

Neisser, U. (1976) Cognition and Reality: Principles and Implications of Cognitive Psychology, San Francisco: W. H. Freeman and Co.

Nelson, J.A., Crookes, R.J., and Simons, S. (1990) 'On obtaining the fractal dimension of a 3D-cluster from its projection on a plane - application to smoke agglomerates', Journal of Physics D-Applied Physics 14, 465-468.

Nelson, T.R., West, B.J. and Goldberger, A.L. (1990) 'The fractal lung - universal and species-related scaling patterns', Experientia 15, 251- 254.

Nelson, T.R. and Jaggard, D.L. (1990) 'Fractals in the imaging sciences - Introduction', Journal of the Optical Society of America A - Optics and Image Science 7, 1052-1052.

Noorden, K.G. von (1971) 'Past-pointing in paralytic strabismus', American Journal of Ophthalmology 71, 27-33.

Obert, M., Pfeifer, P., and Sernetz, M. (1990) 'Microbial growth patterns described by fractal geometry', Journal of Bacteriology 172, 1180-1185.

Olton, D.S. (1978) 'Characteristics of spatial memory', in S.H. Hulse, H. Fowler, and W.K. Honig (eds.), Cognitive processes in animal behavior, Hillsdale, NJ: Lawrence Erlbaum.

Olton, D.S. (1982) 'Spatially organized behaviors of animals: Behavioral and neurological studies', in M. Potegal (ed.), Spatial abilities: development and physiological foundations, New York: Academic Press.

Ott, D., Eckmiller, R., and Bock, O. (1987) 'A head-mounted device for measurement of pointing to visual targets without seeing the pointing arm', Vision Research 27, 307-309.

Peterson, I (1989) The Mathematical Tourist, W.H. Freeman: New York

Pegels, C.C. and Sekar, C. (1989) 'Determining strategic groups using multidimensional scaling', Interfaces 19, 47-57.

Piech, M.A., and Piech, K.R. (1990) 'Fingerprints and fractal terrain', Mathematical Geology 22, 457-485.

Plomin, R. (1986) Development, genetics, and psychology, Hillsdale, NJ: Lawrence Erlbaum.

Plomin, R., DeFries, J. C., and McClearn, G. E. (1990) Behavioral genetics, New York: Freeman.

Poincare, H., (1900) La valeur de la science. Paris: e. Flammarion [The value of science: Authorized translation with an introd. by G. B. Halsted (1912), with a special prefatory essay, New York: Dover Publications, 1958].

Polzella, D.J. and Reid, G.B. (1989) 'Multidimensional scaling analysis of simulated air combat maneuvering performance data', Aviation Space and Environmental Medicine 60, 141-144.

Potegal, M. (1982) Spatial abilities: development and physiological foundations, New York: Academic Press.

Presson, C.C., DeLange, N., and Hazelrigg, M.D. (1989) 'Orientation specificity in spatial memory: What makes a path different from a map of the path?', Journal of Experimental Psychology: Learning, Memory, and Cognition 15, 887-897.

Price, D. de S. (1975) Science since Babylon, enlarged edition, New Haven: Yale University Press.

Raven, J. C., Court, J. H., and Raven, J. (1985) A manual for Raven's Progressive Matrices and vocabulary scales, London: H. K. Lewis.

Riddoch, G. (1917) 'Dissociation of visual perceptions due to occipital injuries, with special reference to appreciation of movement', Brain 40, 15-57.

Rosch, E. (1973) 'On the internal structure of perceptual and semantic categories', in T.E. Moore (ed.), Cognitive Development and the Acquisition of Language, New York: Academic Press.

Rosch, E (1978) 'Principles of categorization', in E. Rosch and B.B Lloyd (eds.), Cognition and Categorization, Hillsdale, NJ: Erlbaum.

Rosenfield, I. (1988) The Invention of Memory: A New View of the Brain, Basic Books, New York, New York.

Schone, H. (1984) Spatial orientation: the spatial control of behavior in animals and man (translated by Camilla Strausfeld), Princeton, NJ: Princeton University Press.

Seilacher, A. (1967) 'Fossil behavior', Scientific American 217, 72-80.

Sholl, M.J. (1987) 'Cognitive maps as orienting schemata', Journal of Experimental Psychology: Learning, Memory, and Cognition 13, 615-628.

Shepard, R. N. (1966) 'Metric structures in ordinal data', Journal of Mathematical Psychology 3, 287-315.

Shepard, R. N. (1974) 'Representation of structure in similarity data: problems and prospects', Psychometrika 39, 373-421.

Shepard, R. N. and Metzler, J. (1971) 'Mental rotation of three-dimensional objects', Science 171, 701-703.

Skarda, C.A. and Freeman, W.J. (1987) 'How brains make chaos in order to make sense of the world', Behavioral and Brain Sciences 10, 161-195

Spearman, C. (1904) '"General intelligence" objectively determined and measured', American Journal of Psychology 15, 201-293.

Stiles-Davis, J., Kritchevsky, M. and Bellugi J. (1988) Spatial cognition: brain bases and development, Hillsdale, NJ: Erlbaum Associates.

Spencer, C.J S., Blades, M., and Morsley, K. (1989) The child in the physical environment: the development of spatial knowledge and cognition, Chichester: Wiley.

Takahashi, M. (1989) 'A fractal model of chromosomes and chromosomal DNA replication', Journal of Theoretical Biology 141, 117-136.

Teich, M.C. (1989) 'Fractal character of the auditory neural spike train', IEEE Transactions on Biomedical Engineering 36, 150-160.

Thomas, J.A. and Stock, W.A. (1988) 'The concept of happiness: a multidimensional scaling investigation', International Journal of Aging and Human Development 27, 141-154.

Thomson, G. H. (1941) The factorial analysis of human ability, London: University of London Press.

Thorndike, R. L., Hagen, E. P., and Sattler, J. M. (1986) The Stanford-Binet Intelligence Scale: Fourth edition: Technical manual, Chicago: Riverside Publishing Co.

Thurstone, L. L., and Thurstone, T. G. (1941) Factorial studies of intelligence, Chicago: University of Chicago Press.

Turcotte, D.L. (1989) 'Fractals in geology and geophysics', Pure and Applied Geophysics 131, 171-196.

van der Mass, H.L.J, Verschure, P.F.M.J., and Molenaar, P.C.M. (1990) 'A note on chaotic behavior in simple neural networks', Neural Networks 3, 119- 122.

Vernon, P. E. (1961) The structure of human abilities (2nd ed.), London: Methuen.

Vernon, P. E. (1979) Intelligence: Heredity and environment, San Francisco: Freeman.

Voss, R.F. (1989) 'Random fractals - self-affinity in noise, music, mountains, and clouds', Physica D 38, 362-371.

Wakefield, J. A., Jr., and Goad, N. A. (1982) Psychological differences: Causes, consequences, and uses in education and guidance, San Diego: EdITS.

Wang, X.W., Dong, L., and Xiong, L.Y. (1990) 'The change of fractal dimensionality in the recovery and recrystallisation process', Journal of Physics-Condensed Matter 2, 3879-3884.

Watson, A. (1978) 'A Riemann geometric explanation of the visual illusions and figural after effects', in E. Leeuwenberg and H. Buffart (eds.), Formal Theories of Visual Perception, Wiley: New York.

Wechsler, D. (1974) Manual for the Wechsler Intelligence Scale for Children—Revised, New York: Psychological Corp.

Wechsler, D. (1981) WAIS-R manual: Wechsler Adult Intelligence Scale—Revised, San Antonio, TX: Psychological Corp.

Wehner, R. and Menzel, R. (1990) 'Do insects have cognitive maps?', Annual Review of Neuroscience 13, 403-14.

Wen, Z., Deng, Z.-G., Liu, Y.-Z., and Xia, X.-Y. (1989) 'The fractal dimension in the large-scale distribution of galaxies with different luminosities', Astronomy and Astrophysics 219, 1-6.

West, B.J. (1990) 'Physiology in fractal dimensions - error tolerance', Annals of Biomedical Engineering 18, 135-149.

Xu, N. and Xu, J.H. (1988) 'The fractal dimension of EEG as a physical measure of conscious human brain activities', Bulletin of Mathematical Biology 50(5):559-65.

Young, F.W. (1987) 'Multidimensional scaling: history, theory, and applications', R.M. Hamer (ed.), Hillsdale, NJ: Erlbaum Associates.

Zimbardo, G., Veltri, P., and Mangeney, A. (1990) 'Photon propagation in the solar corona - ergodic and fractal regimes', Astronomy and Astrophysics 232, 215-224.

PATH FINDING IN FREE SPACE USING SINUSOIDAL TRANSFORMS: III

IAN PRATT
Department of Computer Science
University of Manchester
Manchester M13 9PL,
United Kingdom

ABSTRACT. This paper investigates the application of a system of spatial representation, called the sinusoidal representation scheme, to the problem of finding a path between two points in a plane strewn with a finite collection of convex obstacles. The main thrust of the argument is that the sinusoidal representation scheme permits the ready extraction of a restricted set of candidates for the shortest path between two given points. The structure of the paper is as follows. First, we define a restricted set of paths between a given pair of points which is guaranteed to include all optimal paths. Secondly, we present in detail the sinusoidal representation scheme. Finally, we show how the aforementioned set of paths can be read off from the sinusoidal representation of the scene in question.

1. Introduction

Consider the problem of finding the shortest path between two points in a space strewn with obstacles. It often happens, when people attempt this task, that they can select a few likely candidates with great ease, but cannot confidently decide which of these candidates is really the shortest. Thus, given the arrangement depicted in fig. 1, three paths at once stand out as sensible routes between points X and Y; but it is less easy to decide, simply by looking, which of these three is the shortest.

This paper presents a method for finding paths between points in a two-dimensional space strewn with convex obstacles. Our method, though it does not aspire to simulate human performance, shares with human geometrical intuition an ability to select a few likely candidates rather than to compute a single optimum path. The structure of the paper is as follows. First, we define a restricted set of candidate paths between a given pair of points which is guaranteed to include all optimal paths. Secondly, we present a technique for representing convex regions of the plane using periodic signals—the *sinusoidal representation scheme*. Finally, we show how the aforementioned set of paths can be read off from the sinusoidal representation of the scene in question. Thus, the paper draws together two quite separate ideas: first, an idea about how to identify a set of plausible-looking candidates for the shortest path between two points; second, an idea about how to represent shapes in the plane. We show how the proposed system of spatial representation facilitates the identification of a collection of sensible routes between two points.

D. M. Mark and A. U. Frank (eds.), Cognitive and Linguistic Aspects of Geographic Space, 219–233.

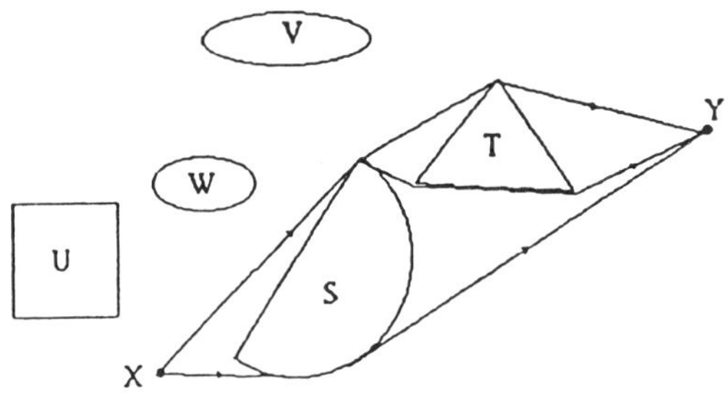

Figure 1: A path finding problem

2. The Path Finding Problem

Our task in this section will be to characterise a set of paths between two points in arrangements such as that of fig. 1 which is guaranteed to contain all optimal paths, yet which does not contain too many blatantly sub-optimal paths. We begin with a basic observation about optimal paths.

Observation 1 *If p is an optimal path from point X to point Y around a collection Ω of obstacles, then p consists of a sequence of line segments interspersed with segments (possibly single points) of the boundaries of the obstacles in Ω. Each of the aforementioned line segments is tangent at each endpoint either to some obstacle in Ω or to the point-regions X and Y.*[1]

In fig. 1, for example, the bottom path from X to Y starts with a line segment tangent to the point-region X and to the obstacle S, then has a segment of S's boundary, and finishes with a line segment tangent to S and to Y. In the same figure, the middle path starts with a line segment tangent to both X and S, visits S at a single point, has a line segment tangent to S and to T, proceeds along a segment of T's boundary, and finishes with a line segment tangent to T and to Y. A good way to think about these paths is to imagine a piece of elastic stretched taut between X and Y and around the relevant obstacles in Ω. We call a line segment which is tangent at its endpoints to a pair of regions, a *joint tangent* to those regions.

Another notion useful for path finding is that of the *orientation* of a directed joint tangent about a region. In a path p from X to Y, the joint tangents which link one obstacle to the next in the path all have a direction: that is, they run *from* one region *to* another. Now any directed line which is tangent to a convex region has a well-defined orientation—clockwise or anticlockwise—about that region (assuming that the region has a non-empty interior), as shown in fig. 2. Thus, any directed joint tangent will have a well-defined orientation about both of the obstacles it visits (excepting the point-regions X and Y). For example, in the bottom path in fig. 1, the directed joint tangents to and from S both have anticlockwise orientation about S; in the middle path, by contrast, they both have clockwise orientation

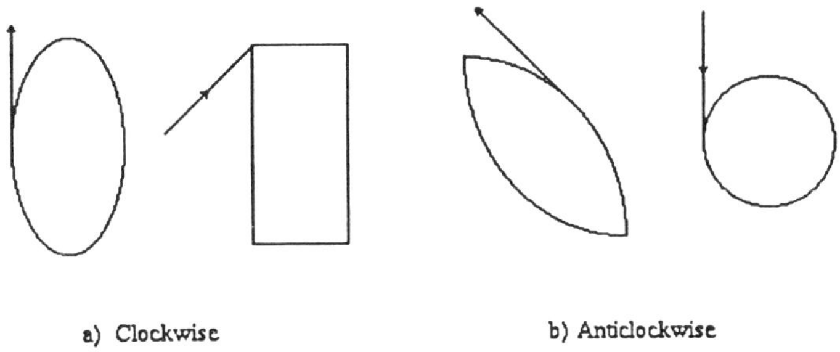

Figure 2: Clockwise and anticlockwise orientation

about S. Henceforth, all obstacles in Ω will be assumed to have non-empty interiors, so that the orientation of joint tangents involving those obstacles is well-defined.

The notion of orientation is important because of the following observation.

Observation 2 *Let p be a path which visits obstacles $X = O_1, \ldots, O_n = Y$, so that, for $i = 2, \ldots, n-1$, p contains a directed joint tangent from O_{i-1} to O_i and another directed joint tangent from O_i to O_{i+1}. Then, for p to be optimal, those two directed tangents must have the* same *orientation about O_i.*

Observations 1 and 2 limit the paths under consideration to a relatively small collection. Nevertheless, they fail to eliminate many clearly suboptimal paths. It is clear, for example, that a path from X to Y in fig. 1 which skirts the left of U and proceeds around the top of V cannot be optimal, even if it does obey the conditions set by observations 1 and 2. The notions developed in the remainder of this section help to deal with such cases. Specifically, we develop a sufficient, but not necessary, criterion for detecting sub-optimal paths between two points in a plane strewn with convex obstacles. This criterion, it is important to notice, does not rely on the numerical determination of path lengths.

The idea behind the criterion is illustrated by fig. 3. Fig. 3a shows two paths, p and p', between points X and Y; the obstacles these paths pass around are not shown.[2] If we imagine elastic bands stretched taut along these paths, and held in place by the obstacles they pass around, then it is clear that removing all obstacles in the region enclosed by p and p' will allow band p to snap back to the line of band p', but not *vice versa*. It follows that path p must be longer than path p'. Similar remarks apply to fig. 3b. The sub-optimality criterion developed below is a generalisation of this judgement.

In seeing thus that p is longer than p' in figs. 3a and 3b, we are relying on two important notions. The first is that of a path's being *clockwise definite* or *anticlockwise definite*. The second is that of one path's being on the *clockwise side* or on the *anticlockwise side* of another.

Although fig. 3a does not show the obstacles which p' passes around, it is clear that p' must visit them all with clockwise orientation. This is because p' is always veering to the right, and so must be held in place by obstacles lying only on its right hand side. Similarly, in fig. 3b, p' must visit all its obstacles with anticlockwise orientation, since it is always veering to the left. This observation prompts the following definition:

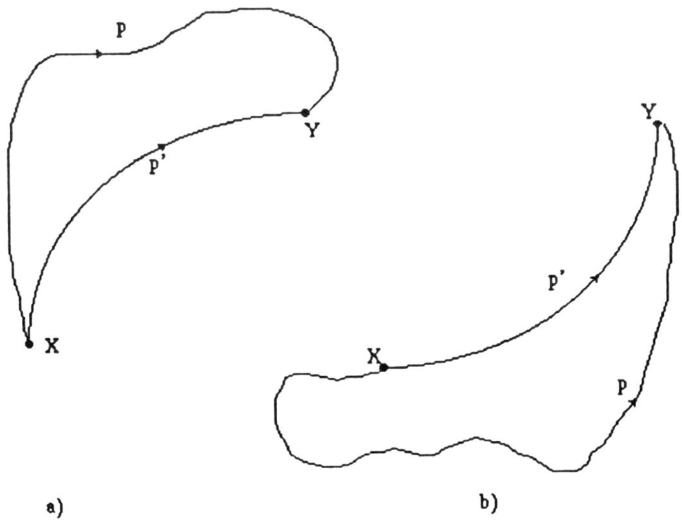

Figure 3: Obviously sub-optimal paths

Definition 1 *Let p be a path from point X to point Y. We say that p is* clockwise definite *if, for any obstacle O encountered by p strictly between X and Y, p passes clockwise around O. We say that p is* anticlockwise definite *if, for any obstacle O encountered by p strictly between X and Y, p passes anticlockwise around O.*

We now turn to the notion of one path's being on the *clockwise side* or on the *anticlockwise side* of another. Consider the loop p^* formed by first following p from X to Y and then p' (in the reverse direction) from Y back to X. In fig. 3a, p^* winds clockwise; in fig. 3b, p^* winds anticlockwise. Accordingly, we define:

Definition 2 *Suppose p and p′ are two paths from point X to point Y which cross over neither themselves nor each other, and which have only their endpoints in common. Let p^* be the path formed by first following p, and then following p′ in the reverse direction, so that p^* is a simple circuit. If p^* winds clockwise, then we say that p is* on the clockwise side *of p′. Otherwise, if p^* winds anticlockwise, then we say that p is* on the anticlockwise side *of p′.*

Notice that, provided p and p' do not cross over themselves or each other, and have only their endpoints in common, p^* is always defined and winds either clockwise or anticlockwise. Thus, either p lies on the clockwise side of p' (and so p' on the anticlockwise side of p), or *vice versa.* Fig. 3 shows cases where p lies on a) the clockwise, and b) the anticlockwise side of p'.

We can now state the general criterion of suboptimality which we use in the case of fig. 3.

Observation 3 *Let p and p′ be paths from point X to point Y which cross neither themselves nor each other, and which have only their endpoints in common. If p′ is* clockwise definite *and p* lies on the clockwise side *of p′, then p is longer than p′ (and is hence sub-optimal). Likewise, if p′ is* anticlockwise definite *and p* lies on the anticlockwise side *of p′, then p is longer than p′ (and is hence sub-optimal).*

3. Representing Convex Regions using Sinusoidal Transforms

In the sequel, we show how the sets of paths satisfying the above criterion can be easily generated using the *sinusoidal representation scheme.* This section sets forth the fundamentals of that scheme; the next discusses its application to path finding.

We start with the observation that there is a natural one-to-one correspondence between the set of sinusoidal oscillations with some fixed frequency—say, $\omega = 1$—and the set of vectors in two-dimensional space. Given a vector $\vec{v}$ whose polar co-ordinates are (a, ϕ), we can associate with $\vec{v}$ the sine wave

$$w_{\vec{v}}(t) = a\sin(t + \phi)$$

of frequency 1, amplitude a and phase-lead ϕ. In geometrical terms, $w_{\vec{v}}(t)$ plots the height above the origin of the endpoint of $\vec{v}$ as $\vec{v}$ is rotated anticlockwise about the origin with an angular velocity of 1 radian per time unit. We shall say that $w_{\vec{v}}(t)$ is the sine wave *corresponding to* the vector $\vec{v}$.

The key feature of this correspondence is that, given two vectors, the *superposition* of their corresponding sinusoidal waves yields a third sinusoidal wave, also of frequency 1, corresponding to the vector sum of the original two vectors.

Again, more formally:

Theorem 1 *Let $\vec{v}_1$, $\vec{v}_2$ be vectors having polar co-ordinates (a_1, ϕ_1), (a_2, ϕ_2) respectively, and let $w_{\vec{v}_1}(t) = a_1\sin(t+\phi_1)$, $w_{\vec{v}_2}(t) = a_2\sin(t+\phi_2)$ be the corresponding sinusoidal waves of frequency 1. If $w(t)$ is the pointwise sum $w_{\vec{v}_1}(t) + w_{\vec{v}_2}(t)$, and if $\vec{v}$ is the vector sum $\vec{v}_1 + \vec{v}_2$, then $w(t)$ is also a sinusoidal wave of frequency 1, $w(t) = w_{\vec{v}}(t) = a\sin(t + \phi)$, where (a, ϕ) are the polar coordinates of the vector $\vec{v}$.*

Now this method of representing vectors—or, equivalently, points—can be extended to arbitrary convex regions of the plane. Consider a convex region R such as the triangle, quadrilateral or circle in figs. 4a, 4b and 4c respectively. Let us imagine R being rotated anticlockwise about the origin at a constant angular velocity of 1 radian per time unit. If, for every time t, we plot the maximum and the minimum heights of all the points in R as R is rotated, we obtain a pair of periodic signals $\langle a(t), b(t)\rangle$ as shown for the three cases in fig. 4.

More formally:

Definition 3 *If R is a convex set of vectors, define the* sinusoidal hull of *R to be the pair of functions: $\Sigma_R(t) = \langle a_R(t), b_R(t)\rangle$ defined by:*

$$a_R(t) = \sup_{\vec{v}\in R} w_{\vec{v}}(t) \qquad b_R(t) = \inf_{\vec{v}\in R} w_{\vec{v}}(t).$$

If R is a region consisting of a single point, $R = \{\vec{v}\}$, then $a_R(t) = b_R(t) = w_{\vec{v}}(t)$.

It turns out that this mapping from convex regions of the plane to pairs of periodic signals is: (i) always defined, (ii) one–one, but (iii) not onto. Thus, certain pairs of periodic signals can be used to represent convex shapes in the plane.[3] If convex regions—including point-regions—are represented by their corresponding sinusoidal hulls, we can test for the membership of a point in a region using the following result.

Definition 4 *If $\Sigma(t) = \langle a(t), b(t)\rangle$, is a sinusoidal hull (period 2π), and if $\sigma(t)$ is a periodic function (period 2π), we write $\sigma(t) \preceq \Sigma$ (read σ* is contained in *Σ) iff $\forall t,\ a(t) \geq \sigma(t)$.*

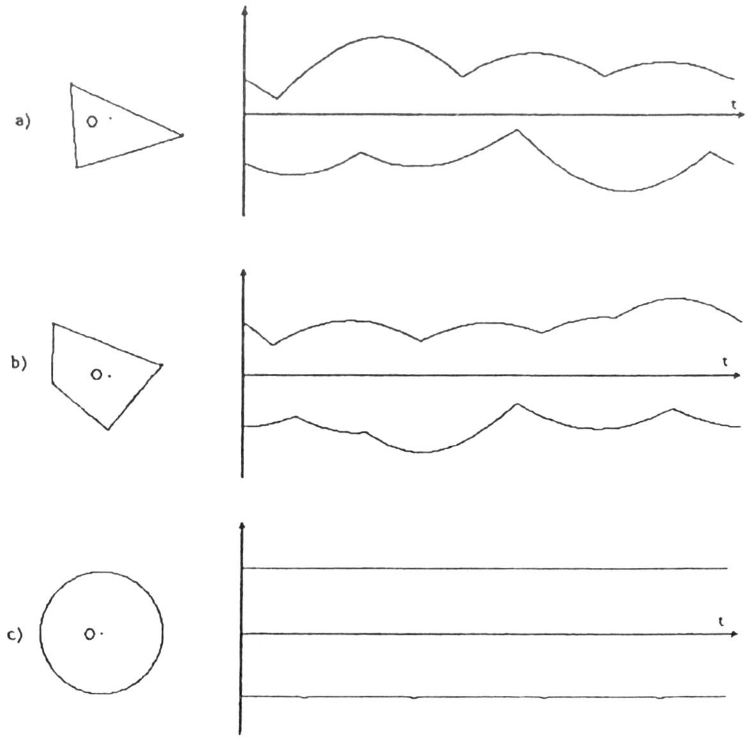

Figure 4: Some convex regions and their sinusoidal representations

Theorem 2 *If $\vec{v}$ is any point in the plane and R a convex region, then*

$$\vec{v} \in R \text{ iff } w_{\vec{v}}(t) \preceq \Sigma_R$$

To test whether a point is in a region, all we need is a device which tests whether the signal representing the point has a smaller magnitude than the a-signal representing the shape, over a time period of duration 2π units.

The sinusoidal representational scheme has a number of useful properties. In particular, suppose we want to translate a convex set S by a vector $\vec{v}$, where S is represented by the signals $\langle a(t),\ b(t)\rangle$ and $\vec{v}$ by the signal $w_{\vec{v}}(t)$. Since $a(t)$ and $b(t)$ are the supremum and infimum of all the waves $w_{\vec{u}}$ for $\vec{u} \in S$, respectively, then theorem 1 ensures that the translated convex set will be represented by $\langle a(t) + w_{\vec{v}}(t),\ b(t) + w_{\vec{v}}(t)\rangle$. Hence, the translation of a convex regions S by a vector $\vec{v}$ can easily be effected by means of piping the signals representing S and the signal representing $\vec{v}$ to appropriately arranged units that simply superpose incoming signals, so yielding signals representing the translated shape. Similarly, the rotation anticlockwise through an angle θ of a convex region about the origin can be effected by delaying the signals $a(t)$ and $b(t)$ by an amount of time $(2\pi - \theta)$. Formally:

Theorem 3 *Let S be a region represented by signals $\langle a_S(t), b_S(t)\rangle$, $\vec{v}$ be a vector represented by the sine wave $w_{\vec{v}}$ and θ be a real number. Let $S_{\vec{v}}$ denote the region obtained by translating S by $\vec{v}$, and S_θ the region obtained by rotating S anticlockwise about the origin through θ radians. Then $S_{\vec{v}}$ is represented by the signals $\langle a_{S_{\vec{v}}}(t), b_{S_{\vec{v}}}(t)\rangle$, where:*

$$a_{S_{\vec{v}}}(t) = a_S(t) + w_{\vec{v}}(t) \qquad b_{S_{\vec{v}}}(t) = b_S(t) + w_{\vec{v}}(t).$$

And S_θ is represented by the signals $\langle a_{S_\theta}(t), b_{S_\theta}(t)\rangle$, where:

$$a_{S_\theta}(t) = a_S(t+\theta) \qquad b_{S_\theta}(t) = b_S(t+\theta)$$

Another useful property of the sinusoidal system is the ability to determine whether two regions have non-empty intersection or are contained one within the other.

Definition 5 *Let $\Sigma_1 = \langle a_1, b_1\rangle$, $\Sigma_2 = \langle a_2, b_2\rangle$ be sinusoidal hulls, with period 2π. Then we define:*

1. *$\Sigma_1 \bowtie \Sigma_2$ (read: Σ_1 and Σ_2* intersect*) if $\forall t\ a_1(t) \geq b_2(t)$.*
2. *$\Sigma_1 \sqsubseteq \Sigma_2$ (read: Σ_1 is* contained in *Σ_2) if $\forall t\ a_2(t) \geq a_1(t) \geq b_1(t) \geq b_2(t)$.*

If the Σ_1 and Σ_2 are sinusoidal hulls, then the above notions of intersection and inclusion correspond to the familiar notions of intersection and inclusion for the regions they represent. Formally:

Theorem 4 *If Σ_S and Σ_T are sinusoidal hulls corresponding to convex regions S and T, respectively, then*

1. *$\Sigma_S \bowtie \Sigma_T$ iff $S \cap T \neq \emptyset$.*
2. *$\Sigma_S \sqsubseteq \Sigma_T$ iff $S \subseteq T$.*

The sinusoidal representation scheme facilitates a number of geometrical constructions, transformations and tests, some of which are described in (Pratt, 1988a; Pratt, 1988c). Here, we concentrate on the problem of path finding using sinusoidal representations.

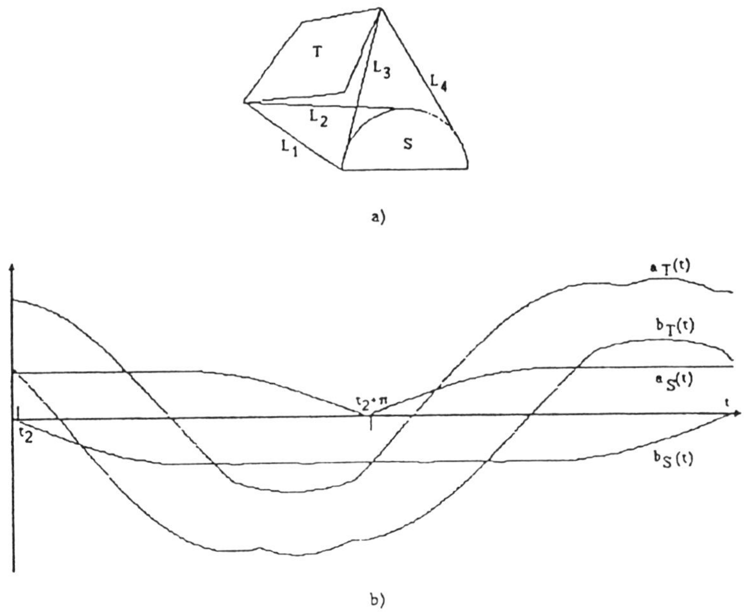

Figure 5: The four joint tangents involving two obstacles

4. Path Finding using Sinusoidal Transforms

As we have already observed, any optimal path between X and Y in a plane strewn with convex obstacles consists of a sequence of joint tangents interspersed with segments of the obstacles' boundaries. Furthermore, in any optimal path, successive joint tangents have the same orientation about their common obstacle. It transpires that joint tangents have a particularly simple representation on the sinusoidal scheme, as we shall now see.

Fig. 5a shows two convex regions, S and T, together with the four joint tangents between them. Consider the joint tangent L_2. If we imagine S, T and L_2 rotating together anticlockwise about the origin, we see that L_2 will be horizontal at times t_2 and $t_2 + \pi$, say. We also see that, at time t_2, the b-signal of T equals the a-signal of S, and at time $t_2 + \pi$, the b-signal of S equals the a-signal of T. Moreover, at t_2, the a-signal of S becomes greater than the b-signal of T, whilst at time $t_2 + \pi$, the a-signal of T becomes greater than the b-signal of S. Let us say that, in the former case, the a-signal of S *eclipses* the b-signal of T, whilst in the latter, the a-signal of T *eclipses* the b-signal of S. Thus, the joint tangent L_2 corresponds to a pair of signal-crossings involving signals belonging to S and T: one in which a signal belonging to S eclipses a signal belonging to T, the other in which a signal belonging to T eclipses a signal belonging to S. As for L_2, so for the other joint tangents. Each joint tangent corresponds to a pair of signal-crossings involving the signals representing S and T: one in which a signal belonging to S eclipses a signal belonging to T, the other in which a signal belonging to T eclipses a signal belonging to S.

Of course, for the purposes of route planning, we will need to specify the *direction of travel* along a joint tangent between two regions: that is, we will need to specify which of the two regions in question our route visits first. The above observations make it possible to adopt the following convention. We can take a signal-crossing at which a signal belonging to T eclipses a signal belonging to S to represent a joint tangent directed *from* S *to* T, and a signal-crossing at which a signal belonging to S eclipses a signal belonging to T to represent a joint tangent directed *from* T *to* S. Under this convention, we have a one-to-one correspondence between the eight *directed* joint tangents between S and T and the eight signal-crossings involving Σ_S and Σ_T, with the direction of travel always being taken to run from the obstacle represented by the eclipsed signal to that represented by the eclipsing signal.

Suppose, then, we have a signal-crossing involving signals belonging to regions S and T, representing a directed joint tangent between these two regions. The question arises as to whether we can determine its orientation about these regions merely by looking at the signals in question. The answer is straightforward. Consider the two directed tangents along L_2 of fig. 5a: one running from T to S, the other from S to T. Notice that the directed joint tangent from T to S has *clockwise* orientation about S and *anticlockwise* orientation about T, whilst the directed joint tangent from S to T has *clockwise* orientation about T and *anticlockwise* orientation about S. Notice also that the directed joint tangent from T to S corresponds, under the convention adopted above, to a signal-crossing involving the a-signal of S and the b-signal of T, whilst the directed joint tangent from S to T corresponds to a signal-crossing involving the b-signal of S and the a-signal of T. The observation generalises: a directed joint tangent has *clockwise* orientation about a region if its corresponding signal-crossing involves the a-signal belonging to that region, and *anticlockwise* orientation if its corresponding signal-crossing involves the b-signal belonging to that region.

Let us summarise what we have established so far. An optimal path p in free space between X and Y, and avoiding a collection of obstacles Ω, must consist of a sequence of joint tangents between the regions in $\{X, Y\} \cup \Omega$ interspersed with segments (possibly single points) of those regions' boundaries. Moreover, successive joint tangents along such a path must have the same orientation about their common region. Translating these facts into the language of sinusoidal signals, an optimal path can be thought of as a journey in transform space along some subset of the signals representing the regions in $\{X, Y\} \cup \Omega$. Each crossing from one signal to the next (which eclipses it) represents a joint tangent between the corresponding regions; each portion of the journey along a signal represents a portion (possibly a single point) of the boundary of the corresponding obstacle.

Therefore, we may represent paths as sequences of signals

$$\Pi = [\sigma_1, t_1, \sigma_2, t_2, \ldots, t_{n-1}, \sigma_n]$$

where $\sigma_1(t)$ is the sine wave representing X, $\sigma_n(t)$ the sine wave representing Y, $\sigma_i(t)$ $(1 < i < n)$ either the a- or b- signal of some obstacle $O_i \in \Omega$, and t_i $(1 \leq i < n)$ a time at which $\sigma_{i+1}(t)$ eclipses $\sigma_i(t)$. We should think of Π as hopping from signal σ_i to signal σ_{i+1} at time t_i in transform space where the latter signal eclipses (becomes greater than) the former.

Finally, if $\sigma_i(t)$ is a signal visited by some path Π (with $1 < i < n$), then the difference between the times t_{i-1} and t_i at which Π joins and leaves the signal $\sigma_i(t)$ has a geometrical interpretation. Suppose $\sigma_i(t)$ is an a-signal. Then $t_i - t_{i-1}$ is equal modulo 2π to the angle, measured in a clockwise direction, between the two directed joint tangents represented by the transitions to and from σ_i. In other words, the length of time which Π spends on the signal $\sigma_i(t)$ is equal, modulo 2π, to the clockwise angle through which path winds as it traverses the boundary of the obstacle O_i to which $\sigma_i(t)$ belongs. Since we need never be concerned with paths that wind through more than a complete revolution about an obstacle (such paths could hardly be optimal), we might as well insist that all transitions happen at the earliest possible opportunity: in other words that $t_i - t_{i-1}$ lies in the interval $(0, 2\pi)$. Thus, we think of the path Π as moving forwards along a-signals for a time interval in the range $(0, 2\pi)$ until a suitable transition point to another signal.

When $\sigma_i(t)$ is a b-signal, then $t_i - t_{i-1}$ is equal, modulo 2π, to *minus* the angle, measured in an anticlockwise direction, between the two directed joint tangents represented by the transitions to and from $\sigma_i(t)$. For this reason, it is easier if we think of the path Π as moving *backwards* along b-signals. Again, since we need never be concerned with paths that wind through more than a complete revolution about an obstacle, we might as well insist that the next transition happens at the earliest possible opportunity (remembering that we are proceeding backwards along the b-signal): in other words that $t_i - t_{i-1}$ lies in the interval $(-2\pi, 0)$. Thus, we think of the path Π as moving backwards along b-signals for a time interval in the range $(0, 2\pi)$ until a suitable transition point to another signal.

It should be stressed that this convention of taking a sinusoidal representation of a path to move forward along a-signals, and backward along b-signals is to be taken as a way of conceptualising the problem of route planning within the sinusoidal scheme, rather than as an implementation decision. Implementations in which obstacles are represented by signal-emitting devices are not precluded, of course: the convention merely aids in formulating a criterion for eliminating certain sub-optimal routes, as we shall see later. Henceforth then, let us take a sinusoidal representation of a path

$$\Pi = [\sigma_1, t_1, \sigma_2, t_2, ..., t_{n-1}, \sigma_n] \tag{1}$$

to have, for $1 \leq i < n$,

$$\begin{aligned} t_i - t_{i-1} &\in (0, 2\pi) \text{ if } \sigma_i(t) \text{ is an } a\text{-signal} \\ t_i - t_{i-1} &\in (-2\pi, 0) \text{ if } \sigma_i(t) \text{ is a } b\text{-signal.} \end{aligned}$$

Figure 6a shows a path from point X to point Y passing clockwise about S and anticlockwise about T. Figure 6b shows the corresponding sinusoidal representation: starting anywhere on $w_X(t)$, we jump to $a_S(t)$ at the point labelled L_1, corresponding to the joint tangent connecting X and S; then we proceed along $a_S(t)$ until the point L_2 when we jump to the signal $b_T(t)$; proceeding backwards along $b_T(t)$ (because it is a b-signal), we arrive at the point L_3, when we jump onto our destination signal $w_Y(t)$. The signals $b_S(t)$ and $a_T(t)$ have been omitted from figure 6b for clarity.

Of course, not all sequences of signals of the form (1) correspond to free paths between X and Y, since many joint tangents will be obstructed by other obstacles in Ω. We need a condition for determining, using the sinusoidal representation, when a joint tangent is obstructed by an obstacle.

Fig. 7a shows the directed joint tangent L from O_i to O_{i+1} with clockwise orientation about O_i and anticlockwise orientation about O_{i+1}. Here, L is obstructed by a third obstacle, O. Fig. 7b shows the sinusoidal representation of L, where $\sigma_i(t)$ (in this case, the a-signal of O_i) is eclipsed by $\sigma_{i+1}(t)$ (in this case, the b-signal of O_{i+1}). The a- and b-signals of O are also shown. Fig. 7c shows, in schematic form, the pattern of signal-crossings in the

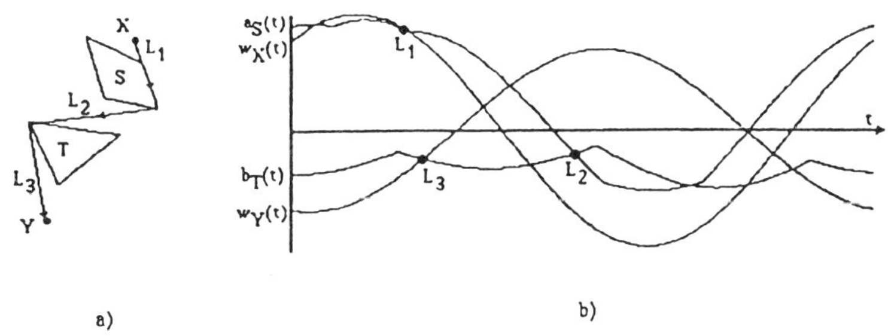

Figure 6: The sinusoidal representation of a path

relevant region: the signals $\sigma_i(t)$ and $\sigma_{i+1}(t)$ form an X-shape, with $a(t)$ cutting both upper branches of the X, and $b(t)$ cutting both lower branches. Simple geometrical considerations suffice to show that this pattern is characteristic of the fact that O intersects the joint tangent involving $\sigma_i(t)$ and $\sigma_{i+1}(t)$. Formally:

Theorem 5 *Let $\sigma_i(t)$ and $\sigma_{i+1}(t)$, be signals belonging to regions O_i and O_{i+1} respectively, so that the eclipsing of $\sigma_i(t)$ by $\sigma_{i+1}(t)$ represents a joint tangent L from O_i to O_{i+1}. Let O be an obstacle not intersecting O_i or O_{i+1}, with sinusoidal representation $\Sigma = \langle a(t), b(t) \rangle$.*

Let ${}_{\sigma_i}t_a$ be any time at which $a(t)$ eclipses $\sigma_i(t)$; let ${}_bt_{\sigma_{i+1}}$ be any time at which $\sigma_{i+1}(t)$ eclipses $b(t)$. Then O intersects L if and only if the following two conditions hold:
1. After ${}_{\sigma_i}t_a$, $\sigma_{i+1}(t)$ eclipses $\sigma_i(t)$ before $\sigma_{i+1}(t)$ eclipses $a(t)$.
2. After ${}_bt_{\sigma_{i+1}}$, $\sigma_i(t)$ is eclipsed by $\sigma_{i+1}(t)$ before $\sigma_i(t)$ is eclipsed by $b(t)$.

This concludes the basic embedding of the route planning problem within the sinusoidal representation scheme. In seeking an optimal path from X to Y avoiding obstacles Ω, we can look for a suitable sequence of signals of the form (1) such that, for $i = 1, \ldots, n-1$, there is no $O \in \Omega$ such that σ_i, σ_{i+1} and O satisfy the relevant conditions of theorem 5. The set of paths corresponding to all such sequences will contain all shortest paths.

Our final task is to show how to exploit the criterion of sub-optimality contained in observation 3 within the sinusoidal representation scheme. That criterion requires two geometric notions: firstly, that of a path's being *clockwise/anticlockwise definite*, and secondly, that of one path's *lying on the clockwise/anticlockwise side* of another path. It is trivial that clockwise definite (anticlockwise definite) paths are identifiable as paths involving only a-signals (only b-signals). The next result shows that the relationships of clockwise side and anticlockwise side have remarkably simple sinusoidal correlates.

Theorem 6 *Let p and p' be two paths from point X to point Y which cross over neither themselves nor each other, and which meet only at their endpoints. Let the sinusoidal representations of p and p' be*

$$\Pi = [\sigma_1, t_1, \sigma_2, \ldots, \sigma_{n-1}, t_{n-1}, \sigma_n]$$

and

$$\Pi' = [\sigma'_1, t'_1, \sigma'_2, \ldots, \sigma'_{n'-1}, t'_{n'-1}, \sigma'_{n'}]$$

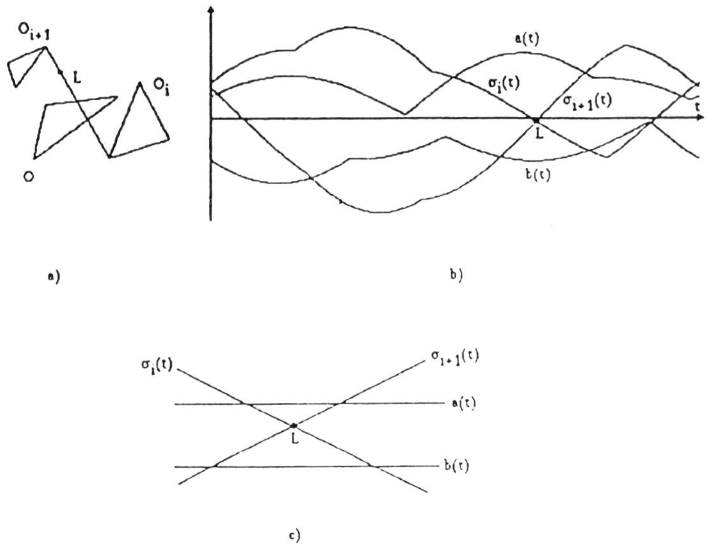

Figure 7: Obstruction of a joint tangent

respectively, with $\sigma_1 = \sigma'_1 = w_X(t)$ *and* $\sigma_n = \sigma'_{n'} = w_Y(t)$.

Then p *lies on the clockwise side of* p' *if and only if*

$$t_{n-1} - t_1 > t'_{n'-1} - t'_1. \tag{2}$$

The times t_1 and t_{n-1} indicate when Π leaves its initial signal and arrives at its final signal, respectively. Thus, the quantity $t_{n-1} - t_1$ denotes, as it were, the 'journey time' of Π. Of course, this 'journey time' has nothing to do with actual path length: if Π involves b-signals, its 'journey time' may be negative! Theorem 6 states that one path lies on the clockwise side of another just in case it has the greater journey time. Geometrically, the 'journey time' $t_{n-1} - t_1$ equals the net angle through which the path p turns, using a clockwise-is-positive convention. Thus, theorem 6 alternatively states that, if p and p' satisfy the conditions of definition 2, p^* winds clockwise if and only if the net angle of p is greater than that of p' (whence the theorem can easily be proved). The important point about this result is that it enables the relations of clockwise- and anticlockwise-side to be established merely by observing and comparing the times of a few signal-crossings.

If a path p from X to Y passes through intermediate points A and B, in such a way that the section of p between A and B is suboptimal, then, clearly, p is suboptimal. In particular, p is suboptimal if there exists a clockwise definite path between A and B lying on the anticlockwise side of p between A and B, or if there exists an anticlockwise definite path between A and B lying on the clockwise side of p between A and B. The detection of such paths using sinusoidal representations proceeds as above. A sinusoidal representation Π of a path involving signals σ_i and σ_j can be seen to be suboptimal if there is a path consisting just of a-signals which does not cross over itself and which goes between sections of σ_i and σ_j covered by Π with a shorter 'journey time' than Π does (where the journey time of a path between two signals is measured from the time at which it leaves the first signal to the time at which it joins the second signal). Similarly, Π can be seen to be suboptimal if there is a path consisting just of b-signals which does not cross over itself and which goes between sections of σ_i and σ_j covered by Π with a greater 'journey time' than Π does. Note that, in this latter case, the journey time of an anticlockwise definite path is negative: thus Π will be ruled suboptimal if its journey time between σ_i and σ_j is a larger negative number than the journey time of the anticlockwise definite path in question.

Thus, a sinusoidally represented path Π can be seen to be suboptimal if it involves two signals, σ_i and σ_j, such that we can find a definite path which does not cross itself, and which gets from σ_i to σ_j 'quicker' than Π does in the sense made precise by theorem 6. By generating minimal definite paths which do not cross over themselves (a straightforward matter), it is thus possible to declare many paths to be suboptimal. For example, all the paths in fig. 1 except the three shown can be rejected on the above criterion.

5. Relation to Other Work

The technique of using sinusoidal transforms for a variety of spatial reasoning tasks is developed in (Pratt, 1988a). Sinusoidal transforms are most closely related to Radon transforms (Leavers, 1988), though the latter represent shapes as a function of two variables, whereas the former require only a function of one variable. There is not the space here to engage in a detailed comparison of the two representation schemes.

The present method of finding paths in a cluttered plane is both a generalisation and a restriction of the method of *V-graphs* presented in, for example (Lozano-Péres and Wesley, 1979): a generalization, because the the V-graph method is confined to polygonal obstacles whereas the present method works in exactly the same way for both polygonal and curvilinear obstacles; a restriction, because the present method is confined to convex obstacles whereas the V-graph method can cope with re-entrant polygons. When only convex polygonal obstacles are involved, the present method yields a subset of the paths in the V-graph, yet is still guaranteed to include all optimal paths.

Another path finding method which can yield paths close to those generated by the present method is that given by the SPLA algorithm described in (Chubb, 1987). Chubb's algorithm uses an iterative process to converge on a set of forced paths between two points. However, Chubb does not use transformed space, and does not have the benefit of easy identification of joint tangents. Chubb's method aspires to work for non-convex obstacles.

The path finding problem addressed here concentrates on the simple case where the moving object is a point, and where there is no uncertainty as to the positions and shapes of the obstacles. The problem of path finding for non-point obstacles is discussed in, e.g. in (Brooks, 1982; Lozano-Péres and Wesley, 1979). Path finding under uncertainty is discussed in (McDermott and Davis, 1984).

6. Conclusions

The main purpose of this paper has been to demonstrate the utility of the sinusoidal representation scheme for a simple version of the path finding problem. The approach has been by way of picking out a set of sensible paths between a given pair of points in a crowded plane—a set which is guaranteed to contain all optimal paths—and of showing how such a set can be easily characterised within the sinusoidal representation scheme. Thus, the results presented here allow only the determination of a set of plausible candidates for being the optimal path. In particular, no mention has been made of computing the lengths of paths.

Once this set of sensible paths has been produced, its members can then be compared, if necessary, by determining exact path lengths, or, if possible, by some computationally less intensive method. The fact that the method proposed here does not result in the selection of a single optimum path does not mean that it is devoid of utility: after all, *people* are bad at obtaining, by eye, accurate determinations of arc-length, but they are certainly able to find near-optimal paths in relatively simple environments.

In showing how to obtain a set of sensible paths between two points, we have gone some way—but only some way—towards matching the ability of human spatial intuition to identify near-optimal paths. In many cases, the set of paths satisfying the above-presented condition contains clearly sub-optimal candidates, so that there remains a need for heuristics which can further whittle down such sets, and which are at the same time amenable to easy expression in the sinusoidal representation scheme. Such heuristics, however, have yet to be developed.

One final point. A striking feature of the sub-optimality criterion just described is that the critical notions of *clockwise/anticlockwise definite* and *clockwise/anticlockwise side* are projective invariants. This fact is important because information about the dispositions of objects in one's environment can often come in the form of perspective views (i.e. projective transformations) of the layout of those objects. A projectively invariant criterion of sub-optimality thus holds out the prospect of circumventing the explicit construction of a plan view of one's environment. However, full discussion of these ramifications would take us too far afield to be included in this paper.

7. Acknowledgements

This paper describes an improved approach to the problems tackled in (Pratt, 1988b) and (Pratt, 1989). Some excerpts from (Pratt, 1988b) are reproduced here verbatim, with the kind permission of the publisher.

8. Notes

1. For brevity, I shall treat X and Y sometimes as *points*, and sometimes as *point-regions*—i.e. sets containing just one point. A point of terminology: I say that a line Λ is *tangent* to a convex region S if $S \cap \Lambda \neq \emptyset$ and if Λ does not divide S into two regions. According to this usage, there are many tangents to a region at a kink in its boundary.

2. Fig. 3a and fig. 3b must be regarded as approximations in the following respect: in both cases p and p' would be made up of straight line segments interspersed with curved portions of obstacle boundaries, and could therefore not be continuously curving. But this detail can be ignored for present purposes.

3. In fact, the mapping from R to $a_R(t)$ also has the properties (i)–(iii); the signal $a_R(t)$ can therefore represent R just as well as the pair of signals $\langle a_R(t), b_R(t) \rangle$ can. However, it

is computationally convenient to keep both signals around.

The author may be contacted via electronic mail at ipratt@cs.man.ac.uk

9. References

Brooks, R. (1982), 'Solving the Find-Path Problem by Good Representation of Free Space', Proceedings of AAAI, 381–386.

Chubb, D. (1987) 'An Introduction and Analysis of a Straight Line Path Algorithm for Use in Binary Domains', In A. Kak and Su-shing Chen (eds.), Spatial Reasoning and Multi-Sensor Fusion: Proceedings of the 1987 Workshop, Morgan Kaufman, Los Altos, CA, 220–229.

Leavers, V. (1988) 'Use of the Radon Transform as a Method of Extracting Symbolic Representations of Shape in Two Dimensions', In Proceedings of the Fourth Alvey Vision Conference, Manchester, U.K., 273–280.

Lozano-Pérez, T. and Wesley, M.A. (1979) 'An Algorithm for Planning Collision-Free Paths Among Polyhedral Obstacles', Communications of the Association for Computing Machinery 22(10), 560–570.

McDermott, D. and Davis, E. (1984) 'Planning Routes through Uncertain Territory', Artificial Intelligence 22, 107–156.

Pratt, I. (1988a) 'Spatial Reasoning using Sinusoidal Oscillations', In Proceedings of the Tenth Annual Conference of the Cognitive Science Society, Montréal, Canada, 219–225.

Pratt, I. Pratt, I. (1988b) 'Path Finding in Free Space using Sinusoidal Transforms', In A.G. Cohn (ed.) Proceedings of the Seventh Conference of the Society for the Study of Artificial Intelligence and Simulation of Behaviour, Pitman, London, 127–136.

Pratt, I. (1988c) 'Spatial Reasoning and Route Planning using Sinusoidal Transforms', Technical Report UMCS-89-8-2, Department of Computer Science, University of Manchester.

Pratt, I. (1989) 'Path Finding in Free Space using Sinusoidal Transforms II', AISB Quarterly Newsletter 69, 26–37.

SECTION 4: CARTOGRAPHIC PERSPECTIVES

There are many definitions of cartography, mapping, and maps. Prototypically, however, cartography is involved with the concept that maps represent geographic space, usually graphically. This focus on the map as a pictorial or graphic representation has always been central to cartography, but has become even more important recently, because maps have been freed of the necessity to act as storage media for geographic information. Geographic information systems have been developed for the storage and retrieval of such information. But just as the data-storage role of maps diminishes with the introduction of computerized GISs, computer graphics and "scientific visualization" open up new opportunities for cartographic display. Efforts to understand "map grammar", the syntax and semantics of maps and map symbols, become more important. The first three papers in this chapter investigate different aspects of studying "cartography as language" and its relation to text, linguistics, and semiotics. The fourth paper provides a link between maps and image processing.

Grant Head starts with an extensive and detailed overview of the contributions of various researchers, stressing the contribution of linguistic and semiotic theory to the understanding of cartographic rules. An immediate problem is caused by the widely varying terminology used by researchers applying semiotic methods to maps. This is due in part to different theoretical points of view, but also is due to different translations and interpretations in a body of literature distributed over time and in many different journals and disciplines. Head compares some definitions and finds agreement in essential points and differences in details. It seems unfortunate (and unnecessary) that the terminology is so varied, considering that it expresses very similar concepts. In his conclusions, Head proposes that we say "a *sign* has a *content* and *expression*; the link between these is the semantic link or *code*". There is less agreement on the smaller graphic units and different sets are proposed. The smallest units are often separately disarticulated into expression units with no content and content units with no expression.

In the second paper, **Hansgeorg Schlichtmann** investigates the specific semiotic point of view, and concentrates on how maps communicate planimetric information. Necessarily, he starts with different kinds of spatial objects and how we conceptualize them. Somewhat surprisingly, a cartographic view results in more spatial concepts than are contained in most current GISs. Furthermore, the spatial concepts that Schlichtmann identifies differ from those in the GISs. Schlichtmann then goes on to discuss the notion of fidelity of a map to the world, and some origins of "infidelities".

Dalia Varanka next considers the more direct interplay of map and text, primarily concentrating on early English atlases. She follows a "map-as-text" concept, where the map is seen as a compilation and expression of the map-maker's personal experience. She describes a project in which she will collect data about the cartographic and the textual elements in atlases and examine their interrelationships. The research is an attempt, in a post-modern, deconstructionist tradition, to place the cartographic expression in the social context, understanding maps not only as truthful representations (at least in intent) of some external reality, but also as a product of the social, cultural, personal situation of the map maker.

In the last paper in this section, **Geoffrey Edwards** bridges the gap between cartography and image interpretation, through an examination of aerial photographs and satellite images. He uses some of the concepts presented by Head and Schlichtmann in their contributions, namely the construction of complex objects from simpler ones. Applying a language paradigm to cartography is more obviously justified, since cartographic signs are produced by people, expressing their interpretation of geographic space. Remotely-sensed images are the product of a physical processes that records (mostly) light reflectance; however, their interpretation is still a human activity. Edwards investigates the methods people seem to use to interpret such images and finds striking parallels to the methods used for map interpretation. His concern is primarily with the aggregation of visual elements into higher order, complex objects. Results of an informal experiment suggest three levels of structuring, going from the field of view to large object scale to a textural scale. The interesting problem is the verbal expression of spatial relations and shape. The ability of natural language and human reasoning to bridge several levels of detail is very important in image processing, as an image contains several levels of detail, which are interpreted individually and then related to the other levels. Edwards' analysis stresses the importance of working on multiple levels of resolution, allowing aggregation and pointing to multiple sources of meaning.

MAPPING AS LANGUAGE OR SEMIOTIC SYSTEM: REVIEW AND COMMENT

C. GRANT HEAD
Department of Geography
Wilfrid Laurier University
Waterloo, Ontario Canada N2L 3C5

1. Introduction

1.1 MAPPING

A large part of the core of what geographers do is *mapping*. By mapping I do not mean merely the drafting of maps with pens or with the computer but, to turn Robinson and Petchenik's noun "mapper" into a verb, mapping is *the act of conceiving spatial relationships in the milieu.* "Whether those relationships involve the hills and valleys of the topography, the familiar nodes and corridors of the city dwellers, or the invisible shoals and reefs of the ship's pilot" they observe, "the conception of things in spatial relationship is the critical operation...." (Robinson and Petchenik 1976, p. 17). Not all of what geographers do, of course, is mapping. Much attention is given also to *process*, to the understanding of the manner in which things of the milieu—things physical, cultural or economic—evolve and interact, but in the final analysis most geographers would probably admit that it is the knowing what to map so as to reveal significant spatial patterns or the explanation of spatial patterns that appear from mapping that seems to define the essence of their discipline. Mapping—this focus upon spatial relationships—seems almost always to be the core of creating geographical *information* from mere spatial *data*.

Mapping involves the use of the map, but it is helpful to think of the map as functioning in at least three basic ways. First, it may act as a form of *storage for data about the milieu*. Second, it may act as a *surrogate for the milieu* during study—during mapping. Thirdly, it may act as a *medium of transmission of spatial concepts*. Like all thought or communication, mapping is selective, it does not present all of reality all at once. And the map, as its vehicle, can represent it in an infinite variety of ways. Some maps will hold a wide selection of data, apparently emphasizing none more than others; some maps will transmit clear spatial messages even to the most passive mind.

Because the electronic Geographical Information System traffics in spatial data and in spatial relationships, it makes much use of the map. The data from paper maps are digitized into electronic databases that replicate the storage function of the map. Although there are other query and analysis methods in GIS, the map is often used to present a selection of the spatial data from the electronic database to the human user of the system. The user searches these presentations for potentially meaningful spatial relationships and then clarifies these by manipulating either the data selected or the graphics of the map. "Mapping" is thus very much like thought and creative writing, specialized for spatial relationships.

The craft of encoding aspects of our milieu into map storage—the art and science of cartographic production—has ancient roots and is no stranger to technological innovation. Yet the understanding of the use of maps as a medium to facilitate human thought and communication about spatial relationships is a relatively recent concern and along this path we have made surprisingly little progress. To use a geographical metaphor: the path is not clearly cut, the night is dark, we have no torch, and we have no map.

At this meeting we all recognize that language is relevant to geographic space. Some of us, it seems, are interested in how most efficiently to encode spatial data and relationships within an

D. M. Mark and A. U. Frank (eds.), Cognitive and Linguistic Aspects of Geographic Space, 237–262.

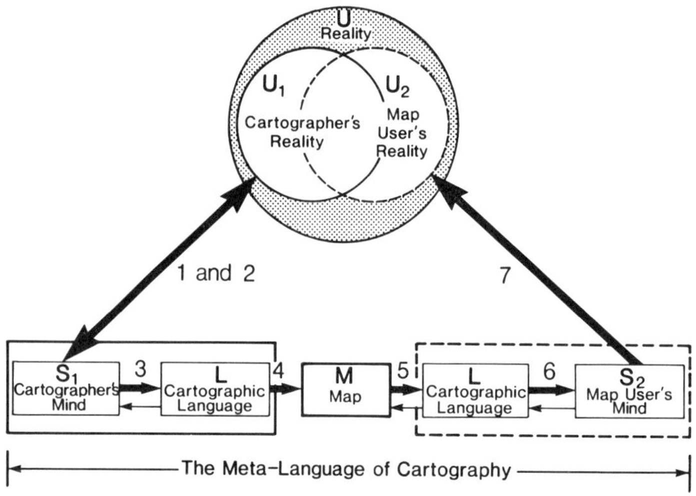

Figure 1: Kolacny's Cartographic Communication Model (as simplified by Robinson and Petchenik, 1976 p. 29).

electronic database; others in how best to correlate our variety of natural human languages to consistently query the database. This paper presents, in part, the concerns of those who see the map as the medium in a system of communication specializing in spatial relations. The specific point of view of this paper is that this system of communication specializing in spatial relations can be most fruitfully studied within the paradigm of semiotics or of language.

1.2 PARADIGMS FOR RESEARCH INTO CARTOGRAPHIC COMMUNICATION

Despite the long history of cartography, it has been but forty years since academic cartographers began to think explicitly of the map as a "functional graphic display", and began to consider how the efficiency of that functionality might be measured, experimented with, and improved. Much of the first work was in the psychophysical testing of the graphic parameters of individual map symbols.

We know a great deal now about the response of the average map user to variations in size of point symbols (especially circles), about perceptions of changes in value (lightness) of grey-tones used for choropleth maps, and even about the reactions of map readers to the complex meanings, both qualitative and quantitative, that are imbedded in the use of color (examples of the work that informed us on these issues can be seen in Flannery, 1956; Jenks and Knos, 1961; Cuff, 1972; and Gilmartin and Shelton, 1989). Yet cartographers are fully aware that these psychophysical studies dealt with only bits of the full map and that we seem not to be developing a consistent and operational theory of the map as information communicator. For a time in the 1970s it was popular to apply the information communication theory originally developed by Shannon and Weaver (Shannon and Weaver, 1949). This application took many diagrammatic and explanatory forms (e.g., Figures 1 and 2) all of which saw the map as a channel transmitting information from sender to receiver. The sender formulated the message and encoded it, the receiver decoded it into a set of information. The focus of research could then be upon efficiency of the channel, aiming at increased capacity and decreased loss in transmission. It came to be recognized, however, that map readers often seemed to get things from map reading that were not consciously designed-in by the cartographer, and this made measurement of information loss a fuzzy business. In addition, it was never fully understood how to quantify the information itself. Lately, there has been a tendency to retreat from such dalliance with science, and to stress cartography as, once again, an art (for example, Keates, 1984). But in the meantime there has also been a counter-current that has looked to language and semiotics as offering the broad general paradigm that will provide the consistent and operational theory in which to place the map as information communicator.

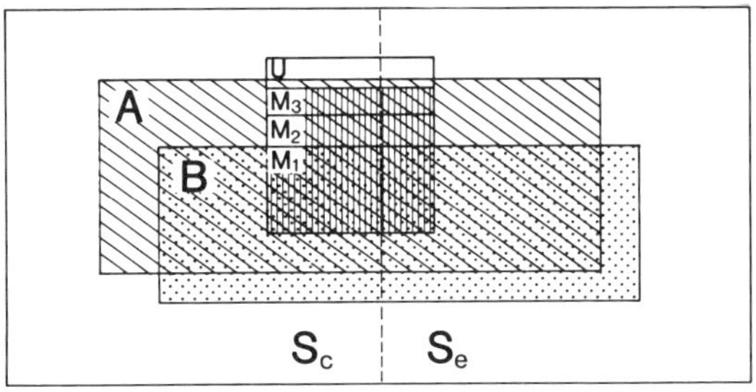

In the terminology, Cr refers to the cartographer and Pt to the percipient.
Diagram elements are as follows:

S	=	Geographic space, the milieu
S_c	=	Correct concept of the milieu
S_e	=	Erroneous concept of the milieu
A	=	Concept of the milieu held by Cr
B	=	Concept of the milieu held by Pt
M	=	Map prepared by Cr and viewed by Pt
M_1	=	Fraction of M previously conceived by Pt
M_2	=	Fraction of M not previously conceived by Pt and newly comprehended by him: an indirect increment
M_3	=	Fraction of M not comprehended by Pt
U	=	Increase in conception of S by Pt not directly portrayed by M but which occurs as a consequence of M: an indirect increment

Figure 2: Cognitive Elements in Cartographic Communication (after Robinson and Petchenik, 1976, p. 34).

2. Language, Semiotics, and Maps

2.1 DEFINITIONS

Those of us who believe that we will find in language or semiotics the theory for cartographic communication have not been consistent in our struggle to use the analogy. Some of us speak of language but use semiotic terminology and concepts, while others reject the strict linguistic paradigm to use semiotics but find it heuristically useful to discuss cartography as it compares to language. To be more consistent in terminology and in concepts must now be an early goal so that we may build more cumulatively upon each other's work. Some definitions of semiotics, language, and natural language should be provided at the outset, but even here we find much disagreement. The simplest definition of semiotics is perhaps "the study of sign systems", or the study of semiosis, "the process by which empirical subjects communicate, communication processes being made possible by the organization of signification systems" (Eco, 1976, p.316). Language is but one of these systems, often set very much apart by particular qualifiers, for example aspects of its structure such as "double articulation". Natural languages are set apart from formal languages in that the latter are completely rule determined; natural languages evolve, sometimes unsystematically. Finally, before we begin to trace the rambling path that cartographic theorists have cut through language and semiotics it will be useful to note its most crucial indicator, the distinction between content and expression. The *sign* has two components. It has a physical form, a mark on paper or a sound: an *expression*. And it has a meaning, a concept it conveys: a *content*.

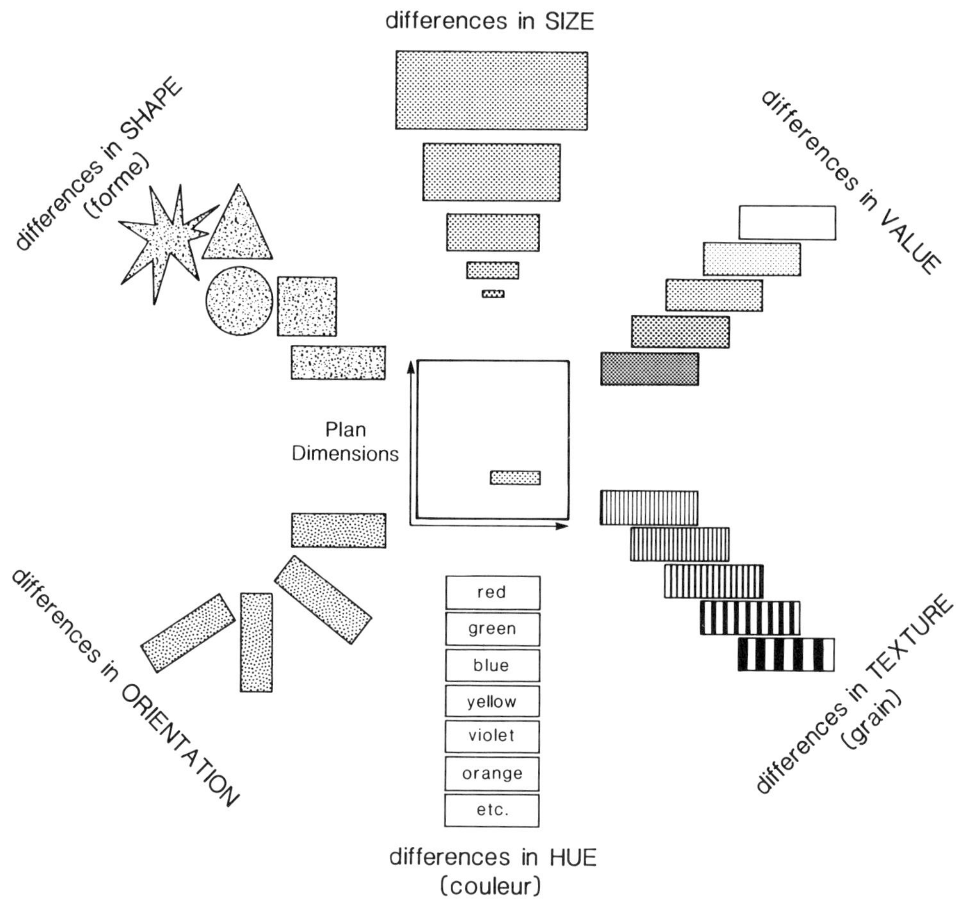

Figure 3: Bertin's Visual Variables (Bertin, 1967, p. 43).

2.2 REVIEW OF CRUCIAL LITERATURE

2.2.1 *Bertin's Sémiologie Graphique.* Jacques Bertin's *Sémiologie Graphique* appeared in French in 1967 but not in other languages until many years later. It was a landmark study, striking directly and deeply at the relations between expression and content. At the heart of the system he proposed seven "visual variables" (see Figure 3), manners in which a simple graphic symbol may be manipulated so as to clearly change perception and thus to allow the presentation of a clearly-differentiated message item. The seven variables included the plan location of the symbol, although we must recognize that this has obvious limitations in mapping.

Bertin stressed that each visual variable was either more or less suited to particular types of content (Figure 4). The variable size is the only one he recommended to express *interval-scaled* or *ratio-scaled* quantitative data. Circle area, for example, can be scaled through a constant to portray precisely a city population; this can be decoded for a particular city through the use of the map legend, or can be read as a ratio measure by comparing the circles for different cities on the map itself. The visual variables of size, value and texture can effectively portray *ordinal* data. We often use them to represent the different traffic capacities of highways; though we use value heavily to signify intervally-scaled data on choropleth maps, Bertin does not recommend it for this task. All the other visual variables except form can be used to create clearly-perceivable *classes* from nominally-scaled data. Thus, we might signify all forest land uses in one color. Finally, Bertin notes that any visual variable can be used for mere *nominal* differentiation although with size and value their abilities to transmit ordinal and quantitative messages over-ride our perceptions of mere nominal differences.

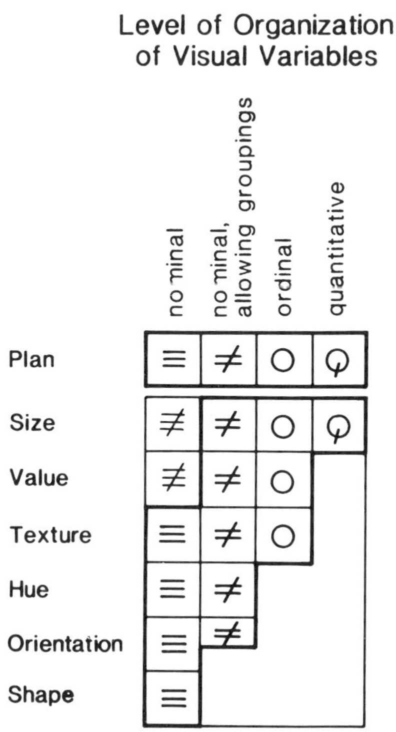

Figure 4: Suitable uses for each graphic variable (Bertin, 1967, p. 69).

The visual variables of Bertin were (and remain) important. They systematized much of the approach to cartography's content-expression link at the simple "map symbol" level and set the psychophysical studies within a clear framework.

2.2.2 *Basic concepts but basic problems: the 1970s*. Michael Dacey (United States) as early as 1970 struck clearly at today's concerns with an exploration of "...an information system that provides a [wide] variety of geographical information to a diverse group of users" (Dacey, 1970, p. 73). For introductory purposes, Dacey viewed a GIS as "anything that functions like a map in communicating geographically ordered information to users of the system" (p.72). Since, "by definition", Dacey maintained, "all communication is by language, identification of the functions and properties of these languages is prerequisite to the successful description of a geographic information system" (p.73). His object in this short paper was "to go beyond the bare observation that maps have a linguistic basis to clarify some of the ways in which maps and other formulations of geographic information constitute a language and then indicate how a linguistic approach contributes to description and analysis of the information content of a geographic information system" (p.72). He was interested it seems not only in the traditional printed map but also in all "things that function like a map", such as textual material, systems of equations, sketches and drawings (p.72). Linguistic concepts are appropriate to all of these, he notes, "because they take into account the meaning of geographical information but are largely unaffected by the structure and form of alternative symbolic expressions of geographic information" (p.72). Thus, central to Dacey's concepts was the recognition of both content and expression components of sign. Dacey used the sign concept of Morris (1946) defining the sign as "the smallest unit that designates" and that which is pointed out as the "designation" of the sign. He defined *language* as "an institutionalized collection of signs that have common designations to members of the community using these signs" (p.73).

The greatest problem in applying common semiotic or linguistic models to maps, Dacey observed, would be that "while the serial order of language is commonly contained within the ordering concepts of 'before' and 'after', the ordering of map symbols is multi-directional" (pp.74-75). A language for the transmission of geographic information must be, then, a two-dimensional language. Dacey suggested the application of certain linguistic models such as the "context-free phrase structure language or immediate constituent analysis" (p.76) being developed in the early 1960s by Chomsky. At the time of his paper Dacey was attempting to develop a language for the syntactical description of a variety of polygon figures. His work, he noted, provided partial solutions to the definition of *picture primitives*, the expression of *picture syntax* and the generalization of the notion of concatenation to two dimensions. Problems yet to be explored included "the manner of describing the spatial relations that obtain between expressions that are components of complex pictures", and the procedures "for conducting syntactical analysis of pictures" (pp.77-78). He speculated on the nature of the "object" for geography. Whereas "objects" for most languages are characterized by non-spatial attributes, "a primary concern for geographic information is the spatial attributes of the things occurring in a specified time-space region" (p.78). The theory of semiotics, Dacey maintained, "leaves open many problems concerning the specification of the semantic system for a language whose domain includes individuals whose positional attributes are as significant as are the non-positional attributes" (p.78). These problems of the "geographic object" or *unit* and of the role of two-dimensional space would continue to cause difficulties in the development of the cartographic language or sign system paradigm.

During the 1970s, the International Cartographic Association's "Commission on Cartographic Communication" continued the investigation. Representative of the work of members of this commission are papers by Board (United Kingdom, 1973) Morrison (United States, 1974) and Ratajski (Poland, 1975, 1976). The first two papers focussed largely upon the semantic links between map content and map expression, and they confined themselves mainly to lower-level, simpler, map "symbols", be they point, line or area (for example, see Table 1). There are close parallels here with the work of Bertin.

Table 1: Morrison's "Proposed Syntactical Structure"
(Morrison, 1974, p. 125)

Possible Symbol Dimensions Available for Symbolization

Data Element Category	Available Symbol Dimension		Additional possible dimensions (not recommended)
Point Symbols	Black & White	Color	
nominal scale	shape pattern orientation pattern arrangement	hue	pattern texture
ordinal, interval or ratio scales	size pattern texture	value intensity	shape
Line Symbols			
nominal scale	shape	hue	pattern orientation pattern arrangement
ordinal, interval or ratio scales	size pattern texture	value intensity	shape
Area/Volume Symbols			
nominal scale	pattern orientation pattern arrangement	hue	pattern texture
ordinal, interval or ratio scales	pattern texture	value intensity	

Lech Ratajski saw cartographic marks such as the lower two in Figure 5 (which he calls "signs") as equivalent to word-size in the natural language hierarchy. Like words in natural language, they can be subdivided into morphemes (top three in Figure 5). In these cartographic signs he championed the use of primary and secondary morphemes—"guiding signs" that denote the general class of the sign content (e.g."processing" or "extraction") and "secondary signs" which add more specific denotation (e.g. "crude oil") (Ratajski, 1975, p. 6).

While the focus of the above work had been on lower-level cartographic symbols rather than higher-level complexes of map marks and the manners in which these complexes are achieved and used, certain significant aspects of the linguistic analogy were noted. One was again the problem of the linearity of natural language as against the two-dimensional presentational mode of the cartographic system. Dacey (1970), Aslanikashvili (1974), Robinson and Petchenik (1976), and Ratajski (1975, p. 76) all commented upon this problem during the 1970s: the spatial arrangement of marks on maps reflects closely the spatial arrangement of items on the ground. Thus, it is easy to see "the syntactic structure of the object language of cartography as a reflection of a structure of concrete space" (Ratajski, 1975, p. 10 referring to Aslanikashvili) or, in Ratajski's own words, "positions of words-signs depend solely on positions of the facts in reality represented by a map" (Ratajski, 1975, p. 12).

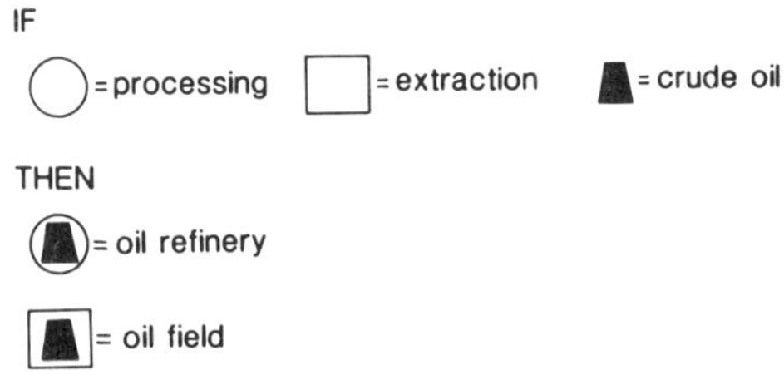

Figure 5: Ratajski's (1976) cartographic signs.

But the eventual resolution of this seemingly key difference between the map symbol system and language may lie in the fact noted by Board that in cartography (in parallel with language) "the signs employed represent geographical concepts not the real world itself" (Board, 1973, p. 230). By 1978 Pravda had further drawn attention to the mark-referent-concept triangle (Pravda, 1978, pp. 1-2). *Geographical concepts* are human mental constructs that although they may be derived from the observation of geographical reality (the referent) are not that reality itself. Ratajski in his 1975 paper notes how these geographical concepts may influence the map reading process. If the map user has developed from geographical experience the concept that "every river has its own valley", then as he uses a map that denotes only a river, the map implies "the existence of a valley, the rise of a terrain as it becomes more distant from the river..." (Ratajski, 1975, p. 11). Other geographical concepts arise from the experience of map reading itself. From extended map use, "a whole atlas of mental maps is left in a map-user's mind", and this atlas is put to use to enable the user to recognize on other maps the shapes of Europe, Great Britain and so forth (Ratajski, 1975, p. 13). These mental maps if known and used as part of the design of new (real) maps can play a key role in assisting the recognition of map syntax: "They play the same role in a map as punctuation, paragraphs, bold-face type or setting out in the written language" (Ratajski, 1975, pp. 13-14). Finally, based partly upon experimental research, Ratajski suggested another significant aspect of the map communication process: "A map is never read 'from cover to cover'". Map reading is a *process*, a sequence of concentrations upon "...parts which include the feature in demand..." and of comparison of map images with elements of our mental maps (Ratajski, 1975, p. 16). This was to be a particularly important concept, since it runs counter to the widely-held generalization that maps are seen "all at once" and thus function in a manner fundamentally different from language (Robinson, 1978, pp. 5-7).

In this early period of thinking within the semiotic/language paradigm, therefore, cartographers had expanded their notion of sign or symbol to include content as well as expression, had recognized that the map reader not merely the map maker is active in linking content with expression, and implied, at least, that the map reading process is sequential and, essentially, cognitive. But they had also tended to continue to see the location of map marks as reflections mainly of globe space and thus arranged very differently from marks in natural language.

2.2.2 *Advances in detail, continued problems: late 1970s to present.* Jan Pravda (Czechoslovakia) in papers of 1978 and 1982, and Macioch (Poland, 1982) particularly enumerated the units of the map language and their ability to combine to form complex signs. Cartographic signs have "...in the thinking the same function as have the lexemes and syntagms in the natural language from which the percipients of the map is able to determine whole spoken phrases, written sentences and compact texts, i.e. both simple and complex sets of ideas" (Pravda, 1978, p. 2). Pravda defined cartographemes, cartomorphemes, and cartosyntagma. Examples appear in one of his figures, redrawn here as Figure 6.

As is the grapheme in the natural languages, the *cartographeme* is the "...most elementary and visually perceivable graphic formation with a mainly graphic function, i.e it has an individual graphic motive within the framework of a two-dimensional graphematic space and has also the ability to arrange into cartomorphems. It is not a bearer of an independent significance /sensual,

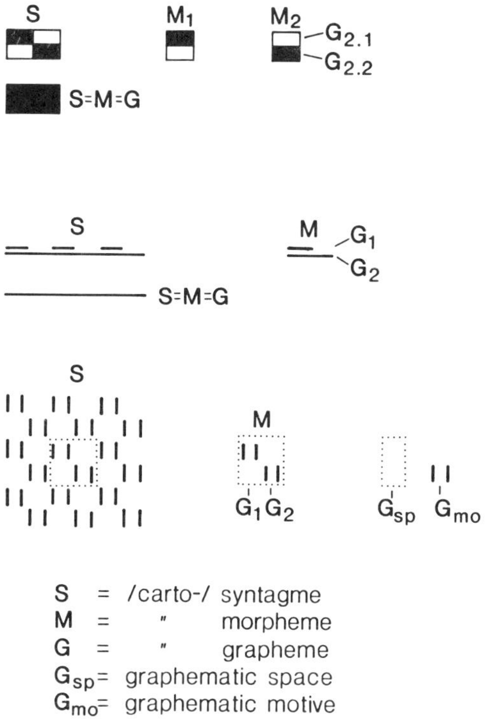

Figure 6: Pravda's (1978) Language Units.

semantic content/ but only a material part /component/ of the cartomorpheme and together with it of the cartosyntagma" (Pravda, 1978, p. 4).

A *cartomorpheme* is a "graphic formation which developed from the combination of cartographemes", "able to become a bearer of the significance component or of a part of the significance or even of a selective significance if the concept is complex" (p.4).

The *cartosyntagma* Pravda defines as "every independent cartographic sign", "a graphic conventional or symbolic significate of an arbitrary /even abstract/ concept"; "it fulfills its full function only in a mathematically defined or otherwise logically defined space of the map" (p.4).

The word or lexeme that we might expect to find by pursuing the analogy with natural language, he notes, is not a unit in cartographic language "...because there exists no logical dependence of the cartographic sign on the one- or multiverbal concept" (p.4). He does not elaborate.

Pravda in these papers made only brief mention of the rules that might govern the combination of signs into syntagmes—the syntax of cartography. Syntax lies within the various mapping approaches, he suggests, such as in dot mapping, choropleth mapping, isolines, and so forth. Again, he does not elaborate.

A. A. Lyutyy (Soviet Union) appears to have been developing his concepts of cartographic language concurrently with Pravda. In a 1984 paper he describes map *syntax* as "...models of combinations of symbols on maps", "combinations of symbols in the sense of grammatical categories, or the forms of correct expression in the language" (Lyutyy, 1984, p. 115). He illustrates (Figure 7) how some 21 combinations of "graphic devices" derived from six basic

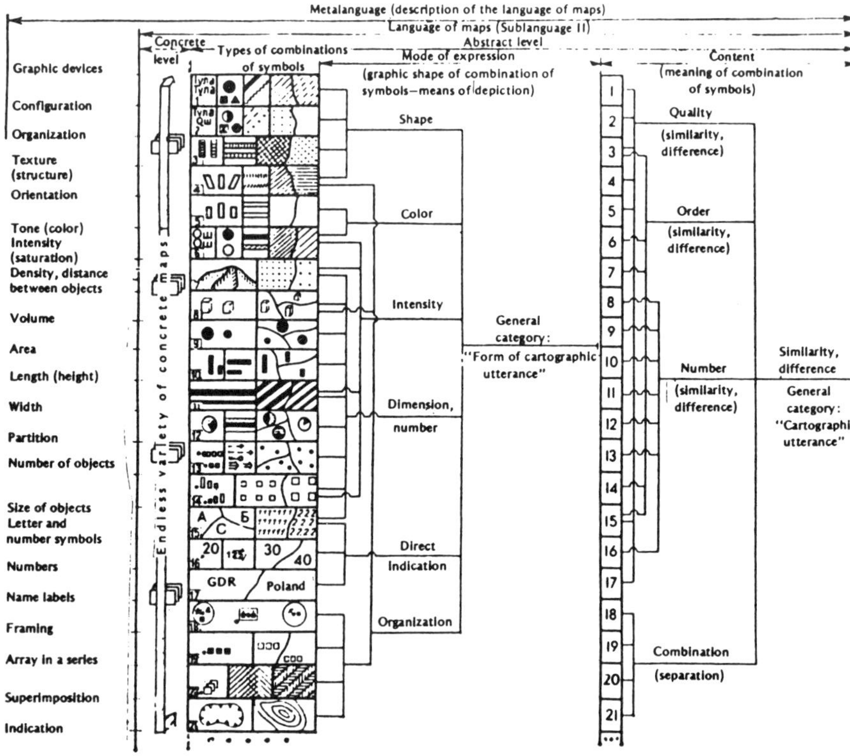

Figure 7: Lyutyy's (1984) Sublanguage II.

"modes of expression" (similar to Bertin's "visual variables") can be used to express nominal, ordinal or interval content (p.114).

The most original aspect of this important paper is the recognition of what Lyutyy calls two "sublanguages" in cartography. His Sublanguage II is characterized by the lower-level expression units that are found typically in the map legend. "It is through this sublanguage that the nature of mapped objects is represented: the general category of the mapped object (soil, plants, rocks, sectors of the economy), the specific type of object (the name of the soil, the plant, etc.), its qualitative and quantitative characteristics, its qualitative-quantitative structure, and the way it functions" (pp.107-108). In current GIS terminology we might call this the expression of the "attribute" data. It is this sublanguage to which he is referring in Figure 8. "But", he goes on, "there is still another linguistic apparatus working in maps. And that is the grid of geographical co-ordinates..." (p. 108) or other major geographical elements such as hydrography (p.110). Sublanguage I, therefore, is used to "locate mapped objects on the surface of the earth and to represent their mutual arrangements and their spatial shapes and orientation" (p.109). The grid of coordinates alone is the extreme case of this sublanguage, bordering upon "the language of mathematical graphs" (p. 110).

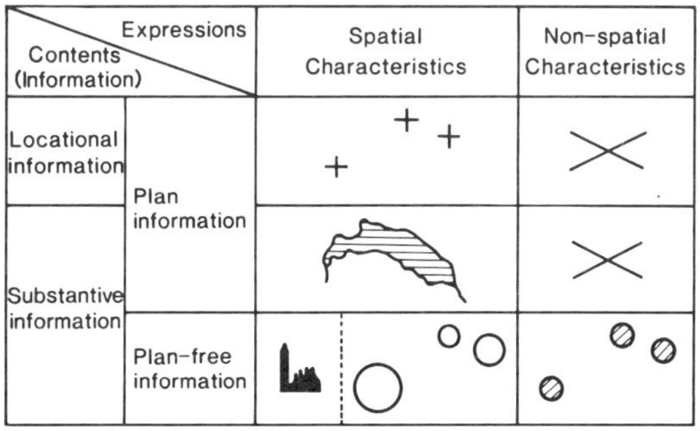

Figure 8: Schlichtmann's Plan-free, Plan, and Locational Information (Schlichtmann, 1984, p. 24).

2.2.3 *Summary thus far...* Much of the cartographic communication research work of the 1960s and 1970s in Anglo-America was of a psychophysical nature, experimenting with subject response to graphic parameters of individual cartographic symbols such as circle size and aspects of color. The cartographic language work of Bertin, Board, Morrison and Ratajski tended to emphasize the organized use in map design of this semantic link between symbol and meaning at the level of individual map symbols. Ratajski, Pravda and Macioch developed the notion of units, applying linguistic unit concepts to cartographic symbols and symbol-groups. The importance of "mental maps" and geographical concepts as "human mental constructs" began to be noted, particularly by Ratajski. The role of these concepts thus joined the role of the homology of map space with globe space in the process of chunking map signs into supersigns. The syntax of the cartographic language had not been much detailed by these authors, and when it was referred to it was seen either as the mathematical ordering of items in map space or the style of map expression adopted, such as choropleth, contour, or dot mapping. Lyutyy proposed two sublanguages, one dealing with the subject matter of the map, the other with the spatial context in which the first was imbedded. Unfortunately, this work found its way only slowly into the widely-read Anglo-American literature. During the period that it was available only in French, Bertin's work was virtually unknown in the United States; Ratajski's and Pravda's major works on cartographic language were available in English only in typescript. Dacey's overview work was published in a small-circulation provincial geographical journal.

3. Three Recent Position Papers: Two Crossovers from Semiotics, One from Psycholinguistics and Human Information Processing

3.1 SCHLICHTMANN: CHARACTERISTIC TRAITS OF THE SEMIOTIC SYSTEM "MAP SYMBOLISM"

Hansgeorg Schlichtmann (Canada) in this 1985 paper brings to maps a careful reading of the theoretical framework of the semioticians Buyssens, Prieto and (particularly) Eco. It is a tightly-worded paper and, although no summary can do justice to its contents, we will try.

First, Schlichtmann brings into the cartographic literature the semiotic terminology that a *sign* is composed of two parts: a conceptual item, a meaning or *content* on the one hand; and a perceivable item, an *expression* or sign vehicle or form on the other. "Expression and content, " he notes, are coupled by *code*: "a set of correlating rules or conventions". "Without a code there is no sign" (Schlichtmann, 1985, p. 23). Secondly, he clarifies the "composite nature of map symbolism": map symbols carry both spatial and non-spatial components of meaning. Most of the graphic elements of the symbol—its hue, value, size, etc.—carry what Schlichtmann calls *plan-free information.* An example is a point symbol for a power plant (red circle) of 10 million horsepower capacity (area 1 sq. cm.). The symbol also carries *locational information* by virtue of its placement on the map. And if the symbol had a shape that mimicked its geographic shape (as often do the symbols for large urban places) it would also carry what he calls *plan information* (p. 24). See Figure 8 for some diagrammatic clarification.

A third major set of points deals with the *units* of content or expression. Schlichtmann starts with *macrosigns* —minimal acts of *message* conveyance or reception. The macrosign in language is the sentence ("in the everyday and intuitive sense of the term"). In map symbolism, a macrosign is defined as a "complex which contains one location and one plan position"—a *localized sign* (p. 25). See Figure 9. Macrosigns may be combined into complexes which he calls *texts*, or can be decomposed into signs of a lower order, into *minimal signs*—signs whose expressions can no longer be broken into smaller units that still hold content. The square in Figure 9—a macrosign—can be decomposed as in Table 2. In this paper Schlichtmann holds that minimal signs cannot be broken down—as they can be in natural languages—into smaller "distinctive" units on the expression side (pp.25-26). He thus does not recognize cartographic graphemes nor distinctive features below them.

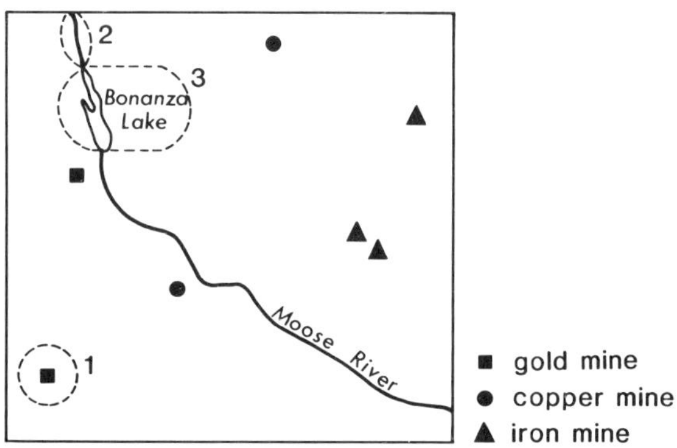

Figure 9: Schlichtmann's Localized Sign (Schlichtmann, 1984, p. 24). Fictitious map. The expressions of three localized signs are enclosed in dashed lines. Examples 1 and 2 have one segment each, a localized sign normally contains one segment which indicates the location (in example 3 the lake symbol).

Schlichtmann treats three aspects of the links between expression and content of map signs: motivation, singular transcription and gradual distinction (pp. 25-26). *Motivation* occurs when a content characteristic is reflected in the expression. Pictorial map marks are "motivated"; so too are colors used in an ordered series to reflect homologously content items that are so ordered (as, for example, temperatures). Geometric shapes used in point symbols on the other hand do not usually reflect a characteristic of the content, but are *arbitrary.*

Table 2: The Components of an Example Macrosign ("Localized Sign").

expression level	(x + y)	+	point implantation	+	segment	+	square shape
content level	($\phi + \lambda$)	+	point information	+	mine	+	gold

Whereas it has been said that "a map mark means only what a cartographer designates it to mean" (Robinson, 1978, p.6), it is in fact not quite so *singular* in its transcription. It is bound by many *global* cartographic conventions, by codes of efficient transcription such as those developed by Bertin, and by the codes established by the legend supplied for each map. The legend may reveal that the set of codes to be used on a particular map is non-standard, but once these codes are established by the legend of that map its individual marks are set, their transcription of content is no longer "singular" (Schlichtmann, pp. 26-27).

Many aspects of map marks (for example, areas of point symbols) can vary continuously: differences from one symbol to another can be *gradual*. In language, however, the distinction of expressions is differential: two very similar expressions are seen as either the same or different, not slight variations one of the other. Schlichtmann notes that in spoken language it is the listener who must decide whether the word spoken is /bed/ or /bad/. In cartography, however, he notes the mapmaker may treat continuous qualitative variables as discrete, and design his expression so that "only so many terms are directly rendered as can be coupled with distinguishable expression values" (p. 50); for example, we tend to use only five to seven variations of value and saturation in a choropleth scale. He does not comment on the possibility that the map user may also treat continuously-varying graphics as discrete, as is implicit in the word "distinguishable".

Schlichtmann's discussion of the syntax of macrosigns and their combination beyond into texts is perhaps the most significant part of his paper. He identifies two kinds of syntax. The *local syntax* refers to that which governs relations of the components of the localized sign, and thus is not complicated by spatial arrangements (p. 28). He has elaborated on this local syntax in a separate paper (Schlichtmann, 1988). The second type of syntax he calls *supralocal* and as it governs relations of localized signs within texts it involves spatial arrangements. Schlichtmann recognizes that although "The map arrangement of the expressions of localized signs in map space always models globe-spatial arrangements", yet this is "not a sufficient condition for text constitution" (p. 28). The reader has a strong role in creating chunks of text. Texts like all semiotic units at the level of minimal signs and above are *units of content as well as expression* and the map user must interpret "specific spatial relations as characteristic of an *ensemble* of specific objects" (p.28). It is the map reader who must bring to the map a concept of this compound geographical object: the map marks alone are insufficient to create a text. And during different uses of the same map driven by different user needs or different user knowledge: "several conceptual foci can coexist and can place a localized sign into different texts which overlap" (p. 28).

3.2 WOOD AND FELS: "DESIGNS ON SIGNS"

Denis Wood (United States) and John Fels (Canada) in this 1986 paper cover semiotic aspects of maps from the smallest sign to (taking inspiration from the semiotics of Roland Barthes) the meanings of the map as a whole. Their work focuses upon *codes*, the links between expression and content: "...it is the code that makes the sign, not the mark" (Wood and Fels, 1986, p. 68).

Table 3. Wood and Fels' Codes

Intrasignificant Codes	**Extrasignificant Codes**
Iconic	Thematic
Linguistic	Topic
Tectonic	Historical
Temporal	Rhetorical
Presentational	Utilitarian

They identify ten codes (Table 3). Five of these "the map exploits"; they operate "within" the map at the level, they say, "of language" (p. 68). These *codes of intrasignification* are similar (though different in name) to those recognized by earlier authors: the iconic, the linguistic, the tectonic and the temporal. The *iconic code* operates to link content and expression largely within localized signs (and may be motivated, arbitrary or conventional); the *linguistic code* explains the manner in which cartographers use map lettering (the authors use "linguistic" here in the narrow sense). To Wood and Fels the *tectonic code* provides the "*syntactical superstructure*" "to the iconic and linguistic codes. The tectonic code "configures graphic space in a particular relation to geodesic space ..."(p. 80). It operates through map projection and map scale. Iconic and linguistic codes allow us to signify "what", the tectonic code "where". *Temporal codes* are the way in which maps deal with time; they do so through two sub-codes: "tense" (most maps, of course record a past condition even as they are made) and "duration" (the amount of time recorded on the map) (pp. 82-86). Finally in intrasignificant coding is the *presentational code* by which all the signs of the previous four codes are "ordered, arranged, organized" on the map display. "Title, legend box, map image, text, illustrations, inset map images, scale, instructions, charts, apologies, diagrams, photos, explanations, arrows, decorations, color scheme, type faces are all chosen, layered, structured to achieve speech: coherent, articulate discourse" (p. 69).

And as the style of language can establish a message as strongly as can the explicit content of the presentation, the presentational codes of cartography assist in creating meanings for maps beyond those taken consciously. This new level of meaning is governed by *codes of extrasignification* which "govern the appropriation of entire maps as sign vehicles for social and political expression—of values, goals, esthetics and status—as the means of modern myth" (p. 54). Whereas the map exploits the intrasignificant codes to assist its communication, the extrasignificant codes by contrast "exploit the map" for their own ends. The authors identify five such codes (Table 3). The *thematic code* establishes the true subject of the map (which may not be the stated one); the *topical code* ("topos"—place) binds the rather abstract "spaces" governed by the tectonic code into real world "places". The *historical code* exploits the signs of the temporal code for its own ends, and the *rhetorical code*, "having consumed the presentational code" (p. 70), governs how the map is seen within the culture—and it may be seen subconsciously very differently than it appears on the surface. The *utilitarian code* specifies how the map is actually put to use—to navigate, to manifest rights of possession, to decorate—"the anthropology of cartography" (p. 72). All these codes operate upon the map and its elements to move it into a new level of signification: the map as myth.

In addition to extending the semiotics of maps into their meanings as myth, the paper of Wood and Fels deals also with certain mechanical problems touched by earlier writers. They recognize four sign units that vary in size from what they call "elemental" (similar to the "localized sign" of Schlichtmann), to "presentational", at the level of the full map sheet (pp. 90-99); see Table 4 for a summary. They maintain that the smallest unit of both form and meaning, their *elemental sign*, must always include as one constituent meaning the concept "distinct geographical entity", be it concrete (a town for example) or abstract (1, 000 pigs for example) (pp. 90-91).

Finally, Wood and Fels make the point that the hierarchy of map units should not be transposed directly into the map reading process as the necessary sequence through which users acquire meaning. Rather, the reader can begin with signs of any size, and can move up or down the hierarchy at will. The map thus "*unfolds* [meanings] *through a cycle of interpretation* in which it is continually torn down and rebuilt" (p. 88). The map user "...creates a wealth of meaning by selecting and subdividing, combining and recombining its terms in an effort to comprehend and understand" (p.88).

3.3 HEAD: "THE MAP AS NATURAL LANGUAGE: A PARADIGM FOR UNDERSTANDING"

My own position, presented in the paper of the above title in 1984, a revision of a discussion paper of 1981, is similar to the above-mentioned works in its employment of many of the language and semiotic concepts. It recognizes units in the map language, smaller ones of which can be "chunked" into larger. It recognizes the content and expression components of signs, calling them "concept" and "surface" levels of language or "signifié" and "signifiant" components of sign. Where its strikes a different focus from many of the previous papers, however, is its insistence throughout upon the role of the map reader. "The process of map reading, like the process of

Table 4: Units of Language or Sign Systems, and of the Cartographic Language or Sign System, according to Various Schools and Authors

STRUCTURAL LINGUISTICS

	seme (content)				
distinctive feature	*grapheme* (expression)	*morpheme*	*lexeme*	*syntagme*	*proposition*
	phoneme				

RATAJSKI

feature/parameter	*morpheme*	*cartographic sign* (lexis?)
(visual variables in various implantation modes = 15 types)	- primary sign represents a class - secondary sign represents member of class - by location, can be specific - can carry semantic value eg.shape - arbitrary - drawn from reality	- represents a feature of reality - must be situated within a system of spatial reference

PRAVDA

graphematic motif *graphematic space*	*cartographeme*	*cartomorpheme*		*cartosyntagma*
	(visual variable)	minimal unit of both form and significance	(no lexeme, for "no logical dependence of the cartographic sign on the one- or multi-verbal concept")	cartomorpheme situated in logically-defined map space - represents a feature of reality

LYUTYY

cartographic (letter) *("parts of an alphabet")*	*cartographic lexeme*	*cartographic symbol*	*cartographic utterance*
subelemental graphic symbols without meaning of their own eg, a line segment, pattern of texture.	*a cartographic symbol divorced from actual representation*	*used on the map*	*combination of symbols on a map*

SAUSHKIN and IL'YINA

conventional symbol (I)	conventional symbol (II)	*sentence*	map/text
- the most elementary combination of symbol and mapped object that does not carry additional graphic content - the object can be a model of a natural or social phenomenon (river, inhabited place) or a model of a process or regionalization. - at "letter" level.	- as conv. symb. I, with additional information regarding arrangement in space or other attribute - at "word" level	- a combination of conventional models (symbols?)	

(Table 4, continued)

HEAD

distinctive feature	*seme* (grapheme missing)	*morpheme*	*word* (lexeme)	*syntagme*	proposition (sentence)

EASTMAN

graphic variable	*graphic sign*	graphic symbol	graphic variable
- the basic unit of cartographic communication: the range of perceivable differences that graphic marks can exhibit, consisting of variations in location, hue, value, etc.	minimal unit of meaning. -may be discrete in form, indeed individual instances are always discrete, though they may be drawn from a continuum repertoire	contains a meaningful position (see Schlichtmann's "localized sign")	arises from relationships between map symbols. Eastman implies that the relationships are on both surface and meaning levels

ECO (not all semiotic systems have all units - see his Theory of Semiotics pp. 232-234

semantic marker parameter	*sememe* (content) *figura* (expression)	*sign*	*supersign* content is proposition	*text*

SCHLICHTMANN

segment?	*minimal sign*	*macro sign* *"localized sign"*	*text*
	minimal units that have both content and expression whose vehicles can no longer be broken down into limits which are themselves expressions which convey meaning	contains one location and one plan position - at sentence level	- combination of localized signs - examples suggest that it can be of sub-map size

WOOD and FELS

elemental sign	*sign system*	*synthetic sign*	*(map)*
- legend symbol placed within geographic space - sign of geographic feature - refers to a "distinct geographical entity"	a set of similar elemental signs extensive in the space of the map image	two or more sign systems related	

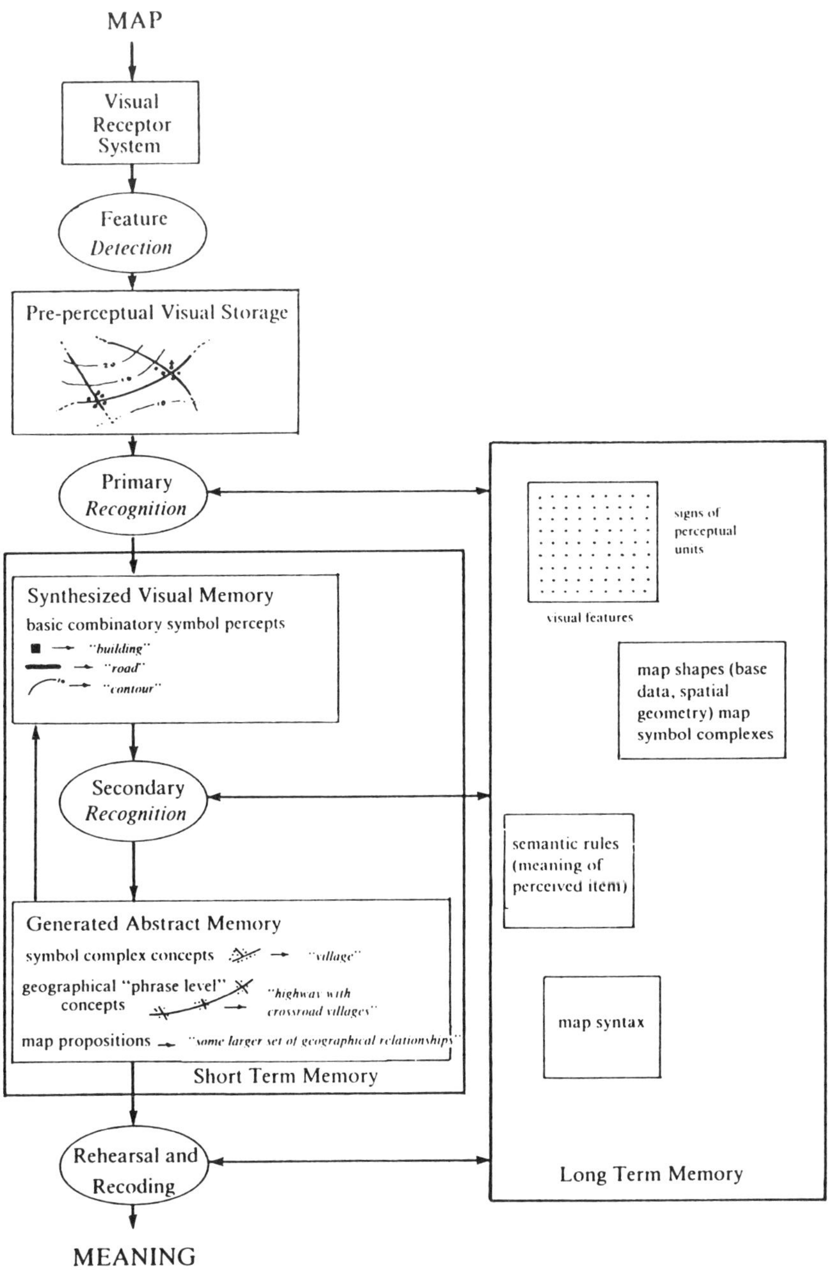

Figure 10: Head's Human Processing of Maps (Head, 1984, p. 24).

reading of text, requires the existence of structures in the reader's mind that are, at the least, equally as important as the marks on paper" (p. 2). Thus this paper gives much more attention than other than other language or semiotic papers to the *process* of map reading and to the *reader's role* in that process. Figure 10 suggests how the map reading process can be fitted neatly within a human information processing model of the reading of printed text. Crucial to this process is the reader's role in holding the codes for sign recognition and identification. Figure 11 stresses that it is the map *reader* that defines which map marks or groupings of map marks are endowed with meaning and thus are components of signs. Although this paper does tend to present a case for upwards chunking of low-level signs into higher, it also points out that as the more efficient text readers "...transfer into STM [Short Term Memory] whole *words*, even phrases, not just letters" (p .4), so too do the more efficient map readers recognize and utilize larger map signs such as "hill", "town", "oil field", etc., rather than using only the minimal signs such as one sinuous brown line (contour), one small black square (building), or one small black circle (oil well). In the reading of a topographic map for landscape visualization, for example, the user "with a developed lexicon of geographic items—'drumlins, kames and outwash plains'—will be better equipped to move quickly and effortlessly through the successive stages of the visualization process" (p. 12). "What we look for is at least as important as what is there" (p. 10). Consider Figure 11. It is almost a raster-structured visual display with no graphic cues for word or sentence breaks. Those of us who use Latin script will re-orient it and those of us who read English will, with some effort, be able to read and make sense of it, and for many Canadians it is a part of National myth. But for some it will be mere marks on paper. The expression level of sign is but a catalyst that releases meaning within the reader's mind.

asforyouwhostandonthethresholdofli
fewithalonghorizonopenbeforeyoufor
alongcareerofusefulnesstoyournative
landifyouwillpermitmeafteralonglifeish
allremindyouthatalreadymanyproblem
srisebeforeyouproblemsofracedivisio
nproblemsofcreeddifferencesproblem
sofeconomicconflictproblemsofnation
aldutyandnationalaspirationletmetelly

Figure 11: Alphabetic field (after Head, 1984, p. 10)

Another way in which we can use the natural language analogy as a pathway to understanding map communication is in considering map messages at the level of texts, and in applying the human information processing and psycholinguistic notion of the *proposition* to the problem of syntax. The paper suggests that the basic map communication is of the nature "what is where". This can be a very simple communication such as "the post office is at the corner of King and Queen Streets" or it may be more complex. From a map or map set of ethnicity and religion, landform and land uses, in a nineteenth century Canadian city, an efficient reader trained in historical geography might be able to see relatively quickly that "Roman Catholic Irish are intermixed in a random pattern with Methodist English within a northeastern sector (which has near its center-city edge an industrial concentration) and is relatively low-lying and cut by sluggish creeks, whereas Anglican English and Presbyterian Scots are in two separate but adjacent clusters in a southwest sector (which has near its center-city edge a commercial concentration) and is located on a knoll". This latter message requires the map reader to relate the individual ethnicity and religion symbols one to another to form recognizable patterns; to relate contour lines one to another to create signs for hill or flats, and to relate these to hydrographic information; to see groupings of similar sets of religion and ethnicity patterns as sectors; to group map symbols representing specific types of industry and specific types of commercial functions into clusters; and then to relate all of these clusters, patterns, groupings or "chunks" one to another.

The paper proposes that the basic schema that drives all this search is that which underlies the natural language *proposition* (pp.14-19). In Anderson and Bowers' "Human Associative Memory" model the proposition is composed of subject, predicate and context. "The Map as Natural Language" paper applies this concept to simple "niveau élémentaire" map reading and to the more complex "niveau d'ensemble" map reading (Figure 12).

In a final section, Head's paper notes how this paradigm of maps as natural language, through its potential for cross-over from the much more widely-researched disciplines of psycholinguistics, human information processing, and reading comprehension and teaching, might assist our research and teaching in cartography (pp. 19-25). First, it focuses attention upon the importance of message content, while many cartographers tend to focus upon expression. Secondly, it may give us insights into better ways of teaching map use, perhaps using the methods of teaching English as a Second Language. And thirdly, as it is maintained that cartographic communication is sufficiently similar to communication through natural language, it will allow us to transfer fairly directly much experimental work from these other disciplines.

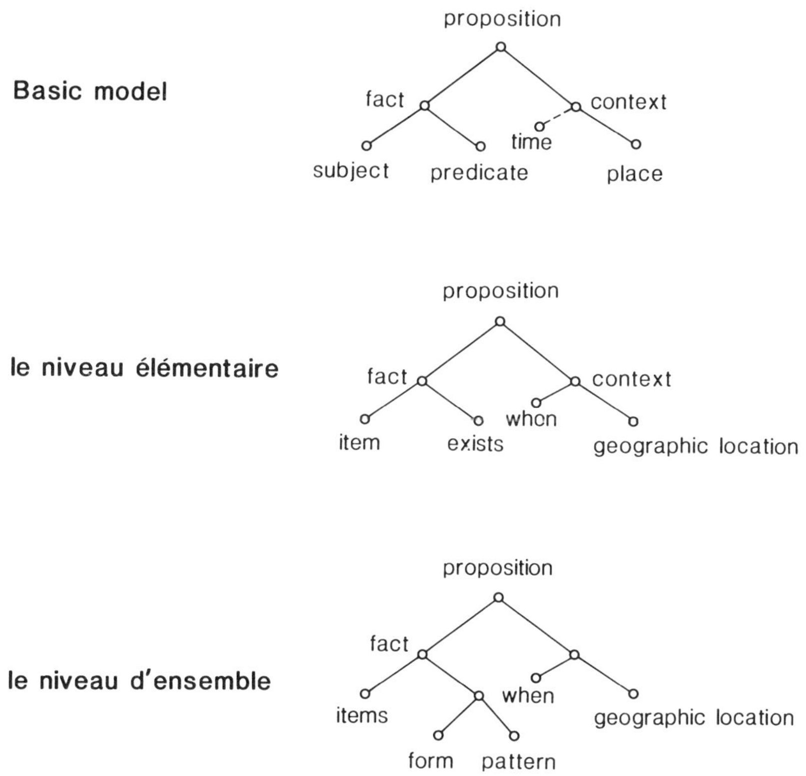

Figure 12: Head's Basic and Derived Propositions (Head, 1984, p. 18).

4. Perspective

4.1 SUMMARY

During the past two decades a broad range of terminology and concepts has been drawn into cartography from semiotics and linguistics. Indeed, the range has probably been too broad, and a number of key concerns in the application of these terms and concepts to cartography have remained untreated or unresolved. We will comment on some of them below.

4.2 AGREEMENTS AND VARIATIONS

4.2.1 *Basics: Definitions.*

- *Content, expression and code.* Linguists and semioticians speak variously of the basic unit of meaning or communication as the morpheme, the lexeme or word, the syntagme or phrase, the proposition or sentence, the sign, the text and others. Whatever sign it is, it is recognized as having two parts, again variously known as meaning, signifié, concept or content on the one hand, and form, signifiant, surface level or expression on the other. For clarity of scholarly communication, I suggest we might agree to use the terminology that a *sign* has *content* and *expression* (in English), *signifié* and *signifiant* (in French). The link between these is the semantic link or code. The word *code* seems most used thus far in the cartographic literature.

- *Sign units.* Table 4 is a summary of a variety of approaches to sign units in the cartographic language or sign system, along with two approaches in structural linguistics and semiotics. The terminology varies considerably as do the correlations between units of particular sizes and the size of the smallest meaningful unit of form. Which set we choose as a basis for a cartographic model is open for discussion. Let us note, however, three aspects of these unit hierarchies.

  1. The semiotic model specifies a series of units of increasing size each of which can be articulated or disarticulated. Some have both content and expression components, some at the lower end may have either content or expression and not both, but these units that have only one component are not present in all signs systems.
  2. The linguistic model also has units that can be articulated or disarticulated, but have a key smallest unit of both content and expression. Articulation proceeds above this unit with units that have *both* content and expression components, below this the units are separately disarticulated into expression units with no content on the one hand and content units with no expression on the other. This "double articulation" is by some linguists seen as one required characteristic of any sign system to be called language.
  3. At least two semiotic-based cartographic theorists require that the smallest cartographic unit of both content and expression must include placement in geographic space as one content element.

  Again, to advance our research we need to agree upon units and their definitions, probably on functional grounds. Can we adopt the semiotic terminology, as it appears to be the most broadly-based? Whatever we adopt, the roles of two-dimensional space and of the reader will also be relevant factors.

- *Syntax.* Cartographic theorists using the linguistic or semiotic paradigms show little agreement on the notion of syntax. Some see it as operating largely on the expression level, and find difficulty in applying the model of a linearly-presented language to a medium that in part of its expression level at least models the two-dimensional globe space. Others clearly recognize that, if syntax is to bind units that have both expression and content, it too must involve contents, and that significant contents and structuring schemata are brought to the map use process by the user.

4.2.2 *The Role of Two-dimensional Space.* Cartographers wrestling with the linguistic analogy have always found the two-dimensional nature of the map language expression a particularly slippery problem. "The elements of the cartographic language however possess a significant peculiarity that they—unlike the verbal sign—do not arrange in a ...linear...sequence...". "Due to the mathematical base the cartographic signs have a logical spatial location..." (Pravda, 1978, p. 2). In other words the location of the expressions on a map is dependent upon the globe location of their referents. "The scale transformation and the projection or rotation transformation are the basic syntactical rules for all (ordinary) maps (Blaut, 1987, p. 29). "While linguistic and iconic codes access the semantic field of geographic knowledge, the tectonic code [which configures graphic space in a particular relation to geodesic space] provides their syntactical superstructure" (Wood and Fels, 1986, p. 20). "Symbols acquire cartographic meaning only as part of a certain 'grammatical structure', namely the mathematical base of the map" (Saushkin and Il'yina, 1985, p. 7). Lyutyy in fact sees this operation as defining his Sublanguage I whose role is to "order the map symbols in accordance with the location of the mapped objects on the surface of the earth" (Lyutyy, 1984, p. 109).

Many of the same authors who point out this strong link between globe space and map space note also that reader assembly of signs into supersigns is not wholly dependent upon these spatial relations. Thus, while Ratajski notes "...positions of words-signs depend solely on positions of the facts in the reality represented by a map", yet he suggests that neither the map maker nor the map reader use only the positions of these map symbols in their message creation. "Treating the syntax of a map as an active process, that is in categories of data transmission, we have to take into account the cartographer's conscious formulation of a message and the map user's apprehension of a transmission." The map reader brings knowledge to the map that bears upon how this spatial data is used: signs on a map "obtain a semantic value only if a map user is familiar with similar arrangements of relations in reality..."(Ratajski, 1975, p. 13).

Schlichtmann extends Ratajski's comments, noting that although "the arrangement of the expressions of localized signs in map space always models globe-spatial arrangements", he too goes on to emphasize that "the process of combining localized signs into texts will require the map user to "have or develop a concept of the compound object which contains the [spatial] relations in question as diagnostic or pertinent traits" (Schlichtmann, 1985, p. 28). "A map text is created by interpretation, or a focus, which 'filters out' some of the macrosigns on a map and organizes them into a larger unit "(p. 28). A map reader can interpret, "at different moments", Schlichtmann notes, "'valley bottom with winding river, oxbow lakes and backswamp' and 'valley bottom with river crossed by causeway with bridge'" (p. 28). The clue to this differing approach to the role of two-dimensional space is in Schlichtmann's comment that "Two or more signs combine by virtue of relations that hold between their contents"(p. 27) not merely by relations between the signs' expressions which reflect the sign referents' global-spatial arrangements. Sign *content* and thus by extension *syntax* are heavily influenced by reader knowledge. The map's localized signs may thus indeed be positioned in a reflection of globe-space, but it is the map reader who selects individual localized signs to build texts and at this point the spatial arrangement of the expression gives precedent to the selection mechanism of the map reader.

4.2.3 *The Role of the Reader*. That map reading is concept-driven as well as data-driven is of course inherent in the paradigm of maps as a medium of semiotics or language. The fact needs reinforcing and elaboration. Much attention needs to be given to "the structures in the reader's mind".

- *Localized signs*. The reader must bring codes to the interpretation of localized signs. He may obtain these codes in any of three ways. They may have been learned in using similar types of maps, in the same way that a natural language text reader will acquire vocabulary from long experience. Secondly, the code may be obtained through the use of the map legend in a manner analogous to the novice stumbling through a foreign language text with the aid of a dictionary; when this method is used exclusively, higher, propositional-level meanings are slow to come as alternative meanings confuse and chunking is awkward. Thirdly, he may obtain the code through the recognition of similarities of the expression with its content. Many of our cartographic expressions are *motivated*; they are chosen so that the average human perceiver senses a homology between one or more parameters of the expression and its contents. Two decades of experimental psychophysical research in cartography has given map makers a good store of information on how to match expression to content for localized signs (Figure 13). Unfortunately, map makers still give too little attention to organizing the *contents* of localized signs before assigning the parameters of expression (but see Schlichtmann, 1988).

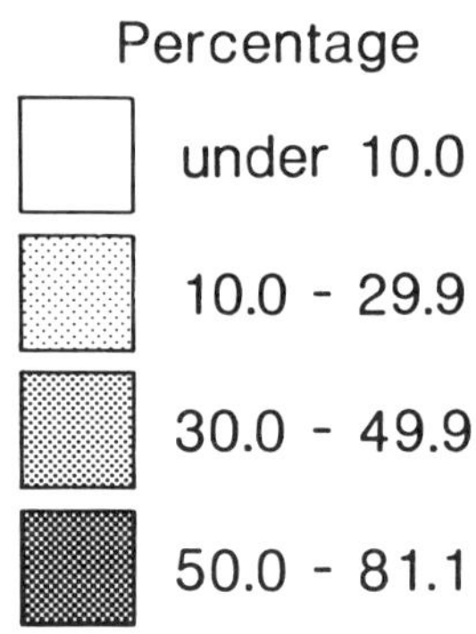

Figure 13: Matching Expression to Content, a Graded Series Example.

- *Signs into supersigns; supersigns into signs*. The *chunking* of localized signs into compound supersigns can be both data-driven and concept-driven. We presume that all the parameters of figure-ground such as articulation, closed figure and so forth operate here. Contrast with the rest of the visual field, as when one set of localized signs is in a low value or high saturation of a hue, can be a powerful factor in forcing a perceived association between these signs. These factors can be effectively utilized when a map designer has a particular content message to drive through. With many maps, however, the message has either been unclear to the author or designer, or the map has been designed to be an efficient *store* of much data rather than to be an efficient *communicator*. In these cases it is the map user who must work hard to create the supersigns. One of the essential map functions is to allow data reduction—by the maker and by the user—so that *spatial relationships*, geographical information, can be discovered and explored. The user then, must apply the cognitive cycle (Figure 14) in striving to recognize what he considers significant spatial relationships. Certain localized signs or portions of such signs are selected and combined with others to create supersigns. Once created, each supersign can become a sign in the further process of combination upwards into texts. Conversely, at some points in the map-using process the texts or supersigns are also broken down into signs. When and how all this proceeds will depend upon the experience and strategies of the map user, as emphasized by Wood and Fels.

- *Map texts*. It is Head's suggestion that map texts are based upon a propositional form. The basic proposition *fact-context* applied to mapping is more specifically of the nature *what-where*: the city is located at 2.56x, 4.7y; or at 135 degrees W, 48 degrees N; or at the big bend of the river. This is using one localized sign—a dot for the city—for *what*, and another series of localized signs—location markers—for *where*. The text might be compound: the high land is to the left and to the right of the route I am taking through the pass—four propositions. To build each of these simple propositions of the *what-where* form, however, a more complex proposition is usually required in which the *predicate* component of *fact* is composed of a series of simple what-where statements. A whole series of localized signs (item and location) is recognized and the spatial relations and other content relations between these are recognized as characteristic of some learned supersign. To expand the above example of a route and a pass, the simplest supersign might be <the route>, built from the expression of a /road symbol/, with

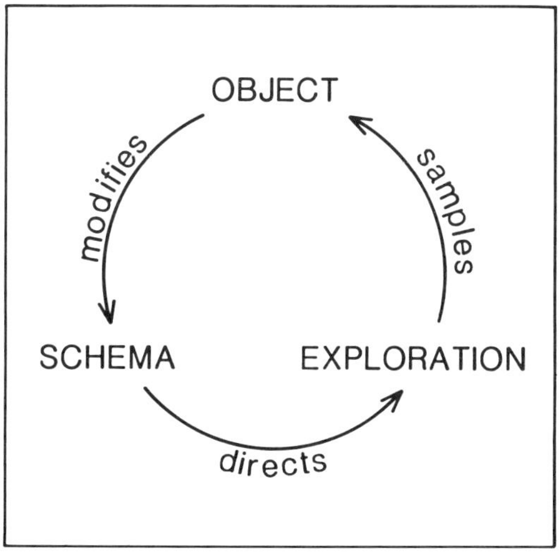

Figure 14: Ulrich Neisser's Cognitive Cycle.

a direction of use attached to it by the map user: <line from xy to xy to xy; has a movement attribute, vector a>. Two other supersigns might be built from the /contour lines and their characteristic spatial relationships/ to form the concept <high land at location xy> which takes the connotation <avoid>. The <pass> is the final supersign, built again from /contour lines and their characteristic spatial relationships> and with a connotation <spatial co-incidence with route>. Each of these fact-context propositions are chunked into new "subjects" of a larger proposition that puts (<route> in conjunction with <pass>) which is centered between two <avoid>s, which in turn may be chunked to <proceed xy>;<avoid xy, xy>. The *fact-context* or *what-where* proposition seems unavoidable.

4.2.4 *Learning the language.* Human languages like English or French or Spanish are learned either as mother tongues or as Second Languages. To learn a language as a mother tongue is a long evolutionary process and involves much of one's childhood if not one's lifetime. Specialized teaching can speed up the acquisition of second languages. Mapping as seen here has enough similarities with human language to be teachable as a second language. There are many methods used in such second language instruction, but they all combine in various ways and various proportions vocabulary, grammar and practice. We need in mapping to teach the lexicon of signs and supersigns (remembering the dual nature of the "sign") the basic syntax, and the combinations that can be generated from these basics. We then can apply these in graded series of exercises—in both map-making and map-using—and in problem-solving in "immersion" situations using complex real-life mapping examples. Mere twenty-minute lessons along these lines, teaching lexicons and teaching thematic map syntax have, in controlled experimental situations, raised map comprehension levels dramatically (Head and Elgood, 1988; Saku, 1990). The language or semiotic paradigm promises, therefore, to be not only theoretically interesting but of practical utility.

5. Conclusions

The basic conclusions from this review must be that cartographic theorists exploiting the semiotic and language paradigms for the last two decades have still only scratched the surface. There is, as yet, little agreement upon terminology and the details of the concepts. There are reasons for this. The literature in semiology, linguistics and human information processing is complex and often contradictory. Theoretical cartographers are few, widely scattered, and publish in several languages.

Recently, however, more interchange is being pursued and there is an effort to adopt mutually-acceptable terminology and conceptual approaches. This advanced study institute is one opportunity for such interchange. The semiotics and language paradigms will both have value for cartographic communication research. Since semiotics does indeed cover all communication systems, it gives us a broad frame of reference. There are significant differences between cartographic communication and human natural language narrowly conceived and we can perhaps progress more quickly if we adopt more of the semiotic *terminology and concepts*, referring them to language only where appropriate.

Language research is more detailed than semiotics in its consideration of *process*. Thus in dealing with *combinatorics and meaning at levels beyond the minimal sign*, where the role of the map reader is large, perhaps we can learn most by employing the linguistic paradigm. More specifically, the knowledge that cartography has built up over the last three decades on the role of psychophysics links well with the semantic theory of the construction of intuitively-readable minimal signs. On the other hand we have hardly begun to explore in any ordered sense the processes of extracting higher-level meanings from maps; for this the research methodologies and approaches of psycho-linguistics, human information processing, reading comprehension and teaching of second languages have much to offer.

What we will learn by applying these paradigms can be put to use in both the way we make maps and the way we use maps. The electronic Geographical Information System requires its operator to both make and use, so the results of our work will be doubly important.

6. Bibliography

Anderson, J. R., and Bower, G. H. (1973) Human Associative Memory, New York.

Aslanikashvili, A. F. (1974) Metakartografiya, Tbilisi.

Bertin, Jacques (1967) Sémiologie Graphique, Paris.

Blaut, James M. (1987) 'Notes Toward A Theory of Mapping Behaviour', Children's Environments Quarterly 4(4), 27-34.

Board, Christopher (1973) 'Cartographic Communication and Standardization', International Yearbook of Cartography 13, 229-238.

Board, Christopher (1984) 'Higher Order Map-Using Tasks: Geographical Lessons in Danger of Being Forgotten', in Board (ed). New Insights in Cartographic Communication, Cartographica Monograph 31, Univ. of Toronto, 85-97

Cuff, David (1972) 'Value versus Chroma in Colour Schemes on Quantitative Map', The Canadian Cartographer 9, 134-140.

Dacey, Michael (1970) 'Linguistic Aspects of Maps and Geographic Information', Ontario Geography 5, 71-80.

Eastman, Ronald (1987) 'Graphic Syntax and Expert Systems for Map Design', Technical Papers, American Congress on Surveying and Mapping 47, 87-96.

Eco, Umberto (1976) A Theory of Semiotics, Bloomington, Indiana.

Eco, Umberto (1979) The Role of the Reader, Bloomington, Indiana.

Flannery, James (1956) The Graduated Circle: a Description, Analysis and Evaluation of a Quantitative Map Symbol, Ph.D. Dissertation, University of Wisconsin, Madison.

Gerber, Rodney V. (1981) 'Competence and Performance in Cartographic Language', The Cartographic Journal 18(2), 104-111.

Gilmartin, Patricia, and Shelton, Elisabeth (1989) 'Choropleth Maps on High Resolution CRTs: the Effects of Number of Classes and Hue on Communication', Cartographica 26(2), 40-52.

Harley, J. B. (1989) 'Deconstructing the Map', Cartographica 26(2), 1-20; and responses to this article in Cartographica 26(3/4), 89-121.

Head, C. Grant (1984) 'The Map as Natural Language: a Paradigm for Understanding', in C. Board (ed.), New Insights in Cartographic Communication, Cartographica Monograph 31, Univ. of Toronto, 1-32.

Head, C. Grant, and Elgood, Richard (1988) 'Impacts of Teaching on Cartographic Language Acquisition: Applying the Linguistic Paradigm to the Process of Map Use', in Head (ed.), Report on the First North American Working Seminar on the Concepts of Cartographic Language, typescript, Wilfrid Laurier University, Waterloo, Canada, 16 pp.

Keates, J. S. (1984) 'The Cartographic Art', in Board (ed.), New Insights in Cartographic Communication, Cartographica Monograph 31, Univ. of Toronto, 37-43.

Lyutyy, A. A. (1984) 'The Language of Maps and its Principal Features', Mapping Sciences and Remote Sensing 21, 103-117 (translated from Voprosy Geografii, 122/123, 40-56).

Lyutyy, A. A. (1985) 'On the Essence of the Language of the Map', Mapping Sciences and Remote Sensing 23 (1986), 127-139 (translated from Geografiya I Prirodnyye Resursy, 1985, 3, 142-151).

Macioch, Andrzej (1982) 'Theoretical Foundations for the Construction of a System of Cartographic Signs', The Polish Cartography, 25-34.

Morris, Charles William (1946) Signs, Language and Behaviour, New York.

Morrison, Joel (1974) 'A Theoretical Framework for Cartographic Generalization with Emphasis on the Process of Symbolization', International Yearbook of Cartography 14, 115-127.

Morrison, Joel (1978) 'The Implications of the Ideas of two Psychologists to the Work of the ICA Commission on Cartographic Communication', International Yearbook of Cartography 18, 58-64.

Pravda, Jan (1978) Map Language: a Logical Graphic System, typescript, 6pp.

Pravda, Jan (1982) 'Kartografia a Kartograficy Jazyk (Cartography and Cartographic Language)', Geograficky Casopis 34(4), 326-335.

Ratajski, Lech (1975) Some Aspects of the Grammar of Map Language in Terms of Cartographic Communication, typescript, 18 pp.

Ratajski, Lech (1976) 'Pewne Aspekty Gramatyki Jezyka Mapy', Polski Przeglad Kartgraficzny 8 (2), 49-58.

Robinson, Arthur H. (1978) 'The Uniqueness of the Map', American Cartographer 5, 5-7.

Robinson, Arthur H., and Petchenik, Barbara Bartz (1976) The Nature of Maps, Chicago.

Saku, James (1990) The Impact of Teaching Cartographic Lexicon and of Geographic Experience on Map Use, Unpublished M.A. thesis, Wilfrid Laurier University, Waterloo, Canada.

Saushkin, Yu G., and Il'yina, L. N. (1984) 'Conversations in the Language of Maps', Soviet Geography 26 (1985), 1-10 (translated from Izvestiya Akademii Nauk SSSR. seriya geograficheskaya, 1984(2), 88-95).

Schlichtmann, Hansgeorg (1984) 'Discussion of C. Grant Head "The Map as Natural Language: a Paradigm for Understanding"', in C. Board (ed.), New Insights in Cartographic Communication, Cartographica Monograph 31, Univ. of Toronto, 33-36.

Schlichtmann, Hansgeorg (1984) 'Characteristic Traits of the Semiotic System "Map Symbolism"', The Cartographic Journal 22, 23-30.

Schlichtmann, Hansgeorg (1988) 'Conceptual Field Structures in Thematic Mapping', in Head (ed.), Report on the First Working Seminar on the Concepts of Cartographic Language, typescript, Wilfrid Laurier University, Waterloo, Canada.

Shannon, Claude E., and Weaver, Warren (1949) The Mathematical Theory of Communication, Urbana, Illinois.

Wood, Denis, and Fels, John (1986) 'Designs on Signs: Myth and Meaning in Maps', Cartographica 23(3), 54-103.

PLAN INFORMATION AND ITS RETRIEVAL IN MAP INTERPRETATION: THE VIEW FROM SEMIOTICS

HANSGEORG SCHLICHTMANN
Department of Geography
University of Regina
Regina, Canada

ABSTRACT. In mapping we convey information about space-occupying and localized objects, alternatively called places. Information is available through signs. It consists of concepts; concepts function as sign contents; these are conveyed by expressions; and expressions are linked to contents by virtue of rules. The information has plan components and plan-free components, which are intertwined. The discussion focuses on the plan components of the information, their conveyance, their retrieval in map interpretation, and the limits of such retrieval. Items of the plan information, shortly called content traits, are divided into internal and external traits. There are three classes of internal traits: cover form (a place is conceptualized as a unitary object, an occurrence range, or a field section), boundary form (from distinct to completely fuzzy), and characteristics of place definition. External traits are values of the variables of dimensionality, location, size, shape, and orientation. Content traits may or may not be formally expressed. Further, they may or may not be implied by object concepts. Where they are formally expressed, they are usually replicated as expression traits through the transformation of point coordinates. The resulting signs are images, *i.e.* motivated signs.

One generally wants to get factually adequate information from a map. Since map images are not completely faithful (*e.g.* for reasons of geometry), interpretation includes an assessment of the fidelity of a map. Motivation permits to infer content traits by virtue of similitude rules. Since these rules presuppose a faithful graphic rendering, they must be "relativized" by the knowledge of infidelities. It is shown what kinds of infidelities exist and how and to what extent the map user can know of them. In the interpretation process, infidelities may be mentally corrected, may only be recognized or assumed to be present, or may go unnoticed. The map interpreter assembles and integrates information from various sources. Map interpretation is a process of reasoning or inference which may be quite involved. It is based on rules which are learnt. Patterns of interpretative reasoning are exemplified; they all involve inferences which take account of infidelities. Apart from simple errors, three sources of such infidelities are identified: effects of coordinate transformations, method-produced misplacement, and shifts between differentially distinguished expression traits.

1. Introduction

Whatever else we do in terrestrial mapping, we represent the space of the earth's surface by means of the map space. To be more exact, it is the two-dimensional space of the globe surface which is so represented, because the map space can accommodate only two dimensions.

A map provides information about objects in space. This information has spatial and nonspatial components. Since we deal with two-dimensional spaces and since the map presents a plan view,

D. M. Mark and A. U. Frank (eds.), Cognitive and Linguistic Aspects of Geographic Space, 263–284.

I shall rather speak of plan information and plan-free information, respectively. This article focuses on the plan components of the information and on and their conveyance. Plan information is all information which relates, in some way, to the two-dimensional space of the globe surface.

But in what way? There are many and disparate aspects under which space is seen in mapping. Cartographic research has gone a long way in disentangling them [1]. But a complete and coherent systematics does not yet exist and may not exist for some time to come. Thus also the conceptual order to be developed below is tentative and likely to require revision. The same holds, incidentally, for a number of technical terms which will have to be suggested.

As for the questions to be answered here, three come to mind immediately:

1. what kinds of plan information are present,
2. how are they conveyed, and
3. how does the map user retrieve them in the interpretation process?

A fourth question will arise during the discussion:

4. within what limits can one interpret?

Not all points can be developed in detail. For simplicity, I shall concentrate on places and static characteristics and generally leave aside movements and relations between places (except for locations). To forestall misunderstandings: I am dealing with maps on paper (rather than with geographic information systems), with their uses by human interpreters, and mainly with those uses which involve visualization.

When we make and interpret maps, we create and use signs. This is a truism. We can study maps in the conceptual framework of semiotics, the field of inquiry devoted to signs, sign systems, and processes of sign use (communication, interpretation, modelling of reality). There are several schools of thought in semiotics, and I shall basically follow Eco (1976). I am concerned with map symbolism, the type of sign system used in mapping (Schlichtmann, 1985), also termed "map language", "cartographic language", and the like. An introduction to map semiotics is given by Head (this volume). In the present context, only three basic points need be restated.

1. A sign consists of a conceptual item - a (sign) content or meaning - and a perceivable item - an expression;
2. Expressions and contents are coupled by a set of correlating rules or conventions, *i.e.* by a code (without which there are no signs);
3. An object or event to which an individual sign is applied is its referent. In cartography, contents are collectively called "(cartographic) information" (*Multilingual Dictionary*, entry 444.0). The terms "content(s)", "meaning", "information" will be used interchangeably. "Transcription" is a cartographer's term for the expression of a content by a visible item. Contents will be cited in single quotation marks, expressions in slashes.

2. Places and Space

Before one can map, one must identify objects to be mapped. Each object occupies a piece of space (even if this piece is only a point). Therefore the object is also a place. Being an object and being a place are here taken as two aspects of the same phenomenon. I shall use the terms "object" and "place" interchangeably, depending on which aspect is of interest. A place has attributes or traits.

A place or object, along with its traits, is the referent of an individual sign at some systematic rank. Referents are states of the world. They are "given" to us through the concepts which we have of them. When we talk about the world, these concepts function as sign contents.

In what follows, I shall usually deal not with objects and their attributes but with the concepts which we have of them. Put differently, I shall usually deal with objects and their attributes as conceptualized. For stylistic reasons I shall sometimes abbreviate and simply speak of objects and traits, if the exact sense is clear from the context.

Defining objects in order to map them amounts to dissecting the continuum of reality. This holds for both aspects, "objectness" and "placeness". A piece of space with its "fill" is subsumed under one (complex) concept and is mentally singled out, i.e. set apart from its surroundings (Schlichtmann, 1989a, 5-6).

What then is one place? Any piece of space that we see fit to regard as a unit under some unifying concept. We can conceptualize in different ways, even at the same time. For the purpose of semiotic analysis one must try to conceptualize in such a way that one can identify basic units which correspond to sentences in natural languages. It is possible to do so. The units so identified are individual localized signs (Schlichtmann, 1985, 25; 1989a, 5-6)[2]. But how does one isolate such a unit? Two operational criteria suggest themselves:

- a notional (content-related) criterion: the object concept is the concept of a place as a whole, independent of any differentiation (example: 'county');
- a graphic (expression-related) criterion: a place is the counterpart of a locator, *i.e.* of a unitary graphic entry that indicates a location (example: area symbol which indicates a county).

As will soon become clear, it is convenient to rely on the graphic criterion.

So much about places. Now what of the space within which they are found? Cartographers conceptualize space in two ways. Either space is dissected into (a finite set of) discrete segments which are points and aggregates of points (*i.e.* lines and areas). Then a plan-free trait is associated with such a space segment as a whole. Example: 'city'. Alternatively, space is treated as continuous, containing an infinite number of points. Then each point is associated with an item of information. Example: a linear place corresponding to a profile or contour, where an elevation value is recorded at every point. A set of adjacent points together with the values assigned to such points will be called a field.

We shall return to this fundamental distinction (subsection 3.2.2). At present we only ask how it relates to the singling-out of places. In the first case space is discretized, one can easily conceive of an individual place, and both of the above operational criteria apply. But what if space is treated as continuous? Does a profile symbol or an isoline correspond to one place or do its many points correspond to different places? In principle, both conceptions are tenable: we have places at two levels. But when we look for a basic unit (as specified before), we see:

- by the notional criterion, the basic places are points;
- by the graphic criterion, a basic place is a point aggregate.

We obviously must choose between the two criteria. To keep the description simple and uniform, we opt for the graphic one. We describe a point aggregate as one place but recognize that this place is inhomogeneous. Such an "aggregate place" is a field or a section of a field.

The present section provides a background for the main argument. I had to develop this background in some detail in order to make clear the nature of the localized sign. The problem is this: the localized sign (more exactly: its expression) is discrete, but space may be discretized or continuous. In the latter case, the continuity of space and the discreteness of the sign must be operationally reconciled in the interest of a uniform and reasonably simple description.

3. The Plan Information and its Transcription

As argued above, we map places or objects as conceptualized. Object concepts are verbalized in ordinary or professional language. They exist before mapping or else are created for the purpose of mapping. When we map an object, we get an expression on the map face, the expression of a localized sign. This expression is linked to a bundle of content components—alternatively called content features or content traits—which is structured by relations. It can be analyzed.

Within such a bundle of conceptual units, we distinguish plan traits and plan-free traits. Suppose a forest is mapped. The object concept 'forest' is an item of the plan-free information. The associated concepts of the location, shape, and size of the forest are plan traits. This is correct as far as it goes. But the two kinds of information are intertwined. 'Forest', even though a plan-free concept, contains components which relate to two-dimensional space, such as 'area' and 'collective of trees with intervening space'. Let us say that such components are implied. "Implied" has a loose, non-technical sense here: a concept is implied if it is conventionally released or called to mind through another concept ("conventionally" meaning "on the basis of a social competence of organizing reality")[3]. The set of implied concepts, incidentally, may be quite large in individual cases. It includes such notions as may appear in a dictionary definition of a word or phrase, but it also includes notions which originate from factual or world experience, as long as these are conventionally called to mind.

3.1 OVERVIEW

What kinds of plan information are there, and how is such information available to the map user?

3.1.1. *Kinds of Plan Information.* As for the first question, I tentatively distinguish four classes of plan information.

1. A mapped object can be seen as covering a piece of space. Intuitively, a lake and the distribution areal of a plant species cover space differently. The difference relates to the disposition of space in the object's interior. Relevant plan traits are subsumed under the term "cover form". They are discussed in subsection 3.2.
2. A mapped place has a boundary which is more or less distinct. This is a matter of the disposition of space at the boundary of the place. I speak of boundary form. It is discussed in subsection 3.3.

3. The cartographer singles out or defines a place as a spatial unit. A place as defined may or may not conform to the spatial distribution of it pertinent "fill". To exemplify: areas where a plant species occurs may be determined by two procedures of place definition:

   - One traces an outline around each set of specimens which are reasonably close together. This way one maps one or more distribution areas (and a residual space). The place definition is contingent on the spatial distribution of the "fill" of the place(s).
   - One defines a set of squares and afterwards records for each whether or not the plant species occurs in it. The place definition is independent of the spatial distribution of the "fill".

   Place definition encompasses several kinds of characteristics which come into view when one distinguishes methods of thematic mapping. In view of space constraints, place definition will not be discussed further.

   Characteristics in classes 1 to 3 relate directly to a place. They are, so to speak, internal to it. For want of a better term I call them internal plan traits. Such meaning components are implied by object concepts.

4. The fourth class encompasses characteristics like location and shape and is discussed in subsection 3.4. Often one can abstract from them (*e.g.* imagining an object without its location or shape); they may be seen as added to the object. I provisionally call them external plan traits. Such meaning components may or may not be implied by object concepts.

3.1.2. *Information Conveyance.* The second question: how are individual items of plan information available to the map user?

Before answering this question, it is necessary to introduce the concept of locator. The expression of a localized sign contains one symbol or more than one. There is usually one mark which indicates the location of the mapped object: it functions as a locator[4]. Other marks, where present, are complements: they do not indicate locations. That is, a complement refers to the same place as the locator, not to a separate place. In Figure 1a, the area symbols are locators, and the circles function as complements.

Items of the plan information, now, may be formally expressed or implied. They may be formally expressed by various means, in the most important and most frequent case by graphic characteristics of the locators. These characteristics are expression traits. Since they involve the disposition of plan space on the map, I also call them plan traits (or graphic plan traits)[5]. Examples of implied traits have already been given. We add: what is implied are not only properties and not only components of the concept of the mapped object. Implication may also point to concepts of other objects and to spatial (and other) relations between objects. For example, in geomorphology 'pingo' implies 'Arctic' (another object or place) and 'in (the Arctic)' (a relation). So, if a map shows a pingo, one may look for other things which are known to exist in the Arctic[6].

An individual item of plan information, then, is formally expressed or implied, and it may be both. (Also, it may not be conveyed at all and thus be unavailable to the map user.) As indicated before, an internal trait is implied by definition; an external trait may or may not be be implied.

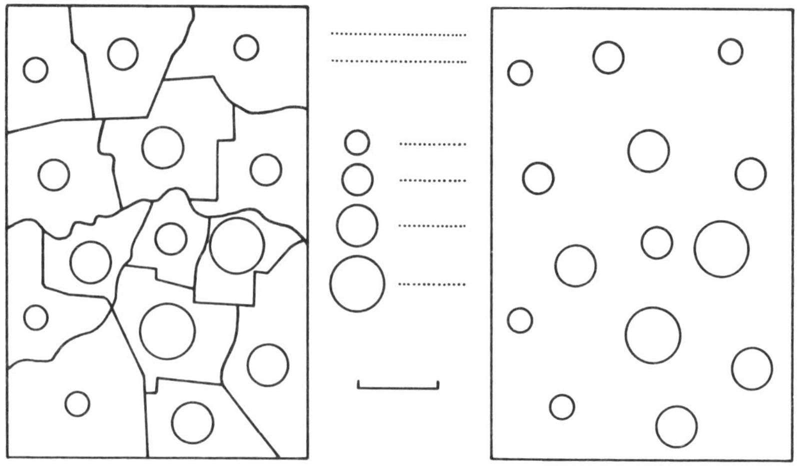

Figure 1. Portion of a fictitious map: "City X - number of inhabitants by wards at date Y". (a) Area symbols serve as locators. (b) The locators have been omitted: area-to-point shift. (After Schlichtmann, 1979, p. 84. Reprinted by permission.)

3.2. INTERNAL TRAITS: COVER FORM

3.2.1. *Basic Distinctions*. Taking one's clues from a systematics of mapping methods, one can distinguish three basic cover forms. Objects or places as conceptualized are unitary objects, occurrence ranges, or field sections.

- Unitary object, *e.g.* a bridge. It is primarily seen as one thing which occupies space, not as a piece of space in which other objects are found. It may be homogeneous or consist of parts (see subsection 3.2.4). In terms of its dimensionality in two-dimensional space, a unitary object is a point, line, or area place.
- Range of occurrence of dispersed objects, *e.g.* the distribution areal of balsam poplar or a township containing 17 farmsteads. It contains objects of the same kind and space segments devoid of these objects. In terms of dimensionality it is an area place. When the objects are counted, the range functions as a reference segment for quantities.
- Field section, *e.g.* a space section within which each point is associated with a separate value of an information variable. It contains many points or only one. It corresponds to a slope-shade mark, a profile, or indeed a spot elevation. That is, a field section may have any of the three dimensionalities.

3.2.2. *Comments*. Here two comments must be added. First, places can sometimes be be thought of in alternative ways. To exemplify: the territory of a county may be seen alternatively as the occurrence range of X inhabitants or as a piece of space that is, as a whole, associated with the plan-free information 'X inhabitants'[7]. Second, the distinction between field section and other cover forms is fundamental but often poorly explained in the literature. It follows from the fact that there are two space conceptions, as noted above. Some cartographers cover this distinction by speaking of discrete and continuous phenomena (*e.g.* Cuff and Mattson, 1982, 15) or of smooth and nonsmooth phenomena (Robinson *et al.*, 1985, 108). It is clearer to speak not of phenomena but of spatial distributions of the values that are associated with places. A distribution is smooth if such values are associated with all points in a delimitable piece of space which contains more than one point; otherwise it is nonsmooth. Since each point carries a value, one can sample values at certain points in a field and interpolate those at other points according to a specific code component (Schlichtmann, 1979, 92). A smooth distribution gives rise to a field. A place within a field which is identified by one symbol on the map is a field section. A field section may, of course, be identical with a whole field (*e.g.* the high-water line of the flood of year X on one side of valley Y).

3.2.3. *Cover Form and Symbols*. Finally, how do we know about cover form? It is always implied. Is it also visually expressed? At first glance one might not expect so: the referent of a localized sign is treated as spatially homogeneous, and the locator should be visually undifferentiated. But this holds only where space is discretized. A field section, as we have seen, is not homogeneous. In the profile case, values associated with points differ between points; since all these values are mapped (in principle at least), the symbol is of necessity visually differentiated. In the isoline case, the values do not differ between points; they are still all mapped, but the symbol is visually undifferentiated.

3.2.4. *More Detailed Distinctions*. The concept of cover form pertains to the object-internal space disposition. One can characterize this disposition in more detail. Concepts of unitary objects can then be assigned to at least four types:

1. Type 'lake': homogeneous object.
2. Type 'farmstead': object composed of parts, usually of different kinds (here of buildings and yard spaces).
3. Type 'forest': collective of dispersed objects of the same kind with intervening space.
4. Type '(deposit of) debris': mass of particles.

For a cartographer, these four types most easily come to mind. The conceptual space disposition within places has been worked out in greater detail by linguists, *e.g.* by Leisi as early as 1953 (now Leisi, 1975) and more recently by Talmy (1983, 1988). Such linguistic works may yield further insights in the ways in which map makers and users think about terrestrial space[8].

Likewise one can make further distinctions for occurrence ranges and field sections. These distinctions, however, are based on criteria of place definition, which have been excluded from the discussion.

3.3. INTERNAL TRAITS: BOUNDARY FORM

Places as conceptualized are discrete, but their boundaries are not necessarily distinct. At least one will separate distinct and indistinct boundaries[9], but at a closer view one must recognize degrees of distinctness, ranging between the extremes of the strict boundary of a cadastral lot and the completely fuzzy boundary of a marine region like the Sargasso Sea.

Boundary form is always implied. In addition, degrees of boundary distinctness can be formally indicated by solid lines, dashed lines, transition fringes (*e.g.* between facies on a geological map), assuming the map scale is large enough. For visual and other reasons, one generally delimits by solid lines, whether or not the boundary is distinct. In the case of extreme fuzziness, boundaries are not indicated; this is familiar from named but undelimited regions as shown on general maps ("Sierra de Gredos", "Sargasso Sea"). In such cases there is no locator, and location is only loosely and indirectly indicated through the placement of the name.

3.4. EXTERNAL TRAITS

The external plan traits by which we characterize objects belong to five classes or are values of five variables: dimensionality[10], location, size, shape, and orientation. They are a heterogeneous set, and their subsumption under a common heading is provisional. External content traits are in most cases formally expressed by expression traits which are carried by locators. (They may also be implied.) More exactly, they are automatically replicated through the transformation of point coordinates[11]. The relevant expression traits are values of five variables: implantation, position, size, shape, and orientation. Because of the replication, content and expression variables correspond strictly, as is obvious from Table 1.

Content variables	Expression variables
dimensionality	implantation
location	position
size	size
shape	shape
orientation	orientation

Table 1. External traits of the plan information and their transcription: content variables and corresponding expression variables.

A few comments may make Table 1 clearer. The expression characteristics have two functions in map symbolism: they may serve to convey plan as well as plan-free information. Where the first function is pertinent, as in the present discussion, one could more strictly speak of plan position, plan shape, etc. As for location and position, it would be more exact to use these words in the plural. A location or position is assigned relative to a reference system, and there are so many kinds of location and position as there are reference systems (coordinate-based or otherwise). One or more of such systems may be made pertinent in map making and interpretation. Orientation and shape are not distinguished once and for all (Bunge, 1966, 69). Recognizing orientation as a separate kind of trait, distinct from shape, is senseful only for elongated or asymmetrical objects and symbols. Orientation presupposes a reference frame with respect to which an object is oriented. Again there are so many kinds of orientation as there are pertinent reference frames. For example, lakes may be described as oriented with respect to compass points, barns with respect to a rural road (*i.e.* to another object).

4. Concepts and Facts, Motivation and Fidelity

So far an overview of the plan information and its conveyance has been given. We must now turn to map interpretation, *i.e.* the retrieval and processing of information by the map user. Although interpretation has more than one aim, I shall here concentrate on the aim of extracting factual knowledge from a map.

Here a new problem comes into view. When we interpret, we manipulate sign contents (or concepts), and these must be kept apart from referents (or facts). But often such a manipulation of contents cannot be fully understood unless we take referents into account. This problem arises when we deal with motivation and with factual adequacy (fidelity).

First, the matter of motivation. A map can be seen as an assemblage of plan images of objects (at least if one focuses on the plan component of map symbolism)[12]. Now, in an image the relation between content and expression is not arbitrary but motivated, i.e. images are motivated signs. Somewhat simplified, content traits are replicated as expression traits, the content of an image determines its expression, and the expression permits to infer the content. *E.g.* the shape of a specific lake symbol permits to infer the contents 'dammed reservoir' and 'Lake Volta'.

Motivation is notoriously difficult to account for. The general problem is this:

- we have images of objects;
- their expressions contain traits which are somehow derived from the objects;
- how then can they reappear as content traits, *i.e.* as conceptual items?

At least for maps in the strict sense (*Multilingual Dictionary*, entry 811.1), an explanation can be developed on the basis of Eco's theory of sign production (Eco, 1976, 195-200, 245-261, 306 f.), if one reconstructs the processes of sign production and sign interpretation. This topic, however, cannot be developed here (for a partial account see Schlichtmann, 1985, p. 26). For the present purpose, we summarize the main points:

- There are mechanisms by virtue of which observable object attributes can be understood to reappear as sign contents.
- Expression traits can be understood as replicated content traits.
- Contents are inferred as replicated by virtue of rules. Extending Eco's term, I call them similitude rules (Eco, 1976, 252).
- Some observable object attributes are learnt as diagnostic of their objects (*e.g.* the shape of an alluvial fan or that of the continent of Africa). This way they recur as content features which are implied by object concepts.

Second, we must take account of fidelity.

- An image has a content, *i.e.* the concept of an object.
- Ideally, this concept is factually adequate (a notion to be made more precise below). Or, focusing on the graphic rendering, one can also say that the transcription of a concept, or the representation of an object, is factually adequate or faithful. Similitude rules, introduced above, presuppose a completely faithful rendering. We then can retrieve the content of an image by reading its expression at face value.
- If one takes a map as a source of factual knowledge, one obviously looks for such a faithful rendering. But maps misrepresent to a greater or lesser degree, *e.g.* for reasons of geometry,

hence there are infidelities or deviations from faithful transcription. Consequently similitude rules are not enough for interpretation; they must be "relativized" or limited by the knowledge of misrepresentation. This limiting knowledge may also be conceived of as a set of rules, usually of a global character.

- Such knowledge permits to assess the fidelity of a map. "Fidelity" is here taken to mean the extent to which the graphic rendering is faithful.

But what is the yardstick of fidelity? We loosely say that we assess fidelity by going back to the facts. But facts are already conceptualized, so we are really going back to specific conceptions of facts. These conceptions have been found "good enough" and serve as a standard. For the plan information, at least for the external traits, we have a standard which is universal[13]: the earth's surface is imagined as seen from vertically above at all points, and all objects selected for mapping[14] are seen at their true sizes, shapes, etc. A large-scale topographic map comes close to meeting this standard (except, of course, that it is reduced). To the extent that a map satisfies such a standard, similitude rules are sufficient for its interpretation. Finally, although one can devise measures of fidelity, let us concentrate here on a global visual assessment.

5. Infidelities and Metasemiotic Information

One interpretation task, then, is to assess the fidelity of the map. It looms large in retrieving the plan information. Thus one must understand what deviations from faithful transcription can be expected and how we know of them.

5.1. THE NATURE OF INFIDELITIES

5.1.1. *Kinds*. There are the following kinds of infidelities in the transcription of the plan information:

1. A trait is not transcribed. If a town is mapped as a point, the shape of the town area is obviously not automatically replicated, and usually it is not replicated at all[15].
2. A trait is formally rendered but not faithfully replicated. We can describe this situation as follows: a "correct" trait is replaced by another trait of the same kind or, rather, from the same paradigm[16]. Such replacements are of two types, depending on the way in which graphic items are distinguished (Schlichtmann, 1985, 27):

- Type "shape distortion": the traits are gradually distinguished; in other words, they are values of a gradual variable (here the variable of shape). That is, between two specific shapes there are many possible intervening shapes.
- Type "point for area": the traits are differentially distinguished, or they are values of a differential variable. We have either a point or an area, but there are no transitions.

Since most relevant variables are gradual, most infidelities are matters of degree. This degree ranges from negligible to intolerable, *i.e.* a representation is fully correct, approximately correct, or wrong. Whether a representation is still approximately correct or already wrong depends on the interpreter's requirements. Take municipalities in a European country. If we want to study the development of the territorial units, their shapes as mapped must not be strongly simplified; if we just use municipalities as reference areas for mapping statistical data, shapes need not be shown in great detail, and a relatively great simplification is tolerable. But whether infidelities are important or negligible, in any case the map user should know about them.

5.1.2. *Origins.* As for the origins of infidelities, one most readily thinks of coordinate transformation with its three component processes: projection, reduction, and choice of map orientation. But deviations from faithful transcription can arise at all stages of map making. That is, they may also be engendered by generalization, mapping methods, and drawing, and likewise by procedures of data procurement (surveying, field observation, and data compilation from man-made sources)[17].

5.1.3. *Description.* To describe infidelities one adopts a heuristic two-step procedure:

- first one assumes that the transcription is faithful, so that the interpreter can rely on similitude rules and read at face value;
- then one searches for information on deviations from faithful transcription.

  In map interpretation we often take the first step only, *i.e.* we assume fidelity. This is easy to understand: what we see on the map is very suggestive. It is more difficult to be aware of infidelities and to take the second step.

5.1.4. *Recognizing Infidelities and Correcting for Them.* Assume an object is not faithfully imaged, and we know it. Sometimes we can mentally correct for an infidelity because we already have learnt the relevant trait, and this trait as learnt sufficiently fits the facts. For example, we have seen Lake X from the air and the image of Jamaica on a globe or, more likely, on a map with negligible distortion. But sometimes we cannot correct; we can only recognize or assume that there is a misrepresentation (and perhaps roughly estimate its degree). Then our interpretation is indeterminate. More exactly, there are four situations, characterized by decreasing degrees of knowledge (columns 1-4 in Table 2). The reader may supply examples, thinking of the various origins of infidelities as mentioned above.

	1	2	3	4
Known or assumed for good reason: there is an infidelity	+	+	+	-
The infidelity is known	+	+	-	NA
The infidelity can be mentally corrected	+	-	NA	NA

Table 2. Degrees of knowledge of infidelities (increasing knowledge toward the right).

Column 4 covers not only the case of the inexperienced map user. Even a trained interpreter may be unaware of infidelities. Much relevant knowledge is factual (or world knowledge), and factual knowledge may well be incomplete. Suppose, for example, the outline of a rock unit on a geological map is in error. The map user will not notice the error, unless he has better factual information (perhaps from a field check) or unless the map message is implausible.

5.2. REFERENTIAL AND METASEMIOTIC INFORMATION

Where, finally, do we find the information to assess fidelity? Here I must take a broader look and present some background.

Depending on the uses to which signs are put, one can distinguish several kinds of meaning or information. Here we need only deal with two of them: referential and metasemiotic information[18]. Referential information is information about referents, *i.e.* states of the world. Metasemiotic information, in the widest sense, is information about the sign system that one uses. In more detail, it relates to concept formation in the referential information, to the ways in which reality is subsumed under concepts, and to the code(s) by virtue of which concepts are conveyed through expressions[19]. The annotation of a map mainly serves to provide metasemiotic information. Only a part of such information is of interest here: that part which relates to the fidelity of the (referential) plan information as conveyed by the map. I call it fidelity information.

Where do we find the referential and metasemiotic information, respectively? As we have seen already, an individual item of the (referential) plan information, if it is available, is accessible in two basic ways: it is formally expressed or implied by an object concept (or both). In addition, some items are tacit background knowledge about the world[20].

As for fidelity information, matters are less straightforward.

- Sometimes fidelity information is expressed on the map face and then pertains to individual places. On geological maps, for example, one indicates by an expression trait whether a fault line or its location has been observed in the field or only assumed (the line on the map is either solid or dashed). Here there is a sign which conveys information about a specific other sign.
- More often fidelity is only globally indicated by a comment in the annotation (*e.g.* "Mercator's projection"). Here there is a sign which conveys information about a set of other signs.
- Too frequently there is no indication at all (*e.g.* the projection statement is omitted). If so, fidelity information can only be available as tacit background knowledge. In many cases one must recognize, by whatever clues, a certain man-made object as a map[21]. The concept 'map' implies features like 'projected', 'reduced', 'generalized'. These meaning components serve as a metasemiotic frame for relativizing the face-value information conveyed by symbol traits. And because of the global character of the fidelity information we do not learn how to correct for a deviation in reading a specific symbol; we only learn that some deviation must be expected. Therefore the interpreting is limited by some indeterminacy.

6. Interpretation: Infidelities and Patterns of Reasoning

Let a map interpreter look for information about a particular place. Then he proceeds just as in understanding a sentence in a language. He retrieves information from several sources: the face-value meanings of symbol traits, content features implied in the concept of the place, and implied meanings released by the context (in linguistics also called co-text)[22], further metasemiotic information. These components are inputs in the interpretation process. The map user assembles and integrates information items, reconciles incompatible meanings, and establishes limits to his interpretation (although some limits may go unrecognized). Interpretation, then, far from being straightforward reading, involves reasoning or inference. Interpretation, like transcription, is governed by rules. These are learnt, although they are not always easy to make explicit.

In what follows, I shall exemplify major patterns of interpretative reasoning (limiting myself, of course, to the plan information). As before, I concentrate on the interpretation aim of extracting factual knowledge. Simple face-value reading is left aside; the more complicated cases all involve reasoning about fidelity (which is why the fidelity problem had to be dealt with beforehand). Not surprisingly, one will find two types of reasoning patterns:

- infidelities are mentally corrected by interpretative choices between incompatible meanings;
- infidelities are recognized but cannot be corrected, thus they set limits to the interpretation, making it somewhat indeterminate.

Infidelities are engendered in at least four ways:

1. commission of errors in the strict sense, mainly in surveying and drafting;
2. transformation of point coordinates;
3. method-produced misplacement, *i.e.* misrepresentation of location;
4. shifts between differentially distinguished expression traits.

Errors are systematic or, more often, unsystematic. The other three kinds of manipulation produce systematic deviations from faithful transcription. For simplicity, the effects of generalization will not be considered. Some of them will find their places under the four preceding headings; others may reveal further sources of infidelities. We leave errors aside and turn to the effects of coordinate transformations.

6.1. EFFECTS OF COORDINATE TRANSFORMATIONS

Coordinate transformations have the following well-known effects:

- reduction (a matter of absolute size);
- distortion of shape, orientation, and relative size;
- change of implantation;
- loss of information as related to change of implantation.

By way of example, we look at reduction-induced change of implantation. Let a city be represented by a point locator rather than an area locator (*i.e.* 'area' is not replicated). Now, dimensionality is always implied (for a cartographer at least). Thus the sign content 'city' contains the feature 'area'. We get two incompatible messages:

- 'point' (from the symbol trait /point/),
- 'area' (from the content 'city' through the annotation). The conflict between the two incompatible information items is resolved by an interpretative choice: we interpret as 'area'.

This choice is governed by two rules (which must be learnt):

1. The mark retains all object dimensions that can be legibly replicated at the map scale.
2. Where two dimensionality terms are incompatible, read dimensionality as implied.

The second interpretation rule holds generally for misrepresentations engendered by coordinate transformations: in case of conflicting information, the implied meaning is taken as correct.

The first interpretation rule corresponds to a transcription rule which aims at keeping the reduction-induced infidelity to a minimum. One must assume such a directive of strict transcription in the absence of information to the contrary. But even in large-scale topographic mapping this directive may be relaxed or, rather, overridden by rules which give the map maker some leeway. They are laid down in the compilation instructions of the topographic services.

6.2. METHOD-PRODUCED MISPLACEMENT

As Morrison has shown, the transcription of location may be subject to method-produced error (Morrison, 1971). I prefer to speak of method-produced misplacement of symbols. Such misplacement is independent of coordinate transformations. Rather, it is brought about by a method of recording or mapping spatial data. The position of a locator, then, is in part an artifact of the method and no faithful expression of the corresponding location.

By way of example, we consider the construction of isometric lines in the mapping of fields. (These are lines of equal quantitative values, where such values are smoothly distributed; see subsection 3.2.2). Specifically we may think of positioning hand-interpolated elevation contours (before the advent of photogrammetric contouring) and of interpolating unreduced temperature values. To get at least a rough idea of possible misplacement, one must not only know about various inputs into the interpolation process (Morrison, 1971) but also about characteristics of the spatial distribution of elevation, unreduced temperature, etc. For example, unreduced temperature is influenced by elevation. If there is a high mountain slope midway between two stations, isotherms will crowd here. An experienced map user might remember that the map author may or may not have taken account of the influence of elevation. (For further pitfalls in maps with isometric lines see Witt, 1970, coll. 323-346.)

In contrast to the preceding case, the interpreter here cannot correct for infidelities. He can only keep in mind that they exist.

6.3. SHIFTS BETWEEN PLAN TRAITS

In subsection 5.1.1, some deviations from faithful transcription have been described as replacements of traits of the same kind. Now let a "correct" shape be replaced by another shape. Shape traits are gradually distinguished, and the replacement is often not obvious. Instead, let an area be mapped as a point. Implantations are differentially distinguished, and for such traits the replacement is a major and obvious change. If such a replacement is independent of the coordinate transformation, I call it a shift between differentially distinguished graphic plan traits or a shift for short[23].

Through such a shift an expression traits becomes an artifact of a procedure. In this sense shifts also belong to the method-produced misrepresentations. In detecting them, the map user again resorts to inferences. Shifts underlie certain methods of thematic mapping (mainly the dot and isopleth methods).

The major types of manipulations under this heading involve:

- scale-independent shifts between point and area implantation,
- shifts between the mapping of smooth and nonsmooth distributions[24].

6.3.1. *Shifts: Area-to-Point and Point-to-Area.* Figure 1a (in subsection 3.1.2) is a portion of a fictitious map entitled "City X - number of inhabitants by wards at date Y". Area symbols are locators and represent wards; circles are complements; circle size expresses number of inhabitants[25].

Figure 1b is derived from Figure 1a by the deletion of the area symbols, *i.e.* the locators. The title remains the same. Now some additional reasoning is required. It is informally summarized in Table 3. Inputs to the interpretation process are listed in the left column, inferences and outputs in the right. Notations are explained in the caption.

Inputs	Inferences, outputs
ET: /point/ A: 'ward'	'point' 'ward' —> 'area' I: 'area' supersedes 'point'
ETs: /positions/ A: 'wards'	'non-contiguous places' 'wards' —> 'contiguous' I: 'contiguous places' supersedes 'non-contiguous places'
ET: /point/ } A: 'ward' —> 'area' } ETs: positions not contiguous (circles do not overlap)	/point/ imposed by reduction /point/ not imposed by reduction I: area-to-point shift

Table 3. Area-to-point shift. (ET: expression trait; A: annotation; —>: implies; I: inference.)

In the three parts of the table we make the following inferences. The mapped places are areas (wards). They are contiguous; thus the map furnishes a rough idea of their relative sizes. At the chosen scale, contiguous areas could have been legibly mapped (otherwise the circles would strongly overlap). Thus, point implantation is not imposed by reduction but is scale-independent. Therefore the mapping procedure involves an area-to-point shift as a separate manipulation. By virtue of the relation of contiguity, the map user can perform the correct inference and can recognize the shift.

The area-to-point shift has another effect. On Figure 1a the point symbols are complements; on Figure 1b they remain complements and must be read as such. The locators have been deleted, therefore the locations of the wards are not strictly expressed, although the positions of the circles permit to guess them in a loose way.

The area-to-point shift is characteristic of the dot method. Here one defines areas, each containing the same number of items, assigns to each area a point symbol which expresses a unit value, and omits the area locators[26]. In discussing a dot map used in teaching, Jenks presents indirect evidence that the shift must be explicitly recognized by the map user. He reports: "...one of my students, a trained and experienced teacher,...had always assumed, although it had seemed odd to her, that there was a herd of exactly 25 cows located under each dot" (Jenks, 1970, 179).

So much on the area-to-point shift. The reverse manipulation, the point-to-area shift, is exemplified by Thiessen polygons used as reference segments. It will not be discussed.

6.3.2. *Shifts: Nonsmooth-to-Smooth and Smooth-to-Nonsmooth.* The nonsmooth-to-smooth shift underlies the isopleth method. (Here nonsmooth distributions are treated as if they were smooth; the requisite lines of equal value are called isopleths.) Take a set of small towns in an overseas area of 19th-century agricultural colonization. For each town take the date of establishment and construct an isopleth map with the title "Area X - spread of towns over period Y". The distribution of the dates is nonsmooth: first, the towns are not points (even though, at the map scale, they are represented by points); second, the dates relate to towns and not to the countryside between them (by virtue of /towns/ in the title). The dates on the isopleths are not "real", the map author has constructed them in order to demonstrate a development in space.

What happens here in spatial terms? The towns, treated as points, make up a complete and finite set of places. This set is taken as a sample out of the infinite set of all points in a definable area, so that interpolations are possible. In other words, a nonsmooth distribution is mapped as if it were smooth. Since the title refers to towns, the map user can realize that the original distribution is nonsmooth, contrary to the message conveyed by the marks on the map face.

Now take a more familiar case: an isopleth map of the population densities in contiguous counties. Here two shifts take place in sequence:

- first, magnitudes associated with areas are assigned to points (area-to-point shift);
- subsequently, this complete and finite set of points is treated as noted before.

The reverse shift, smooth-to-nonsmooth, is rare. A decidedly odd example: one constructs Thiessen polygons around meteorological stations, takes precipitation values measured at these stations, and associates them with the corresponding Thiessen polygons. An point-to-area shift is automatically present.

6.3.3. *Assessment of Shifts.* Although these procedures have a legitimate place in mapping and map-based research, they are nevertheless profound manipulations which one can re-think only at the expense of much interpretation effort. They may produce map images which are far removed from "reality", more exactly: very different from images which can safely be read according to similitude rules. Consequently they may mislead, given the suggestiveness of the visual. The nonsmooth-to-smooth shift is the major case in point. It entails a strong and fundamental misrepresentation, quite apart from additional problems which are linked to specific interpolation procedures. Thus isopleth maps are notoriously controversial (*e.g.* Witt, 1972, 13 f.). A prudent cartographer will balance the advantage of a shift against its disadvantage.

7. Summary and Perspective

Plan information, often also called spatial information, is that part of the cartographic information which relates to the two-dimensional globe space. Contrary to simplifying ideas, it encompasses far more than location, let alone location as described by point coordinates. Rather, there are several plan-related characteristics (and they are partly bound up with each other). They have not yet been sufficiently separated and systematized. Thus the overview which I have presented is tentative. It is also incomplete: I have not, or not fully, covered place definition, movements, and relations between places.

Within these limits, I have discussed the plan information about places which is contained in maps, the ways in which it is conveyed, and its retrieval in map interpretation. Such retrieval includes the assessment of the fidelity of the graphic transcription.

The main results are summarized below.

1. The contents which collectively constitute the plan information are the conceptualized plan traits of objects or places. Four groups of such traits have been distinguished:

- cover form (internal traits);
- boundary form (internal traits);
- place definition (internal traits, not discussed);
- external traits (dimensionality, location, size, shape, orientation).

2. Plan information about a place may or may not be formally expressed (usually by plan traits of locator symbols). Further, it may or may not be implied by the object concept.

3. A major aim of map interpretation is the extraction of information about facts. In view of this aim it was necessary to show, albeit briefly, how facts (referents) are taken into account in handling concepts (sign contents). This point relates to the representation of objects by images (motivated signs) and to the fidelity of the graphic rendering. Owing to motivation, object attributes recur as content components, content traits are replicated as expression traits, and the former can be inferred from the latter through similitude rules. But maps misrepresent: the replication is more or less faithful (with respect to a defined standard), hence there are infidelities or deviations from faithful rendering. One task of interpretation is to assess fidelity; this is done on the basis of a stock of metasemiotic information. In the interpretation process, infidelities:

- may be mentally corrected,
- may only be recognized or suspected without being capable of being corrected,
- may go unnoticed.

Knowledge of infidelities limits or relativizes the interpretation.

4. In interpretation, the map user assembles and integrates information from various sources, including fidelity information. Map interpretation is a process of reasoning or inference which may be quite involved. Like transcription, it is based on rules which are learnt. Complications in the retrieval of the plan information are due to deviations from faithful transcription. Apart from simple errors, three sources of such deviations have been identified:

- effects of coordinate transformations,
- method-produced misplacement, and
- shifts between differentially distinguished plan traits.

Finally, for the sake of perspective, let us note how the present paper relates to the general theme of this collection. It obviously deals with geographic space, and it deals with cognition. Mapping and map use involve cognitive operations: we conceptualize reality, organize and represent knowledge, process information, and draw inferences. In short, the map is "a tool of thinking" (Pravda, 1987, 19). Its cognitive import is now widely recognized (see *e.g.* Salishchev, 1982; Pravda, 1987; Moore, 1989). In this respect, maps are better understood if they are studied in a semiotic framework, since semioticians have, for a considerable time, developed models of the organization of our conceptual universe and of the processes of expressing and retrieving information (Bouissac, 1985; Gardin, 1985; more detailed Eco, 1984). Thus semiotics is one of the fields of inquiry which contribute to the current lively research into the working of the human mind.

8. Notes

1. Such research has had various aims: to devise a typology of the cartographic information and of its transcription (Kelnhofer, 1971), to provide a catalogue of mapping methods (especially in Soviet cartography, *e.g.* Salishchev, 1976, 62-108; for an example in English see Klawe, 1970, 46-48), or to understand conceptual operations of map makers (Hsu, 1979) and of map users.
2. Individual localized signs correspond to individual acts of message conveyance and are the smallest complete entries on a map. Moore speaks of cartographic statements (Moore, 1989, 360 f., 424-427).
3. One distinguishes denotations or immediate meanings and connotations or mediated meanings, i.e. meanings mediated by a more basic meaning (Eco, 1976, 54-57). Correspondingly, in analyzing a bundle of content components, one should, more exactly, separate denotative and connotative ones (*ibid.*, 84-86, 145 f.). Now, whether a content component is denoted or connoted can be difficult to decide (*ibid.*), and there is no need to do so in the present discussion. Therefore I summarily speak of implied components.
4. As will be seen later, locators are sometimes omitted.
5. In some cases, plan information is expressed by plan-free expression traits or by lettering marks.
6. Conversely, such implications may be released by a spatial context as replicated on a map. An example (sand in front of a glacier terminus) is analyzed by Schlichtmann (1979, 89 f.). 'Sand' is here shown to imply 'outwash' (and not 'dune' or 'tailings') because it occurs in the spatial context 'glacier' (*i.e.* by virtue of a contextual selection).
7. In this sense Kelnhofer notes that his distinction between homogeneous and structured (internally differentiated) phenomena does not lead to mutually exclusive classes (Kelnhofer 1971, v. 1, 30). Alternative conceptualization is even more conspicuous in the plan-free information (where very few authors have recognized it); see Schlichtmann, 1989b, 18-25.
8. The cases 2-4 correspond to Leisi's individuatives, group collectives, and mass words (Leisi, 1975, 28-35). Homogeneous objects correspond to individuatives. (Leisi's terms, which have here been rendered in English, do not relate to concepts but to the nouns of which such

concepts are contents.) The distinction between case 1 and cases 2-3 corresponds to Talmy's uniplex-multiplex distinction (Talmy, 1988, 176-182).

9. Strictly this point holds for area places only. For lines and points, interior and boundary coincide, and the boundary is distinct.
10. The previous distinction between extent and dimensionality (Schlichtmann, 1979, 86) is now abandoned, and "extent" is replaced by "dimensionality".
11. In semiotic terms, coordinate transformation is a sign process. Point coordinates on the globe surface are contents. They are expressed by map coordinates, if only for the purpose of map construction. The equations which couple both sets of coordinates constitute the code.
12. This point has been stressed by several authors. Salishchev incorporates it is his map definition,
    stating that geographical maps are "*obrazno-znakovyye izobrazheniya*", translatable as "representations by means of signs which are images" (Salishchev, 1976, 9).
13. In the case of the plan-free information, by contrast, one often must set up standards which are relative to topics or purposes.
14. This selection is not under discussion.
15. If traits are not automatically replicated, one can, instead, replicate them consciously. Take a set of villages of rectangular shape and different sizes and orientations, mapped by means of point symbols. Even though one must use point symbols, one can draw them as rectangles with the requisite orientations (example: Warkentin, 1959, Figure 1 on p. 346) and even with correct relative sizes. The three characteristics, in their rendering, are, so to speak, separated from the locations. Incidentally, one also could transcribe plan traits of the information (other than location) by formal characteristics of a complement rather than of a locator.
16. The notion of paradigm or substitution class is one of the basic concepts in the analysis of sign systems. A paradigm has as its members all items which can be intersubstituted in a given context.
17. Thus, before using a map as a source in research, one must subject it to source evaluation. This need is especially obvious in historical geography.
18. For language, Jakobson (1960) has set up several functions, including the referential and metalingual one. Child (1984) has applied Jakobson's function systematics to maps. In analogy to the functions, some semanticians speak of referential etc. meaning. As I am not dealing with language, I prefer "metasemiotic" to "metalingual". My "information" is synonymous with "meaning".
19. Jakobson relates the metalingual function specifically to the code (1960, 356). I use "metasemiotic" in a wider sense. Subsumption of facts under concepts is here taken to include the collection and organization of data.
20. For example, the tacit knowledge that this specific locator refers to a specific place and that this place exists on the earth's surface. Such background knowledge seems to be more important in interpreting for plan-free information than for plan information.
21. Different clues are required to recognize a cartogram or a plan sketch.
22. But circumstances of message reception, important in the case of spoken language, do not appear to be of systematic importance.
23. Previously called "substitution of locational data" or "data substitution" for short (Schlichtmann, 1979, 92 f.). These terms are unsatisfactory: one substitutes expressions rather than data, and there are several processes which could be called substitutions. Therefore I now use "shift" for a specific type of replacement. Where a more general term suffices, "substitution" remains acceptable.

24. Changing the point of view and taking the expressions at face value, one could also speak of shifts between sign contents, *i.e.* between dimensionalities and between conceptualized distributions (the latter replacements could be seen as shifts between cover forms). In this view (which is also applicable to other cases discussed in the present section) a shift would be a reconceptualization. This point will not be pursued further.
25. It is assumed that the map user knows the following points: (1) As the circle is a complement, its implantation is meaningless. Thus point implantation does not indicate that inhabitants are systematically concentrated in the middle of each ward. (2) The circles have standard positions relative to the locators (in this case they are placed in the visual centers). Therefore the position of a circle does not indicate any central tendency in the distribution of the inhabitants. (For a different situation see the following note.) For simplicity, these two points will not be considered in the following derivation.
26. But there is an important difference from the preceding case. In the strict variant of the method, a dot is the locator of a center of gravity in a spatial distribution. A dot, then, is a complement with respect to the reference area (as is a circle in Figure 1a) but a locator with respect to the center of gravity (unlike a circle in Figure 1a). It retains this locator function after the shift.

9. Acknowledgements

Participation in the NATO Advanced Study Institute was made possible by financial aid from the Staff Travel Fund of the Dean of Arts, University of Regina, and, through the Director of the Advanced Study Institute, from the NATO conference grant. This assistance is gratefully acknowledged.

10. References

Bouissac, P. (1985) 'The Potential Role of Semiotics for the Advancement of Knowledge', Recherches Semiotiques / Semiotic Inquiry 5(4), 339-346.

Bunge, W. (1966) Theoretical Geography, 2nd. ed., Lund Studies in Geography, Ser. C, 1, Gleerup, Lund, Sweden.

Child, J. C. (1984) Creating a World: The Poetics of Cartography, Unpublished Ph.D. dissertation, Dept. of Geography, University of Washington.

Cuff, D. J. and Mattson, M. T. (1982) Thematic Maps, Their Design and Production, Methuen, New York.

Eco, U. (1976) A Theory of Semiotics, Indiana University Press, Bloomington and London.

Eco, U. (1984) Semiotics and the Philosophy of Language, Indiana University Press, Bloomington and London.

Gardin, J.-C. (1985) 'Semiotics and Artificial Intelligence: the Operationalization of Semiotic Research in the Humanities', Recherches Semiotiques / Semiotic Inquiry 5(4), 368-375.

Head, C. G. (1991) Mapping as language or semiotic system: Review and comment. In D. M. Mark and A. U. Frank (eds.), Cognitive and Linguistic Aspects of Geographic Space, Kluwer, Dordrecht.

Hsu, M.-L. (1979) 'The Cartographer's Conceptual Process and Thematic Symbolization', The American Cartographer 6, 117-127.

International Cartographic Association - Commission II - (1973) Multilingual Dictionary of Technical Terms in Cartography, Wiesbaden: Steiner.

Jakobson, R. (1960) 'Closing Statement: Linguistics and Poetics', in Sebeok, T. A. (ed.), Style in Language, The MIT Press, Cambridge, Mass., 350-377.

Jenks, G. F. (1970) 'Conceptual and Perceptual Error in Thematic Mapping', Proceedings, American Congress on Surveying and Mapping, 175-188.

Kelnhofer, F. (1971) Beitrage zur Systematik und allgemeinen Strukturlehre der thematischen Kartographie, erganzt durch Anwendungsbeispiele aus der Kartographie des Bevolkerungswesens, 2 vols., Graz: Hermann Bohlaus Nachf (Vienna, Cologne), Forschungen zur Theoretischen Kartographie 1.

Klawe, J. J. (1973) 'Population Mapping', The Canadian Cartographer 10(1), 44-50.

Leisi, E. (1975) Der Wortinhalt. Seine Struktur im Deutschen und Englischen, 5th ed., Quelle und Meyer, Heidelberg [First edition 1953].

Moore, R. W. (1989) Cognition, Semiotics, and the Cartographic Statement, Unpublished Ph.D. dissertation, Dept. of Geography, University of Kansas.

Morrison, J. L. (1971) 'Method-produced Error in Isarithmic Mapping', American Congress on Surveying and Mapping, Technical Monograph CA-5, Washington, DC.

Multilingual Dictionary of Technical Terms in Cartography, See International Cartographic Association.

Pravda, J. (1987) 'Semiologicke, jazykove a logicke aspekty mapy', Geograficky Casopis 39(1), 3-22.

Robinson, A. H., Sale, R. D., Morrison, J. L., and Muehrcke, P. C. (1984) Elements of Cartography, Wiley, New York.

Salishchev, K. A. (1976) Kartovedeniye, Moscow: Izdatel'stvo Moskovskogo Universiteta.

Salishchev, K. A. (1982) Theoretical Issues of Cartography in the 80s, Moscow: Moscow State University.

Schlichtmann, H. (1979) 'Codes in Map Communication', The Canadian Cartographer 16, 81-97.

Schlichtmann, H. (1985) 'Characteristic Traits of the Semiotic System 'Map Symbolism', The Cartographic Journal 22, 23-30.

Schlichtmann, H. (1989a) 'Map Semiotics: Concepts, Terms, and some Problems', in Head, C. G. (ed.), Report on the First North American Working Seminar on the Concepts of Cartographic Language, Wilfrid Laurier University, Waterloo, Ontario, February 12-14, 1988, [Waterloo: C.G. Head], 6 pp. [separately paginated].

Schlichtmann, H. (1989b) 'Conceptual Field Structures in Thematic Mapping', in Head, C.G. (ed.), Report on the First North American Working Seminar on the Concepts of Cartographic Language, Wilfrid Laurier University, Waterloo, Ontario, February 12-14, 1988, [Waterloo: C.G. Head], 27 pp. [separately paginated].

Talmy, L. (1983) 'How Language structures Space', in Pick, H. and Acredolo, L. (eds.), Spatial Orientation. Theory, Research, and Application, Plenum Press, New York, 225-282.

Talmy, L. (1988) 'The Relation of Grammar to Cognition', in Rudzka-Ostyn, B. (ed.), Topics in Cognitive Linguistics, Benjamins, Amsterdam, 165-205.

Warkentin, J. (1959) 'Mennonite Agricultural Settlements of Southern Manitoba', Geographical Review 49, 342-368.

Witt, W. (1970) Thematische Kartographie. Methoden und Probleme, Tendenzen und Aufgaben, 2nd ed., Hannover: Gebruder Janecke. Veroffentlichungen der Akademie fur Raumforschung und Landesplanung, Abhandlungen 49.

Witt, W. (1972) 'Ungeloste Probleme in der thematischen Kartographie', International Yearbook of Cartography 12, 11-27.

AN APPROACH TO MAP/TEXT INTERRELATIONSHIPS

DALIA VARANKA
Department of Geography
University of Wisconsin
Milwaukee, Wisconsin 53201 USA

ABSTRACT. After a short theoretical review, placing map/text interrelationships into contemporary cartographic theory, a quantitative approach is outlined to organize and structure information at five stages of analysis: the publication as a whole, the maps, the texts, and observed interrelationships between maps and texts. A discussion of meaning to be derived from related materials is included in the method. The data are to be interpreted in contexts of theory, content, and social practice.

1. Introduction

Any document consisting of both maps and accompanying geographical descriptions or explanatory texts employ the characteristics of both. The main objective of the research approach described in this paper is to see how both maps and text are interrelated in the structure and "discourse" of publications as a whole. The premise is that the texts are decisive to any reading of the maps and that they lend substance to the abstract and highly codified cartographic images. Without the text the map gives a minimum of information but with it they take on new meanings that are crucial to understanding the geographical attitudes and thoughts held by their users. Thus the arrangements of maps with texts have a critical, yet subtle effect on the nature of geographical information.

The research in progress is a Ph.D dissertation examining the earliest English world atlases. The contemporary conventions of prose and maps which developed then are being examined within the context of their philosophical foundations. To do this, the characteristics of the collected evidence of maps and texts in a selection of atlases will be identified, and where possible, the logic and meaning interpreted. This paper will describe in general terms the approach by which this will be done. Then I will see how well contemporary cartographic theory explains these systems of representation by testing our modern rationalizations of cartographic design. It is hypothesized that map and text interrelationships are not neutral representations, but an interplay between the world and society in terms of its own questioning, so that neither media can be separated from its social context. It is hoped that the research will contribute to a broader examination of what cartographers try to do and what view cartography has of itself.

At a theoretical level, a method for the study of systematic map and text arrangements will complement literature that has only recently begun to appear investigating general image/word relationships (Mitchell, 1986). The two representational forms are traditionally studied separately, but their prolific occurrence together has called for new investigation into this problem. At the most

D. M. Mark and A. U. Frank (eds.), Cognitive and Linguistic Aspects of Geographic Space, 285–294.

general level, their coordination is a vital issue in human communication, including cartography (Bann and Hunt, 1988). Theoretical opinions by modern cartographers about the relative roles of map and text in cartographic communication are split. Some have proposed that the map is independent of text (Eckert, 1908). On the other hand, J. K. Wright argued that the map never loses its dependance on, or at least it receives aid from, some textual source or other (Wright, 1944). The contemporary study of map/text interrelationships falls within the "map as text" movement, which has emerged within the last few years. The study of the map as a text draws upon four bases: that the map can then be interpreted within humanistic lines of inquiry, that the symbolic rather than analytic qualities of the map can be accounted for, that the map be studied within the demands of its context, and that the graphic quality of the map is combined with words to create a variety of meanings (Harley, forthcoming; Pickles, forthcoming). It is this last point which draws our attention to map/text interrelationships.

In pragmatic terms, cartographers have expressed the need for information to be represented in the most effective way--whether by map, prose, or table--and that this decision must be made in an informed way (Guelke, 1976). Separately, the two media share some similar qualities; map-reading can generate verbal statements, and language can structures spatial imagery such as concepts of way-finding. It is commonly agreed that one media fills in meaning more specifically or quickly for the other. Studies of image/word relationships appear in the literature of psychology at least as early as the 1950s. Results of psychological experiments about the learning efficacy of maps and texts together conclude that maps provided with a passage of geographic text help learn the reader learn the content presented in the prose (Gilmartin, 1982, 1986; Davis and Hunkins, 1968). Still, conclusions about the interactive properties of these representational forms remain undefined. A version of dual-coding theory has been presented by Kulhavy, Lee, and Caterino (1985) which they call the conjoint retention hypothesis. In conjoint retention, the independent coding of spatial and linguistic information facilitates recall due to increased accessibility of appropriate retrieval cues. Furthermore, information stored in both codes may be retrieved from either system to activate the other, thus providing multiple avenues for recall of map/text information.

2. Map Language

Early investigation of the interrelationships between maps and texts were made by the development of an analogy between the two media called map language. As a dominant movement, it makes it earliest appearance in articles by the central European authors, as part of a larger movement called cartographic communication theory. Kolacny (1969) calls map language the system of map symbols and rules for their use. Ratajski's model of cartographical transmission makes a statement emphasizing "correctness" of map language in the design and reading stages (1973). The actual concepts of map language receive more attention in an article by Meine (1977), in which he discusses the establishment of a standardized "cartographic alphabet" for applied design tasks. Building on the premise that the logic and design of maps varies with the cartographer's experience and with inconsistent traditions, scientifically derived cartographic symbols should be formulated which would conventionalize map language. Meine proposes some first-step approaches to the analysis of a map language, but no solutions. The paper argues for the applicability of linguistic models of syntactics, semantics, and pragmatics to graphic sign systems, since both are branches of semiotics.

The other main direction that research extended the map language analogy was to cognitive studies. Morrison (1976) calls the process by which cartographic communication occurs the cartographic language. The idea of language is at first used generally, almost metaphorically, but then pressed for semantic and grammatical substance which Morrison argues can be formulated through psychological/physiological research, a direction continued by Head (1984), Eastman (1986), and others. In these studies, cognitive processing models and a model of reading/writing were applied to map user research. This helped clarify some of the problematic aspects of the analogy, and map language began to resemble a working model applicable to practices of graphic design and map-using education.

Writing in 1982, John Keates noted that acceptance of the model by the cartographic community was split. Problems of cartographic language focused particularly on the analysis of units and the establishment of syntax, although the specific analyses and definitions of the map/language analogy vary. The literature using or discussing map language (which is sometimes referred to as a model) is extensive, and still being used today. Only a few of the ideas on these themes will be considered here.

An example of the more elemental controversies could be the criticisms raised by Robinson and Petchenik (1976) that no alphabet exists for graphic language. The signs, the number of which are infinite, are continuous, not segmented, and have no meaning, but depend upon an agreed context for interpretation. Head responded that the eye fragments the image in scanning it. The map symbols are equivalent to morphemes, with nothing equivalent to phonemes or graphemes. Cognition of these symbols is not purely data driven; the reader's intentions in these tasks achieve an effect similar to the process of combining units of language.

As to the problem of syntax, Eastman's view was that the variable symbolism is not anarchy, but the very basis for employing syntax. The order of graphic symbols is due to geographical control, yet still analogous to language at the level of a reader's cognitive model for recognition. Maps have understandable meaning or concept of syntax below the surface level signs and these can be discovered through an examination of the user and his expectations. "However, instead of superimposing a syntax upon our units of communication (such as with the use of order in the reading of text), we embed that syntax within the definition of the unit itself" (Eastman, 1986). An example Eastman mentions is symbol size hierarchies. This would mean that a reading order of depth (from global to local) may have importance for capturing the underlying content organization of maps--the syntax.

Research employing cognitive models were empty of any directly semiotic statements. The fact that semiotics provides a more direct relationship between maps and language than cognition was emphasized by Schlichtmann, in a response to the article by Head (Schlichtmann, 1984). Schlichtmann argued that because semantic comprehensiveness or independent patterning of its expression are constrained on maps to a spatial component, they cannot be regarded as a natural language, but semiotics can help the study of maps as a language. Schlichtmann suggests some techniques for pursuing such research.

A system of semiotic classification of cartographic signs is presented in detail by Wood and Fels (1986). Drawing from the analysis of myth by Roland Barthes, Wood and Fels argue that intrasignificant map codes are at the level of language and serve to support extrasignificant codes, a broader level they call the myth of the map. (The system includes linguistic codes, refered to as explicit map/text interrelationships later in this essay.) In addition to the classification of codes, four levels of signification, at which the map generates and structures the interrelating signing processes are explained: elemental, systemic, synthetic, and presentational. Neither the vocabulary used by the authors nor their concepts really fall within the mainstream map language terms, yet their aim, to analyze in detail the meaning embedded in the map, was a similar one. The article differs in that it extends map language into the context of its use, or its "myth". This brings Wood and Fels to conclude that the map is a text.

Head also arrived at this by the analysis of intelligence and cognitive linguistics. For him, communication can be thought of as attempts to convey assertions about the world. At a conceptual level, these assertions can be referred to as proposition, associations between elementary ideas or concepts. A proposition has a subject and a predicate, a predicate can have a relation and an object. The subject and predicate form the fact component, but it also has the context component of time and space. On maps, the fact component is served by symbols, title, and other things. The context component is created by grid, base map, notes, and such. At a higher order, the landscape of the map may be read as a text. In most general terms, the graphic nature of cartography can be considered the substitution of one set of signs for another and a subset of the symbols by which we experience ourselves and the world—thought.

The map language analogy served an important role in the cartographic communication theory of the 1970s, and furthered the investigation of meaning behind the map image. The movement determined that it is both possible and desirable to establish a correspondence between map symbols and reality. In contrast, the view of the map as text, the method still uses the elements of the map image as the basis of analysis, but the approach has shifted in viewpoint. The correspondence in the map as text is between the map image and world view (Kearney, 1984). I propose that issues raised in the research of cognitive processing and mental imagery, as described by Pylyshyn (1981), are the same in the shift in cartographic perspective from map language to map as text, but extended to the creation of images as documents.

This debate, briefly, is whether mental imagery is driven by external sensory stimulation and is not cognitively penetrable, or if the imagery is largely a result of cognitive experience. Cartographic communication theory assumes that map-making is a product of the cartographer's cognitive process, so this debate has a direct impact on the formation of map imagery. Map language, whether as a theory, analogy, or model, has been built on the assumption that the map image is a mirror of nature, and thus less vulnerable to human manipulation. The map as text idea views the map as a compilation and expression of the map-maker's personal experience.

3. The Analysis of the Atlases

The essence of the research method which is being applied to the English world atlases involves interrogating a representative sample of atlases, selected by date and format from the general corpus, in terms of a set of standard questions. This information is entered into a digital database prior to further analysis and interpretation. The entries consist of single words or codes entered under separate headings and will be processed using computer software which allows variable arrangement and quantitative summary. The bulk data will be manipulated to find connections, patterns, and organizing principles of the atlases. In this way, the general characteristics of a genre of work can be detected despite individual deviations from the trend.

The entries of the database document four stages of analysis, A to D: the atlases in general, the maps, the texts, and observed map/text interrelationships. Maps and texts are separated for purposes of analysis, although both media share many qualities and are read together in the atlas. The database will also be used to assemble relevant material derived from "primary source materials," to be used as evidence toward the establishment of conventional symbolic meanings which were crucial for understanding the atlases.

3.1 STAGE A: GENERAL CHARACTERISTICS OF THE ATLAS

Social circumstances and the technology of production are viewed as "external power" practiced upon the atlases as a whole (Harley, 1990). To approach and analyze these influences, the information entered for each atlas is:

1. Carto-bibliographic detail
   a) principal author's name;
   b) title of atlas;
   c) year of publication;
   d) edition of atlas;
   e) name and place of publisher.

2. Attributes of physical form that impose constraints on the key cartographic variables of generalization and hierarchy
   a) size;
   b) format;
   c) mode of technological production;
   d) The collation and ratios of text to maps.

3. Publication history
   a) profile of patron or commercial sponsorship;
   b) inference of objectives: targeted readership;
   c) dissemination of the atlas to libraries and customer.

4. Atlas map and text publication processes sharing a number of basic steps
   a) data on publication personnel;
   b) use of common data sources;
   c) a preliminary overview of geographic content;
   d) the publishing process;
   e) readership education.

3.2 STAGE B: ANALYZING THE MAPS

The techniques of analysis to be performed upon maps as documents indicative of contemporary culture and scientific thought are:

1. Information sources and compilation

2. Scale and its attributed generalization of subject matter
   a) The coverage and number of maps ranging in scale from world to continental, regional and local;
   b) Observations made from maps enlarged or reduced photographically to a common scale;
   c) Comparisons made between maps on a section-by-section basis of a standard grid superimposed on each map of a common or roughly equivalent scale.

3. An inventory of the map content within an analytical framework of graphic qualities and criteria
   a) identify conventional signs of landscape features;
   b) classify signs as standardized depictions of unique elements.

The technique outlined above furnishes the empirical evidence for an analysis of maps within certain terms of styles of the contemporary period and place. The style of the map is determined by the map logic, and is studied as a reflection of contemporary thought and creation of an image reflecting belief, but shaping behavior as well (see a discussion of the "internal power" of maps in Harley, 1990). The assumption validating a comparison of prose style to map is that the philosophies influencing these styles did not originate with either one, but in the general attitudes regarding knowledge of the time. Since ideas can be expressed by either word or image or both, it is the expression of meaning, inherent in the form, which is the basis of comparison.

1. Maps are tested for style at different levels of complexity
   a) The individual signs and their relationship to each other serve as vocabulary. The patterns of these relationships are considered expressions of spatial visualization and cognition;
   b) Decoration is considered figurative language for description or interpretation. This approach follows the work of Panofsky (1939), as modified by Harley (1983);
   c) Concepts of cartographic syntax now current in the cartographic literature (Head, 1984, Eastman, 1986) will be tried out on the maps. If these approaches, after experimentation, are not appropriate to this research, a concept of syntax will be determined from the evidence, or subsumed in the analysis of point d);
   d) map theme and over-all presentation.

2. Theoretical beliefs and conventions relating map-making and geography. These social aspects are compared to the empirical evidence.

3.3 STAGE C: ANALYZING THE TEXTS

Four techniques are involved in analyzing the texts:

1. The organization of the text, the arrangement of headings and subheadings, is the expression of the geographical scope of the atlas and the geopolitics of its representation.

2. A simplified form of Salience Analysis (developed by James Blaut, and presented in Lucarelli, 1987) will be used to quantify the space allotted to particular text themes. Based on the fact that the differential allocation of space to different topics denote deliberate policy decisions by authors or publishers, the themes treated briefly or not present will be hypothesized to be less important than the ones that gain more attention. The method is as follows:
   a) The total number of content pages is determined by counting all textual pages of the section;
   b) A line count of each content page is made, excluding tables, graphs, and maps;
   c) The number and percentage of textual lines devoted to different themes will then be calculated.
3. The texts will be carefully read to establish the operation of valuative language as an indicator of ideology. The scale of analysis under consideration includes a word, a phrase, and, at most, a sentence. Three types of textual units will be noted and entered into the database:

a) the valuative unit per se, which expresses a judgement in itself;
b) cognitive textual units with valuative overtone, i.e. apparently neutral textual units that convey positive or negative valence to the reader;
c) standardized classes of units, such as the index;
d) frequency count of other words.

4. I will analyze texts in terms of style for the aims mentioned in 2. of *Analyzing the maps* (above). Particular attention will be paid to qualities of language structuring space (Talmy, 1990; Herskovits, 1986).
   a) Study of nouns, verbs, and adjectives will search for different sorts of stylistic functions within the sentence;
   b) Figurative language is considered to consist mainly of metaphors. This analysis will focus on the qualities of the "whole" idea conveyed by the metaphor;
   c) Elements of syntax and theme are closely associated. Syntax will be tested for adherence to philosophical norms by its communicative qualities.

3.4 STAGE D: OBSERVED INTERRELATIONSHIPS BETWEEN MAPS AND TEXT

Implicit, explicit, and theoretical interrelationships are found in atlases. Data for the first two are discussed below.

1. Database entries for explicit interrelationships defined as the presence of the elements in both media include:
   a. place names;
   b. written legends relating to symbology;
   c. notation in the form of short descriptions and statements on the map;
   d. cross-referencing between map and text.

The conventional signs of landscape features determined from the inventory of stage B will be categorized and compared to the index categories.

2. Implicit relationships will be entered into the database by the idea or meaning linking them. A fundamental aim of this study is to identify the 'messages' (of both map and text) and their clarity, tension, or obfuscation in the form of their presentation and translation between map and text or vice versa. Does one media expand the presentation in a way that the other is unable to, or is the information simply repeated? What about inconsistencies or contradiction in information?

Some interrelationships may not be recorded in either the map or the text, but would be revealed from the independent evidence of the primary sources, or in the genealogical similarities between the sources employed in map and text, respectively.

The database will be used, alongside 'soft' insights from the primary sources, to lend itself to forms of 'hard' analysis and statistical presentation. Interpretation will be aided by mathematical, graphing, and mapping techniques to be accomplished at several levels of penetration. Together with the qualitative insights, these procedures will allow new generalizations to be made about the nature of the atlases and their significance in the culture of the period. An example of a more complex manipulation would build on trends revealed by the simple analysis to construct relationships such as 'families' of atlas maps, filiated by related characteristics, the particular perceptions of different regions or the ideological aspects of geography they embody.

4. Meaning in its Social Context

Careful attention aimed toward determining the meaning behind map/text interrelationships should be consistently maintained. It is expected that different map and prose meanings exist at several levels of realism ranging from the purely factual to the abstract meanings registered in the unconscious of the reader. General interpretations are probably adequate for some aspects of representation such as lettering, emblems, and the design of conventional signs, but symbolic meaning is both more ambiguous and elusive. No simple set of techniques exists for recovering the symbolic layer of meaning in the atlases, but it involves recourse to both graphic and written sources contemporary with their production. The conjunction of other evidence may indicate certain cultural associations that have no surface expression in the atlas. Applications of cognitive and linguistic models are acceptable and may also prove useful, but the social record of the period and place of the atlas must take priority over our own interpretations.

The main advantage of using a social record is that the cultural influences are more open for objective interpretation. The economy and technology of atlas production, as well as the available source material, has rational limits which modifies the images. Those images are further influenced by authors, editors, and the editorial concepts that they followed. As map use passes out of control of the original maker's intentions, it proliferates in a variety of social contexts. Through maps and text, atlases rely upon concepts as well as facts associated with land and territory to communicate effectively. These concepts structure the ideas and uses to which they are applied by the readers. In turn, the reception of the atlases may account for continuity or innovation in their content.

5. Acknowledgements

The research described in this paper was significantly furthered by my participation at the NATO Advanced Study Institute "Cognitive and Linguistic Aspects of Geographical Space," supported by NATO and by a travel grant from the National Science Foundation. Thanks are extended to the ASI Director David Mark, and as well to J. B. Harley, University of Wisconsin-Milwaukee, for his suggestions towards the research; to J. B. Harley and John Pickles, University of Kentucky, for sharing unpublished papers with me; and to Mark Blades, University of Sheffield, for relevant references.

6. References

Bann, Steven, and Hunt, John Dixon (eds.) (1988) 'Maps and mapping', Word and Image 4.

Davis, O. L., and Hunkins, F. P. (1968) 'The usefulness of a map with geographic text', Journal of Geography 67, 362-366.

Eastman, J. (1985) 'Cognitive models and cartographic design research', Cartographic Journal 22, 95-101.

Eckert, M. (1908) 'On the nature of maps and map logic', Bulletin, American Geographical Society 40, 344-351.

Gilmartin, Patricia P. (1982) 'The instructional efficacy of maps in geographic text', Journal of Geography 81, 145-150.

Gilmartin, Patricia P. (1986) 'Maps, mental imagery and gender in the recall of geographical information', American Cartographer 13, 335-344.

Guelke, Leonard (1976) 'Cartographic communication and geographic understanding', The Canadian Cartographer 13, 107-122.

Harley, J. B. (1989) 'Deconstructing the map', Cartographica 26, 1-20.

Harley, J. B. (1983) 'Meaning and Ambiguity in Tudor Cartography', in Sarah Tyacke (ed.), English Map-making 1500-1650, Historical Essays, The British Library Board, London.

Harley, J. B. (forthcoming) 'The Myth of the Great Divide: Art, Science, and Text in the History of Cartography', Imago Mundi.

Head, C. Grant (1984) 'The map as natural language: A paradigm for understanding', Cartographica 21, 1-32.

Herskovits, Annette (1986) Language and Spatial Cognition: an interdisciplinary study of the prepositions in English, Studies in Natural Language Processing, University Press, Cambridge.

Kearney, Michael (1984) World View, Chandler and Sharp Publishers, Inc., Novato, California.

Keates, John (1982) Understanding Maps, John Wiley and Sons, New York.

Kolacny, A. (1969) 'Cartographic information--a fundamental concept and term in modern cartography', The Cartographic Journal 6, 47-49.

Kulhavy, R. W., Lee, J. B., and Caterino, L. C. (1985) 'Conjoint retention of maps and related discourse', Contemporary Educational Psychology 10, 28-37.

Lucarelli, Helena (1987) 'Eurocentricism in Brazilian Geography Textbooks', Unpublished M.A. thesis. University of Illinois at Chicago.

Meine, K-H. (1974) 'Cartographic communication links and a cartographic alphabet', Cartographica, monograph 19, 92-110.

Mitchell, W. J. T. (1986) Iconology; Image, Text, Ideology, The University of Chicago Press, Chicago.

Morrison, Joel L. (1976) 'The science of cartography and its essential processes', International Yearbook for Cartography 16, 84-97.

Panofsky, E. (1939) Studies in Iconology, Humanistic Themes in the Art of the Renaissance, Oxford University Press.

Petchenik, Barbara B. (1975) 'Cognition in Cartography', Proceedings, International Symposium on Computer-Assisted Cartography, 183-193.

Pickles, John (forthcoming) 'Text, Hermeneutics and Propaganda Maps', in Barnes T. and Duncan, J. (eds.), Languages of Landscape: Text, Metaphor and Discourse, Routledge and Kegan Paul.

Pylyshyn, Zenon (1981) 'The imagery debate: analogue media versus tacit knowledge', Psychological Review 87, 16-45.

Ratajski, Lech. (1973) 'The research structure of theoretical cartography', International Yearbook for Cartography 13, 217-227.

Robinson, A. H. and Petchenik, B. B. (1976) The Nature of Maps, The University of Chicago Press, Chicago.

Schlichtmann, H. (1984) 'Discussion of C. Grant Head, The map as natural language: a paradigm for understanding', Cartographica 21, 33-36.

Talmy, Leonard (1990) 'How Language Structures Space', Paper presented at Cognitive and Linguistic Aspects of Geographical Space: NATO Advanced Study Institute, Las Navas del Marqués, Spain.

Wood, D., and Fels, J. (1986) 'Designs on signs: myth and meaning in maps', Cartographica 23, 54-103.

Wood, Denis (1987) 'Pleasure in the Idea: The Atlas as Narrative Form', in Carswell, R.J.B., de Leeuw, G.J.A., and Waters, N. M. (eds.), Atlases for Schools: Design Principles and Curriculum Perspectives, Cartographica 24.

Wright, J. K. (1944) 'Map makers are human', Geographical Review 32, 527-544.

SPATIAL KNOWLEDGE FOR IMAGE UNDERSTANDING

GEOFFREY EDWARDS
Centre de géomatique
Département des sciences géodésiques et télédétection
Université Laval, Ste-Foy,
Québec G1K 7P4 Canada

ABSTRACT. Both computer-assisted and fully automated image understanding in remote sensing require the ability to extract and merge spatial knowledge in a variety of forms. In this article some of the processes involved in extracting, representing and interpreting spatial knowledge in images, and in maps as they relate to images, are examined. In particular, a "naming" of image units perceived as single entities is found to be the preferred mode of describing shape information when both map and image interpreters communicate their findings. Furthermore, the interpreter appears to construct a top-down hierarchical structure of embeddings and relations, each node of which consists of a single "sentence". The structure of this cartographic sentence is determined by the attentional field of the interpreter, and consists of an attentional frame, an object/texture combination (subject) and a set of spatial and topological referents (predicate) to other objects. The analysis of this nested structure provides insight into appropriate strategies for automated image and map interpretation as well as providing clues to building natural language interfaces with GIS.

1. Introduction

Automated scene understanding requires combining image analysis, symbolic and cartographic processing and propositional knowledge in new ways. Digital remote sensing has traditionally been focussed on the spectral information contained in images, via pixel-based image analysis. In recent years, the inclusion of spatial variability in the form of texture has been increasingly adopted. Most recently, cartographic information has been used to further enhance the information obtained from automated processing. However, the high sensitivity of spectral information to atmospheric conditions and illumination geometry, combined with the complexity of the Earth's surface, creates a barrier to using these analyses directly for universal scene understanding. That is to say, one cannot point to a particular pixel and, based solely on its spectral characteristics, unambiguously determine its content.

The inclusion of other information implicit in the scene is hence of fundamental importance for reliable image understanding. The human perceptual system uses a variety of information to identify objects and "understand" a scene, including the identification of regions with similar spectral and textural properties, the shapes of such regions, the relationships between a region and its neighbors, information on scales present in the image, and knowledge about physical connections between scene elements. Indeed a human interpreter is usually able to determine global image content without recourse to other information than what he or she carries in his or her

D. M. Mark and A. U. Frank (eds.), Cognitive and Linguistic Aspects of Geographic Space, 295–307.

own head. We are still unable to duplicate this feat using a computer. However, for detailed photointerpretation, an expert must also rely on other cartographic information, on standardized procedures (e.g. an interpretation "key"), or on both (Ryerson, 1989).

In this paper, several aspects of scene understanding are explored. A triple perspective is employed: observations about the human perceptual process (supplemented by references to the literature of perception), a review of work on written texts describing scenes, and general background on the application of automated image analysis techniques applied to remotely sensed images. Although the discussion centers on the analysis of remotely sensed imagery, many of the examples used offer insight on general spatial representation issues related to GIS operation and design.

2. A Sampling of Image Understanding Tasks

In order to provide a context for the analysis to follow, a number of example tasks have been selected. These include locating specific entities with known characteristics, the extraction of statistical information from images and maps and the accurate mapping of complex scenes.

The recognition of target entities is a "detection" task, equivalent to an existence proof in mathematics. It does not mean a great deal can be said about the target entities, it only implies one can detect them and locate them. This kind of task is often found in military applications, where the presence and location of planes or tanks needs to be determined. In non-military applications, one might be interested in locating buildings in a scene, or some particular type of object, such as flooded forest around a lake, indicating the presence of beaver dams. This kind of tasks usually involves relatively "low-level" algorithms, although they may need to sift through a lot of data, and hence need to be fairly efficient.

A second task, that of extracting statistical information about image entities which may be combined with other information for a variety of deductive tasks, has been the traditional goal of remote sensing analysis. That is, the user supplies information on a region of interest and the software is then called on to determine statistical properties for the purposes of inventory, modelling or prediction. For example, obtaining the percent of fields in a particular administrative zone which have been planted with a variety of different crops is a typical inventory task. The precise location and shape of the fields is not necessarily relevant information, or may be available just on a visual display.

The third kind of task is the mapping of complex scene objects. This requires first of all that we know what we mean by "scene objects". For instance, is a field the arable surface inside a given boundary, or does it include the ditches, fences, tractor lanes, trees and shrubs, etc. which might also be present? Furthermore, since the goal is to produce maps, we must understand how cartographic information is represented and manipulated. At the present time no complete theory of cartographic representation exists (Head, 1991). Indeed, the whole study of cartographic representation has been sorely neglected (Head, 1991). Leaving this to one side for the moment (i.e. assuming we are able to determine the nature of the objects we wish to extract), we must then proceed to the identification of these objects. This is also not a trivial task. The image must be broken into identifiable regions. Traditional pixel-based classification, used, for example, for the classification task outlined above, does not readily generate such regions. Post processing must be performed in order to convert neighborhoods of individual pixels into regions. "Segmentation" of the image done in this way tends to be very sensitive to noise, since it is based on the classification of single pixels only. More sophisticated segmentation algorithms utilize the spatial context of pixels at an earlier stage in the processing, allowing neighborhoods to be formed based on a variety of criteria and not just individual spectral values.

In such region-based segmentation, complex scene objects can be built up from single segments using a "mental" model or schema which specifies their structure. Hence, for example, a city park is not a single unit, but is actually composed of some subset of lawns, playing fields, individual trees, woods, glades, flower gardens, lakes, ponds, open water, streams, rivers, waterfalls, fountains, river banks, beaches, bridges, tennis courts, playgrounds, buildings, fences, etc. Even in low resolution satellite imagery, moderately sized parks will be made up of several of these components. Scene recognition software must be able to put such units together and give the total an object name ("a park"). Some other examples of complex objects include:

- an agricultural field (plants, rows, drainage ditches, access route, fences, rocks, isolated trees, etc.),
- a tree (leaves, shadows, branches, soil, flowers, fruit, etc.),
- a stand of trees (individual trees, a road, a stream, bare ground, shrubs, grass, etc.),
- a farm (house, barns, other small buildings, roads, driveway, lawns, fences, fields),
- a river zone (flowing water, still water, river bed, river banks, rocks, islands, river spillover, dams, open spaces),
- a university (woods, isolated trees, lawns, playing fields, roads, walkways, sidewalks, parking lots, large buildings, small buildings, tarmac, gardens, etc.).

The task of recognizing such complex entities can be performed by a human interpreter relatively easily. Even a university complex can be distinguished from other institutions based on subtle clues such as building placement, building sizes and walkways, etc. A great deal of effort is currently being spent on understanding how the mental models and processes underlying such recognition function in the human brain or, alternatively, how they can be implemented on a computer. Most of this effort is oriented, however, towards obtaining information from visual scenes of a factory floor or other, relatively closed environments. The problems of photointerpretation of remotely sensed images is different than those associated with such scenes, because the kinds of "objects" sought are more cartographic abstractions than they are real, physical objects. Another difference is that, rather than filling in a three-dimensional model, the information obtained from remotely sensed images is essentially confined to a thin two-dimensional membrane.

In the next section, the question of cartographic representation and the nature of cartographic entities is taken up in more detail.

3. Cartographic Representation

In the previous section, it was noted that the kinds of "scene objects" which are extracted from images are, essentially, cartographic abstractions. Hence, in order to be able to examine the image interpretation process, a good grounding in the theory of cartographic representation is necessary. As has been already indicated, this is difficult, since no such full theory exists. However, some ground theory exists. For instance, it has been recognized that one of the fundamental functions of maps is to communicate spatial concepts (Head, 1991). A secondary function is as a storage device for spatial data, and a third, derivative function, is to act as a surrogate for the "terrain" (Head, 1991). The idea of maps as communication media for information has led to a skeletal theory of cartographic representation based on semiotics and/or language. A number of insights have been gained from this: (1) that map elements are signs, and hence are composed of content, graphic expression and a code that links the two; (2) that maps are two dimensional systems of such signs; (3) that map symbols are signs representing abstract geographical concepts and not the real world directly; (4) that maps are not "read" all-at-once but rather as a sequential process; and (5) that cartographic entities are structured in hierarchical complexes. In particular, the recognition that there are basic units and more complex units obtained by combining basic units has been a convergent result of several studies.

Most of the attempts evoke a close structural similarity to sentence building in language. This leads to a hierarchy whose structure involves five levels: a basic indivisible morpheme, a word or lexeme level, a syntagme level and a propositional or sentence level. Propositions can then be combined to form "texts". There is disagreement, however, on the exact correspondences between these levels and cartographic structures.

Another point of interest brought up by several of these studies (and reviewed by Head, 1991) is the extent to which cartographic meaning is determined by the map reader as well as the map-maker. The map reader takes the basic elements in the map and generates new relationships of which the map-maker may not even have been aware. In this sense, the map is not only more like language than a formal code, it is more like a poem than it is akin to prose. Furthermore, the map-reading process appears to occur through a chunking of low-level signs into higher level signs, and sometimes, also, the converse. Finally, this chunking is concretized by assigning names to higher level chunks. Indeed, the experienced map reader has a whole dictionary of names for grouped entities which he or she carries around her head, allowing her to read and understand maps more quickly and effectively. These dictionaries are frequently domain dependent.

In the next section, an attempt to obtain clues for automating image interpretation is examined. This study arose from several considerations. The author has been developing an integrated environment which combines image segmentation, a geographic information system and knowledge based reasoning (Edwards, 1990). While this system can be used "as is" for traditional remote sensing tasks such as classification for inventory purposes, the environment provides an ideal base for building more advanced automated image interpretation functionality —provided a procedure for extracting meaningful entities from the segmented data can be found. The starting place for image interpretation of these data is with the segmented image. Furthermore, while most efforts to automate image interpretation over the past ten years or so have been based on an internal approach to visual perception, combining low-level and high-level processing in a single algorithmic structure, an alternative approach is to require that the intermediate stages of visual processing be communicable. The actual configuration of human visual processing in terms of one or the other of these approaches is not yet known. There is evidence, however, that at least the naming of spatial elements, and perhaps other linguistic processes, occurs at fairly low levels of visual processing in a separate visual processing channel linked to short-term memory (Haber and Hershenson, 1974; Head, 1984). This suggests that insight on how visual interpretation tasks could be implemented can be obtained from studying how a human interpreter names and describes entities obtained from intermediate visual processing.

4. Codes for Complex Shapes

The experiment consisted (a) of showing several subjects some line drawings similar to those obtained from image segmentation and asking for verbal descriptions, and (b) of showing the same subjects an actual partition and asking for a partial scene description. The experiment was meant to be, at this stage, fairly informal. It was not a rigorously constructed experiment of the kind carried out by psychologists to study human perceptual response, but merely an attempt to get a feeling of the kinds of visual processing revealed by verbal descriptions of abstract drawings.

One of the more obvious parameters used to describe or represent objects is their shape, and this aspect was targeted by the study for particular attention. Surprisingly, shape is still not widely used in image analysis, outside of character recognition. Where it is used, it is frequently implemented via mathematical formulations, such as moment analysis, and not using the more "intuitive" shape criteria most of us are used to: shape descriptors such as "rectangular", "oblong", "ovoid", etc. Shape descriptors which are used, for example, include "compactness", "asymmetry", the ratio of length over width, the ratio of perimeter over surface area, and so on. Frequently, shape parameters are selected on the basis of their convenience in terms of ease and cost of computation, or on the basis of some desired property, such as invariance with regards to rotation or scale changes. Such shape information is encoded as a number and used with respect to a user-defined threshold, or as a parameter in a classification process. Rather than describe such computational variants on shape here, it is of interest to look closer at how humans use shape to identify or describe objects.

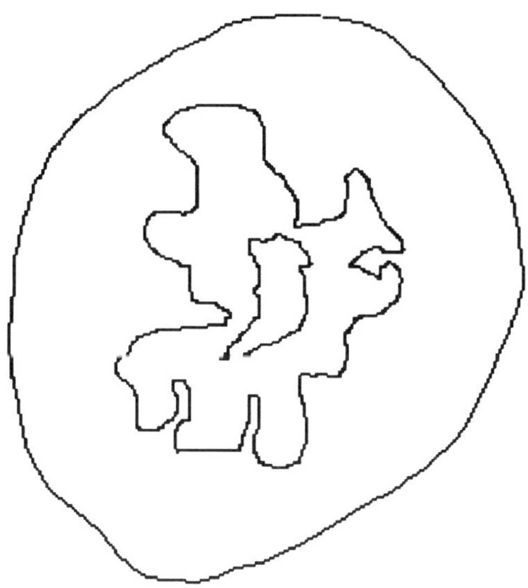

Figure 1: A simple abstract line drawing

Here are some textual descriptions of a simple abstract shape (Figure 1). Person A chose the following descriptions: "a flower", "an eye", "a chain of protuberances around a circular center with an irregular contour, the whole surrounded by a large circle" and "an island". For individual parts, the same speaker spoke of "an irregular ovoid form in the center", "a closed space" and "a closed curve in the center". Speaker B chose, instead: "a three-layered pudding", "a broken egg" and "a central turd shape separated from a circular egg by a liquid barrier".

Several observations can be readily made, based on these descriptions. First of all, the descriptions of shape are characterized by a strong use of comparison with other, known (and named) images, a form of comparison not unlike metaphor, except that it is a mapping between graphical elements rather than linguistic structures. This is hence a roundabout way of assigning "names" to chunks of the image. Secondly, each person tends to use a series of such descriptions, each of which captures a different subset of the shape's characteristics. A "complete" description is provided only by the set of such descriptions.

A second example, this time of a more complex abstract shape (Figure 2) led to the following descriptions: "a geographic map", "a jigsaw puzzle" and "a patchwork quilt". When the scribes were told there are four basic units present in the image, the following additional descriptions emerged: "four squares in a larger square", "four fields", "there's a basic structure that suggests four squares" and "four segments form a square-like shape".

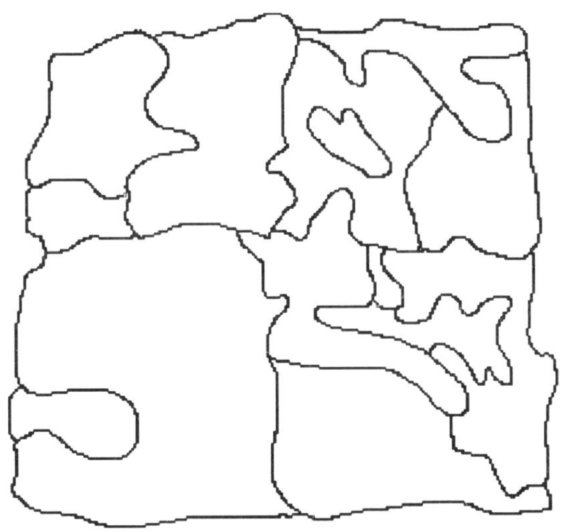

Figure 2: A more complex abstract shape

First impressions of this example seemed to focus on the overall type of pattern. Only after the existence of four units was specified did the observer recognize the frame, but once specified, one observer remarked that there was a "basic structure" present. This seems to concord with evidence that the detection threshold of an observer can be reduced by giving him pre-information (Röhler, 1988).

It is emphasized that this extended use of comparison with named images to describe these shapes is fundamentally different than the approach taken by most computational analyses of shape. It is also fundamentally similar to the naming of chunks in map reading, as observed by Head and others. While in map reading there is a "vocabulary" of clearly defined and labelled complex entities, in abstract drawings the interpreter appears to chunk visual information into entities which are named through association with a set of prototype objects of known shape characteristics. We might call these associative forms "associative codes", in opposition to the "explicit names" or well-defined labels used in map interpretation. The use of such "iconic names or codes" seems to be the preferred mode of describing shape among human speakers. Even in the more technical descriptions, similar codes are present: the use of the word "protuberances" and the idea of a "chain" in the first example. On the other hand, explicit shape-reserved words such as "irregular form", "ovoid", "closed curve" and "square-like" are also present in some of these descriptions. There appears, indeed, to be two modes of description operating side by side, one (the preferred mode) which relates (frequently complex) image units to associative or explicit iconic codes and the other which attempts to break the image down into elemental parts and the relations between them. This is very similar to the semiotic structure observed in the cartographic interpretation process.

These two modes of description are well characterized by recent experiments on right and left brain specialization (Heller, 1990):

> When both hemispheres confront the same spatial task, the right hemisphere takes control of the person's response ... the strategy adopted by the right hemisphere for [many] spatial tasks is fundamentally different from that of the left hemisphere, not just a more adept variant ... The right hemisphere ... is skilled at grasping a general form — the gestalt — of an object. The left hemisphere is more successful at apprehending objects that can readily be analyzed into simpler features, easy to count or name: corners, edges, indentations.

These studies are focussed on developing a more complex view of right and left brain specialization. For the past ten years, popular opinion has tended to associate spatial tasks with the right brain hemisphere and language tasks with the left hemisphere. More recently studies show that both hemispheres are capable of both types of task, but each hemisphere operates differently than the other. Hence, with regards to language (Heller, 1990):

> ... though left-hemisphere damage can severely hamper reading, right-hemisphere damage can affect the process in subtler ways. Howard Gardner of the Boston University School of Medicine has shown that the right hemisphere adds depth — layers and metalayers of meaning — to the material. Right brain-damaged patients have trouble understanding puns and double entendres, proverbs, metaphors and tongue-in-cheek comments.

Hence recourse to metaphor-like functionality and to perception as gestalt are right hemisphere tasks, whereas analytical language competence and feature breakdown of images appear to be left brain specializations. The experiments indicate that the right brain "takes control" for spatial tasks, hence favoring a metaphorical and "chunking" approach to description, as well as a gestalt approach to perception.

How can these observations be related to the task of interpreting a real scene? Figure 3 shows the results of applying an image segmentation algorithm to such a scene in remote sensing. The spectral information has been suppressed: all that is shown is the set of "segments" resulting from a particular image partition. Two regions (labelled A and B) have been isolated (see small inset).

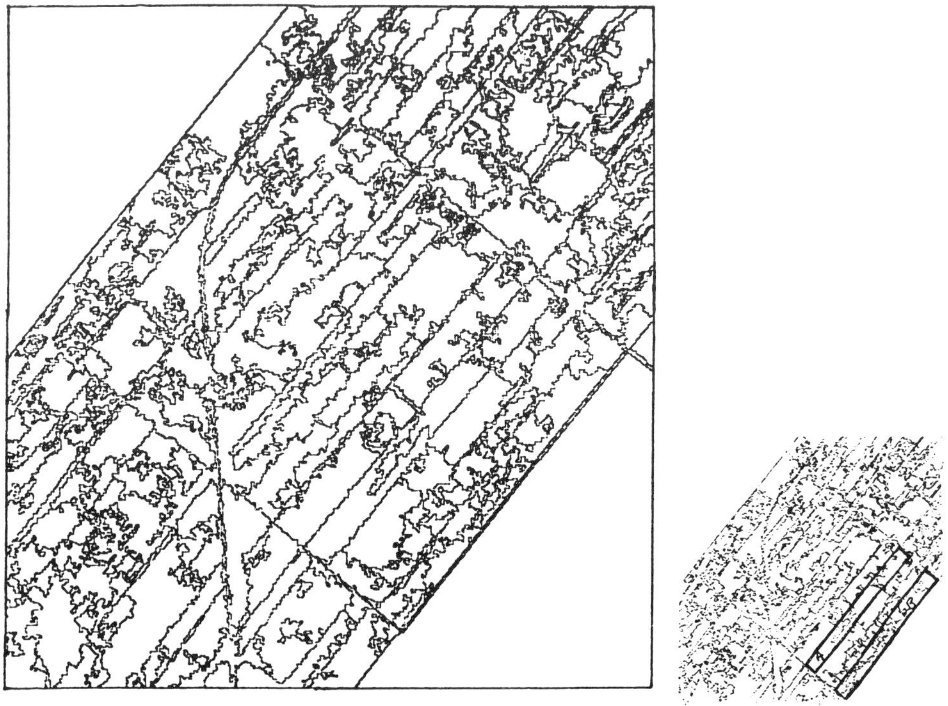

Figure 3. Results from the segmentation of a complex scene. The data used for the segmentation were a synthetic aperture radar image acquired over an agricultural site near Drummondville, Québec. The reduced image in the lower right shows the two regions for which descriptions were obtained.

Two individuals were asked to describe these scenes. One individual "knew" the image, while the other was ignorant of the image content. First, the observer who had knowledge of the scene content:

> Field A is long and skinny, divided into three major sections, each about the same length. It is a rectangular shape. The middle section intrudes on the first (west-most) with a long skinny spike. Between the second and third sections, there is a small, narrow irregular region which appears to extend along the north east side of the third section a small distance. Field B is about the same size and shape is Field A. Unlike the latter, however, it is broken into numerous small and irregular regions, especially near the middle of the field (lengthwise). The west end is divided into several long and thin regions which run parallel to each other. The east end is broken into larger regions which straddle the entire width of the field, even if the dividers are not straight or regular.

The scribe who had no knowledge:

> Field A is an oblong form in the middle of which is drawn a parallelogram. An arm emerges from the north-west end of the parallelogram in a straight line, even though the arm's sides are rather irregular. The east end consists of a double line, also irregular, which splits from its center further east, describing partially the letter M. Field B contains an wave-like structure more easily discernable at the beginning and end of the corridor. In the middle section, many small islands form from several closed curves, suggesting a multitude of directions but dominated by the global structure of the field as a whole.

These descriptions confirm the observations made earlier, as well as adding new material to explore. Associative codes are used for units perceived as being a single entity, although these may be quite complex. Hence, for example, the extension described variously as a "spike" or an "arm". However, in some cases, such codes are not used. Hence, for example, the first scribe describes "a small, narrow irregular region" between two zones. Interestingly enough, this scribe, on re-reading his text, came up with the word "isthmus" for this same feature, and wanted to change the text correspondingly. Hence it would seem that if no name, whether associative or explicit, "comes to mind", an analytical description or feature breakdown is provided; otherwise, an name is preferred.

Furthermore, the first scribe uses less iconic coding than the second, and those codes he uses tend to be landscape oriented ("field", "isthmus", etc.) because this is how he perceives the image: as a representation of an agricultural landscape. The second scribe uses more iconic codes, and her codes are "drawing" oriented ("the letter M", "a parallelogram", "wave-like", etc.), because this is how she perceives the image: as a drawing. It would seem that domain competent interpreters favor a more analytical description, while less knowledgeable speakers use more associative codes. This suggests that explicit codes, when available, are preferred over associative codes.

Finally, the shape descriptions of Figure 3 are nested. The scribe began with a description of the frame, then looked for "large-scale" regularities or structures within the frame, compared similarities between these structures, then differences, and then selected one of the structures as a new frame. The process stopped either at the simplest units, or at a unit which was identified with an explicit or associative code. If the process is similar to that of map reading, we can expect this sequence of events to vary with the individual. Indeed, just as for map reading, the nested structure is not seen all-at-once. It is "navigated" within a limited but moving attentional focus. This attentional field appears to be characterized by three levels of awareness, a frame or field of view (Level -1), a large scale regularity (Level 0), and a perception of fine or intermediate scale regularity (Level +1). Interestingly enough, the "or" in the last level appears to be an exclusive "or". If both fine scale "texture" and intermediate scale "structure" are present, the individual appears to "see" one or the other, but not both (unless she changes her framing to a different level). In addition, the "fixing of attention" on the object scale (Level 0) seems to leave the texture scale (Level 1) as a gestalt-like perception. Hence the recursion process stops either when the third level is perceived to be absent or is too blurry to see any finer detail, or when it is recognized as a known pattern.

This tri-level attentional field has characteristics similar to those in the figure/ground segregation as found by Hebb (1949) and modified by Hochberg (1971a, 1971b). Hebb discussed three aspects of this segregation: a primitive, sensorially determined unity, a nonsensory unity affected by experience and an identity function, also affected by experience. According to Hebb, the primitive unity is a direct product of the pattern of sensory excitation and coding strategies of the visual nervous system, and is constituted by an area of homogeneous color and intensity with a sharp boundary which is perceived as a single unified entity distinct from the surround. The nonsensory unity characteristic was derived from the observation that some boundaries are not determined by luminosity discontinuities in the visual field. The classic wife versus mother-in-law image (Figure 4) was used to illustrate this point, highlighting the fact that nonsensory boundary allocations are affected by our previous experience.

Figure 4: The wife and mother-in-law reversible figure (taken from Haber and Hershenson, 1974)

Finally, Hebb notes that both the naming of the figure and the observation that the figure is similar and/or dissimilar to other figures or falls into a category of figures are functions of identity, which appear to occur via spontaneous association upon first exposure to the object. Indeed, as pointed out by Hebb, recall and subsequent recognition of objects are dramatically aided by such identification. Identity as defined by Hebb represents properties of association or memory inherent in perception. Hebb also observed that such naming and/or association is heavily affected by our accumulated experience and also that it can change with time.

Although some of the structures found by Hebb have been modified by later research (among other things, Hebb's "primal unity" does not seem to be obtained from a single eye fixation but must be accumulated over several fixations), the main structure appears to hold (Hochberg, 1971a,b). The nonsensory unity discussed by Hebb appears to be a "reframing" function, where the scale and focus of the field of attention is shifted. The primitive unity can be equated with the large scale structure (Level 0) discussed above, and the association of iconic codes with shape and color are part of the naming or identity function.

This overview leads to a view of image entities as a hierarchical structure of embeddings which is navigated serially by an interpreter via a moving attentional field. The attentional field limits the number of elements which an algorithm must combine at one time in order to build useful intermediate structural models. Furthermore, since the attentional field in language is the sentence, the mapping of the sentence to this attentional frame seems to be a natural correspondence. The (cartographic) sentence then consists of a frame, an object located with respect to external referents (concepts of direction, distance and connection), and a set of embedded structures. The subject of the cartographic sentence is the object with its internal structure, while the predicate is identified with the spatial and topological referents. The "navigation" process then appears to consist of sentence building: looking for namable "chunks" (shapes or sub-structures which can be associated with an explicit or associative code) and identifying the spatial and topological relations which connect them to other structures. The entire structure of chunks+relations may then be assigned a name (or code), thus becoming a new "subject", and then nested within a larger scale of relations, just as complex objects described in one sentence can be abducted into another sentence through a single word referent. An indefinite number of levels could be nested in this way. This is fortunate, since the number of levels which may arise when combining image segmentation and interpretation at several resolutions may involve six or more levels of detail.

Hence the "map" produced by the intermediate stages of image interpretation outlined above may be thought of as a multi-tiered hierarchy of embeddings of different levels of detail, going from individual branch clusters in trees or individual plants in agriculture, right up to entire continents. Traditional paper maps and most digital maps today skip levels in this hierarchy and mix levels of detail in a single presentation. In this sense, much of the information in a map is hidden or implied. This information must be made explicit in automated image interpretation. Even so, the actual navigation process remains to be worked out (why one sub-structure should be selected rather than another? etc.). A top-down approach seems to be suggested, where some form of overall identification of large-scale regions helps provide context for unambiguously identifying smaller entities. This is further supported for image interpretation by noting that low resolution imagery tends to provide better identification than high resolution imagery, at least for certain tasks (Latty *et al.*, 1985). More work on appropriate associative codes for abstract shapes needs to be done and, in addition, these ideas need to be tested out in a real system.

5. Applying Attentional Field Based Structural Decomposition

The nested structure of feature-based shape descriptors, iconic codes and spatial and topological relations discussed in the previous two sections provides a framework for carrying out automated image interpretation. Hierarchical region-based image segmentation algorithms (Beaulieu and Goldberg, 1989; Lifshitz and Pizer, 1990) generate automatically nested structures which are similar to those which the human visual system seems to pick out. Feature-based shape descriptors for image segments are also being exploited by a number of researchers.

The use of associative codes for shape description is an alternate approach which is worth exploring, both for GIS-human interfaces and for automated image interpretation. The use of such codes may help reduce ambiguities when GIS users attempt to distinguish cartographic objects which are identified with the same explicit codes. Associative codes may also provide a convenient and efficient way of encapsulating shape information in highly irregular objects, such as are frequently found in remotely sensed scenes. They may also aid in rule-based reasoning about image primitives, in constructing knowledge schemas or frames for image interpretation and in building cartographic parsers such as attempted by McGranaghan (1990). Indeed, iconic codes, both explicit and associative, may serve as one of the needed bridges between spatially-coded knowledge (maps) and propositionally-coded knowledge (texts) about scenes.

Furthermore, a natural language interface to a GIS should take into consideration both the attentional frames by which human interpreters interact with maps and the nested construction and deconstruction process used to describe maps and scenes. Explicit recognition of levels of detail in the map digitizing process (or in assigning themes or levels to cartographic entities) may also decrease some of the confusion which seems to be involved in interpreting natural language descriptions.

In a typical remotely sensed image, about four levels of detail seem to be present. For example, in a Landsat TM scene, different principal use zones will be seen (urban, agriculture, forestry, alpine, river delta, etc.), followed by sectors (woods, fields, primary roads in an agricultural zone), inside of which will be found land units: fields, buildings and secondary roads. Inside these structures will be the texture level of road surfaces, crops, etc., since pixel size is 30 meters. In digital airborne imagery, the highest level will be sectors, followed by land units, and then land objects (hedges, ditches, crops, access lanes). The texture level in these images will be row patches and soil patches (for a pixel size of 50 cm.). In current GIS, the number of levels also rarely exceeds four. The human visual field encompasses six levels, however, and the full range of interest for GIS extends over at least eight levels, from individual crop plants to entire continents. Remote sensing capabilities are also extending to better spatial coverage of the Earth, leading to more levels at the top, both for airborne and satellite systems. It is not clear that existing GIS can deal effectively with maps or images over the entire range of scales of interest. A semiotic structure may provide the necessary framework.

6. Conclusions

In this paper, descriptions of intermediate results from image segmentation as well as abstract line drawings are studied in order to determine some general guidelines for automating the image interpretation process. The interpretation of abstract figures appears to resemble closely map reading, except that sets of associative prototypes which represent different aspects of the entity are used instead of explicit names to label complex entities. Furthermore, the named structures are nested into larger structures through the medium of a changing attentional field, which limits the number of elements (and levels) which are processed at one time. Studies in human visual perception have revealed a number of characteristics of this attentional field, including evidence for the primacy of naming the figure, as well as details on the nature of the frame, the figure and the ground.

The interpretation process consists of constructing and naming more complex entities out of simpler ones, one step at a time. Named structures may also be broken down into simpler structures and the relations between them. In a parallel with semiotics theory, the unit for constructing complex entities from structures one level of detail below is identified with the sentence. Cartographic entities are hence associated with the subject and the relations between such structures with the predicate.

A number of possible applications for results of this study are also presented, including scene interpretation, internal GIS structure and natural language interfaces for GIS. It is hoped that this paper, as well as the entire ASI conference, will help stimulate studies along similar lines and reveal new approaches to spatial representation and analysis.

6. Bibliography

Beaulieu, J.-M. and M. Goldberg (1989) 'Hierarchy in Picture Segmentation: A Stepwise Optimization Approach', IEEE Transactions on Pattern Analysis and Machine Intelligence 11, 150-163.

Edwards, G. (1990) 'Image Segmentation, Cartographic Information and Knowledge-Based Reasoning: Getting the Mixture Right', International Geoscience and Remote Sensing Symposium IGARSS '90, Washington, D.C.

Haber, R.N. and M. Hershenson (1974) The Psychology of Visual Perception, Holt, Rinehart and Winston, Inc., New York, 398 pp.

Head, C.G. (1984) 'The Map as Natural Language: A Paradigm for Understanding', New Insights in Cartographic Communication, Cartographica Monograph 31, Toronto, 1-32.

Head, C. G. (1991) 'Mapping as Language or Semiotic System: Review and Comment', in D. M. Mark and A. U. Frank (eds.), Cognitive and Linguistic Aspects of Geographic Space, Kluwer, Dordrecht.

Hebb, D.O. (1949) The Organization of Behaviour, John Wiley and Sons, New York.

Heller, Wendy (1990) 'Of One Mind: Second Thoughts About The Brain's Dual Nature', The Sciences, May/June, 40 ff.

Hochberg, J. (1971a) 'Perception: I. Color and shape', in Woodworth and Schlosberg (ed.), Experimental Psychology, Holt, Rinehart and Wilson, Inc., New York, 395-474.

Hochberg, J. (1971b) 'Perception: II. Space and movement', in Woodworth and Schlosberg (eds.), Experimental Psychology, Holt, Rinehart and Wilson, Inc., New York, 475-550.

Latty, R.S., R. Nelson, B. Markham, D. Williams, D. Toll and J. Irons (1985) 'Performance Comparisons Between Information Extraction Techniques Using Variable Spatial Resolution Data', Photogrammetric Engineering and Remote Sensing 51, 1459-1470.

Lifshitz, L.M. and S.M. Pizer (1990) 'A Multiresolution Hierarchical Approach to Image Segmentation Based on Intensity Extrema', IEEE Transactions on Pattern Analysis and Machine Intelligence 12, 529-540.

McGranaghan, M. (1991) 'Matching Representations of Geographical Locations', in D. M. Mark and A. U. Frank (eds.), Cognitive and Linguistic Aspects of Geographic Space, Kluwer, Dordrecht..

Röhler, R. (1988) 'Contribution of Colour to Pattern Recognition, Processing Structures for Perception and Action', Sonderforschungsbereiche, 133-161.

Ryerson, R. (1989) 'Image Interpretation Concerns for the 1990s and Lessons from the Past', Photogrammetric Engineering and Remote Sensing 55, 1427-1430.

SECTION 5: FORMAL TREATMENT OF SPACE IN MATHEMATICS

It is difficult to discuss spatial concepts in natural languages, because such discussion is essentially circular—we can hardly avoid using the same terms and concepts to describe what we are describing. The description and the described become the same, and definitional statements may be reduced to tautologies, understandable only if one knows what the terms are supposed to mean in the first place. This section includes some contributions to the effort to provide formal mechanisms for addressing this problem.

Formalization is always based on a set of symbols and rules for their manipulation, described as equality and substitution rules. If a particular combination of symbols is equal to some other combination of symbols, one can be replaced by the other one. At this level, "meaning" is associated neither with the symbols nor with the operations—it is only a mathematical statement, with no particular relation to any reality. But the formalist can then explain an interpretation for the symbols and operations, linking them to percepts of our experience, stating that the results of the 'real' operations applied to the "real" objects have equivalent results to the symbolic operations applied to the symbolic objects. This correspondence can become so compelling that "educated" people can easily accept something like Euclidean geometry as if it were a property of the world, rather than a set of mathematical axioms. In his book *Computation and Cognition*, Zenon Pylyshyn commented on the changing relation between geometry and the "real world":

> "Plane geometry is an outstanding example of how the acceptance, as a view of reality, of what was once a tool for calculation made a fundamental difference in science. The Egyptians were familiar with geometry and used it widely in surveying and building. The Greeks later developed geometry into an exquisite formal instrument. For the Egyptians, geometry was a method of calculation—like a system of ciphers—whereas for the Greeks it was a way of demonstrating the perfect order of that platonic world of which the observable one is just a shadow. It would be two millennia before Galileo began the transformation that resulted in the view so commonplace (and mundane) today that virtually no vestige remains of the Aristotelean ideas about natural places and motions. We conceive of space as a completely empty, infinite, three-dimensional, isotropic, disembodied receptacle, distinct from the earth or any object that might be located on the earth, one that is capable of housing not only things but also such incorporeal mathematical entities as points and infinite straight lines. Such a strange idea—especially if it were taken to describe something that occurs in this world—was unthinkable before the seventeenth century; yet not even Galileo fully accepted the idea of such a world as real. For him, a "straight line" was still bound to the earth's surface. Not until Newton was the task of "geometrization of the world" (to use Butterfield's 1957 phrase) completed. The transformation that led to reification of geometry, though basically one of attitude and perception rather than of empirical observation, profoundly affected the course of science." (Pylyshyn, 1984, pp. xiv-xv)

The correspondence between mathematics and experience is useful in two respects: in an engineering sense, we can rely on the symbolic operations to predict the outcome of equivalent "operations" in the world, without trying them. People do this all the time, from the analysis of very dangerous situations that cannot be tried (for example, modeling and predicting accidents in nuclear power plants) to the mundane prediction of the total cost of an order in a restaurant. Physics is useful for engineering because the mapping between the formulae and the world is a close one. But the correspondence between mathematics and experience is also useful in the way that the symbolic descriptions can be communicated without assuming any preexisting understanding of reality, because they are self-contained and by themselves do not refer to anything. The assumed interpretation in terms of real objects can then be compared with one's own experience. The formalism either determines what part of our experience is meant, or selects a subset of our experience. The formalism then serves as a definition of what is meant.

It is interesting and significant that the latter sense of the meaning of formal mathematics is very close to the essence of experientialism—it assumes that people by and large have essentially the same basic bodily experiences. Language then names some of these experiences, such as Mark Johnson's image schemata, and through social interaction, children learn which experiences relate to which words (symbols), trying to adapt their usage of terminology to the usage they see in their environments. The major difference between natural language and mathematical formalization is that language, as it is used, presents a flow of more-or-less consistent examples, whereas in the formalization, we try to give the governing rules as "typical examples".

Mathematics is a science that has dealt with the manipulation of symbols, and has constructed various consistent, complex systems of symbol manipulation, including algebra, geometry, topology, and several others. **John Herring**'s paper details first the basic concept of formal models, used here to avoid the circularity of natural language description, and puts these concepts in an object-oriented framework. He gives a systematic overview of the parts of mathematics that are relevant to Geographic Information Systems and the topic of spatial languages in particular—a very unusual collection. But the similarities between mathematical abstractions and natural language should not be surprising, since each is a product of human awareness of geometric and numerical experience.

The second paper in this section, by **Kelly Chan** and **Dana Tomlin**, complements this viewpoint by extending the discussion of geometry to a spatial occupancy view, rather than an object view. They remind us that delimiting objects or regions in space is a fundamental problem in geography, but is often impossible to achieve. Instead, Chan and Tomlin discuss the construction of an algebra to treat space with properties, without needing to differentiate objects in this space. They associate values for properties directly with location, and structure a cartographic model in map layers of discrete resolution elements; each layer contains a data value for a specific theme for each location. A model contains a layer for each theme, such as forest type, land value, or zoning regulations. Tomlin's map algebra contains operations for combining two layers and computing a new one, according to a variety specific rules. The operators can be categorized into one of four groups, which the authors call "local", "zonal", "incremental", and "focal", and all expressions in the map algebra are of the form

newlayer := function_of (oldlayer1[,oldlayer2,...]).

Map algebra has been applied mainly in raster or grid cell GISs, but has also been generalized to map layers as collections of irregularly-shape resolution elements.

Christian Freska's paper picks up a more cognitive stream, and first points to the fact that space and spatial relations can be experienced through different sensory channels—visual, auditory, olfactory, etc. This may be the reason that people seem to take spatial experience as being more "real" than other experience, and use spatial structure as a base for metaphors to organize other conceptual material. This leads to questions regarding which spatial relations should be considered, that is, which ones are worth abstracting from. There is some evidence that qualitative information is what people abstract from scenes, and that they use such information later for spatial reasoning. One important reason for this is that people usually deal with a context in which only a very small number of objects are visible and in need of differentiation. The qualitative observation is sufficient to identify an individual. A very good example is the case of the relative position of two objects—the tree in front of the house, the person behind the counter, etc. Allen analyzed the relations between two temporal intervals and found that there are 13 different relations between two such intervals. In two or three dimensional space, there may be considerably more relations possible between extended objects, depending on the kind of objects.

Daniel Hernández' paper examines a special case of the use of spatial relations—the deduction of the relative positions of objects in a three-dimensional world based on observations from different points of view. The individual views are encoded in an extension of the relations Allen proposed for temporal data. He then shows how such relative position must be transformed to account for differences in reference frames and points of view. Values from these relations can be combined in cases like "A is before B" and "B is before C" being combined in order to conclude that "A is before C". Some combinations yield a single relation, others a disjunction of several. Because it is possible to combine knowledge from several viewpoints, the relations can be further restricted and the position of an object can be determined as precisely as is necessary. On the other hand, if only coarse information is available, the computations are simpler, but yield less precise results.

The paper by **Matthew McGranaghan** discusses a problem which is in some ways the inverse of this: given a verbal description of a location, determine that location. The problem is motivated by the practical problem of determining coordinates (for mapping or for GIS data integration) for the locations of specimens in a herbarium. For each specimen, there is a label, which gives a terse verbal description of the location at which the specimen was found. Can this description be matched with the features in a cartographic database? The obvious first step is to use the place names and match them with a place names database. This will not work for labels which do not include names found in the gazetteer, and is only a starting point if the description of location is substantially qualified, for example, "2 km north of Hillsville on Smithman trail." McGranaghan analyzes the digital cartographic data distributed by the US Geological Survey regarding their suitability to this task. He also analyzes the use of prepositions in his sample and compares them with the use types Herskovits has given for English spatial prepositions. Most important for herbarium labels are geomorphic features which can potentially be extracted from the Digital Elevation Model data file from USGS, but they are not linked to the toponymy stored in the other files. In order for a GIS to support this application, it would have to represent geography differently than do current cartographically-oriented databases and GISs. He considers the alternative of a frame based representation that would contain explicit representations of neighborhood, topological relations, etc.

In the final paper in this section, **Michael Worboys** extends the formal, mathematical framework beyond the spatial and shows how temporal data could be integrated in a GIS. Numerous user groups and applications demand the extension of GISs to include temporal knowledge. Some of the more obvious examples include archaeology, anthropology, cadastral mapping, global change, history, and geology. Worboys presents the relevant theory from modal logic, which includes different types of temporal reasoning. Temporal aspects of GIS, including both statistical approaches and integrated space-time reasoning, will form a very important part of the research agenda for GIS. We expect that efforts to link cognitive and linguistic aspects of temporal data with formal aspects of the problem will prove productive.

Reference:

Pylyshyn, Zenon W., 1984, *Computation and Cognition: Toward a Foundation for Cognitive Science*. Cambridge, Massachusetts: MIT Press.

THE MATHEMATICAL MODELING OF SPATIAL AND NON-SPATIAL INFORMATION IN GEOGRAPHIC INFORMATION SYSTEMS

JOHN R. HERRING
Intergraph Corporation
Huntsville, Alabama 35894-0001 USA

ABSTRACT. Geographic Information Systems (GIS) use a variety of approaches to model spatial information and the data processing associated to spatial analysis. Each of these primitive data models has its own set of inherent strengths and weaknesses which determine how its users view the spatial world and reason about it. Further, each GIS application has developed its own jargon to describe complex spatial interactions not usually addressed in natural languages. To study how humans reason about space should require us to examine these logical approaches to spatial reasoning, especially those which extend natural language concepts with their own more application specific jargon. To extend this study to encompass methods for automated spatial reasoning in a manner applicable to existing GIS environments and applications requires a logically consistent framework for the models themselves, for the manners in which they interact, and for the spatial concepts that they can represent.

1. Introduction

To understand automated spatial reasoning in a Geographic Information System (GIS), one must understand how such a system might store and manipulate data. In a standard Relational Data Base (RDB), this understanding is straight forward, since the capabilities of the data base are fully defined by the relational model itself. Relational systems can claim a "completeness" for their model, and for the access methods provided for it (as implemented in their query language) by arguing equivalence to the theory of mathematical relations upon which the model is based.

The GIS implementor is not so lucky. There is no universally accepted mathematical model for geographic information, and thus no universally accepted concept of a complete GIS language. The RDB's relational model completeness does not apply here, since that "completeness" is a function of the relational model as compared to the theory of mathematical relations, and not of any geographic model. Furthermore, the completeness of the relational model is a syntactic or structural one, applied to the accessibility of the data stored in the database, not to the capability of the database to represent a complete set information types or a complete semantic model for spatial reasoning. As one would guess, this lack of universal model has lead various implementations off into separate non-convergent paths. This paper will not try to bring these paths together, as much as it will compare their separate underlying design philosophies as shown by their models.

D. M. Mark and A. U. Frank (eds.), Cognitive and Linguistic Aspects of Geographic Space, 313–350.

To begin to study or to build an acceptable picture of combined spatial/non-spatial models, we have to go back to some very basic first principles.

2. Data and Process Modeling

The purpose of any GIS is to model the real world in some manner. The system model provides a framework that mimics the interactions of real world objects. The user fills this framework with data collected in a manner consistent with the framework. The model acts upon this data and answers questions, draws conclusions or makes predictions about the real world which can then be applied to a variety of decision processes.

This type of system makes two fundamental assumptions:

1) the model and the data represented within it somehow reflects the pertinent facts about the state of the real world; and
2) the internal processes and relationships of the model somehow correspond to the processes and relationships in the real world.

If these assumptions hold, it is assumed that inferences from the model can be directly translated into inferences about the real world. If this is the case, the model is useful in analysis or predictions, as long as the constraints of the similarities in 1 and 2 above are not exceeded.

The question then arises as to what constitutes a "model", and what makes two "models" similar enough so that facts about one can be taken as conclusions about the other. Several logically equivalent variations of the definitions of a model exist. For the purposes of this paper, I will adopt an entity-operator-algebra definition. To describe a model, we must specify:

1) what objects are covered by the model,
2) what are the valid operations on those objects and relationships between them, and
3) what are the rules that govern these actions and relationships, their interactions with one another, as well as their interactions with the objects.

2.1 BASIC STRUCTURE

For our purposes, a "model" consists of:

1) a collection of entity domains (classes of objects).
2) a collection of operator domains on and between those domains (classes of operations and relations).
3) an algebra relating the actions of those operators (expressed as rules of calculation and composition).

The entity domains supply the semantic subsystem of the model with nouns, more or less concrete objects upon which the model will act and about which it will reason.

The operator domains supply the system with verbs, including both actions and descriptions. The active operators supply the model capability to transition from state to state through the modification of the entities or their description. The existence of active operators is a fundamental distinction between models. Models with active operators are dynamic in the sense that they model change, either in a predictive (into the future) or analytic (into the past, or interpolative) manner. Models without active operators (other than those the system needs for edit and capture) are called static or descriptive models. Descriptive operators supply the system with the capability to describe the entities or represent relationships between two or more entities. Obviously, any dynamic model contains a smaller descriptive one. In fact, several, possibly incompatible, dynamic models may share a single descriptive base.

The algebra supplies the model with the sets of rules under which the operators and entities can act and interact. The division between entities and operators is a formal one that we can impose at our convenience to aid in the translation of semantics. For example, "the road is concrete" can be interpreted in several different model-theoretic manners, depending upon which concepts we wish to define as operations and which we wish to define as entities. Taking "concrete" as a noun, we have "is" as a relations associating objects with their composition:

is (road, concrete) = TRUE

Taking a slightly different view of the world, we can use "is concrete" as an operator that distinguishes entities on their composition:

is_concrete(road) = TRUE

A third approach, which may expand upon the meaning of the sentence a bit, is to apply an operator "has a surface of" to road entities:

surface_of(road) = concrete

In our discussions to follow, we will assume that the algebra contains rules that allows us to define the equivalence of forms as the above example demonstrates. We could in fact define all domains as entity domains, and introduce an evaluation operator between corresponding domains. For example, the relational model can be viewed as containing entity domains for each attribute type, for each tuple type definable as a member of a relation, for the relations, and for truth values (Boolean TRUE, FALSE). The existence of a tuple in a relation is an operator mapping the tuples and relations into Booleans:

is_in: tuples x relations → { TRUE, FALSE }.

An alternate but equivalent description allows the relations to be considered as operators on the tuples, and then thought of as an entity-domain themselves. Using this approach the following equation defines for each relation an equivalent boolean operator on tuples:

is_in(tuple, relation) = relation(tuple)

The algebra of the model defines the operations and how they interact. For example the "is_in" operator above can be universally used to define a boolean operator for each relation which is equivalent to it. As another example, if our relational model is to contain an "entity integrity" rule, then there would be a concept of a key projection for each relation, and a concept of a NULL tuple for each tuple entity domain, that would satisfy the following identity in the algebra:

for every "relation" in "the relation entity-domain"
is_in (NULL, key (relation)) = FALSE

The operator algebra again blurs the distinction between operator and entity, since we can think of the operator algebra as operators on the entity-domain of operators. In fact this is exactly how the concepts of "views" in a relational data base are defined; a query (as an operator on a set of tables) is equated with the table it would produce, if it were executed.

In a system for spatial reasoning, we would define the entity domains as being the sets of objects capable of being reasoned about, and a set of relations (operators to {TRUE, FALSE}) corresponding to the relations those entities might exhibit. For example, reasoning about books on a shelf, there would be a entity domain of "books", and relations for "left-of" and "right-of". The operator algebra would be able to make statements concerning to the relations like "if Dante is left of Shakespeare, and Shakespeare is left of Voltaire, then Dante is left of Voltaire". The algebra could also be used to define non-primitive relations so that we could say that "if Dante is left of every other book on the self, then it is at the left end of the shelf" or "if Dante is left of Shakespeare and there is no other book right of Dante which is also left of Shakespeare, then Dante is immediately left of Shakespeare".

2.2 MORPHISMS (STRUCTURE PRESERVING CORRESPONDENCES)

The other problem discussed above is how to draw inferences about the character of one model from the character of another model. This leads to the other key concept in our algebraic approach to models, the concept of a morphism between models. A morphism is a mapping between models, which establishes a correspondence between entities and operators from the two models in such a way as to assure that inferences drawn in the one model apply to the corresponding objects, actions and relations in the other. More formally, suppose that we have two models, A and B: $A=\{E_A,P_A,A_A\}$, the entities, operations and algebraic rules for A respectively, and $B=\{E_B,P_B,A_B\}$. A morphism "m" from A to B (written $m{:}A \rightarrow B$) is a triple of mappings

$m_E: E_A \rightarrow E_B$
$m_P: P_A \rightarrow P_B$
$m_A: A_A \rightarrow A_B$

such that the internal structure of the entity-domains, operators and operator-algebra is preserved under the mappings. Since the context of an equation is usually sufficient to deduce the distinction between entity, operator or algebra, the various forms of m can be written without a subscript. This preservation of structure is usually symbolized by what are called commutativity diagrams. Such a diagram consist of a set of vertexes joined by arrows. At each vertex of the diagram is an entity-domain of the model in question. Each arrow joining two of the domains represents an operator of the model. The implication of such a diagram is that any path between two vertices is equivalent (that is, the functions represented by the arrows commute).

Pairs of diagrams, one from each of the two models joined by a morphism, can be joined into a larger diagram by the application of the appropriate function from the morphism for each vertex and for each arrow. Thus in our model of models, the concept of a morphism as an operator mapping one model to another means that the following diagram commutes for every operator in $p \in P_A$:

$$\begin{array}{ccc} E_A & - m_E \rightarrow & E_B \\ \downarrow p & & \downarrow m_P(p) \\ E_A & - m_E \rightarrow & E_B \end{array}$$

In this diagram the left hand column is a diagram in model A, the right hand column is a diagram in model B. The ladder of horizontal functions from the morphism shows the relation between the two subdiagrams.

In our bookshelf example above, we could take the model consisting of the integers 1 through n (if there are n books) and use "$<$" to correspond to "to the left of". Assuming that the data is collected and we are told that "Dante"→1, "Shakespeare"→2, "Twain"→3 "Cervantes" →4 and "Voltaire" →5. Then our morphism says "the Dante is left-most on the shelf" because "$1 < n$ for every other n mapped to (via the morphism) by a book in the shelf model". The advantages are obvious. Once the morphism is established, any query as to the order of the books on the shelf can be answered by investigating the relationship of their target numbers. Also, the morphism has produced an equivalence between the real world objects and a domain easily represented, analyzed, and manipulated by a software system (integers and the order relation "$<$").

In defining a morphism, it is necessary only to define the target for a generating set of entities, operations and algebraic rules. A generating set consists of a set of entities, operations, and algebraic rules, that through repeated application would generate the entire model. In general, models must have a finite generating set to be amenable to implementation; a finite set of basic concepts that can be implemented and used in combination to generate essentially infinite models. For example, a generating set for the integers is {1, + (binary addition), - (unary minus)}, since all other integers and integer operations can be written (using some induction) as combinations of these primitives such as in the following formulae:

$2 = 1 + 1, \quad 3 = 1 + 2, \quad \text{etc.}$
$m + (-m) = 0, \quad n - m = n + (-m)$
$n + m = (n + 1) + (m - 1)$
$n \times 1 = n, \quad n \times m = n + (n \times (m-1))$
$n < n + 1, \quad \text{if } (k < n, \text{ and } n < m) \text{ then } (k < m)$

In our bookshelf model it is not necessary to specify a "right-of" relation or an "end-shelf" operations since these can be defined using left-of and simple logical statements:

x right-of y $\equiv$ y left-of x
x left-end-of-shelf $\equiv$ for every y, x left-of y

Using the concept of the generating set we have m(right-of) = "$>$". Further, we can define an operator for the bookshelf model to correspond to a $m^{-1}(()+1)$ by seeing how "()+1" is related to "$<$" in the integers. Compare the following statements:

$x+1 = y$ if and only if $x < y$ and for every $z \neq y$, if $x < z$ then $y < z$

x is immediately left of y if and only if
x is left of y and for every $z \neq y$, if x is left of z they y is left of z

Thus the relationship that maps to "()+1" is "immediately left of".

As with other mappings, a morphism can act as an embedding if it does not collapse diverse items (m is one-to-one). In our bookshelf example, the morphism embeds the book model into the integers. As long as the logic applied within the integers deals only with the part of the system mapped to by the morphism or items generated by this set (entities={1,2,3,4,5}, operations={$<$}), conclusions drawn from the integers are applicable to the bookshelf. Proofs using other symbols (such as $>$, $+$, $-$, $\times$) in the integers are valid on the bookshelf if we can extend the morphism to include the operators used, as was done with the ()+1 operator above. This is how we can derive a theorem that says:

(x is immediately left of y) if and only if (m(x) + 1 = m(y))

If m is one-to-one, and maps to every entity and every operator in each domain (m is onto), then an inverse mapping m^{-1} exist and m (m^{-1}()) = m^{-1}(m()) = identity(), and we say that A and B are logically equivalent (isomorphic, or structurally equal). One would think that isomorphic models are rare, but in fact they are quite common (as we shall see). More importantly, the existence of an isomorphism (which is easily proven if one can derive a generating set for the mapped model) means that processes in one model mimic directly processes in the other.

2.3 CONCEPTUAL, USER AND IMPLEMENTATION MODELS

In building a software/hardware system, the designer, either consciously or subconsciously works with at least three models and a set of associated morphisms. The first model is the conceptual model, which describes the designer's logical view of the system. The second model is the implementation model, which describes the manner in which the system is actually implemented. Third, and often neglected, is the user's model, the model that describes how the user perceives and operates within the system (which may or may not be the conceptual model).

The conceptual model generally comes first. Then, the system designer constructs an implementation plan consisting of the implementation model with a morphism from the conceptual model. Finally, the external interfaces are implemented, each based upon a user's perception of the conceptual model. This gives us a triangle of models and associated morphisms:

conceptualization ⟵ d – user-view
↓ i   i(d())
implementation ⟵

The triangle is closed by the composite morphism i(d()). For the purposes of this paper, we will refer to "i()" as the "implementation" or "design embedding", since it must be an embedding to preserve the complete conceptual model, and it is usually realized through a system or software implementation design. We will call "d()" the "data definition embedding", the "data dictionary", or simply "schema", since it is usually implemented through a "data definition language", a "data dictionary", or "database schema". We can use these little semantic tricks to embed our theory in the software engineering language as it is used today.

It must be noted that these model names are not absolute, but are relative to one another. For example, at one layer of the software, the conceptual model may be a relational database model, while the implementation model deals with the various input/output and data manipulation techniques associated with this model. While at a higher level, an application, such as a natural language interface (NLI) uses the RDB as its implementation, and a more semantically appropriate model for a natural language. Note that in such an application, the pivotal role of the designer would be the automation of the morphism (interpreter and parser) between the NLI front- end and the RDB back-end.

3. Category Theory - Tensor Sums

What has been described so far is a category theory of models, see Casti, 1989; Barr and Wells, 1990; Herring, 1990b; and Cartan, 1956. Category theory is the study of structure-preserving morphisms between "algebraic" structures, especially concerned with the types of information and inferences that can be preserved under such mappings. We have seen how this preservation plays a role in system design. The tensor sum allows us, based upon common concepts, to fuse multiple models (and therefore, multiple designs) into a single model (and single design). One way to put two models together is simply to union their parts. Given a pair of models A and B then we can declare a new model $A \oplus B$ as:

$$\text{entity-domains}(A \oplus B) = \text{entity-domains}(A) \cup \text{entity-domains}(B)$$
$$\text{operators}(A \oplus B) = \text{operators}(A) \cup \text{operators}(B)$$
$$\text{algebra}(A \oplus B) = \text{algebra}(A) \cup \text{algebra}(B)$$

In this way we have put A and B into a single model. But, internally, A and B still maintained separate identities. A-entities and B-entities are separate. A-operations and B-operations are separate. And no valid expression in the combined algebra contains both A-operations and B-operations. This is a very loosely coupled system, and formally is equivalent to two separate systems, even though they may share a common implementation level design.

A method of joining models generating a more tightly-coupled result is the tensor sum. This sum identifies the common elements, both in terms of entities and operations, from the two models as mapping into common elements in the sum of the models. This sort of coupling allows for a free flow of technique from one category to the next. We have already been using a natural tensor sum in that statements from formal logic have been used in some of our "proofs", even though we have not included the entities and operations of formal logic as a integral part of the model definitions. Formally, to define the tensor sum, we require a third model C (representing the submodel common to the two systems) and two injectives $i:C \rightarrow A$ and $j:C \rightarrow B$. In $A \oplus B$, we define an equivalence relation that says for every c in C, $i(c)=j(c)$. This gives us the following diagram:

$$\begin{array}{ccc} C & \xrightarrow{j} & B \\ \downarrow i & & \downarrow \\ A & \longrightarrow & A \oplus_C B \end{array}$$

In a semantic sense, there is a limitation on the tensor sum of A and B. Suppose that there is a semantic model S which acts as an interpretation of both A and B. The interpretation for A (or B) can be thought of as a morphism from S to A (or B), giving a triad of models:

$$\begin{array}{ccc} S & \xrightarrow{j_S} & B \\ \downarrow i_S & & \\ A & & \end{array}$$

If the tensor sum $A \oplus_C B$ is to make semantic sense, in terms of the embedding of S, then the common interpretation in each model must factor through an embedding of S into C, so that there is a mapping $m:S \rightarrow C$ such that $j_S() = j(m())$ and $i_S() = i(m())$.

$$\begin{array}{ccccc} & & \overset{j_S}{\frown} & & \\ S & \xrightarrow{m} & C & \xrightarrow{j} & B \\ | & & \downarrow i & & \downarrow \\ \llcorner & \xrightarrow{i_S} & A & \longrightarrow & A \oplus_C B \end{array}$$

What this means is that if we have two models that have a common interpretation, then the tensor sum will have that same interpretation if the common submodel does.

4. Some Useful Models

This section describes three standard data models that are in common use in GIS environments: the Entity-Attribute-Relation (ER) model, the Relational Model (RM) and the Object-Oriented Model (OOM). It also describes a logic model for predicate manipulation that is common to all three implementation models. It then shows how the discussion above can be used to join them into a common implementation environment.

A Word on Entity-Domains: In many models, the various entity-domains may in fact be models in their own right. For example, we often include entity domains for integer and real numbers, which are in turn models with entity domains and operators of their own (the integers, "=", "-", "+", "$\times$", etc..). This is not usually a problem to the theory, since we can assume that this internal structure of common entity domains is preserved in the morphisms. We will, however, need to be careful in cases where entity-domains are called by the same name, but are not in fact the same model. A common example of such a "mistake" is the identification of real numbers, and the "real" numbers represented as floating point within computers. The analogy fails in several places. For example, in the real numbers, between any two numbers, there is a third, larger than the first and smaller than the second (infinite precision); while in the floating point numbers (which are finite representations) any two numbers differing only in the low order mantissa bit have no other floating point between them. Accompany that with the finite precision of the arithmetic, and anomalies abound. For example, given a circle and a line that are tangent to one another, if we rotate the configuration in the computer, say by 37°, the line will no longer be tangent (either having fallen into the circle or completely out of it) since the rotation matrix would have used sine of 37° and cosine of 37°, both of which are not exactly representable in any floating point number system, regardless of precision. Further, the larger the magnitude of the coordinates of the circle, the worse the absolute accuracy of the points, and the more noticeable the introduced errors.

4.1 LOGIC AND SET THEORY

The standard logical and set theory model is embedded in the algebra of all of the models used in this paper. This allows us to use standard style proofs. Similar models could be built with fuzzy set theory and fuzzy logic (both of which include a concept of probable truth).

These models contain as entity domains: atomic constants and variables, relations and functions symbols (appropriate to the various domains), statement symbols, logical connectives (AND ("$\wedge$"), OR ("$\vee$"), XOR (exclusive or), NOT ("$\neg$"), IMPLIES ("$\Rightarrow$"), and EQUIVALENT ("$\Leftrightarrow$","$\equiv$", "$\leftrightarrow$", IFF or IF-AND-ONLY-IF) and quantifiers (for-every "$\forall$", and there-exists "$\exists$"), sets, set operators such as intersection "$\cap$", union "$\cup$", difference ("$-$" or "$\backslash$"), set comparisons such as subset "$\subset$", and element of "$\in$", and Booleans (TRUE, FALSE). The operators are: combination and evaluation. The algebra consists of the basic theorems of mathematical logic and set theory and includes the standard inferences such as (A, B are statement variables, X, Y are set variables, and x, y are element variables):

$$(A \Rightarrow B) \equiv ((\neg B) \Rightarrow (\neg A))$$

$$(X \subset Y) \equiv (\forall x\, (x \in X) \Rightarrow (x \in Y))$$

This mass of notation is actually an aid to understanding, not in any Cartesian sense of the "world as mathematics", but in that it allows us to define formal algebraic rules for the manipulation of logical statements which are independent of the peculiarities of language and cultural biases. In fact, formal logic is the oldest example of a model designed to mimic the reasoning patterns of philosophers, so that they might be analyzed for "truth" (set theory is 2000 years younger).

Much of the relational query language is defined in terms of the evaluation of first-order predicates. A predicate is a valid statement from the model's algebra (wff or well-formed-formula) made up of qualifiers, variables, constants, and operators (comparisons, AND, OR, etc.) whose truth or falsity can be determined for any complete set of appropriately assigned values for the unqualified (free) variables. Constants are values from some set of domains, and variables are representations for some unknown or unspecified elements of these same domains. The operations include comparisons for each simple type.

For example, the following are predicates:

"$x = 3$ AND $y > z$"

"$(\exists x)$ ($y \geq x$ AND $x \in X$)"

In the first, given values for the free variables x, y, and z, we can evaluate the truth of the statement. In the second, we need values for y and the set X to evaluate truth; x is already bound by the statement "$(\exists x)\ x \in X$" which in effect defines the possible values of x. The comparison "$\geq$" does limit the values for y and and the elements X to those which can be compared with each other (meaning that y and the elements in X must be from the same entity domain, and that domain must admit "<" as an operator).

4.2 ENTITY ATTRIBUTE RELATION MODEL

The ER model is the simplest of the three models and is used mainly as a manual method for designing data types for software environments, see Bédard, 1989; Stonebreaker, 1986; Ullman, 1988. As the oldest of the three, several variations exist in terms of representational methodology, but here we will refrain from discussing these minor differences and stay with their common capabilities.

The three basic domains of the ER, as you might guess from the title, are the entity, the attribute, and the relation. Entities are divided into types (domains or subdomains). Each entity type is associated with some number of descriptors (called attributes) which associate directly with the entity. Entities (instances of an entity type or domain) can engage in relations with entities of the same or other types in a strictly binary fashion. Through the use of "relation entities", these relations can be of any degree and carry attributes (it is usually in the handling of relational entities that the specific ER formalisms differ). In our terms, any of the ER models is isomorphicaly embedded in (and is therefore logically equivalent to) a model allowing any order of attributed relations. We will assume that our version of the ER model is the most general one.

A couple of the relation types in the ER need special note to aid in our description of the Object Oriented Model (OOM) later: aggregation and generalization.

Generalization, and its inverse specialization, are the "is a" relation between entity types. We say that one entity type is a specialization or subtype of another, if the instances of the first type are also instances of the later. A specialized (or descendant) type has all the attributes, and all the relations of its more general ancestor, but has additional associated information. Generalization is a relationship among entity types, not the instances of those types.

Aggregation is the "is a part of" relation which exist between instances of specified entity types. This relation allows us to represent specific instances of a class as a composition of other entities from the same or different classes, and treat the composite entity as a single object.

The ER model plays another very important role in database technologies: semantic translations. The ER model is the most natural of the currently used data models in that it most closely mimics the noun-verb style of language syntax. As such, ER expressions can be quickly translated into natural language sentences, and given a proper dictionary, sentences can be translated back to the ER model without significant loss of meaning. This is often used in databases to implement a natural language interface via a morphism to the ER model.

4.3 RELATIONAL MODEL

The relational model, RM, is based upon a mathematical formalism called relations, see Date, 1983, 1984, 1986; Codd, 1979, 1988, 1990; Stonebreaker, 1986; and Ullman, 1988, 1989. A mathematical relation is a set of ordered sets (with a fixed number of elements) called tuples, such that each commonly ordered entry in the ordered set comes from a common mathematical domain. The most common usage associates these relations to a rectangular table. Each tuple in the relation corresponds to a row in the table, and the position-to-domains correspondence is associated with a column (called an attribute) of the table.

There are several points in which the RM differs from standard mathematical theory. The most important one is the type of domains allowed by the model. In the mathematical theory, the domains can consist of any mathematically well-defined set of concepts or objects, including other relations, functions, infinite sets, etc. We have already used this concept above in showing how models can be constructed so that operators may operate on operators. The RM domains are generally staticly defined, ruling out the ability to use variable sets as domains (this comes out later in the discussion on RM entity and referential integrity). Furthermore, in the RM, several restrictions are placed on the domains. RM attribute domain entities cannot contain other relations and are not decomposable into smaller entities (invalidating lists, sets, some types of predicate statements and functions as valid domains). RM relations are finite (a necessity due to the nature of the Relational Data Base (RDB) implementations). In practice, the RM domains are generally restricted to numeric types (float, integer, decimal), fixed length character strings, or minor variations such as date or time data types.

The RM then includes a relational algebra inherited from mathematics. Each of these operations takes one or more relations and defines a new relation. The basic operations of this algebra include:

Select: pick a subset of tuples based on a predicate.
":" : relation × predicate → relation
Project: pick a subset of attributes.
"%" : relation × attribute-names → relation
Product: paste tables side by side, every tuple to every tuple.
"×" : relation × relation → relation
Union: paste tables end to end.
"∪" : relation × relation → relation
Difference: tuples in one table not in another (similar) table
"−" : relation × relation → relation

These operations are augmented by some standard composite operators:
Join: select from a product based upon a predicate.
"⋈" : relation × relation × predicate → relation
Intersection: pick common tuples from two similar (equal) tables;
Writable as a combination of union and difference.
"∩" : relation × relation → relation

4.3.1 *RM Query Languages.* The Relational DataBase (RDB) implementation of the RM supply a data access and manipulation language based upon the model and the logic-set-theory model. These languages allow the user to do the following basic operations on a table:

- create table (definitions) in the database,
- alter table (definitions) within the database,
- drop table from the database,
- insert rows into a table,
- delete rows from a table based upon a predicate,
- update rows from a table based upon a predicate,
- select and project rows and columns from a table based upon a predicate whose free variables are bound to the columns of the relation,
- specify an "exists" predicate based upon attribute values, existence of tuples with a relation (subquery), and simple functions of an entire column or set of columns,
- specify a limited number of functions on a single column, such as maximum,minimum, average, count and sum.

A word is needed here on the imbedding of the logic model in the query language. Quantifiers in the logic model apply to domains unless otherwise bound. Quantifiers in the relational query language are always bound to the existence of a tuple in the relation (a version of the "closed-world hypothesis (assumption)" (CWH) - if it is not specified as true in the database, it is false). For example, in logic the following statement is TRUE, regardless of what "y" is meant to be, as long as the underlying entity-domain is non-empty:

$$(\exists y)\ 1 = 1$$

In the relational query language, if y is bound to a NULL-filled column in a relation, or to an empty relation, then the statement is false. Alternatives to the CWH can be supplied by default logic, where the lack of data is defaulted to a reasonable hypothesis.

4.3.2 Integrity Rules. Several integrity rules are generally thought of as being included in the RM, although many RDB implementations ignore them. The two basic integrity rules are entity integrity and referential integrity. Both of these rules operate on the concept of a key, the set of attributes that uniquely identifies a tuple within a relation. The usual semantic meaning of a key is the identity of the real world entity or object to which the tuple attributes apply.

Entity integrity informally means that each tuple of information must reference its real world entity. Formally, this requires that for all relations, the key is not NULL:

entity integrity: $(\forall\, r \in \text{relations})\ (\neg\ (\text{NULL} \in \text{key}(r)))$

Notice that the justification for inclusion of entity integrity into the RM is based upon usage: the key of a relation is the linkage for that relation to the real world, the identifying attribute for the real-world object which is described by the relation.

Referential integrity, informally means that an external reference (a "foreign key") in a tuple must be matched with an entity that exists in the database. Formally this requires that for all references between one relation and another, the target of the reference exists:

referential integrity:

$(\forall\, r1, r2 \in \text{relations})\ (\forall\, A \text{ key-domain})$

$[A = \text{key}(r1) \wedge \text{is\_attribute}(A, r2)] \Rightarrow [(\forall\, a \in A)\ ((a \in r2\%A) \Rightarrow (a \in r1\%A))$

4.3.3 *Codd's Extended Model.* In light of the integrity rules the lack of a concept of "entity" and "reference" has bothered many database designers. One of the better solutions to this is Codd's extended model RM/T (Relational Model - Tasmania), see Date, 1983; and Codd, 1979, 1990. The model defines, in an almost algorithmic fashion, how to embed the full ER model into the RM. Codd extends the concept of entity-domains to include entities whose existence is predicated on the descriptions of or relations between other entities:

Kernel Entities: those which correspond to the real-world objects.

Characteristic Entities: those which qualify or describe other entities.

Associative Entities: those which represent relations between other entities.

He goes on to classify relations and attributes in the database according to how they relate to the description of these entities:

E-relations: relations which establish the existence of an entity-class.

P-relations: relations which establish the properties of an entity-class.

E-attribute: attribute of an E-relation used to identity a participant (instance) of an entity-class; its corresponding domain is used to reference (as surrogates of) this class in all relations.

What Codd has done in this "extended model" is define the morphism that embeds the ER model into a RM model in a standard way. This embedding is not the only one possible between the ER and the RM, but it does exhibit how such a morphism could be formally built.

4.4 OBJECT-ORIENTED MODEL

The object oriented model, OOM, is an extension of the ER model to encompass process modeling, see Cox 1986; Ullman, 1988, 1989; Herring, 1987; White, 1986; Ellis, 1990. The OOM terminology changes "entity" to "object". Each entity type (now called object class) is associated not only with its attributes and relations, but also with a type-specific suite of operators (the operator is a message, while the class-specific implementation is called a method). Relations are still relations, although not specifically a formal part of the model. They can be implemented as either as an object class in their own right or as messages on the appropriate related object classes.

4.4.1 *Overloading - Methods and Messages.* The OOM model is unique among the models described in this paper in that it makes statements about the implementation of its operators. It includes, for each data model class-entity-domain, two associated operator-entity-domains for protocols (messages) and algorithms (methods). A special operator called bind associates message-protocols and classes to methods:

bind: object-class × message-protocol → method

"Bind" takes the class of the object being sent to, and the classes (or entity-data-types) of the parameters in the message, and specifies (binds) them to an algorithm (method). Another operator (send) invokes the algorithm:

send(object, protocol) = bind(class(object),protocol)(object, protocol)

The output of the bind() operator is a method in the class-associated method-domain that operates on the object, given the parameters of the protocol.

Thus, in the OOM, a formal distinction is made between the conceptual level of an operation and the implementation level; a class's operator is defined at two levels: message and method. At the conceptual message level, the input parameters or protocol for the operator are defined. At the implementation method level, the message-protocol is translated into the appropriate actions to be taken on the specific object receiving the message. The protocols are overloaded in the sense that the same message may take several forms depending on the object class and parameter types passed in. A good example of this is that the standard "+" operator in mathematics or most programming languages is the same conceptually for a wide variety of entity-domains (integer, real, vector, matrix, function, tensor, etc.), but actually represents several different algorithms.

4.4.2 *Encapsulation.* Inclusion of the operators with the object definition is referred to as encapsulation. At the implementation level, this means that an object class consists of public and private sections. In the public section, the object class defines all message protocols and instance data (attributes) that are available to the system at large. In the private section, the object class defines internal data and internal messages that are private to the class implementation and not accessible to the system at large.

4.4.3 *Inheritance.* The distribution of message-protocols and method-implementations is spread through the various object classes by a set of rules depending upon the definition dependencies of the various object classes derived from one another through generalization. A specialized class (subclass, or derived class) supports the protocols of the classes from which it descended. For example a coffee cup is a instance of the "coffee cup class". As such, it inherits attributes and operations from the class "container class", such as attributes for "contents", and "capacity" and operation-protocols for "fill", and "empty". It also inherits from the "ceramics class" the operation-protocols for "break". If the methods inherited from the super-class are appropriate, then no further implementation is needed for the protocol. If the methods are somehow inappropriate, or incomplete, the class definition over-rides the method and supplies either augmented or replacement methods as appropriate.

4.5 A HYBRID MODEL

Using the tensor sum above, we can define a hybrid model for the OOM and the RM based upon the common imbedding of the ER. This hybrid OOM $\oplus_{ER}$ RM is an extension of Codd's RM/T. Basically, we have the following identifications (Ø is used to represent "no corresponding structure):

OOM	ER	RM/T
objects	entities	e-tuples (e-relations)
instance data	attributes	attributes in p-relations
relation - objects	relationships	surrogates in p-relations
messages, methods	Ø	Ø
Ø	Ø	relational algebra

Figure 1: Comparison of OOM, ER and RM/T

Other authors have noted these same equivalences used to create this tensor sum, see Andrews 1987.

4.5.1 *Extended Query Language.* Given that the three models are joined, the model can be completed by giving a descriptions of how the two operator algebras (messages-methods and the relational algebra) can be combined.

Incorporation of OOM-object concepts into the relational algebra, as implied by the embedding-morphism, is quite simple: include the domains expressed by object classes (Codd's E-domains). This extends the algebra to the full theory of first-order mathematical relations. Since the concept of the existence of an object is independent of attribute value, the identity of an object would probably be specified through a system-assigned identity system, see Date 1983. Methods for the creation and destruction of objects would have to be supplied for the "insert" and "delete" RM queries (as is done in RM/T).

Incorporation of the OOM method and message concepts into the query syntax is also well understood. Using a function-based model for the RM, we can associate any attribute as a message on the primary-key of the tuple (compare for example the C++ syntax for substructure "object.data" with its syntax for messages "object.message(parameter-list)"). A message syntax in a query would cause that message to be sent (note that object-identities are passable parameters). The message return value used as indicated by the query syntax. A straight "send ... where ..." could also be defined in parallel to, or incorporated into the "update" and "insert".

5. Geometry Models

In this section, I will develop the standard geometric models used in GIS applications and investigate how they can be combined using the model tensor sum. I will also describe some of the uses of geometric models for modeling things other than geometry, the intrinsic geometry of distribution attributes, movement, and change.

5.1 COORDINATE SYSTEMS

The oldest geometric model (2300+ years) and the one still in widest use, is that described in Euclid's *Elements*. In general cartographers used Euclidean methods to project the spherical earth on to flat planes so that mapping could be done (most famous of which is the Mercator projection still widely used today). In the 17th century, Descartes put the model in a more mathematical form by describing geometric objects by coordinates, unifying geometry with functional analysis. Cartographers, were quick to seize Cartesian coordinate systems and the new calculus to analyze old projections and define new ones. What they learned, and what we still need to keep in mind is that any map projection distorts most (probably all) of the following:

a) scale (it is non-constant over the surface of the map, and it usually does not take elevation or other landforms into account),
b) distance (the variable scale confuses it),
c) geodesics (the shortest path is not a straight line),
d) area (the scale problem squared),
f) shape (distorted by both scale and landform).

In designing a geographic spatial system, each of the above can cause problems, in the perceptions of the user if not in fact.

5.2 GEOMETRIC REPRESENTATIONS

The ability to represent the geometry of objects, or to distinguish the extent of the validity of an attribute value is a universal requirements of all GIS systems. There are two basic views taken of this requirement: feature-based and position-based. In a feature-based system, the primitive objects are entities, one of whose attributes is a description of geometric extent. In a position-based system, the primitive objects are the (coordinate) locations with which attributes have been associated. The two views can lead to radically different systems, but their underlying needs to distinguish pieces of geometry are logically equivalent. The universal common denominator is the use of a coordinate system to associate internal numeric structures to real world positions.

In this section, we look at some of the common geometric representation techniques and how a GIS system might use them.

5.3 RASTER OR GRID REPRESENTATION

Raster or grid representations of spatial data consider the fundamental object to be the point (these structures tend to be used exclusively in position-based systems). The underlying geometry of space is divided into adjacent cells (pixel. or picture element, for 2-dimensional spaces and voxel, or volume element, for 3-dimensional spaces) usually of a fixed size and shape in the coordinate system. Each cell is assigned values for a small set of attributes, and this information is compressed, indexed, or both and stored. Although the schemes for storage vary widely, and each has technical merit from an implementation point of view, for the conceptual view that we need here, they are all equivalent.

The major conceptual distinction between different forms of the data is the meaning of the measurement assigned to a cell. A raster data set usually contains an average value for the attribute over the area centered at the cell coordinate, and a gridded data set usually contains a point sample that is the value of the attribute at the cell coordinate.

5.4 VECTOR REPRESENTATION

In a vector representation, objects are represented by key coordinate locations and by an associated algorithm for the logical generation of the full coordinate description (stroking). The standard representational objects are lines, line strings, circular arcs, elliptical arcs, general conic and often other Euclidean constructive objects. Vector structures can be used in either position or feature-based systems, depending on whether the structures are attributed directly, or used as an attribute for aggregated feature-objects.

5.5 FUNCTIONAL REPRESENTATION

Functional geometric descriptions are a generalization of the algorithmic objects in vector representations. Each functional geometric object consists of a domain (parameter space) and a function mapping the domain into the range coordinates system. Domains are usually some subset of a Euclidean space, the most common of which are I (the closed interval from 0 to 1, [0,1]) or some unit cube I^n (the cartesian product of I, n times). Logically, such a function supplies an identification of points in the domain space with their image points in the range space.

The most common of the functional representations, spline functions are based upon a variable weighted average (weight functions) over the domain of some specified set of points in the target coordinate space (poles), see Bartels, 1987; Farin, 1990; Faux, 1979. Formally, for a three-dimensional data set, we have a spline as a function:

$$f : I^n \rightarrow E^3$$

$$f(t) = \sum_{i=1}^{n} w_i(t)\, P_i$$

where t is a vector in I^n, P_i are points in E^3 and $w_i()$ are functions from I^n into I^1 such that:

$$0 \leq w_i(t) \leq 1 \text{ for every } t \in I^n$$

$$\sum_{i=1}^{n} w_i(t) = 1$$

Such a set of weights, "w_i", is called a partition of unity.

Curves can be represented as one-dimensional splines:

$$c : I^1 \rightarrow E^3$$

The continuity and differentiability of the curve are determined by the smoothness of the weight functions. The various derivatives of the curve as a function have exactly what you might expect. c'(t) is a tangent vector to the curve, c"(t) is a normal vector to the curve pointing directly away from the center of curvature, and c'"(t) is a binormal indicating the direction of the torsion of the curve (tendency of the curve to leave a planar surface).

Surfaces can be represented as two-dimensional splines:

$$s : I^2 \rightarrow E^3$$

The images of the domain lines in the surface give a spline grid of constant parameter values. The partial derivatives of a surface spline give us the tangents to the surfaces in the direction of the parameter curves.

Solids (volumes) can be represented in two distinct manners. The most common representation is as a collection of surfaces which form the boundary of the solid. In the second technique, generalizing from the above, we can always consider solids as representable as three-dimensional splines:

$$v : I^3 \rightarrow E^3$$

Such volume splines must usually be almost everywhere one-to-one to prevent the function from collapsing multiple points from the parameter space into single points in the range space (the mathematical equivalent of "spindle, fold and mutilate"). The most common exception to this is where the boundary of the parameter space is collapsed to give us non-rectilinear ranges.

Distributions of attribution can also be addressed by spline and other functional representations, see Prenter 1975. Given an attribute or a set of k attributes, each of which is expressible as a real value, and each of which is a continuous function of space, we can generate spline functions whose first three range coordinates (dimension of target geometry) represent points, and whose k trailing range coordinate values are for the attributes along the spline:

$$f : I^n \rightarrow E^3 \text{(geometry)} \times E^k \text{(attributes)}$$

In using splines or other functional geometric descriptions for the distribution of attribute values, we are creating a tensor sum model that makes no implementation distinction between geometry and other numerically measurable attributes. Such distributions are inherently position-based representations, but they can be used in feature-based systems if the functions are used as feature attributes and not as separate entities.

6. Simplified Geometric Spaces

Several types of non-coordinate geometric spaces are useful in GIS. Each of these spaces simplifies the nature of geometry, limiting the number of objects that must be contended with during data analysis or manipulation, and simplifying the operations. Since they are simplified models, each can be associated with a natural embedding into the standard coordinate geometry model.

6.1 METRIC SPACES

Metric Spaces consist of a set of "points" or objects and a "distance" function that returns a non-negative real number for each pair of objects. Symbolically, a metric space is a pair {X,d} where X is a non-empty set and d is a function on pairs of objects in X, $d:X \times X \rightarrow R$ such that the following are true:

$$(\forall x \in X) \quad d(x,x) = 0$$

$$(\forall x \in X)\,(\forall y \in X) \quad d(x,y) = d(y,x)$$

$$(\forall x \in X)(\forall y \in X)\,(\forall z \in X) \quad d(x,z) \leq d(x,y) + d(y,z)$$

A metric space can be generated from a set of geometric points and the standard Euclidean metric. More interestingly, a metric space can be generated from a set of geometric objects and a metric defined as the distance between those objects; for a metric space (X,d), we can define a metric on the subsets of X, 2^X (so named because the relation between the number of items in X and the number of subsets of X) by:

$$(\forall A \subset X)(\forall B \subset X) \quad d(A, B) = \text{minimum}\ \{ d(a, b) \mid a \in A \ \wedge \ b \in B \}$$

In such a metric space, sets which have non-empty intersections are 0-distance apart:

$$(A \cap B \neq \emptyset) \Rightarrow d(A,B) = 0$$

6.2 TOPOLOGY

The history of topology dates back to the 17th century, and was, at that time, called "analysis situs" (1650), the analysis of space; the study of geometric properties not tied to distance or coordinate measure; coordinate-free (coordinate-independent) geometry. Today, the mathematical field of topology studies the characteristics of geometry (in any dimension) that are invariant under continuous transformations (spatial distortions without singularities - rips or tears). Specifically, the subfield of algebraic topology deals with the algebraic manipulation of symbols which represent geometric configurations and their relationships to one another. It is this symbolic geometry that is used in GIS systems today. The fundamental methodology in algebraic topology (as in category theory) consists of three steps:

1) convert the problem from a geometric statement about geometric configurations, to an algebraic statement about symbols representing those configurations;
2) solve the algebraic form of the problem; and
3) convert the solution back to a geometric statement.

Although the results of algebraic topology are useful in themselves (having implications in other fields of mathematics, as well as in chemistry and physics), for a GIS it is this basic three-step process that leads to fundamental improvements in geographic data manipulation. This theme is repeated many times in topological-GIS algorithms.

Point set topological spaces are based upon the concept of "open" sets. A set is said to be open in a metric space if every point in the set is surrounded by points also in the set, for some slight distance:

$$(A \text{ is open}) \equiv (\forall a \in A)(\exists \epsilon \in R, \epsilon > 0)(\forall x \in X)((d(x, a) < \epsilon) \Rightarrow (x \in A))$$

A topological space removes the concept of "distance" and relies only upon the concept of "open set". Formally, a topological space is a set of points X and a set of subsets of X, $\Im$, such that: ($\emptyset$ is the null or empty set)

$$\emptyset \in \Im$$

$$X \in \Im$$

$$\forall A, B \in \Im, \quad (A \cap B \in \Im)$$

$$\forall S \subset \Im, \quad ((\bigcup S) \in \Im)$$

The last two conditions state that $\Im$ is closed under finite intersections and arbitrary unions. It is quite easy to show that the definition of open in metric spaces produces a topology satisfying the above 4 conditions. With a great deal more effort, it can be shown that most topological spaces can be produced via a metric. In terms of models, this implies a morphism from the topological model to the metric one. Open sets in the topological model are mapped to sets satisfying the metric definition of open given above.

Quite a bit of analysis can be done with point-set topology, see Egenhofer, 1990a, but, the more common usage in GIS is for algebraic topology, see Alexandroff, 1932; Spanier, 1966; Switzer, 1975; Herring, 1987. In algebraic topology, the geometric space is divided into sets, and the topological relations between these sets are reduced to algebraic equations. There are two main technical approaches; simplicial topology and cellular topology.

Use of algebraic topology allows us to define a morphism between a geometric space and a simple algebraic space. Instead of manipulation of large numbers of coordinates describing geometric objects, we can manipulate symbols using a fairly simple algebra. This has several impacts on the GIS implementation. Performance is improved, since the manipulation of the symbols requires less system resources. More importantly, a certain amount of precision in the expression of geometric relations is gained, since the algebra reduces the geometry to a simple set of primitives that, in combination, can be used to express the more complex ideas.

The basic algebraic topology primitives are closure, interior and boundary (in addition to the standard set theory primitives). The point set definitions of these concepts differ slightly from the algebraic topology ones, but they carry a good intuitive feel for the meanings:

$$\text{closure}(A) = \{ x \in X \mid (\forall U \in \Im)(x \in U) \Rightarrow (U \cap A \neq \emptyset) \}$$
$$\text{interior}(A) = \{ x \in X \mid (\exists U \in \Im)(x \in U) \wedge (U \subset A) \}$$
$$\text{boundary}(A) = \{ x \in X \mid (\forall U \in \Im)(x \in U) \Rightarrow (U \cap A \neq \emptyset) \wedge (U \cap (X-A) \neq \emptyset) \}$$

The closure of a set is thus the collection of points within the "universal" set that are "close to" the set ("close to" is defined by the topology). Boundary and interior are a bit simpler, since they correspond very closely to the English words. The interior of a set is the collection of points that are completely surrounded by the set (again where surround is defined by the sets in the topology). The boundary of a set is the set of points within the universal set that are "close to" both the set and its complement. Using set theory, one can prove that:

$$\text{interior}(A) = X - (\text{closure}(X - A))$$
$$\text{closure}(A) = X - (\text{interior}(X - A))$$
$$\text{boundary}(A) = \text{closure}(A) \cap \text{closure}(X - A)$$
$$\text{closure}(A) = \text{interior}(A) \cup \text{boundary}(A)$$
$$\text{closure}(A) = A \cup \text{boundary}(A)$$
$$\text{interior}(A) \cap \text{boundary}(A) = \emptyset$$

This would allow us to define the entire theory using any of the three topological set operators (in combination with classical set theory) as the only primitive. Modern algebraic topology uses the boundary operator (usually written as "$\partial()$"). For the purposes of this paper we will mnemonically write the interior operator as "i()" and the closure operator as "cls()".

6.2.1 *Simplicial Topology.* The simplest of the basic tools of topology is the "simplex", see Alexandroff, 1932. In the most general of terms, an n-dimensional simplex (or n-simplex) is the convex hull of (n+1) points, in general position, embedded in a space of dimension n or greater. Thus, for the modeling of real 3-dimensional space, a 0-simplex is a single point, a 1-

simplex is a line joining two distinct points, a 2-simplex is a filled triangle joining three non-colinear points; a 3-simplex is a solid tetrahedron joining 4 non-coplanar points.

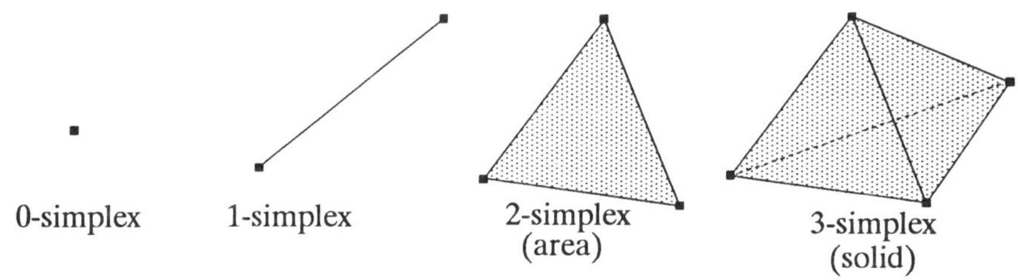

Figure 2: The Simplicies Used in 3-D Simplical Complex

Assuming the n-simplex is defined by the ordered set of n+1 distinct points $a_0, a_1, a_2, \ldots$, an, we can write the simplex as $\langle a_0, a_1, a_2, \ldots, a_n \rangle$. The ordering of the points within a simplex is considered as an orientation. The exchange of any two vertices in the list is said to reverse or negate the orientation (e.g. <...,a,...,b,...>=−<...,b,...,a,...>). Orientation has a natural geometric interpretation. For a 1-simplex, the orientation is a direction of traversal, distinguishing start from end, left from right. In a 2-simplex, orientation can be used to distinguish top from bottom. The top of a 2-simplex is the side from which the ordering of the corners is counter-clockwise (mathematically positive).

The boundary of a n-simplex consists of the union of the (n-1)-simplices that can be obtained by ignoring each of the extrema points one at a time. The boundary of a n-simplex contains (n+1) (n-1)-dimensional simplices; for example, a 3-simplex (tetrahedron, <a,b,c,d>) has four 2-simplex (triangles <b,c,d>, <a,c,d>, <a,b,d>, <a,b,c>) faces.

A n-dimensional simplicial complex is a collection of n-simplices sharing common boundaries. Most geographers are familiar with two-dimensional simplicial complexes in the form of triangulated irregular networks (TIN's) used in terrain modeling, and with one-dimensional simplicial complexes in the form of path-node graph models used in network analysis.

In simplicial topology the boundary operator works on the vector space (over the integers or some other algebraic field) generated by using a "< · >" symbol for each simplex and using orientation reversal as negation (Alexandroff 1932). The boundary of a 0-simplex is the zero vector($\partial(\langle a \rangle) = 0$). The boundary of a 1-simplex is the vector difference of the end node 0-simplex and the start-node 0-simplex ($\partial(\langle a,b \rangle) = \langle b \rangle - \langle a \rangle$). The boundary of a triangle is the closed sum of its oriented edges ($\partial(\langle a,b,c \rangle) = \langle a,b \rangle + \langle b,c \rangle + \langle c,a \rangle = \langle b,c \rangle - \langle a,c \rangle + \langle a,b \rangle$). The general formula is:

$$\partial(\langle a_0, a_1, \ldots, a_n \rangle) = \sum_{i=0}^{n} (-1)^n \langle a_0, \ldots, a_{i-1}, a_{i+1}, \ldots, a_n \rangle$$

The boundary operator "∂", as an algebraic tool, lets us define a symbolic counterpart of the topological boundary defined above. This algorithm becomes very important in the spatial operator discussion below, basically because of how it interacts with unions of simplices. Assuming we have a set S of simplices, then the boundary of the union of these simplices can be written as an algebraic sum of the boundary of the component simplices, as follows:

$$\partial(\bigcup S) = \sum_{s \in S} \partial(s)$$

Given a feature model, where each feature's geometry is expressed as set of topology, this formula, and others derived from it form thc basis for topological spatial analysis.

6.2.2 *Cellular Topology.* A more complicated notion in topology, but one which is nearly equivalent to the simplicial complex, is the cellular complex.

A homeomorphism between two geometric objects, is an invertible function from one to the other, such that both the function and its inverse are continuous. The two objects are said to be homeomorphic (from Greek "homoiomorphos", meaning "of similar form").

This essentially means that there is an identification of the points in the first object to the points in the second that preserves the concept of "near", introduces no new tears, holes or seams. The function is often called a "rubber sheeting", a metaphor for for the deformation of one of the objects into the other. Taking an inflated balloon, we could deform it by applying pressures along it surface, but short of its destruction, any of the shapes so derived would be homeomorphic to any other. Under this sort of association, a basketball is homeomorphic to a football, or to a long tube. But once the ends of the tube are joined, the homeomorphism is destroyed (a new seam was introduced). Although this is intuitively obvious, its proof from the definition is very difficult. A great deal of topological theory consists of establishing characterization theories that allow us to regain, formalize, and deepen this type of geometric intuition.

An n-disc is the set of points in Euclidean n-space with distance from the origin less than or equal to 1. An n-cell is any object homeomorphic to an n-disc (this includes an n-simplex). The surface of the (n+1)-disc (all points at a distance exactly 1.0) is called the n-sphere. The boundary of n-cell is that portion of the n-cell mapped onto by the (n-1)-sphere by any homeomorphism. An n-sphere can be thought of as a collection of n-cells with points along the boundary identified, such as the surface of the earth, a homeomorphic 2-sphere, can be divided into a northern and southern hemisphere, homeomorphic 2-disc, joined along the equator, a homeomorphic 1-sphere.

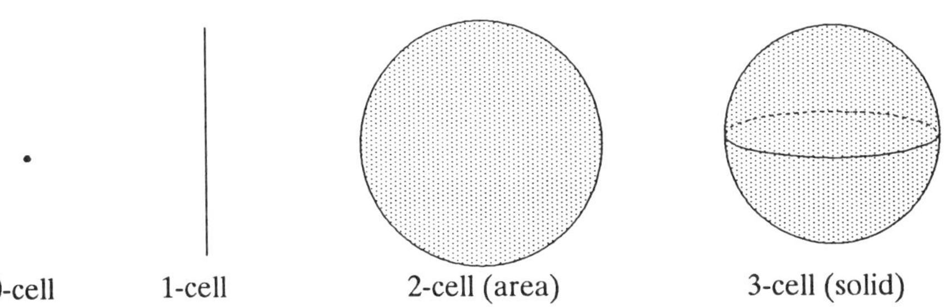

Figure 3: The Cells Used in a 3-D Cellular Complex

An n-dimensional cellular complex (n-complex) is a collection of n-cells, such that portions of their boundary (homeomorphic to (n-1)-cells) have been identified to one another. An n-complex of particular interest is the punctured n-cell, consisting of an n-cell from which a smaller n-cell has been removed. In two dimensions, this is homeomorphic to the annulus (or the cylinder if embedded in 3 space). In three dimensions, this is a ball with a smaller bubble within it. In either case, it is a n-complex representable by an n-cell, with four (n-1)-cells on its boundary, two of which have been identified. In 2 dimensions, it is the square with one pair of opposite sides bent to meet each other. In three dimensions, it is a ball with a stripe around its equator, that has been divided in two and bent to fold back on itself. The complete algebraic machinery from simplicial algebraic topology also exists in cellular algebraic topology, with appropriate interpretation of cells as vectors and orientation.

The big difference between simplices and cells is for visualization and storage. A geometric configuration of the type with which we are concerned is called a topological manifold. A cellular realization of a topological manifold is a cellular complex which is homeomorphic to the manifold. Normally a cellular realization of a topological manifold needs fewer cells than a simplicial realization need simplices. For example, a solid torus needs only one 3-cell, two 2-cells, two 1-cells, and one 0-cell (6 total); for a simplicial realization it would require at least four 3-simplices, eleven 2-simplices, ten 1-simplices and four 0-simplexes (29 total). A solid sphere needs one 3-cell, two 2-cells, one 1-cell, and one 0-cell (5 total), while it would need one 3-simplex, four 2-simplices, six 1-simplices, and four 0-simplices (15 total).

6.2.3 *The Geometric Model as Topology.* The fundamental approach to using these topological structures within a GIS to describe space is to view the underlying coordinate space as a union of disjoint, fundamental topological entities. In a 2-dimensional simplicial implementation, there would be points, open lines, and the interior of triangles. In a 2-dimensional cellular implementation, there would be points, simple curves, and connected areas possible with holes (in some implementations, phantom lines (1-cells) are used to remove holes from the areas and make them 2-cells).

The simplicial approach, as is used at the University of Maine at Orono (Frank and Kuhn, 1986), for surface modeling and similar statistical models has the advantage of simplicity and easy generalization of code to higher and higher dimensions, and is very appropriate for models with a great deal of micro-structure (such as surface modeling and subsurface geology). The cellular approach as used in vector GIS's such as TIGRIS at Intergraph (a feature-based system which aggregates topology to from features, Herring, 1987) and ARC/Info at ESRI (a position-based system) and in exchange structures such as DLG at USGS (generally position-based, but the current plan for DLG-E is feature-based) has the advantage of reduced data set sizes (fewer entities, more coordinates per entity).

Unfortunately, the calculation complexity of three-dimensional cellular topology usually restricts cellular systems to surfaces (carrying z values for the topological entities, but not using them in the topological model), or to limited three-dimensional application (such as finite element modeling in structural engineering). After the model is picked, the topological relations, based upon the boundary relation and its inverse relation (coboundary) are explicitly stored in the data.

6.2.4 *Graph Theory and Networks.* For two-dimensional complexes, called networks or graphs, cellular and simplicial complexes are logically equivalent. Such graphs consist of a set of vertexes (nodes) joined by paths (edges). Each edge generally has been assigned two values associated with the cost of traversing it in each direction(weights). Weighted graphs are susceptible to a variety of analysis, see Bondy, 1976, including:

Shortest Path: the path between two points with the minimal cost (or distance) associated with it.

Optimal Tours: a single path through each of a set of nodes with the minimal total cost (some restrictions may be made on the order of some of the nodes); or variations include the Traveling Salesman problem (visit each node at least once and return to start at minimal cost) and the Chinese Postman problem (visit each edge at least once and return to the start at minimal cost).

Multiple Source Scheduling: a set of paths, which when taken together pass through a set of nodes with the minimal total cost, each path associated with a "home" node from which the path begins and to which the path must return.

Optimal One-Waying: Given a bi-directional graph, find an optimal one way system that will maximize internal connectivity at a minimal cost.

7. The Feature Model

There are two fundamentally different ways of looking at geographic data. The first, and the oldest in terms of computer products, is to think of the data as "attributes of position", a sort of "geometry-first" approach (position-based). The other paradigm is to think of "geometry as another attribute" of the primitive objects, the "feature-based" approach. The geometry-first approach means that the primitive objects in the system are geometric constructions, and that they are associated with attribute descriptors. The feature-based approach means that the primitive objects in the system (the features) can be described in many ways, one of which may be geometric extent or position.

Each of the two approaches have their adherents, usually based upon some processing model. Logically we can define a morphism between the models and make them equivalent, so that for our purposes, we can pick one and allow the morphism of models to translate results between them. From a linguistic point of view, the feature-based model fits better with English ("the house is on the corner", "the ball is on the table", "Madrid is in Spain" all refer to objects as the primary focus). Since we are here to discuss cognition and linguistic aspects of geographic data, I will concentrate on the feature-based approach.

The feature model is constructed on the geometric model by associating each primitive feature with some collection of geometric (or topological) entities. The three simplest features, derived form the Euclidean model, are the point, line and area

If the topological model is used, each of these feature types is built from a homogeneous collection of topological entities (points are composed of only nodes (0-cells), lines are composed of only edges (1-cells), usually connected in chains, and areas are composed of faces, (2-cells, punctured 2-cells or multiply punctured 2-cells).

In most applications, higher levels of feature abstractions are needed. This can be done by allowing features to be composed of basic features or other features. This is useful for representing objects that contain pieces of different dimensions (such as rivers which might be both lines and areas) or associating similar lower-level features into abstractions (as an airport made up of hangers, runways, approach lights, radar beacons, tower, etc.).

8. Spatial Analysis and Spatial Query Languages

Spatial analysis on the above models classically falls into two categories, one generally for position based or raster systems called map algebra, and the other more for feature-based or vector systems called overlay or synthesis.

8.1 MAP ALGEBRA

In a map algebra system, each attribute is represented as a separate grid or raster file (referred to as a layer, generally co-registered so that the pixel centers from each layer overlie one another (Tomlin, 1990). Each layer can be considered as a disjoint union of pixel sets, called regions, zones, categories, or classes, each set containing all pixels having a particular attribute value.

An algebra is defined on the layers via a variety of rules based upon the algebraic combination of co-registered or nearby pixels. For example, an elevation separate could be used to create a slope separate by the comparison of each cell with its neighbors so that the local slope of the land can be estimated.

In addition to the algebra between map layers, map algebra systems supply functions for the generation of neighborhoods of various types. Neighborhoods can be defined via any set of metric relations, and represent a vector formalism embedded in the raster map algebra domain. For example, a neighborhood of all points within 100 meters of a point (the interior of a circle) is a vector geometry concept, but it can be approximated as a set of pixels in the map algebra. We define a morphism "center" that takes pixels to their center, and use the metric structure of the underlying geometry space to define the buffer of a point at a radius as:

$$\text{pixel-buffer}(\text{focus}, 100\ \text{m}) = \{ \text{pixel} \mid d\,(\text{center}(\text{pixel}), \text{center}(\text{focus})) \leq 100\text{m} \}$$

We in essence, defined a morphism "center" from the map algebra model (MA) to the geometric model (G) mapping pixels to the center of their ground extent.

$$\text{center} : MA \rightarrow G$$

Taking the definition of buffer(center, radius) from the geometric model, we pulled back a similar definition by using the set inverse relation $\text{center}^{-1}()$ defined by:

$$(\forall\ X \text{ set of points}) \quad \text{center}^{-1}(X) = \{ p \mid p \in \text{pixels}, \text{center}(p) \in X \}$$

$$\text{pixel-buffer}(\text{focus}, \text{radius}) = \text{center}^{-1}(\text{buffer}(\text{center}(\text{focus}), \text{radius}))$$

In the pixel (map algebra) domain, some of the pixels in the set will have parts lying more than 100 meter from the point, and some out of the set will have parts lying closer than 100 meters. The problem is that the mapping "center" from the pixel model to the geometric model is not an isomorphism (cannot be perfectly inverted). The best that can be done is to pull definitions back from the geometric domain and use these as pixel-domain approximations.

8.2 FEATURE SYNTHESIS

Feature synthesis is the process of taking features which share some underlying geometric extent, and producing (synthezing) new features whose type and attributes are functions of the original features. The usual route to feature synthesis examines overlapping areas of different types and functionally combines the attributes of these areas to derive a set of attributes for the area of common overlap.

Such a process could be considered the vector equivalent of the Map Algebra. For example, suppose that suitability of a particular site for construction is dependent upon the type of soil, the average and extreme soil moisture content, and the slope of the land. Given map layers describing the distribution for each of these attributes, another layer can be produced for "site suitability", where a combining map function has been used that assigns dry, hard, flat soil to a higher suitability factor than wet, soft, sloping soil.

Other approaches to feature synthesis are possible for non-area features, such as the intersection of lines. For example, one might take all points where roads or railroads cross rivers, and in the absence of other knowledge, assume that a bridge exists at those points. In addition, operators other than intersection might be used. The most common of these is the metric "within a given distance". For example, in deciding upon the site suitability of a plant that handles pollutants, the distance and slope to the nearest wetlands may have effect in the emergency planning problems associated with accidental spills. A site to close or in the immediate drainage basin of a major wetland preserve may be considered as totally unsuitable for some types of industrial installations. In general, any operator that compares two or more features and can return a geometric extent can be used for feature synthesis.

8.2.1 *Vector Overlay.* The vector overlay process is the standard approach to these problems in the vector model. A equivalent to the map layer is defined (called a coverage) which specifies disjoint classification areas via vector polygons. An overlay process begins with a set of coverages, and derives, through repeatedly intersecting the polygons from each coverage, a combined attribute polygon. Attributes for the derived polygons are assigned according to the attributes of the ancestor polygons in each of the input coverages.

Unlike the map algebra, vector feature synthesis can be applied to non-area features and spatial operations other than intersection. For example, road linear features can be intersected with slope polygons to synthesize transportation cost features based upon both distance and slope. Furthermore, other spatial operators can be used to determine the synthesis extent (within distance, adjacency, etc).

8.2.2. *Oriented Synthesis.* Some features have orientation as a part of their geometric descriptions, such as one-way streets, streams (direction of flow) and area boundaries. The orientation of the boundary is inherited from the area by a requirement that the area always lie to a particular (usually left) side of the boundary. Other features have attributes that can depend upon which orientation used. Attributes which do not depend upon orientation are symmetric (equal for both directions). Orientation-specific attributes can be either antisymmetric (such as slope) or asymmetric (traversal cost). In an antisymmetric attribute, the value for one orientation is the negative of the same attribute value for the other orientation. In the antisymmetric case only one of the values need be assigned, since the other value is easily calculated. The relation of boundary to area is antisymmetric, since reversal of the boundary direction would change the side of the boundary upon which the area lies. In the asymmetric case, the two values for the attribute need not have any relation to one another. In the asymmetric case, both values usually must be stored. Oriented synthesis involves orientation-sensitive attributes or orientation-sensitive feature construction (such as the construction of area features from oriented boundaries).

8.2.3. *Feature Merge.* After a synthesis, adjacent features, although derived from different input features, may have a sufficiently similar nature to declare them "equal" according to the new criteria. Merging features brings up two problems: 1) when are the features sufficiently similar to justify the merge, and 2) what are the rules for the combination of attribution and geometry for merging features.

The similarity of features falls into two types of criteria, spatial and descriptive.

The spatial criteria is usually determined by the systems fundamental criteria for the feature description. For raster-based or position-based systems, merging is usually out of the question, since here the division of geometry is defined by the system and not by the features. For vector, feature-based system, some degree of connectivity is required for a feature (area and line feature may be required to be connected, orientation of lines may be required to be compatible at common points of intersections). For some oriented features, it may be possible to adjust the orientation of one of the features to match that of the other, but only if any orientation-sensitive attribution is appropriately modified.

Once the geometric criteria is met, the attributes must be checked for compatibility. For most attributes, this is a check for equality (or at least equality within a range). For orientation sensitive attributes, any possible change in orientation necessary for meeting the geometric criteria must be meet. For some attributes that are measurement-related, there may be no criteria at all to be met (length, population, etc.).

Once the merge criteria has been met, attributes for the new feature must be derived or calculated. Non-metric descriptive attributes are usually simply inherited. Attributes implying a global measurement (population, area, length, etc.) are added. Attributes implying an average or normalized measurement (population density, average rainfall, average slope, etc.) are averaged, usually weighted by an area or length measure. Synthesis needs to take all of this into account, and allow for application-specific rules for each merged attribute.

8.3 SPATIAL QUERY LANGUAGES

Now we come to the difficult question: how to integrate all of the above processes into a single system controlled by a single interface. In addition we must decide how to make that interface match a natural language as closely as possible. In an RDB this is done through a query language that supplies a complete set of access tools to all data stored in the relational model (completeness). To extend this concept to a spatial domain (either feature or position oriented) requires us to supply a "complete" set of spatial operators and relations that could be used to augment a relationally complete query language to make it "spatially" complete.

8.3.1. *Spatial Operators and Relations.* For each of the spatial models above, there are a set of implied spatial relations: those relations between entities or sets of entities that can be distinguished by the model's operators. For each model embedding, (such as topology model inside the metric, or the metric inside the coordinate system model) there is an embedding of the smaller model's operations into those of the larger. Thus, the topological relations are embedded in the metric relations, and the metric relations are, in turn, embedded in the coordinate system relations.

The exception to this general rule is the set of operators defined using algebraic topology, which inherits orientation from the coordinate model through the homeomorphisms of simplices (or cells) to standard coordinate structures.

8.3.2. *Topological Operators and Relations.* There have been several attempts to define a complete set of spatial operators. Most of these have been to define a "natural" set of spatial or topological primitives.

A formal topological approach to this problem (Herring, 1989b; Pullar, 1988; Egenhofer, 1990) allows us to obtain a set of spatial operators from a classification of the possible topological relations. This is consistent with the algebraic topology theory which shows that all topological characteristics of a geometric object can be ascertained from its decomposition into cells (or simplices) and the action of the boundary operator upon this decomposition.

8.3.3. *Topological Classification.* In a feature-based topological system, the spatial description of each feature is given as a set of topological cells. Using the formulae defined above for "∂", we can define the boundary, interior and closure of each feature. We will assume that 1-cells and higher dimension cells are open sets; they do not contain their boundary. This makes the cellular decomposition of the boundary and interior a union of disjoint primitive sets. Taking faces, edges, and nodes as names for whatever the appropriate two-, one-, or zero-dimensional object is for a complex, we can write the boundary of area, lines and points as (recall that the sum includes orientation as a sign, so that cancellation is possible):

$$\partial(\text{area}) = \sum_{\text{face} \in \text{area}} \partial(\text{face})$$
$$\partial(\text{line}) = \sum_{\text{edge} \in \text{line}} \partial(\text{edge})$$
$$\partial(\text{point}) = \emptyset$$

The formula for closure is similar, but we introduce the symbol "$\|\cdot\|$" to represent the support of an oriented set, meaning the unoriented underlying geometric-topological objects. Note that repeated applications of $\|\partial()\|$ takes an oriented boundary and then throws the orientation away. Our formula is then:

$$\text{cls}(\text{area}) = \bigcup_{\text{face} \in \text{area}} \text{face} \cup \|\partial(\text{face})\| \cup \|\partial(\|\partial(\text{face})\|)\|$$
$$\text{cls}(\text{line}) = \bigcup_{\text{edge} \in \text{line}} \text{edge} \cup \|\partial(\text{edge})\|$$
$$\text{cls}(\text{point}) = \|\text{point}\|$$

By examining the character of the underlying sets of topological entities and their boundary relations to one another, we can completely determine the spatial nature of the original object (number of components, number of holes, number of self intersections, etc.). Thus, the topological spatial characteristics of any feature can be completely determined by the characteristics of the set of underlying topological objects (this is the fundamental principle in algebraic topology).

We can extend this concept to the interaction of two topologically described objects (assuming that they are described from the same simplicial or cellular complex. Topologically speaking, there is no concept of distance, and so there is only one relationship that can exist between objects whose closures are disjoint (called "disjoint" in Pullar and Egenhofer, 1988). Other relations require that the two topological objects must share common items in the topological sets that define their closure. Logically there are 3 sets for each object entering into the relation: its boundary, its interior, and the union of these two, its closure. Any topologically definable relation can then be defined by examining how these sets interact with one another. This gives us the following 27 sets to examine for two features A and B:

$i(A) \cap i(B)$,	$i(A) - i(B)$,	$i(B) - i(A)$
$\partial(A) \cap i(B)$,	$\partial(A) - i(B)$,	$i(B) - \partial(A)$
$cls(A) \cap i(B)$,	$cls(A) - i(B)$,	$cls(B) - \partial(A)$
$i(A) \cap \partial(B)$,	$i(A) - \partial(B)$,	$\partial(B) - i(A)$
$\partial(A) \cap \partial(B)$,	$\partial(A) - \partial(B)$,	$\partial(B) - \partial(A)$
$cls(A) \cap \partial(B)$,	$cls(A) - \partial(B)$,	$\partial(B) - cls(A)$
$i(A) \cap cls(B)$,	$i(A) - cls(B)$,	$cls(B) - i(A)$
$\partial(A) \cap cls(B)$,	$\partial(A) - cls(B)$,	$cls(B) - \partial(A)$
$cls(A) \cap cls(B)$,	$cls(A) - cls(B)$,	$cls(B) - cls(A)$

Given the relationships between the three operators($\partial()$, i(), and cls()) and between set intersection and difference, it is highly unlikely that all 27 sets will be used in any statement of a relational condition.

8.3.4. *Set Classification Schemes.* Thus, we can reduce the problem of defining topological relationships to the problem of classifying the above sets according to some criteria. Given a set of criteria (containing, say, "c" possibilities) and choosing a collection of the above sets (say, "s" of them), we would have a "complete" set of as many as s^c possible disjoint primitive classifications for the interaction of two input topological sets. The classifications are primitive in that none can be written in terms of any other simpler operators. They are disjoin in that no two classifications can be simultaneously TRUE. They are complete in the sense that given this classification scheme, each geometric configuration possible falls into one of the classifications. For each primitive classification, we can define a relation operator on two inputs that is TRUE if the classification holds and FALSE otherwise.

We begin by specifying a classification scheme for single sets (possibly useful to support unary operators also). The simplest of the classification schemes is to determine which of the above sets is non-empty ("$\neg\emptyset$" or "$\neq \emptyset$") and which are empty ("$\emptyset$" or "$= \emptyset$").

We now pick a collection of the sets above to classify using the chosen schema. Using the examples in Pullar and egenhofer 1988, we choose the pair-wise intersections of the boundaries and interiors of the two topological sets. In this manner, we can define 16 primitive, disjoint operators (4 intersections, 2 possibilities for each $2^4 = 16$). We are not guaranteed that all 16 operators are non-trivial. Pullar and Egenhofer 1988 names the seven such operators that are valid for areas (eight if we separate the two cases for overlap):

Operator	$\partial(A) \cap \partial(B)$	$i(A) \cap i(B)$	$\partial(A) \cap i(B)$	$i(A) \cap \partial(B)$
disjoint	$\emptyset$	$\emptyset$	$\emptyset$	$\emptyset$
meet	$\neg\emptyset$	$\emptyset$	$\emptyset$	$\emptyset$
equal	$\neg\emptyset$	$\neg\emptyset$	$\emptyset$	$\emptyset$
inside	$\emptyset$	$\neg\emptyset$	$\neg\emptyset$	$\emptyset$
contains	$\emptyset$	$\neg\emptyset$	$\emptyset$	$\neg\emptyset$
covers	$\neg\emptyset$	$\neg\emptyset$	$\emptyset$	$\neg\emptyset$
covered-by	$\neg\emptyset$	$\neg\emptyset$	$\neg\emptyset$	$\emptyset$
overlap	$\emptyset$ or $\neg\emptyset$	$\neg\emptyset$	$\neg\emptyset$	$\neg\emptyset$

Figure 4: The Interpretation of Some Common Spatial Operators

The names applied to the operators are a bit misleading. For example, the proof for the equivalence of the "equal" relation and the fundamental concept of being spatially identical depends upon the dimension of A and B both being maximal (in this case, areas in a 2-complex). For lines, the operator defined as equal can simply mean that two lines begin at the same point and in the same direction, but diverge before they end. To distinguish this case from the "equality" relation for areas would require a more complete classification scheme. Going to a other dimension objects, all of the possible 16 operators in this scheme are in fact non-trivial (see Herring 1990c). For example, the following diagram show two lines whose interiors intersect but have all other pairs of sets disjoint:

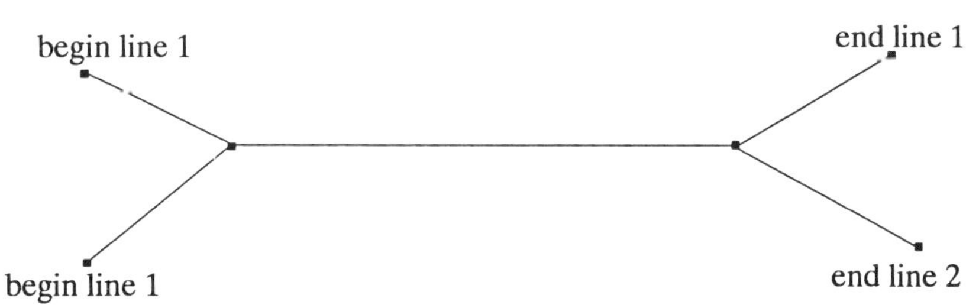

Figure 5:
$\partial(\text{line 1}) \cap \partial(\text{line 2}) = \emptyset$ and $\partial(\text{line 1}) \cap i(\text{line 2}) = \emptyset$ and $i(\text{line 1}) \cap \partial(\text{line 2}) = \emptyset$ and $i(\text{line 1}) \cap i(\text{line 2}) \neq \emptyset$ and $\dim(i(\text{line 1}) \cap i(\text{line 2})) = 1$

Extending to other classification schemes and to other types of objects leads to almost an unlimited number of spatial operators definable in this manner. For example, adding the dimension of the intersection to the above operators extends the logical number to potential 256 relations, of which at least 48 are non-null for simple point, line, and area features (see Herring 1990c). The following diagram shows two intersecting lines whose relation is different from the above example only in the dimension of the intersection:

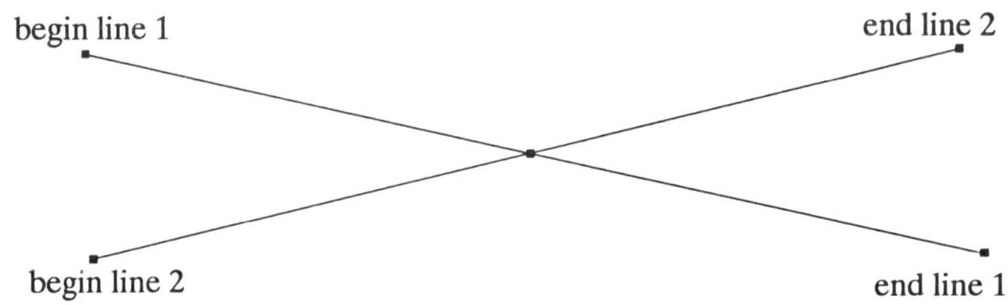

Figure 6:
$\partial(\text{line 1}) \cap \partial(\text{line 2}) = \emptyset$ and $\partial(\text{line 1}) \cap i(\text{line 2}) = \emptyset$ and $i(\text{line 1}) \cap \partial(\text{line 2}) = \emptyset$ and $i(\text{line 1}) \cap i(\text{line 2}) \neq \emptyset$ and $\dim(i(\text{line 1}) \cap i(\text{line 2})) = 0$

Topological invariants include: the number of components of the set, the largest dimension of objects contained within the set, and Euler characteristic, see Spanier 1966, Artin 1969. The three most appropriate set comparison schemes are based upon set intersections, symmetric difference, and set difference. Each one of these can be expanded to use any or all of the the single set classification schemes above on the intersection of the two sets. The most important of these extensions is dimension. Metric invariants include: area, perimeter, length, and diameter (maximum cross-sectional distance). The following two diagrams illustrate how two relations can vary only in a topological invariant (here, the number of connected components, #components(), of the intersection of the closure of the two sets):

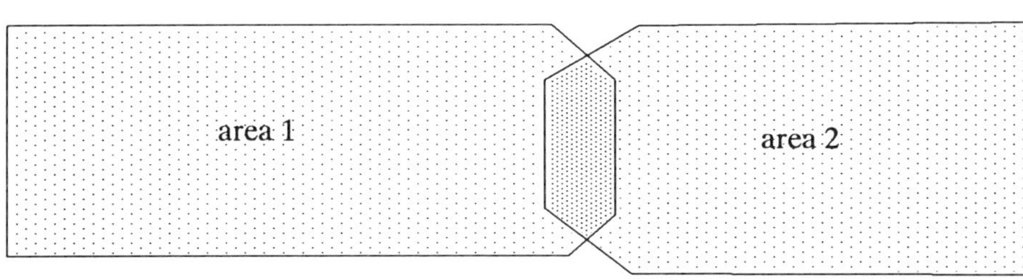

Figure 7: $\text{cls}(\text{area 1}) \cap \text{cls}(\text{area 2}) \neq \emptyset$ and $\#\text{components}(\text{cls}(\text{area 1}) \cap \text{cls}(\text{area 2})) = 1$

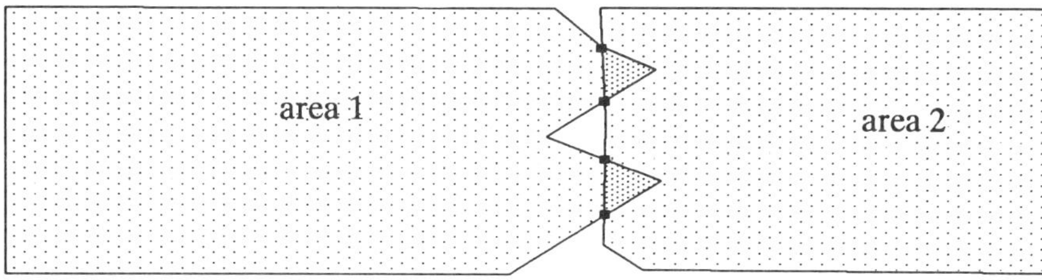

Figure 8: cls(area 1) $\cap$ cls(area 2) $\neq \emptyset$ and #components(cls(area 1) $\cap$ cls(area 2)) = 2

What the geometry-to-algebraic-topology morphism has allowed us to do is to give a non-ambiguous definitions of a set of topology-based spatial binary relations. In the examples below, we will use "dim" as a dimension operator, taking the largest dimension of the objects in a set. Using the relational algebra, we can extend the definition to trinary or higher degrees. For example, we can define "line terminates within area" as:

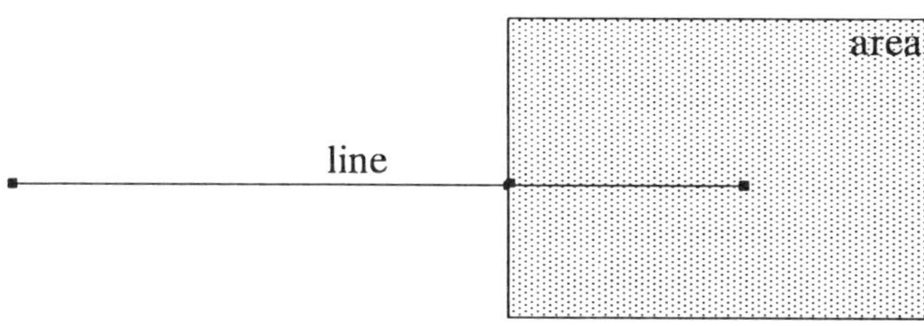

Figure 9:
line terminates within area $=$ (dim (cls(line)$\cap$cls(area)) $=$ 1) $\wedge$ (∂(line)$\cap$cls(area)$\neq\emptyset$)

Using this binary relation, be can define a trinary relation on a line and two areas by logically combining component relations. Thus we can say:

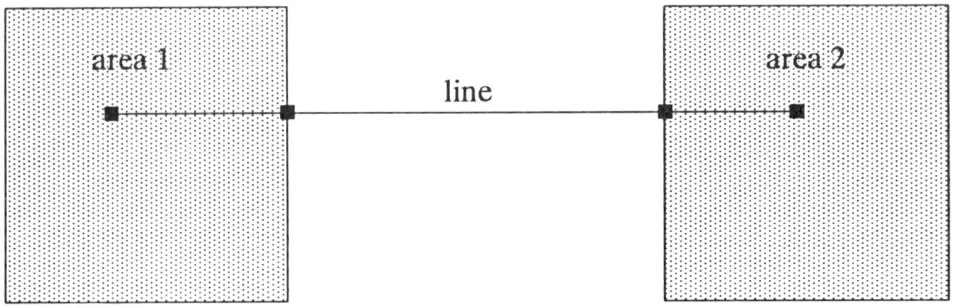

Figure 10: line joins area1 and area2 ≡
line terminates within area1 ∧ line terminates within area2 ∧ area1 and area2 are disjoint

Orientation: The operators above use only the topological nature of the sets involved. By adding orientation into the calculations for set characteristics we can add order and direction to the conditions.

An oriented line inherits the "<" ordering from the ordering of the coordinate stream in its representation (for a vector representation) or from the functional parameterization (for a functional representation). This ordering is reflected in the orientation associated with the underlying topology in the feature description. This in essence defines a morphism from the real numbers under "<" to the line, using a parameterization as the entity mapping. For the boundary of the line, this splits the set $\|\partial()\|$ into a start-point (before all other points) and an end-point (after all others). Denoting this order by "$<_{line}$" for a line, we can define operators that order by specifying the sequence of the underlying support sets (we say that for sets X and Y, $X < Y$ if and only if $(x \in X, y \in Y) \Rightarrow (x < y)$:

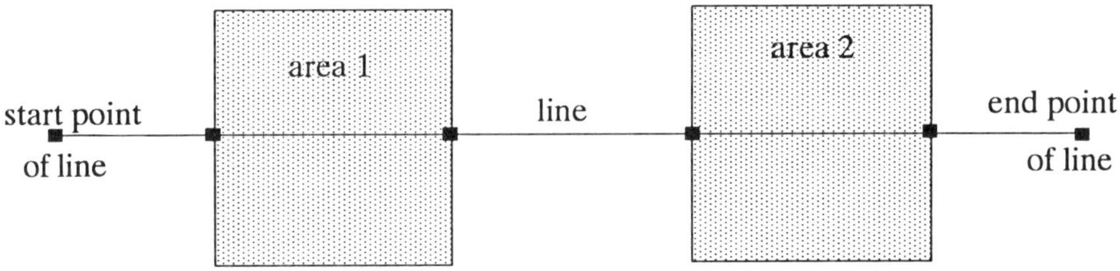

Fiogure 11: area 1 before area 2 along line ≡
$(\text{cls(area 1)} \cap \text{cls(line)}) <_{line} (\text{cls(area 2)} \cap \text{cls(line)})$

In a similar manner to start-point and end-point, a left and right can be assigned to the areas on either side of a line according to the direction of the order. Going back to the $\partial()$ operator defined for complexes, we can define "left of" a line by the following:

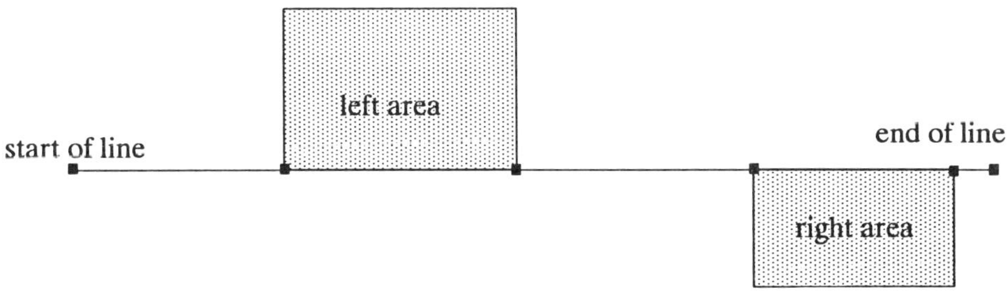

Figure 12: face is left of edge $\Leftrightarrow$ edge $\in \partial$(face)
face is left of line $\Leftrightarrow$ $\exists$ edge $\in$ line such that (face is left of edge)
area is left of line $\Leftrightarrow$ $\exists$ face $\in$ area such that (face is left of line)

For points, the orientation inherited from the lines, using left for exiting lines as positive, gives us a circular ordering (counter-clockwise) of the curves entering and leaving the point.

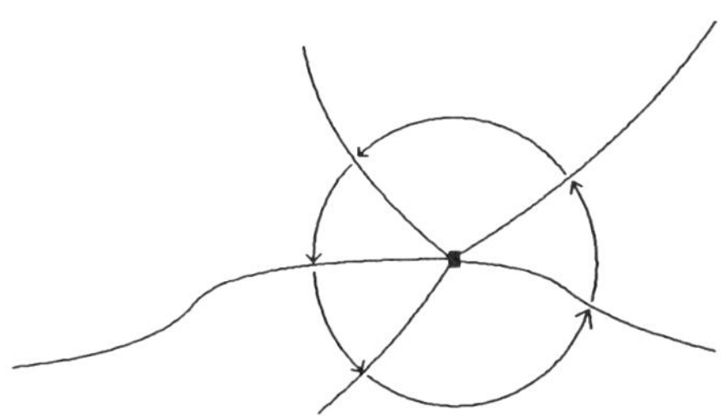

Figure 13: a node point and the circular ordering of its entries and exits

Using this ordering would allow us to distinguish intersections where lines cross from those where they simply touch and diverge. Using the ordering for a line, we can define a node "entry" as the segment of the line directly before the point, and the "exit" as the segment of the line directly after the point. Using the nodes circular order we can define "between" for the entry and exit lines.

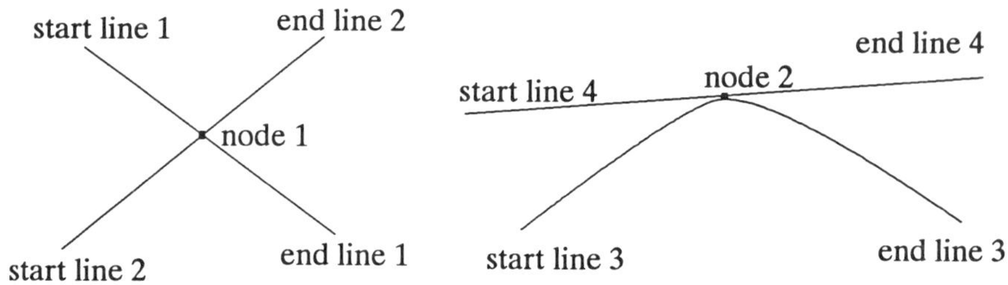

Figure 14:
#segment of line 1 between(entry (line 2), exit (line 2)) = 1
#segment of line 3 between(entry (line 4), exit (line 4)) = 0

8.3.5. *Metric Operators/Relations.* The topological operations above give us a complete set of relations for topological spaces. Since any metric space is also a topological space, the only metric operators that need be added are those which can be shown to be dependent on the metric, but not on the topology. The only such operator is "within distance", applicable to sets and to points (definitions are above). Generally, this operator can be viewed as defining a distance-radius buffer area for each object (called the focus of the buffer), centered upon the object, consisting of all points within a given distance of the feature.

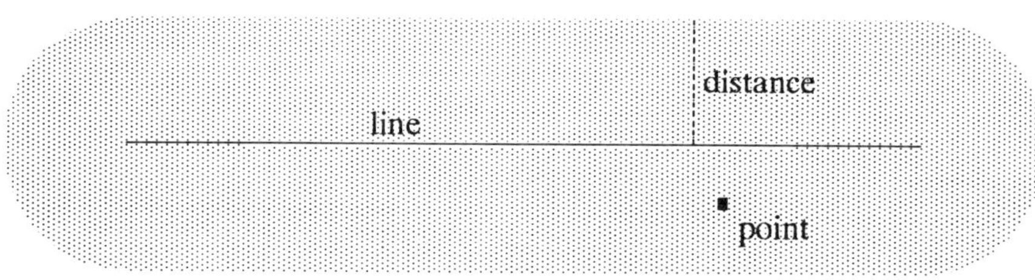

Figure 15:

point within distance of line $\equiv$ point inside of buffer (line, distance)
$\equiv$ i(point) $\subset$ i(buffer (line, distance))

Once the zone at a distance is generated, the resultant area can be used in any other spatial operator, transitively associating the focus of the buffer zone with features interacting with the zone. This in essence reduces most of the metric operators to a set of buffer zone topological operators.

In addition, the metric allows us to define concepts such as shortest path (to use in the generation of geodesics). The property that defines a shortest path (geodesic) between two points is stated in terms of local minimality (recall d(,) is the metric distance):

$\forall$x a point (x $\in$ geodesic(a, b)) $\Leftrightarrow$ (d(a,b) = d(a,x) + d(x,b))

This also gives us a definition of "lies between" for two points:

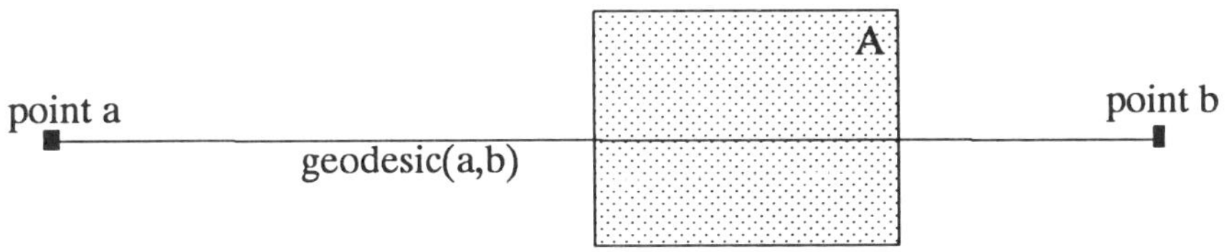

Figure 16: A lies between a and b $\Longleftrightarrow$ cls(A) $\cap$ geodesic(a,b) $\neq \emptyset$

In addition to using the metric to define operators on existing configurations, it can be use to define criteria for the synthesis of new configurations. For example, we can say a set of objects is a δ-cluster if any object is within a distance δ of some other object in the cluster, and any two objects can be joined by a chain of geodesics between points in the cluster that are no longer than δ long. We can further define a δ-polygon to be one whose sides ar no longer than δ. Given a δ-cluster, we can define the δ-hull to be the maximal area δ-polygon through points in the cluster. Using these concepts we can define fuzzy concepts like "among" (actually a "δ-among" for each small real number):

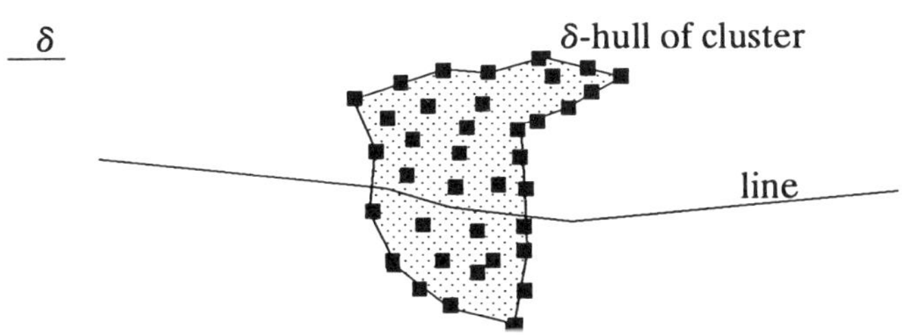

Figure 17: line passes among(cluster, δ) $\equiv$ cls(line) $\cap$ cls(δ-hull(cluster)) $\neq \emptyset$

8.3.6. *Coordinate System Operators/Relations.* Building up through the simpler models to the complete coordinate system, we essentially get back to full geometric configurations, including all of Euclidean geometry. The topology and the metric operators are inherited through morphisms defined by the usual Euclidean distance. In addition to the distance operation, coordinate systems further allow us to determine direction, the size of angles, and to perform differentiation and integration, giving us the ability to measure curve length and feature area.

Using the concept of direction we can define neighborhoods based on direction, or functions of direction and other measurements (see Tomlin 1990). All of these neighborhoods can be used in the same manner we used buffer-zones and δ-hulls above.

The theory for operators and relations that depend absolutely on the coordinate system, and not on the lower-level, simpler models, is the realm of computational geometry. and will not be explored here.

9. Summary

Each model in a GIS comes equipped with a set of concepts that can be used to describe attribute and spatial relations that might exist between spatial objects. By analyzing the logically smaller simpler geometric models, we can categorize all relations representable in such a model. By using the natural morphism embedding of each simple space in the one above it (topology → metric → coordinate space), we can inherit both operator definitions and algorithms. In this manner we can build a complete syntax for the defining of spatial operators.

With the tensor sum of the Object Model (OOM) and the Relational Model (RM), we have a hybrid model capable of encompassing all of the above geometric models as well as a complete attribute system. Using this hybrid, we can design logical query languages that can completely express the complexity of both attribute and spatial relations. Using the hybrid model of the query language, we can build methods to support the spatial relation definition syntax mentioned above.

By using the semantics inherent in the Entity-Attribute-Relation Model (ER) which forms the semantic basis of our OOM, RM tensor sum, and the dictionary of spatial relations, we can construct natural language interfaces to our hybrid spatial model.

Whether or not the implementation of a GIS is compliant with the hybrid model, or capable of extending its query language syntax, we can, by using the morphism theory above, determine the appropriate definition of any of our operators within the GIS's model. Given this ability, we should be able, as an academic community and as an industry, to establish a logically complete set of spatial relations that can be unambiguously defined, and therefore insure consistency of results across systems.

10. References

Alexandroff, P. (1932) Einfachste Grundbegriffe der Topologie, Julius Springer, Berlin; translated by Alan E. Farley (1961) Elementary Concepts of Topology, Dover Publications, New York.

Andrews, T., and Harris, C, (1987) 'Combining Language and Database Advances in an Object-Oriented Development Environment', OOPSLA'87 Proceedings, ACM, October 4-7, 1987, 430-440.

Artin, E., and Braun, H. (1969) Introduction to Algebraic Topology, Charles E. Merrill Publishing Company, Columbus, Ohio.

Barr, M., and Wells, C. (1990) Category Theory for Computing Science, Prentice Hall, New York.

Bartels, R. H., Beatty, J. C., Barsky B. A. (1987) An Introduction to Splines for use in Computer Graphics and Geometric Modeling, Morgan Kaufmann Publishers, Inc., Los Altos, California.

Bédard, Y,, and Paqquette F. (1989) 'Extending Entity/Relationship Formalism for Spatial Information Systems', Proceedings of AutoCarto9, April 1989, Baltimore, Maryland.

Bondy, J. A. and Murty, U. S. R. (1976) Graph Theory with Applications, North Holland, New York

Cartan, H., and Eilenberg S. (1956) Homological Algebra, Princeton University Press, Princeton, New Jersey.

Casti, J. L. (1989) Alternate Realities: Mathematical Models of Nature and Man, John Wiley & Sons, New York.

Codd, E. F. (1988) 'Fatal Flaws in SQL', Datamation, August 15, 1988, Vol. 34, No. 16, 45-48.

Codd, E. F. (1979) 'Extending the Database Relational Model to Capture More Meaning', ACM TODS 4, No. 4, December 1979.

Codd, E. F. (1990) The Relational Model for Database Management: Version2 , Addison-Wesley Publishing Company, Reading, Massachusetts.

Cox, B. J. (1986) Object Oriented Programming: An Evolutionary Approach, Addison-Wesley Publishing Company, Reading, Massachusetts.

Date, C. J. (1983) An Introduction to Database Systems, Vol. II (printed with corrections 1985) Addison-Wesley Publishing Company, Reading Massachusetts.

Date, C. J. (1984) Relational Database: Selected Writings, Addison-Wesley, Reading, Massachusetts.

Date, C. J. (1986) An Introduction to Database Systems, Vol. I, Fourth Edition, Addison-Wesley Publishing Company, Reading Massachusetts.

Egenhofer, M., and Herring, J. R. (1990) 'A Mathematical Framework for the Definition of Topological Relations', Proceedings of the 4th International Symposium on Spatial Data Handling, Zürich, Switzerland, Vol 2, 803-813.

Ellis, M. A., and Stroustrup, B. (1990) The Annotated C++ Reference Manual, Addison-Wesley Publishing Company, Reading, Massachusetts.

Farin, G. (1990) Curves and Surfaces for Computer Aided Geometric Design, Academic Press, Boston, Massachusetts.

Faux, I. D., and Pratt M. J. (1979) Computational Geometry for Design and Manufacture, Ellis Horwood Limited, Chichester, UK.

Frank, Andrew U and Kuhn, Werner (1986) 'Cell Graphs: A Provable correct method for the Storage of Geometry', Proceedings of the Second International Symposium on Spatial Data Handling, Seattle, Washington, 411-436.

Herring, J. R. (1987) 'TIGRIS: Topologically Integrated Geographic Information System', Proceedings of Auto-Carto 8, March 1987, Baltimore, Maryland, 282-291.

Herring, J. R., Larsen, R. C., and Shivakumar, J. (1988) 'Extensions of the SQL Query Language to Support Spatial Analysis in a Topological Data Base', GIS/LIS'88 Proceedings, ACSM, ASP/RS, AAG, URISA, November 30 - December 2, 1988, 741-750.

Herring, J. R. (1989) 'A Fully Integrated Geographic Information System', Proceedings of Auto-Carto 9, April 1989, Baltimore, Maryland, 828-837.

Herring, J. R. (1990a) 'TIGRIS: A Data Model for an Object Oriented Geographic Information System', to appear in Computers and Geoscience, issue on GIS Design Models and Functionality, presented 22 March 1990 at the Midlands Regional Research Laboratory, University of Leicester, UK.

Herring, J. R., (1990b) 'Using Category Theory to Model GIS Applications', Proceedings of the 4th International Symposium on Spatial Data Handling, Zürich, Switzerland, Vol 2, 820-829.

Herring, J. R. (1990) 'A Unified Approach to Binary Spatial Operators', to appear (available from author upon request).

Lefschetz, S. (1975) Applications of Algebraic Topology, Springer-Verlag, New York.

Munkres, J. R. (1963) Elementary Differential Topology, Princeton University Press, Princeton, New Jersey.

Prenter, P. M. (1975) Splines and Variational Methods, John Wiley & Sons, New York, reprinted 1989, Wiley Classics Library.

Pullar, D. and Egenhofer, M. (1988) 'Towards Formal Definitions of Topological Relations Among Spatial Objects', Proceedings of the Third International Symposium on Spatial Data Handling, Sydney, Australia, August 1988, 225-241.

Spanier, E. H. (1966) Algebraic Topology, McGraw-Hill Book Company, New York.

Stonebreaker, M., ed. (1986) The INGRES Papers: Anatomy of a Relational Database System, Addison-Wesley, Reading, Massachusetts.

Switzer, R. M. (1975) Algebraic Topology - Homotopy and Homology, Springer-Verlag, Berlin.

Tomlin, C. D. (1990) Geographic Information Systems and Cartographic Modeling, Prentice Hall, Englewood Cliff's, New Jersey.

Ullman, J. D. (1988) Principles of Database and Knowledge-Base Systems, Vol I, Computer Science Press, Rockville, Maryland.

Ullman, J. D. (1989) Principles of Database and Knowledge-Base Systems, Vol II, Computer Science Press, Rockville, Maryland.

White, E., and Malloy, R., ed. (1986) 'Object Oriented Programming', BYTE, August 1986, 137-233.

MAP ALGEBRA AS A SPATIAL LANGUAGE

KELLY K.L. CHAN
Graduate School of Design
Harvard University
Cambridge, Massachusetts 02138 USA

C. DANA TOMLIN
School of Natural Resources
The Ohio State University
Columbus, Ohio 43210 USA

ABSTRACT. Map Algebra is a language dealing with the analysis and synthesis of spatial information. This language has been developed to realize a land planning methodology commonly known as 'cartographic modelling'. At the heart of cartographic modelling is the idea of decomposing data sets, data-processing capabilities, and data-processing control specifications into elementary components that can then be recombined with ease and flexibility. The language differs significantly from other geometric languages which focus mostly on the organization of spatial data; but Map Algebra makes no pretense that it is not methodologically bounded.

1. Introduction

The theme of the Advanced Study Institute upon which this book is based was the examination of various cognitive and linguistic aspects of geographic space and their implications on our design of geographic information systems. During the discussions at the ASI in Las Navas del Marqués, nonetheless, we found a tendency to disassociate ourselves from the 'geographic space' we try to represent in our systems. The general assumption is that out there are objects to be measured, obstacles to be navigated, and geometric relationships to be recognized. We do not construct objects, cannot rearrange obstacles. In Lakoff's imagery theory, even though his pictures are there only to articulate the theory, the being is nonetheless separated from the image —the metaphorical space happens, somehow, in advance of the reasoning process.

Spatial language, as discussed by the ASI participants, deals principally with the organization of spatial data; the construction of space, the configuration of objects, and the projection of self into a geography are left unsurveyed. With the exception of Maria Cátedra's (1991) paper, none of the presentations seriously considered the manipulation of space, much less the imaginative mode of dealing with geography. Many have proposed languages for the organization of —as opposed to the analysis and synthesis— geographic data include Dacey (1970), Youngman (1978), Shapiro and Haralick (1980), Cox, Aldred, and Rhind (1980), Nyerges (1980), van Roessel (1987) etc; all seek to establish a uniform language for spatial representation. If the premises of geographic information systems are structured only to report a geography which is out there to us, what is the

D. M. Mark and A. U. Frank (eds.), Cognitive and Linguistic Aspects of Geographic Space, 351–360.

relevance of cognitive and linguistic episodes that occur mostly *in situ*. Should a spatial language be limited to geometric language of the type proposed by John Herring (1991)? A geographic entity in a particular geographic information system application may not be identical to some arbitrary physical element and it may not be directly visible. We consider in this paper a 'map algebra', outlined first in Tomlin (1983) as a spatial language for cartographic modelling.

2. Map Algebra

A *map algebra* can generally be defined as the notational expression of a methodology for computation involving cartographic data. Tomlin has used the term in reference to the formal expression of his *cartographic modelling* methodology. This is an integrated set of data processing conventions and capabilities that elegantly addresses a broad range of applications involving the interpretation of maps.

The map algebra associated with cartographic modelling exists as a formal system of symbols, rules governing the formation and transformation of those symbols, and conventions regarding the actual phenomena those symbols represent. As such, it can be regarded as a language. To the extent that the phenomena represented by this language is defined with respect to geographic space, it can further be characterized as a spatial language.

Here, we introduce this language in terms of its syntax and semantics. In doing so, we also contrast map algebra to other formal systems of geographic representation. Our discussion proceeds:

- from the variables of this algebra that are used to represent geographic conditions;
- to the algebraic functions by which those variables can be transformed;
- to the notational expressions by which specific functions are specified.

2.1 MAP ALGEBRAIC VARIABLES

Map Algebra is a one sorted algebra. The complete body of data of a given geographic study area exists as a cartographic model. Each cartographic model is comprised of map layers. Each map layer is a two-dimensional image on which every location is associated with exactly one characteristic. Location is the fundamental spatial entity in cartographic modelling. Unlike the units of data in most relational and object-based systems, locations are not units of "what" but of "where". Although locations can be aggregated into sets of lines, areas and surface features, they remain the elemental units for which attributes are recorded.

Interpretations in the cartographic modelling idiom is facilitated by the fact that values are directly associated with individual locations. Instead of a language which says a region is defined by its enclosure, Map Algebra recognizes it because individual elements say each is belonging to it. This makes it possible to express each map operation in terms of its effect on a single, typical location: the "worm's eye" view. It is this worm's eye view and the function-oriented (as opposed to object-oriented or relation-oriented) structure that distinguishes the cartographic modelling approach from other methodologies.

The cartographic modelling approach associates locations with one another not with declarative statements specifying selected relations but with new entities that are generated by applying selected functions. To interrelate entities that are comprised of multiple locations, each is first disaggregated into a set of individual locations. A function is then applied to those locations to generate new attributes that will ultimately be re-aggregated to characterize the original entities or to form entirely new ones.

Additionally, layers are represented by *titles* (character strings), numerical parameters indicating *orientation* (angular deviation from north) and *resolution* (geographic size of the smallest addressable cartographic location), and a set of one or more *zones*. A zone is a geographic condition associated with a recorded characteristic that distinguishes it from other such conditions. It is represented by a *label* (a character string), a *value* (a numerical quantity), and a set of one or more *locations*. A cartographic model may also contain additional information relating to its geodetic position, its cartographic projection, its size, its origin, its format, and so on. While this information may well be significant, it need not conform to any particular cartographic modelling conventions.

2.2 MAP ALGEBRAIC FUNCTIONS

A cartographic model may include not only data but also (or instead) a set of specified data transformations. In practice, the semantics of a cartographic modelling language must account for four major types of data-processing activity including:

- data preparation (e.g. line digitizing, video scanning, aerial image enhancement and classification, cartographic reprojection, or reformatting of files),
- data presentation (e.g . report generation, map drawing, or visual simulation),
- programming (e.g. error handling, device control, or conditional execution), and
- data interpretation (analysis and synthesis).

Map Algebra, as it currently stands, relates only to 'interpretation'. Map Algebra does not particularly apply to data preparation and presentation, and it does not have a vocabulary for 'evaluation'. Data interpretation here means the understanding of objects in terms of their descriptions. The data interpretation process is generally one in which relationships and/or meanings that are implicit in a set of recorded facts are extracted and expressed in explicit form. Evaluation assigns aesthetic values to particular interpretations.

In Map Algebra, variables exist as map layers that are processed by way of *functions* that are cartographic in nature. Among these are functions to reclassify zones, to combine layers, to calculate distances and directions, to measure sizes, to characterize shapes, to determine lines-of-sight, to simulate dispersion, and so on.

Each of these map algebraic functions accepts one or more existing map layers as its input and generates a new layer as its output. Thus, the output from any one function can be used as input to any other. Sequences of these primitive functions are called *procedures*. Cartographic procedures can be constructed to model complex phenomena such as soil erosion or land development potential.

From the Map Algebra location-specific or *worm's eye* linguistic perspective (as opposed to the *bird's eye* perspective from which traditional cartographic products are normally viewed), the data transforming functions of a cartographic modelling system can be classified into four major types. The four are respectively associated with

- *local*,
- *zonal*,
- *incremental*, and
- *focal*, functions.

Local functions are those that compute a new value for every location as a function of one or more existing values associated with that location. Most do so by applying a familiar mathematical function to each location's value(s) on one or more existing map layers. These local functions are as follows:

LocalArcCosine	computes the arc cosine of each location's value on a specified layer;
LocalArcSine	computes the arc sine of each location's value on a specified layer;
LocalArcTangent	computes the arc tangent of each location's value on a specified layer (or ratio of values from two specified layers);
LocalCombination	computes a value that uniquely identifies the particular combination of existing values that are associated with each location on two or more specified layers;
LocalCosine	computes the cosine of each location's value on a specified layer;
LocalDifference	subtracts from each location's value on one specified layer its value(s) on one or more additional layers;
LocalMajority	computes the most frequent of each location's values on two or more layers;
LocalMaximum	computes the maximum of each location's values on two or more layers;
LocalMean	computes the average of each location's values on two or more specified layers;
LocalMinimum	computes the minimum of each location's values on two or more layers;
LocalMinority	computes the least frequent of each location's values on two or more layers;
LocalProduct	multiplies each location's value on one specified layer by its value(s) on one or more additional layers;
LocalRating	translates the value of each location on an existing layer (or combination of values from two or more layers) at into a new value that is either given as a constant or drawn from the same location on another specified layer;
LocalRatio	divides each location's value on one specified layer by its value(s) on one or more additional layers;
LocalRoot	raises each location's value on one specified layer to the power represented by the reciprocal of its value(s) on one or more additional layers;
LocalSine	computes the sine of each location's value on a specified layer;
LocalSum	adds each location's value on one specified layer to its value(s) on one or more additional layers;
LocalTangent	computes the tangent of each location's value on a specified layer; and
LocalVariety	computes the number of dissimilar values associated with each location on two or more specified layers.

A number of additional functions (e.g. absolute values, logarithms, exponential products, modular differences, medians, factorials) can also be applied to local values. The decision as to whether or not these should be included in a tool box of cartographic modelling functions is largely a matter of anticipated utility and the degree to which such functions can be replicated by combining others. The same is true of the zonal, incremental, and focal functions.

Zonal functions compute a new value for each location as a function of the existing values from a specified layer that are associated not (just) with that location itself but with all locations that occur within its zone on another specified layer. Expressed in terms of existing map layers that are generically entitled ***ThisLayer*** and ***ThatLayer***, respectively, the zonal functions are as follows:

ZonalCombination	computes a value that uniquely identifies the particular combination of ***ThisLayer*** values in each ***ThatLayer*** zone;
ZonalMajority	computes the most frequently occurring ***ThisLayer*** value in each ***ThatLayer*** zone;
ZonalMaximum	computes the maximum ***ThisLayer*** value in each ***ThatLayer*** zone;
ZonalMean	computes the average of all ***ThisLayer*** values in each ***ThatLayer*** zone;
ZonalMinimum	computes the minimum ***ThisLayer*** value in each ***ThatLayer*** zone;
ZonalMinority	computes the least frequently occurring ***ThisLayer*** value in each ***ThatLayer*** zone;
ZonalPercentage	computes the percentage of each ***ThatLayer*** zone that shares its ***ThisLayer*** value;
ZonalPercentile	computes the percentage of each ***ThatLayer*** zone that is of lower ***ThisLayer*** value;
ZonalProduct	multiplies all of the ***ThisLayer*** values in each ***ThatLayer*** zone;
ZonalRanking	computes the ordinal magnitude of ***ThisLayer*** values in each ***ThatLayer*** zone;
ZonalRating	translates a specified combination of ***ThisLayer*** values in each ***ThatLayer*** zone into a new value that is either given as a constant or drawn from the same location on another specified layer;
ZonalSum	adds all of the ***ThisLayer*** values in each ***ThatLayer*** zone; and
ZonalVariety	computes the number of dissimilar ***ThisLayer*** values in each ***ThatLayer*** zone.

The incremental functions characterize each location as an increment of one-, two-, or three-dimensional cartographic form. The size and shape of these increments are inferred from the value(s) of each of location relative to those of its adjacent neighbors on one or more specified layers. The functions performed by these incremental functions are as follows:

IncrementalArea	computes the surface area of each location as part of a two-dimensional pattern on a three-dimensional surface;
IncrementalAspect	computes the compass direction of steepest descent at each location on a three-dimensional surface;
IncrementalDrainage	computes the direction(s) from which surficial drainage would enter each location on a three-dimensional surface;
IncrementalFrontage	computes the amount of boundary associated with each location as part of a two-dimensional pattern on a three-dimensional surface;

IncrementalGradient computes the slope at each location on a three-dimensional surface;

IncrementalLength computes the total length associated with each location as part of a one-dimensional network on a three-dimensional surface;

IncrementalLinkage classifies the form of each location as part of a one-dimensional network;

IncrementalPartition classifies the form of each location as part of a two-dimensional pattern; and

IncrementalVolume computes the subsurface volume beneath each location as part of a two-dimensional pattern on a three-dimensional surface.

Focal functions are those that compute each location's new value as a function of the existing values, distances, and/or directions of neighboring (but not necessarily adjacent) locations on a specified map layer. The distance relationship that establishes each location's neighborhood may be defined in terms of physical separation, travel costs, or inter-visibility. The focal functions are as follows:

FocalBearing computes the compass direction from each location to the nearest of a set of locations defined on a specified layer;

FocalCombination computes a value that uniquely identifies the particular combination of values within each location's neighborhood on a specified map layer;

FocalGravitation computes the inverse-square-distance-weighted average of all values within each location's neighborhood on a specified map layer;

FocalInsularity computes a new value for each location that uniquely matches the new value computed for all other locations within its neighborhood that share its original value on a map layer;

FocalMajority computes the most frequent value within each location's neighborhood on a specified map layer;

FocalMaximum computes the maximum value within each location's neighborhood on a specified map layer;

FocalMean computes the area-weighted average of all values within each location's neighborhood on a specified map layer;

FocalMinimum computes the minimum value within each location's neighborhood on a specified map layer;

FocalMinority computes the least frequent value within each location's neighborhood on a specified map layer;

FocalNeighbor sets each location to the value of the nearest of a set of locations defined on a specified map layer;

FocalPercentage computes the percentage of each location's neighborhood that shares its value on a specified map layer;

FocalPercentile computes the percentage of each location's neighborhood that is of lower value on a specified map layer;

FocalProduct	multiplies the values of all locations within each location's neighborhood on a specified map layer;
FocalProximity	computes the distance between each location and the nearest of a set of locations defined on a specified layer;
FocalRanking	computes the number of zones within each location's neighborhood that are of lower value on a specified map layer;
FocalRating	translates a specified combination zones within each location's neighborhood on a specified layer into a new value that is either given as a constant or drawn from the same location on another specified layer;
FocalSum	adds the values of all locations within each location's neighborhood on a specified map layer; and
FocalVariety	computes the number of dissimilar values within each location's neighborhood on a specified map layer.

2.3 MAP ALGEBRAIC EXPRESSIONS

Map Algebra is essentially an applicative or functional programming language, though there are various generalizations (Chan, 1988; Zhou, 1989). The control of a cartographic modelling system is a matter of specifying particular functions, identifying the map layers to which they apply, and indicating the order in which they are performed. This may be done by way of typed commands, graphic gestures, spoken words, or other forms of communication depending on the particular geographic information system involved.

In order to refer to these various forms of communication in common terms, it is useful to adopt a standard set of notational conventions. These conventions can be expressed in the form of a high-level programming language in which each function is represented by a *statement* in the following generic form:

NEWLAYER = FUNCTION of OLDLAYER HOW

Here, ***NEWLAYER*** is the title to be assigned to a new map layer generated by applying an function called ***FUNCTION*** to and existing layer entitled ***OLDLAYER***. The statement may also include additional modifying phrases, here represented by ***HOW***.

It is the function-oriented (as opposed to object-oriented or relation-oriented) structure that distinguishes the map algebraic approach from other methodologies. Others who have proposed languages for the analysis and synthesis (as opposed to the organization) of geographic data include Tobler (1979), White (1981), Claire (1984), Chan (1988), etc., all of whom seek to characterize spatial relations and/or functions in reference to a unified construct.

3. Conclusion

The cartographic modelling approach to building geographic information systems, as the name indicates, is meant to model the 'surface information' or planimetric information which one can read directly from a map. In the most rudimentary sense, a cartographic model represents a map rather than the geographic theme depicted by the map. The only spatial data type found in a cartographic model is a sample point or picture element: pixel. We do not ask what is where, instead we read directly the characteristic value of a sample point, such as: reflected radiation, soil type, elevation, etc. All other geographic information must be computed from sample points. Objects – sample points – on an image are addressable by their position on an image; that is, we implicitly label them by their physical positions on the cartographic image.

Whereas cartographic models represent an area by sample points, object based models depict the area as composed of distinct and identifiable spatial objects or entities. Albeit these entities can very well be sample points and grid cells. A cartographic based terrain model represents a landscape only by sample elevations; while an object based model, in addition to showing sample elevations, represents the surface characteristics of the study area as a collection of surface features: hills and dales, peaks and pits, ridge lines and course lines, etc, and of spatial relationships: uphill, downhill, leeward, windward, etc.

At the heart of cartographic modelling is the idea of decomposing data sets, data-processing capabilities, and data-processing control specifications into elementary components that can then be recombined with ease and flexibility. This decomposition results in a greater amount of tedium but also a greater amount of clarity and extendibility. While procedures can sometimes become quite involved, their step-by-step structure and simple components make it possible to express even very complex models in a clear and consistent manner. From the brief denotational semantics above, we can appreciate the simple structure of the algebra.

The reason that this specialization is necessary is partly because of the mode in which we deal with a geography. Let us consider the distinction we made above by an example. Suppose we have a land management GIS. When we cartographically highlighted a point, or simply point to a location on a map and indicate that there is a fire, we mean a fire at a certain location in the real world. We are referring to a geographic instance and not to the map or the computer screen. On the other hand, when we color a point on a map and indicate that it is a perfect site in our design for an outdoor summer concert and fireworks display, our reality exists only in our design, on our map. We are referring to the map and not a geographic instance – past, present or future. Philosophical questions aside, efforts on data types and accuracy in recording spatially distributed data in geographic information systems, although fundamental to the data organization in geographic modelling, carry very different significance in this cartographic mode of modelling. Geo-processing languages must handle the descriptive, the interpretative and the imaginative modes of dealing with geography.

4. References

Cátedra, M. (1991) Through the Door: A View of Space from an Anthropological Perspective, in D. M. Mark and A. U. Frank (eds.), Cognitive and Linguistic Aspects of Geographic Space, Kluwer, Dordrecht.

Chan, K. K. L. (1988) Evaluating Descriptive Spatial Models for Prescriptive Inference, unpublished doctoral dissertation, Harvard University.

Claire, R. W. (1982) 'Algorithm Development for Spatial Operators', Proceedings, PECORA 9, 213-221.

Cox, N. J., Aldred, B. K., and Rhind, D. W. (1980) 'A Relational Data Base System and a Proposal for a Geographical Data Type', Geo-Processing 1, 217-229.

Dacey, M. F. (1970) 'Linguistic Aspects of Maps and Geographic Information', Ontario Geography 5, 71-80.

Harley, J. B. (1988) Maps, knowledge and power, in D. Cosgrove and S. Daniels (eds.), The Iconography of Landscape, Cambridge Studies in Historical Geography, Cambridge University Press.

Herring, J. (1991) The Mathematical Modeling of Spatial and Non-Spatial Information in Geographic Information Systems, in D. M. Mark and A. U. Frank (eds.), Cognitive and Linguistic Aspects of Geographic Space, Kluwer, Dordrecht.

Huang, K. S., B. K. Jenkins and A. A. Sawchuk (1989) 'Binary Image Algebra and Optical Cellular Logic Processor Design', Computer Vision, Graphics, and Image Processing 45, 295-345.

McHarg, I. L. (1969) Design with Nature, Natural History Press, New York.

Mitchell, W. J. T. (ed.) (1974) The Language of Images, The University of Chicago Press, Chicago.

Mukerji, C. (1984) 'Visual Language in Science and the Exercise of Power: The case of cartography in early modern Europe', Studies in Visual Communication 10(3), 30-45.

Nyerges, T. L. (1980) Modeling the Structure of Cartographic Information for Query Processing, Unpublished doctoral dissertation, The Ohio State University.

Shapiro, L. G. and Haralick, R. M. (1980) 'A Spatial Data Structure', Geo-Processing 1, 313-337.

Steinitz, C. F., Parker. P., and Jordan, L. (1976) 'Hand-drawn Overlays: Their history and prospective uses', Landscape Architecture 66(8), 444-455.

Tobler, W. R. (1979) Cellular Geography, in S. Gale and G. Olsson (eds.), Philosophy in Geography, Dordrecht, Holland, D. Reidel Publishing Company, 379-386.

Tomlin, C. D. (1975) The Tomlin Subsystem of IMGRID, unpublished master's project report, Department of Landscape Architecture, Graduate School of Design, Harvard University.

Tomlin, C. D. (1983) Digital Cartographic Modeling Techniques in Environmental Planning, unpublished doctoral dissertation, Graduate School Division of Forestry and Environmental Studies, Yale University.

Tomlin, C. D. (1985) The IBM Personal Computer Version of the Map Analysis Package, Laboratory for Computer Graphics and Spatial Analysis, Graduate School of Design, Harvard University.

Tomlin, C. D. (1990) Geographic Information Systems and Cartographic Modeling, Englewood Cliffs, New Jersey, Prentice-Hall.

Tomlin, C. D. and Berry, J. K. (1979) 'A Mathematical Structure for Cartographic Modeling in Environmental Analysis', Proceedings, 39th Symposium of the American Congress on Surveying and Mapping.

Tomlin, C. D. and Tomlin, S. M. (1981) 'An Overlay Mapping Language', in Regional Landscape Planning: proceedings of three educational sessions, American Society of Landscape Architects.

van Roessel, J. W. (1987) 'Design of a Spatial Data Structure using the Relational Normal Form', International Journal of Geographical Information Systems 1(1), 33-50.

Youngman, C. (1978) 'A Linguistic Approach to Map Description', in G. Dutton (ed.), Spatial Semantics: Understanding and interacting with map data, Harvard Laboratory for Computer Graphics and Spatial Analysis, Cambridge, Massachusetts.

White, D. (1985) 'A Taxonomy of Space-time Data Relations', Proceedings, Princeton Conference on Computer Graphics and Transportation Planning.

Woodward, D. (1987) Art and Cartography, The University of Chicago Press, Chicago.

Zhou, Q. (1989) ReMAP Reference Manual, Personal communication.

QUALITATIVE SPATIAL REASONING

CHRISTIAN FREKSA
Institut für Informatik
Technische Universität München
Arcisstr. 21
8000 München 2 Germany

ABSTRACT. Physical space has unique properties which form the basis of fundamental capabilities of cognitive systems. This paper explores some cognitive aspects of perception and knowledge representation and explains why spatial knowledge is of particular interest for cognitive science. It is suggested that 'spatial inference engines' provide the basis for rather general cognitive capabilities inside and outside the spatial domain. The role of abstraction in spatial reasoning and the advantages of qualitative spatial knowledge over quantitative knowledge are discussed. The usefulness of spatial representations with a low degree of abstraction is shown. An example from vision (the aquarium domain) is used to illustrate in which ways knowledge about space may be uncertain or incomplete. Parallels are drawn between the spatial and the temporal domains. A concrete approach for the representation of qualitative spatial knowledge on the basis of 'conceptual neighborhood' is suggested and some potential application areas are mentioned.

1. What is Special about Spatial Knowledge?

Our knowledge about physical space differs from all other knowledge in a very significant way: we can perceive space directly through various channels conveying distinct modalities. Unlike in the case of other perceivable domains, spatial knowledge obtained through one channel can be verified or refuted by perception through the other channels. As a consequence, we are disproportionately confident about what we know about space: we take it *for real*.

More specifically, we perceive spatial dimensions visually, tactilely, acoustically, and even by smell and by temperature sensation. Some animals exploit perception of electrical or magnetic fields for spatial orientation. Within the visual sense, we obtain clues about space through size, brightness, hue, saturation, and texture information; within the acoustic sense, we obtain clues about space through loudness, frequency, and signal/noise ratio. In addition to perceiving space multimodally, we can modify spatial situations by moving objects, we can expose objects to the perceivable influence of physical forces, and we can modify our spatial perception by moving ourselves. In this way, we experience the laws of space as we can experience no other dimension.

In contrast, the perception of other dimensions like color, visual brightness, sound, smell, temperature, etc. relies on single modalities and we cannot directly modify them. This makes the knowledge about these dimensions much more susceptible to perceptional illusion or uncertainty. I will argue that physical space plays a central role in cognition 1) as the domain in which physical events take place, and 2) as reference domain for the interpretation of non-spatial concepts.

D. M. Mark and A. U. Frank (eds.), Cognitive and Linguistic Aspects of Geographic Space, 361–372.

The special cognitive reality of space described above makes the spatial domain particularly suitable as a medium for conveying knowledge, since its properties are universal to different cognitive systems. Thus, the spatial domain can be used particularly well as the source domain for metaphors with a non-perceivable or abstract target domain. In this way, the properties of physical space can be used as a vehicle for conveying non-spatial concepts, provided there exists a mapping from the non-spatial to the spatial domain. Lakoff and Johnson (1980) and Lakoff (1987) present numerous examples for spatial metaphors in natural language.

I propose that our knowledge about the organization of space serves as a 'cognitive interface' between abstract or non-perceptual knowledge and the 'real world'. In other words, we may interpret non-spatial concepts by mentally transforming them into spatial concepts (i.e., *understanding* them in terms of spatial concepts), carrying out mental operations in this 'visualizable' and 'graspable' domain and transforming the result into the original domain. In this way, 'spatial inference engines' may have a much more general function than the term suggests: rather than generalizing by forming a common abstraction for various domains we generalize by forming suitable analogies to a well-understood concrete domain.

2. Spatial Inference Engines

If there is a large class of problems which can be solved well under the specific constraints of the spatial domain, we may want to develop a specialized representation system with an 'inference engine' which is capable of dealing with knowledge about space particularly well. The advantage of a special purpose approach to spatial reasoning is that certain constraints which always hold in the spatial domain do not have to be modeled and verified in each situation anew.

The spatial domain is severely constrained in certain ways in which abstract ideas are not. For example, distinct solid objects never occupy the same physical space and a single physical object always occupies a single contiguous space. The arrangements of objects and locations given by their neighborhood relations are very important features of spatial situations which are relevant in virtually all spatial problem solving tasks. This has to do with the fact that physical objects consist of connected parts ('neighbors') and with the fact that movement in space is only possible between neighboring locations. Therefore we do not want to check in a general representation explicitly if these constraints hold; the constraints should be 'built-in' in the reasoning system as they are built-in in the spatial domain.

For spatial problem solving with a general-purpose theorem prover, for example, such constraints must be explicitly formulated (compare Reiter and Mackworth, 1989). This is due to the extreme degree of abstraction from the original problem which is possible and necessary in a general-purpose representation system. In reasoning about concrete domains like physical space, however, such a high level of abstraction is not desirable since additional effort is required to nullify excessive abstraction as in the example given above. Instead, we are aiming at a representation level which only abstracts from unnecessary details but maintains the important aspects of the domain.

This approach is not new, of course: constructive geometry exploits spatial properties of drawing paper as an analogy to properties in the abstract model domain; cartography utilizes spatial properties of maps as an analogy to spatial properties in the depicted real world; fluid mechanics makes use of self-similarity of space to form an analogy between small scale space and larger scale space when wind tunnel experiments on small scale models are used to predict the behavior of larger scale objects. In addition, neurobiological research has revealed that various areas of the visual cortex are arranged in such a way that certain spatial properties of the external world are preserved in the cortical structures. The crucial question for artificial intelligence in this context is how we can implement a spatial reasoning system which makes direct use of space and its intrinsic properties.

3. The Role of Abstraction for Spatial Reasoning

If we want to maintain properties of space for spatial reasoning, why not solving spatial problems in the real world directly? In fact, there are situations in which this is probably the best approach. For example, for identifying an object in an unusual position or orientation it may be easier to adjust position or orientation than to formalize the object and mentally match the representation against the representation of a template. Also, whenever there is insufficient knowledge for formalization, the *world-based* approach is most promising: rather than using abstract knowledge about the world and its properties, 'real world tasks' can be solved directly under the influence of the world's real properties. Examples are: finding the right key for a lock; testing a vehicle under realistic conditions; diagnosing satellite malfunctions by simulating the malfunctioning process with the aid of a duplicate satellite.

However, in many spatial situations it is too expensive to determine solutions in the real world directly. This is the case when large or heavy objects or large distances are involved. In such cases, we would like to abstract from actual size, weight, or distance; but we have no need to abstract from fundamental spatial properties like spatial arrangement. The usual approach in this situation is to formalize knowledge about the abstract problem and to solve it on a formula manipulation device, a computer. The reason for representing *knowledge* about a domain rather than representing the domain itself is that the world is accessible by formal means only through knowledge. When we are using computers we are bound to using formal approaches to modelling reasoning processes. At this point, however, we do not have formal languages for computers which are suitable for the level of abstraction we described above.

Formal languages like predicate logic initially abstract from all inherent properties of space. Specifying the spatial domain in such an abstract language means constraining the language in order to make the semantics consistent with the target domain 'space'. We are aiming at a language for representing spatial knowledge which leaves us only the expressive freedom we can utilize in the spatial domain ('spatial inference engine' level in Figure 1). In other words, we do not want to specify the laws of space but we want to take them for granted.

Abstraction is very useful in its own right: one main attraction of abstraction is that it liberates representations from insignificant details and focuses on the significant distinctions. This advantage, however, gets partially lost if we go overboard by using formalisms which liberate the representation from significant constraints as well.

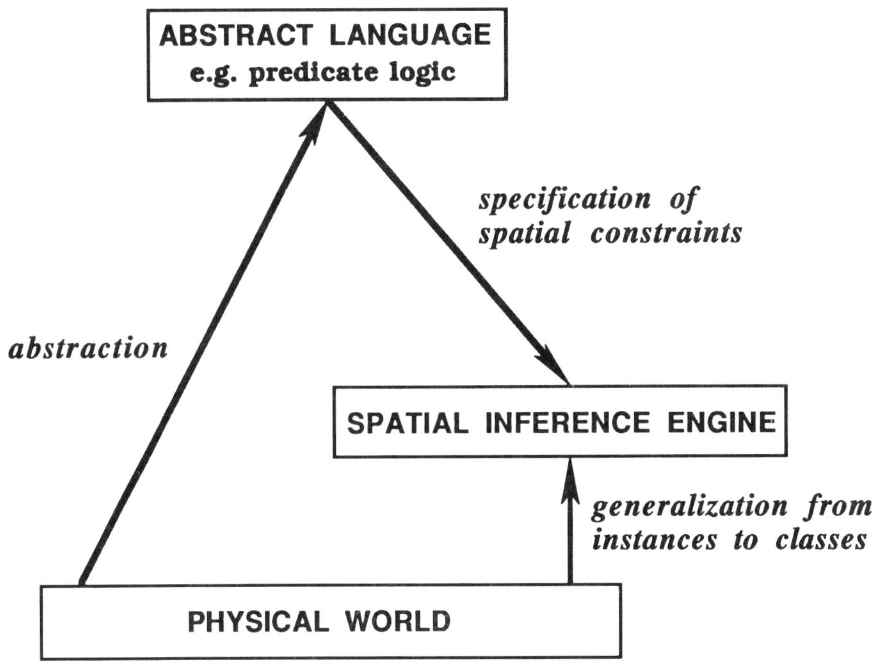

Figure 1. Spatial inference engines can be constructed by constraining very abstract formalization languages or by generalizing over the physical world.

3.1 WHICH PROPERTIES OF REAL SPACE DO WE WANT TO ABSTRACT FROM?

Ideally, we want to represent only knowledge which we will need for solving the tasks which we have to solve and abstract from all other knowledge. Since we typically collect knowledge before we know the specific tasks for which we will use it we can not be so restrictive; but depending on the classes of tasks we want to solve we are usually able to exclude a range of items which we will never consider.

For example, in the domain of spatial knowledge it makes a difference if we represent knowledge in order to construct a precise metric map of some area or if we represent knowledge in order to describe a location to be identified in the real world as in the case of wayfinding. In the first case, we require quantitative knowledge which enables us without reference to the real world to predict precisely at which (absolute) location we will encounter which object. In the second case, we only require enough knowledge to distinguish quantities which are given in the real world and which are available for reference, so a limited amount of qualitative knowledge may suffice.

We are particularly interested in the second case, i.e. the class of identification tasks, since many spatial reasoning problems can be reduced to this class of tasks. In fact, all reconstruction tasks can be structured hierarchically into identification subtasks. We can learn about useful and cognitively adequate abstractions by considering established methods of dealing with spatial knowledge. In the domains of object-, scene-, or route description and identification, for example, we can refer to the long tradition of map production and look which aspects of the spatial domain have been found to be significant in good maps.

For description and identification tasks our culture typically uses linguistic and/or pictorial means to convey knowledge, i.e., we use a natural language description or a depiction of the situation, or preferably a combination of the two. In both approaches, part of the knowledge to be conveyed is used to establish the context of the situation and part of the knowledge actually establishes the target object within that context. A main advantage of combining language and depiction is to reduce the description itself to the identification task level.

3.2 WHICH PROPERTIES OF REAL SPACE DO WE WANT TO MAINTAIN?

For spatial reasoning tasks, the following conceptual properties of spatial configurations appear to be universally useful; we will argue that they should be preserved by the representational structure.

1) Universal domain properties which do not depend on specific situations, for example
   - each object exists exactly once
   - each location coincides with at most one object

   (uniqueness constraints)

2) Properties related to the physics of space (arrangement information):
   - movement in space is possible only between neighboring locations (this property should hold on any level of representation)

   (topology)

3) Properties related to the neighborhood of spatial relations (compare Freksa (1990)). For example, the spatial relations described by 'A is left of B' and 'A touches B on the left' are spatial-conceptual neighbors, since there is a direct transition from one relation to the other when the spatial arrangement is modified; in contrast 'A is right of B' is not a spatial-conceptual neighbor, since other relations have to be traversed in the transition.

   (conceptual structure)

The first two properties are the same properties which constitute the usual semantics of sketch maps of spatial environments (compare Reiter and Mackworth, 1989): each object is depicted only once; different objects are positioned at different locations, and the objects are arranged according to their arrangement in the real world. Other properties of sketch maps are not considered significant, e.g. the thickness of the lines, the exact lengths of the lines, the precise orientation of the lines, the completeness of the representation. The third property (conceptual structure) becomes particularly important for reasoning under uncertainty or in absence of complete knowledge (compare Freksa, 1990).

3.3 QUALITATIVE KNOWLEDGE

Much of the knowledge about time and space is qualitative in nature. Specifically, this is true for visual knowledge about space. Although the retinal image of a visual object is a fully instantiated quantitative image in the sense that specific locations on the retina are stimulated by light of a specific spectrum of wavelengths and intensity, the knowledge about a retinal image that can be retrieved from memory is qualitative. Absolute locations, wavelengths, and intensities cannot be retrieved from memory, in general; rather, comparisons between such features can be performed. Thus, from the representation of the image, the original retinal image cannot be reconstructed in full; merely certain qualitative relations among features within the image or between image features and memory features can be recovered.

Qualitative representations of the kind characterized above share a variety of properties with 'mental images' (see, for example, Block, 1981) which people report about when they describe from memory what they have seen or when they attempt to answer questions on the basis of visual memories:

- depending on familiarity or previous knowledge they will be able to retrieve fewer or more details;
- in the absence of related features, descriptions will be rather vague;
- the accessible information is not sufficient for reconstructing the original image accurately to scale; rather certain qualitative relationships can be reconstructed;
- if enough such qualitative relationships can be recalled, a fairly accurate reconstruction of the original image is possible due to the interaction of constraints;
- quantitative knowledge can be represented by 'anchored' qualitative knowledge ($\int$ measuring);
- qualitative knowledge is robust under transformations.

I would like to suggest that qualitative knowledge of this kind is exactly the type of knowledge we want to represent for spatial reasoning and what we need for solving identification tasks. Most qualitative relations act as constraints which – individually viewed – leave quite a bit of freedom as to the actual quantitative values possible to satisfy the constraints; some qualitative constraints are as informative as quantitative ones, however. For example, the assertion 'a is smaller than b' describes a qualitative constraint which – in a given domain – may be satisfied by a very large number of quantities a, given a certain value for b. In contrast, the qualitative relation described by 'a equals b' constrains the quantity of a to the single value of b. Similarly, the assertion 'k is perpendicular to l' constrains the orientation of a directed path in a 2-dimensional domain to exactly two possible values.

The expressive power of qualitative constraints results from their interaction. For example, the assertions 'a is smaller than b' and 'b is smaller than c' constrain the value of b to the values of the interval]a, c[. In the domain of real values there is still an infinite number of quantities in this interval; however, if we further constrain the assertions to refer to a domain of discrete entities, the qualitative constraints may have the power of selecting small sets of quantities without directly addressing their value.

For solving identification tasks we are not interested in the theoretical specificity of descriptions as obtained by considering the potential set of entities referred to by a descriptor; rather, the *actual* set of entities and values that exist in the specific domain is of interest. Qualitative descriptions are very powerful in this situation: when there are many items to be distinguished there are many relations that can be established for distinguishing them; when there are few, only few are needed for their distinction.

A great advantage of qualitative representations for cognitive systems is the fact that they are independent of specific values and granularities of representation; depending on the situation context and on the granularity of the available knowledge they correspond to more specific or more general entities. In this way, qualitative representations allow for a top-down approach to characterizing situations, in comparison to bottom-up approaches suggested by quantitative representations.

For example, suppose we want to describe objects in terms of their sizes. If we represent sizes quantitatively, we express their values in terms of some absolute unit. Depending on the granularity of the unit chosen, we can distinguish fewer or more classes of sizes; thus, we must know beforehand which size differences we may want to distinguish. If we represent sizes qualitatively, e.g. in terms of the relations 'smaller', 'equal', 'larger', we do not predefine the size differences we will be able to distinguish; depending on the specific needs, we can compose coarser or finer descriptions by cascading relations, i.e., by making fewer or more comparisons of the quality involved. In the absence of a basic reference unit, 'granularity' becomes a relative concept; coarseness of a representation depends on the context of concepts involved.

Qualitative representations structure a domain according to conceptual distinctions corresponding to the way the domain is viewed rather than according to physical quantities the domain is measured by. In other words, qualitative representations do not structure domains homogeneously (i.e., with uniform granularity of physical entities) as quantitative representations do; rather, they focus on the boundaries of concepts: the representation may be viewed as having low resolution for different values corresponding to the same quality and high resolution near the concept boundaries. This is true at least for crisp concepts, i.e., concepts with sharp boundaries. Thus, qualitative representations may be viewed as regions from the viewpoint of quantitative representations and as nodes from a conceptual viewpoint.

4. The Aquarium Metaphor for Spatial Concepts

I will illustrate some properties of perceptual knowledge which we typically encounter when we represent knowledge about the *real world* by introducing the 'aquarium metaphor' for communication about spatial perceptions (compare Freksa, 1980): Two observers A and B look at an aquarium and get excited about the beautiful fishes they see (Figure 2). They do not know the names of the fishes but they want to communicate about particular fish individuals to one another.

The problem observer A is confronted with is to describe the fish in such a way that observer B is able to identify the fish correctly. The situation can be characterized as follows:

- A and B cannot quantitatively locate the fish (because the glass prevents them from putting a ruler inside the aquarium, parallax prevents them from aligning a ruler outside the aquarium, and the movement of the fish prevents them from watching it long enough to precisely determine its position otherwise.
+ However, the observers are not completely uncertain about the position of the fish. Specifically, they have knowledge about its position *relative* to other objects (qualitative knowledge); furthermore, these relative positions do not change arbitrarily when the fishes move; if the relation between two fishes changes at all, a neighboring relation will replace the previous relation (conceptual neighborhood of relations).
- There are various resource limitations: the amounts of time for observation and for object identification are limited, perceptual resolution is limited (coarse knowledge), and the perceivable features are limited (incomplete knowledge).
+ As a consequence, the complexity of the description and identification tasks is limited.

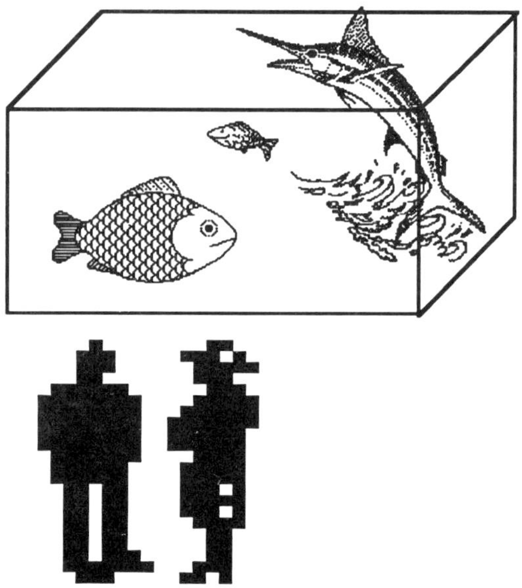

Figure 2. Observer A communicates with observer B about a particular fish in the aquarium.

- – The movement and muddiness of the water may prevent the observers from clearly recognizing the boundaries of objects and relations (fuzzy knowledge).
- \+ They can use fuzzy natural language descriptors which combine conceptually neighboring situations corresponding to their observations (adaptivity of natural language descriptors).
- – The two observers derive their knowledge about the aquarium world from different perceptions, since they are located at different observer positions; as a consequence, some objects and features may be partially or completely hidden from one observer but not from the other and the observers may observe different spatial relations for the same spatial situation (subjective knowledge).
- \+ The knowledge of the two observers is strongly related: gradual changes in the observer position cause at most stepwise changes between neighboring spatial relations; one observer can easily model how the observation of the other observer may deviate from the own observation (conceptual neighborhood of relations).
- \+ The two observers are not only related by the spatial closeness of their positions but also by the common situation context to which they refer. Thus, the object description and identification task is strongly simplified since identification must be possible only in relation to this context (context-aided communication).

In other words, the knowledge about the aquarium world which the two observers can use for object description and identification is incomplete, imprecise, and subjective. Nevertheless, the observers will be able to convey to each other which objects in the aquarium world they are focussing at: they can describe and locate the objects by qualitative means, i.e., they can observe a number of relations between features of the target object and some other objects in the aquarium and convey their knowledge in terms of these relations; they can specify a few features which may suffice to distinguish the target object from the other objects; they will be able to interpret descriptions 'with a grain of salt', i.e., if the description does not completely fit their own conception of the situation they will be able to relax the interpretation in such a way that crucial distinctions are still maintained and less crucial distinctions are tolerated.

The aquarium metaphor emphasizes a few issues which are present in all spatial perception, representation, and identification situations:

1) perceptual knowledge is necessarily limited with regard to resolution, features, completeness, certainty;
2) there is a more or less well-defined context;
3) perceptions are finite;
4) the neighborhood of objects and the conceptual neighborhood of relations between objects provide very useful information for spatial reasoning.

In addition to these perception-related issues, the aquarium metaphor is useful for studying language-related issues of physical space. For example, which properties of space are preserved implicitly, which ones are represented explicitly, and which ones are not represented at all? What are the common features and what are the differences between a linguistic description and, say, a sketch map of the same scene?

Note that the aquarium metaphor is not universal: for example, it does not fit an abstract view of space in which the spatial relations 'right of' and 'left of' may be considered closely related, because they are symmetric. As a perception-oriented construct, the aquarium metaphor is very close to the physics of space, but not to arbitrary abstractions thereof.

In our research, we focus on spatial representation, description, and identification tasks which share the properties discussed in the framework of the aquarium metaphor.

5. The One-Dimensional Case

James Allen (1983) proposed an approach to representing temporal knowledge which is based on qualitative information about time. More specifically, Allen represents relative positions between events represented as intervals. In the linear case of time, any two events have exactly one out of 13 possible mutually exclusive qualitatively distinct relationships to each other. In Figure 3, these relationships are shown for the one-dimensional spatial case: two spatial objects, here a black and a grey fish, are shown in 13 qualitatively different relative positions with regard to the horizontal dimension (the orientation of the objects is ignored).

The thirteen relations are denoted by the labels *before* (<), *after* (>), *during* (d), *contains* (di), *overlaps* (o), *overlapped-by* (oi), *meets* (m), *met-by* (mi), *starts* (s), *started-by* (si), *finishes* (f), *finished-by* (fi), and *equals* (=). In the figure, the 13 cases are arranged according to 'conceptual neighborhood', i.e., relations between which direct transitions are possible when a fish moves or changes its size are located next to each other.

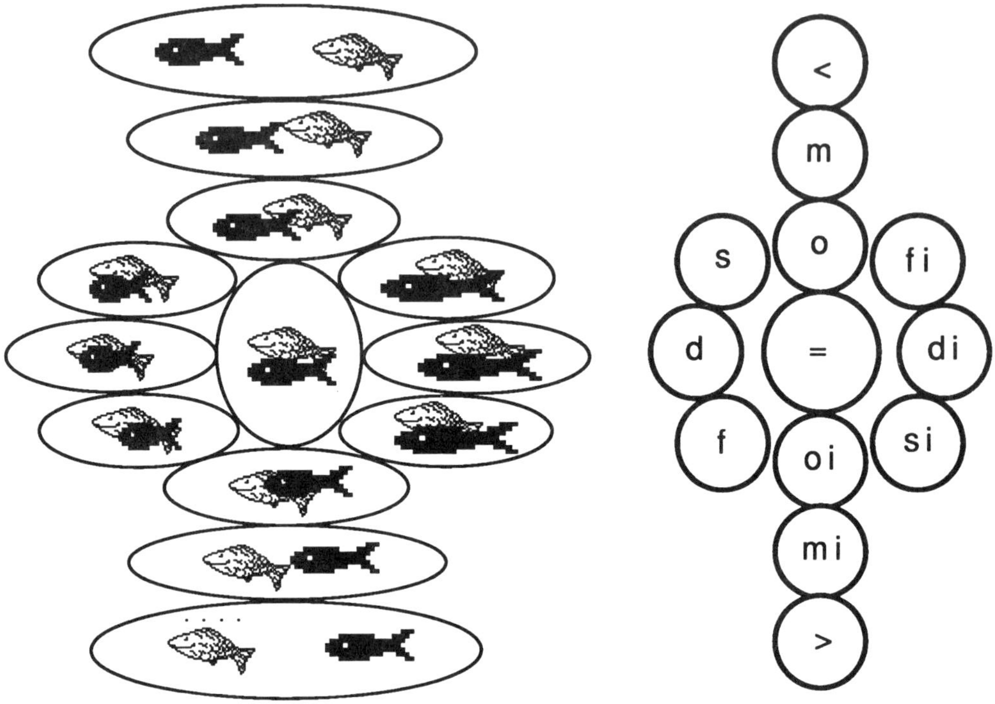

Figure 3: Left: One-dimensional spatial relations between two objects (fishes) arranged according to their conceptual neighborhood. Right: The corresponding labels arranged accordingly.

In Allen's scheme, reasoning is done by computing the set of possible relationships between two entities whose relationships to a third entity must be known. In Freksa (1990), Allen's approach has been generalized to allow for reasoning with incomplete or coarse knowledge about temporal relations. 'Conceptual neighborhood' of relations is the key concept for this generalization. It allows for inferring neighborhoods of relations between objects about which neighborhoods of relations to some other object are known. Interactions of such sets of relations may lead to rather precise inferences. The inferences that can be drawn are not in all cases completely precise – some ambiguity as to which of a neighborhood of relations holds may remain. No quantities are involved in the representation and reasoning can be carried at any level of granularity. Arbitrary levels of granularity can be mixed. The resulting qualitative conclusions are always correct, although – by necessity – they may be rather coarse if translated into quantities. The approach fulfills the requirements for solving identification tasks as discussed in connection with the aquarium metaphor in the previous section. Our aim is now to adapt this approach to higher-dimensional physical space.

6. Higher-Dimensional Space

The spatial domain is considerably more complex than the temporal domain for which the qualitative reasoning scheme originally was developed: the everyday concept of time is one-dimensional and uni-directional; if we give up uni-directionality but add orientation of objects in one dimension, we obtain 26 relations between two objects: each relation in Figure 3 obtains a counterpart where the two fishes face the opposite direction. When moving to 2- or 3-dimensional spaces, in theory more qualitative relations between two objects can be distinguished (compare Güsgen, 1989); however, it is important to determine which of those relations are meaningful in the specific spatial domain under consideration and at which stage of reasoning the distinctions between certain relations become significant. The combination of conceptually neighboring relations at certain stages of reasoning may simplify the representation and/or the reasoning processes considerably (compare Freksa, 1990).

More specifically, we must determine:

1) which type of objects do we want to represent
   a) only solid objects or objects which can penetrate each other?
   b) only convex objects or also concave objects?
   c) blocks, ellipsoids, sectors, or arbitrary shapes?
   d) the objects themselves or views of the objects (e.g. projections)?
2) which type of reference systems we need
   a) deictic, intrinsic (with reference to single objects), extrinsic (with reference to relations between objects)?
   b) are objects viewed as entities without orientation or as entities with orientation; what is the reference frame for the orientation?
3) which type of knowledge organization we need
   a) flat or hierarchical organization?
   b) level-independent or level-specific organization?

Depending on the specific decisions taken, different structures are advantageous and different complexities result. Hernández (1990) presents a representation scheme for 2-D projections of 3-D convex objects with intrinsic, extrinsic, or deictic reference system; the knowledge is hierarchically structured with a level-specific organization ("abstract maps").

7. Application Areas for Qualitative Spatial Reasoning

The paper started off with a motivation for developing spatial inference engines as rather universal vehicles for cognition. In the subsequent chapters we discussed issues specifically relevant to the spatial domain. Once one's attention is focussed on spatial reasoning it appears to be omnipresent. In our research at the Technische Universität München we are presently concerned with reasoning about *real* (i.e. non-metaphorical) space, specifically with description of spatial scenes and with interpretation of scene descriptions with reference to a representation of the scene. This type of reasoning is particularly relevant to route description and to navigation. Spatial planning is another area for which qualitative spatial reasoning is appropriate. Qualitative spatial reasoning also is useful for diagnosis of malfunctions in electrical circuits or in neurology and for Geographic Information Systems (GIS).

8. Acknowledgements

This research is supported by Deutsche Forschungsgemeinschaft under grants Br 609/4-2, Fr 806/1-1, and by Siemens AG. I acknowledge numerous discussions with members and students of the Artificial Intelligence/Cognition research group at the Technische Universität München and with participants of the Advanced Study Institute at Las Navas del Marqués.

9. References

Allen, J. (1983) 'Maintaining Knowledge about Temporal Intervals', CACM 26, 11, 832-843.

Block, N. (ed.) (1981) Imagery, The MIT Press, Cambridge, Massachusetts.

Freksa, C. (1980) 'Communication about Visual Patterns by Means of Fuzzy Characterizations', XXII International Congress of Psychology, Leipzig.

Freksa, C. (1990) 'Temporal Reasoning Based on Semi-intervals', ICSI TR-90-016, International Computer Science Institute, Berkeley, California.

Freksa, C. and Habel, C. (1990) 'Warum interessiert sich die Kognitionsforschung für die Darstellung räumlichen Wissens?', In Freksa, C. and Habel C. (eds.), Repräsentation und Verarbeitung räumlichen Wissens, Springer-Verlag, Berlin.

Güsgen, H. W. (1989) 'Spatial Reasoning Based on Allen's Temporal Logic', ICSI TR-89-049, International Computer Science Institute, Berkeley, California.

Hernández, D. (1991) 'Relative Representation of Spatial Knowledge: the 2-D case', in D. M. Mark and A. U. Frank (eds.), Cognitive and Linguistic Aspects of Geographic Space, Kluwer, Dordrecht.

Lakoff, G. (1987) Women, Fire, and Dangerous Things: What categories reveal about the mind, University of Chicago Press, Chicago.

Lakoff, G. and Johnson, M. (1980) Metaphors We Live By, University of Chicago Press, Chicago.

Reiter, R. and Mackworth, A. (1989) 'A Logical Framework for Depiction and Image Interpretation', Artificial Intelligence 41, 125-155.

RELATIVE REPRESENTATION OF SPATIAL KNOWLEDGE: THE 2-D CASE

DANIEL HERNÁNDEZ
Institut für Informatik
Technische Universität München
Arcisstraße 21
8000 München 2
Germany

ABSTRACT. There have been some straightforward efforts to extend Allen's interval-based temporal logic to spatial dimensions by using Cartesian tuples of relations (Guesgen, 1989). We take a different approach based on a study of the kind of information that best relates two entities in 2-dimensional space qualitatively. The relevant spatial categories turn out to be "projection" and "orientation." We define a small set of spatial relations and stress the importance of making their *reference frames* explicit. Furthermore, we introduce "abstract maps," an analogical representation that inherently reflects the structure of the represented domain, and demonstrate their use in spatial reasoning. This scheme also facilitates "coarse" reasoning and the hierarchical organization of knowledge. These representational issues form the basis for an experimental system to develop "cognitive maps" from 2-D scanned layout plans of buildings.

1. Introduction

Cognitive spatial concepts are *qualitative* in nature. Despite the widespread belief that geometric models are "better," in the sense of being more precise, this preciseness is not present (and not needed!) in cognitive models of space. Even more, due to the ***limited acuity of perception*** (Davis, 1989), geometric models might actually falsify the representation by forcing discrete decisions. We model the qualitativeness of "cognitive space" by using a ***relative*** representation of spatial knowledge based on ***comparative relations*** (over selected spatial dimensions) among objects, and between objects and distinguished reference structures (e.g., landmarks and boundaries). Such a relative representation avoids the falsifying effects of exact geometric representations by "not committing" to all aspects of the situation being represented. By doing so, however, it is also "under-determined" in the sense that it might correspond to many "real" situations. The reason why it still can be effectively used to solve spatial problems is that those problems are always embedded in a particular context. We claim that directly modelling this qualitativeness can lead to more intuitive user interfaces for applications such as Geographic Information Systems, as well as to more efficient ways to handle partial and uncertain spatial information—Freksa (1991) argues this convincingly.

D. M. Mark and A. U. Frank (eds.), Cognitive and Linguistic Aspects of Geographic Space, 373–385.

2. Relative Representations

In a much quoted paper, Allen (1983) introduced an interval-based temporal logic, in which knowledge about time is maintained qualitatively by storing comparative relations between intervals. There have been some straightforward efforts to extend this temporal approach to spatial dimensions by using tuples of relations (Guesgen, 1989). However, while the "cognitive plausibility" of Allen's approach carries over to the one-dimensional spatial case, it gets lost as soon as Cartesian tuples of relations are introduced to handle higher dimensions. People don't go around decomposing the world into three axes and then determining a qualitative relationship for an interval on each of them! We take a different approach based on a study of the kind of information that best relates two entities in 2-dimensional space *qualitatively*.

2.1. 2-DIMENSIONAL SCENES

The type of spatial scenes that we want to model are *2-D projections* of 3-D scenes, i.e., we allow overlapping objects[1] (see for example the layout plan of an office in Fig. 2a). Furthermore, we ignore shape by considering only either "convex" objects or canonically oriented "delineative rectangles" enclosing the objects to be represented. (Size also turns out to play only an implicit role in the kind of relations we consider.)

Determining relations in an n-dimensional space requires an "external" observer in the $(n+1)$st-dimension. This is also true for $n = 3$, since the passage of time, which constitutes the fourth dimension, is necessary to establish relations in 3-D space. For the 2-D case at hand, we assume an external observer that establishes the relations by looking "from above" (3rd dimension) at the 2-D situation. It is important to distinguish the external observer from a possible "point of view" embedded in the scene. Such points of view are used as reference frames in some types of relations (see deictic type of use below).

While in the temporal domain the "beginning" of an interval comes always before its "end" (due to the irreversibility of time), in the spatial domain (even in the one-dimensional case) beginnings and ends depend on intrinsic properties of the objects described and/or on the point of view of the observer. The issue of the reference frame of a spatial relation becomes crucial in two- and higher-dimensional spaces, as we will see shortly.

2.2. PROJECTION AND ORIENTATION

Our goal is to establish qualitative spatial relations between objects in a cognitively plausible way. When comparing the position of two objects in 2D-space there are two relevant dimensions: projection and orientation. By "projection" we mean how the boundaries of the objects relate. It comprises the relations `inclusion`, `overlap`, `tangency` and `disjointness` (Fig. 1). In a quantitative system one would use distance instead. Distance, however, cannot be determined qualitatively, i.e., by comparing just the two objects involved. It requires an external frame with respect to which the positions of the objects being compared must be established first. Thus, in a sense, projection is used in lieu of "qualitative distance." By "orientation" we mean where the objects are placed relative to one another and comprises the relations `front`, `back`, `left`, `right`, `left-back`, `right-back`, `left-front`, `right-front`[2]. Here `front`, `back`, `left` and `right` are taken to be abbreviations for "`front` but not `left-front` nor `right-front`" and so on. That is, `left-back`, for example, is not a finer distinction than say `back` or, with other words, all eight orientations have the same

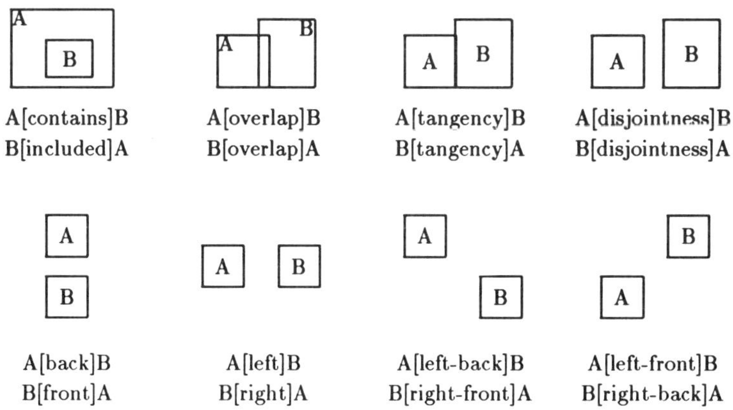

Figure 1. Projection and orientation relations

resolution. We do allow for various levels of granularity, both on projections and orientations, but in hierarchically organized homogenous layers (section 3.2). "Factoring out" orientation reduces the number of projection relations from the 13 used by Allen to five (`inclusion` splits in two—see below) and makes them (except for `inclusion`) *symmetric*, for example in 2-D: A [`overlap`] B = B [`overlap`] A.

`Inclusion` is special in several ways. First, it is asymmetrical, i.e., it does make a difference (in 2-D) if A is `included` in B or if A `contains` B, except, of course, in the case of equality (or rather "perfect overlap") of objects, which is treated here as "mutual inclusion." Second, it precludes orientation, i.e., in our framework if two objects are in an `inclusion` relation it is impossible to establish their relative orientation. Third and most important, it is the interface to the hierarchical organization of abstract maps, since all objects `included` in another one can be manipulated at a higher level by manipulating the parent object. For example, all the objects in a room inherit the orientation of that room with respect to others in the building.

2.3. REFERENCE FRAMES

We have to make explicit not only the *primary object* (the one to be located), the *reference object* (the one in relation to which the primary object is located), and the projection/orientation pair relating them, but also the *reference frame*, i.e., "the orientation that determines the direction in which the primary object is located in relation to the reference object" (Retz-Schmidt, 1988, 95). Given for example a sentence like *The ball is in front of the car*, studies of the use of projective spatial prepositions in natural language reveal three types of reference frames:

intrinsic: when the orientation is given by some inherent property of the reference object (e.g., w.r.t. car front)

extrinsic: when external factors impose an orientation on the reference object (e.g., w.r.t. actual direction of the motion of the car)

deictic: when the orientation is imposed by the point of view from which the reference object is seen (from *within* the scene, e.g., by an internal observer or from the speaker's point of view)

Criteria for determining the intrinsic orientation of objects and places are among others: the characteristic direction of motion or use, the side containing perceptual apparatus, the side characteristically oriented towards the observer, the symmetry of objects.

2.4. EXPRESSING SPATIAL KNOWLEDGE

Now, how are these relations used to express spatial knowledge? We maintain knowledge about the relative position of two objects that could be declaratively stated as follows:

<primary object, [projection,orientation], ref. object, ref. frame>

where the reference frame in respect to which the orientation is determined can be:

1. *implicit*, i.e., the intrinsic orientation of the parent object (the one that includes the reference object) is used as reference. For example, the orientation of the objects in an office is given implicitly by the intrinsic orientation of the office as a whole (which in turn could be determined by such factors as typical entrance or placement of windows).

2. *explicit*, i.e., of the form:

   {type-of-use ⟦, dex-relation ⟦,trans⟧⟧}

   An explicit reference frame must specify its **type-of-use**, which can be one of intrinsic, extrinsic or deictic as explained above, and optionally (indicated by double square brackets), if the **type-of-use** is not **intrinsic**, the **dex-relation**, which relates the external object or observer to the reference object in an intrinsic manner. A further **trans**-formation might be required, to determine the back/front sides of the reference object with respect to the observer ("mirror" transformation: fronts point to each other, backs point in opposite directions; "tandem" transformation: both fronts and both backs point in respective common directions).

The implicit orientation is used as the canonical reference frame. That is, for reasoning purposes, all relations between objects are converted first into the implicit form. This is the most natural representation for most applications (see example below). It corresponds, for example, to the act of an external observer of viewing the layout plan of an office from above such that the labels (assumed to convey the orientation of the office as a whole) are "right side up." On "larger scale" spaces the "preferred" orientation might be given by global directions (north, south, east, west).

2.5. EXAMPLE

As an example consider the following natural language descriptions of spatial relations (Fig. 2a/b):

(1) Petra's desk (A) is to the left of Daniel's desk (B). (Fig. 2b, view I)

(2) Petra's desk (A) is behind Daniel's desk (B) as seen from Christian's (D) office. (Fig. 2b, view II)

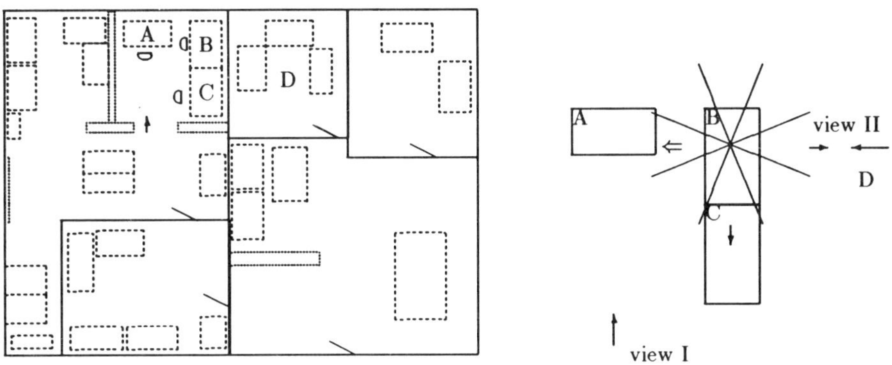

(a) Layout plan of an office (b) Two views on relevant relations

Figure 2. Example of 2-D projection of a scene

The relative spatial information contained in these sentences can be expressed declaratively as:

```
(1) REL1 = <A, [d,l], B>

(2) REL1 = <A, [d,b], B, {deictic, REL2, mirror}>
    REL2 = <D, [t,b], B, {intrinsic}>
```

In (1) no point of view is mentioned, so we assume the scene's implicit orientation which is given by the entrances and the position of the windows (view I in Fig. 2b). To see the difference between implicit (see above) and intrinsic orientations, note that A is to the intrinsic *front* of B, since B's intrinsic front as a desk is the side where the drawers open to or where one typically sits (see double arrow in Fig. 2b). However, if we were to store all relations with an intrinsic reference frame, we would have to transform back and forth between the intrinsic and the implicit orientations for most queries. In (2) a point of view is explicitly mentioned. Here an explicit reference frame is necessary to establish the orientation of the point of view with respect to the reference object in an intrinsic manner.

3. Abstract Maps

We could at this point use general constraint satisfaction mechanisms (see (Mesaguer, 1989) for an overview) to operate on this representation. Note, however, that the relations form a "structural domain": orientations form a circular structure, whereas projections form linear neighborhoods. We would like to use an internal representation that inherently reflects as much as possible the properties of the represented domain (analogical principle). We introduce "abstract maps" containing for each object in a scene a data structure called "rpon" (for relative projection and orientation node), which has the same neighborhood structure of the projection and orientation relations. Such a data structure can be *visualized* as an octagon-shaped figure (Fig. 3).

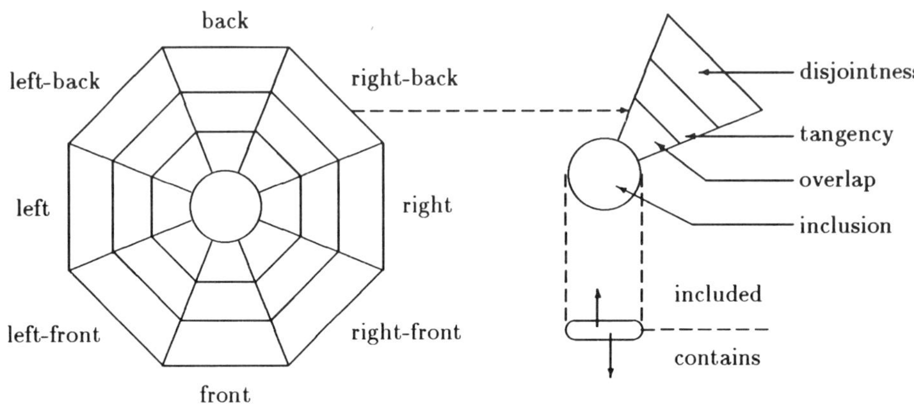

Figure 3. **R**elative **P**rojection and **O**rientation **N**ode

Except for the center, which conveys information on the `inclusion` relation, each of the eight areas corresponding to the eight orientations is subdivided in three slots corresponding to the symmetrical projection relations `overlap`, `tangency` and `disjointness`. A spatial relation between two objects is stored by creating a bidirectional pointer from the projection slot of the given (implicit) orientation of the reference object to the corresponding slot of the "inverse relation" in the primary object. It is important to remember, that rpons represent the neighborhood structure of the relations being used to represent space and *not* the metric of space itself. In particular, even though we tend to *depict* rpons in configurations that resemble the arrangement of the scene being represented (Figs. 4 and 5), every topological deformation thereof would be an equally valid visualization. That is, "internally" only the connections between slots carry information.

3.1. REASONING WITH ABSTRACT MAPS

The obvious advantage of these data structures is that, since they have the same structure as the relational domain they represent, operations such as a change in point of view (section 3.1.1) or the composition of relations (section 3.1.2) can be performed efficiently. The simultaneous effect of multiple constraints (section 3.1.3) is an important factor for the effective use of abstract maps in spatial reasoning.

3.1.1. *Changing the Point of View.* A change in point of view, which affects relations with an explicit deictic type of reference frame, can be easily accomplished by "rotating" the labels of the orientation with respect to the intrinsic one. In the example shown in Fig. 4, A is to the left of B (A [d,l] B) as seen from the original point of view (4a). If the point of view changes to be the one depicted in 4b, A will rather be considered to be left-back of B as indicated by the rotated labels. Note that all relations of a given object to others in the scene are updated simultaneously while keeping projections consistent. So, while B is to the back of C in 4a, their relationship is correctly relabeled in 4b to be "B is right-back of C." Furthermore, this can be done "in parallel" for all objects in a scene.

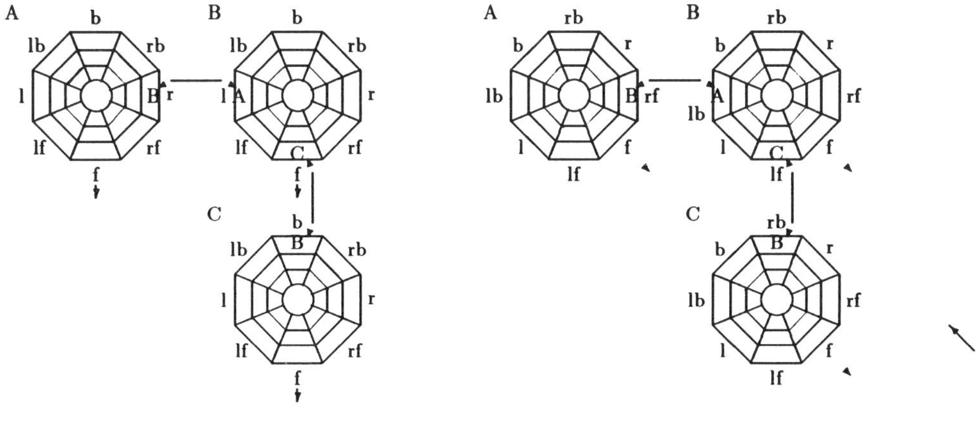

(a) Original point of view

(b) Rotation

Figure 4. Change in point of view as simultaneous rotation of labels

3.1.2. *Computing the Composition of Spatial Relations.* An essential operation in every relational representation is the composition of relations: Given the relation between A and B, on the one hand, and between B and C, on the other, we want to know the relation between A and C. For simplicity we will consider projections and orientations separately. We will first look at simple composition tables and discuss then how the regularities found there lead to efficient ways to compute the composition of relations using abstract maps.

Table 1 shows the projection compositions for an assumed constant orientation. For example, given the projection relations A [o,] B and B [o,] C (the orientation, say `back`, is assumed constant for both given relations *and* for the result!), which can correspond to any of the three (qualitatively) different square configurations shown in the right half of Fig. 5b, the resulting relation between A and C is A [{o,t,d},] C, as can be confirmed by looking at the figures (the notation {o,t,d} means that any of `overlap`, `tangency` or `disjointness` might be the case). Except for the relation pairs involving *only* {i,c} (upper left quarter of the table), there are several interesting regularities:

1. The table is "almost symmetric," i.e., given any pair of relations A [x,] B and B [y,] C such that not both x and y are elements of {i,c}, their composition A [t,] C is the same as the composition of A [y,] B and B [x,] C, except that if an i occurs in any of x, y, t it must be substituted by c and conversely c by i. Example: A [o,] B, B [c,] C $\longrightarrow$ A [{c,o,t,d},] C, so A [i,] B, B [o,] C $\longrightarrow$ A [{i,o,t,d},] C. (For this reason and to

A□B	B□C				
	i	c	o	t	d
c	{i,c,o}	{c}	{c,o}	{c,o,t}	{c,o,t,d}
i	{i}	{i,c,o,t,d}	{i,o,t,d}	{t,d}	{d}
o	{i,o}	{c,o,t,d}	{o,t,d}	{t,d}	{d}
t	{i,o,t}	{t,d}	{t,d}	{d}	{d}
d	{i,o,t,d}	{d}	{d}	{d}	{d}

Table 1. Projection compositions for constant orientation

improve the readability of the table, the vertical order of {i,c} is flipped with respect to the horizontal order.)

2. The relations in the composition are always "conceptual neighbors" (Freksa, 1990).

3. The number of relations in the composition tends to decrease from left to right on all rows except the first one.

Using these regularities, the computation of composition can be implemented as propagation of tokens in a rpon-graph. We will omit the details of the propagation algorithm here. The basic idea is to propagate tokens from both rpons for which the composition is being computed, along the pointers to common known relations, and according to well-defined heuristics, as suggested in Fig. 5a.

Table 2 contains the orientation compositions for disjoint objects (i.e., for constant projection = **disjointness**, which is also assumed for the resulting relation between A and C). Figure 5b shows an example: If A is to the **back** of B and B is to the **left** of C, A can be to the **back**, **left-back** or **left** of C as the three square configurations show. The regularities here are straightforward. Given A [,x] B and B [,y] C, the composition A [,t] C contains all orientations that are "in-between" and including x and y (see Fig. 3) on the "shortest path." The following examples illustrate this rule:

1. If $x = y$ then $t = x = y$, since there are no orientations in-between (distance between orientation = 0).

2. For $x =$ **lb** and $y =$ **r** we obtain $t =${**lb,b,rb,r**} (and *not* {**lb,l,lf,f,rf,r**} which are the orientations between x and y on the longer path "the other way around"). Here, 0 < distance < 4.

3. If x and y are opposites (e.g. **back/front**, **left/right**, **left-back/right-front**) then $t =$ {?}, that is *any* of the orientations apply, since both paths between opposites have the same length (distance = 4).

Again, the relations in the composition are always "conceptual neighbors." Furthermore, the composition is symmetric, i.e., t(x,y)=t(y,x).

Here, again omitting details, the idea is to find the orientations on the common rpon (B in Fig. 5b) according to the rules given above and then, if desired, propagate the result to the rpon representing the reference object for storage.

A□B	B□C b	lb	l	lf	f	rf	r	rb
b	{b}	{b,lb}	{b,lb,l}	{b,lb,l,lf}	{?}	{b,rb,r,rf}	{b,rb,r}	{b,rb}
lb	{lb,b}	{lb}	{lb,l}	{lb,l,lf}	{lb,l,lf,f}	{?}	{lb,b,rb,r}	{lb,b,rb}
l	{l,lb,b}	{l,lb}	{l}	{l,lf}	{l,lf,f}	{l,lf,f,rf}	{?}	{l,lb,b,rb}
lf	{lf,l,lb,b}	{lf,l,lb}	{lf,l}	{lf}	{lf,f}	{lf,f,rf}	{lf,f,rf,r}	{?}
f	{?}	{f,lf,l,lb}	{f,lf,l}	{f,lf}	{f}	{f,rf}	{f,rf,r}	{f,rf,r,rb}
rf	{rf,r,rb,b}	{?}	{rf,f,lf,l}	{rf,f,lf}	{rf,f}	{rf}	{rf,r}	{rf,r,rb}
r	{r,rb,b}	{r,rb,b,lb}	{?}	{r,rf,f,lf}	{r,rf,f}	{r,rf}	{r}	{r,rb}
rb	{rb,b}	{rb,b,lb}	{rb,b,lb,l}	{?}	{rb,r,rf,f}	{rb,r,rf}	{rb,r}	{rb}

Table 2. Orientation compositions for constant projection

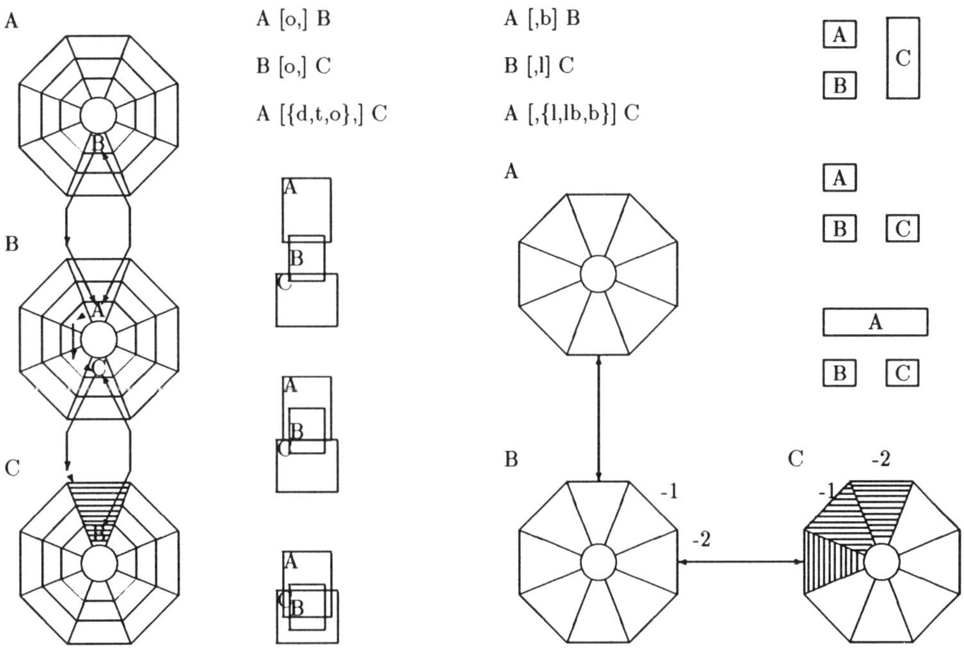

(a) Projections

(b) Orientations

Figure 5. Propagation of tokens in a rpon-graph

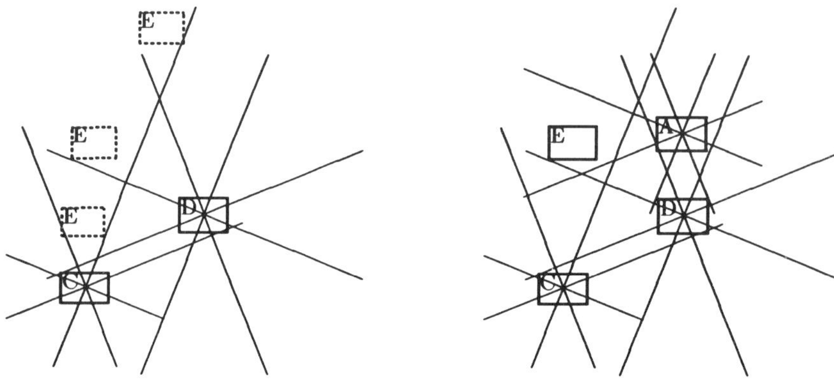

(a) E[]D ambiguous given C[]D, C[]E

(b) Adding E[]A further constrains E[]D

Figure 6. The effect of multiple constraints

3.1.3. *The Effect of Multiple Constraints.* Given the qualitativeness of the relations used, computing their composition seems to lead to a serious "information degradation" in the sense that the resulting relations are increasingly unspecific (see examples above). Note, however, that compositions are not computed in isolation from other relations possibly constraining the one under consideration. Consider for example the situation depicted in Figure 6a. Given the relation between C and D as C [d,lf] D, the fact that E is to the back of C (E [d,b] C) constrains the relation between E and D only to be "somewhere to the left or to the back" (E [d,{l,lb,b}] D). But if we additionally know that E is to the left of A (E [d,l] A), which in turn is to the back of D (A [d,b] D) then E is further constrained to be "left-back" with respect to D (E [d,lb] D).

In general the effects of multiple constraints can lead to an arbitrarily precise description of the relative positions of objects. Of course, the general constraint satisfaction problem (CSP) is known to be computationally intractable as the amount of computation needed to establish consistency in a network of relations grows exponentially with the number of relations. The point of introducing abstract maps with their analogical data structures is to reduce this computational overhead through a pre-structuring of the domain, that limits the number of irrelevant cases considered.

3.2. COARSE REASONING AND HIERARCHICAL ORGANIZATION

Abstract maps also facilitate "coarse" reasoning, i.e., using relations that make less distinctions either to cope with missing information or to simplify reasoning. We could for example only know that the primary object is on the `left` and has `contact` to the reference object. In that case it suffices to distinguish between `left` and `right` in the orientation dimension, and between `contact` and `no-contact` in the projection dimension. In our analogical representation this corresponds to using levels of rpons with less subdivisions (in the visualization of rpons, discs with less subdivisions are superimposed on the original octagon).

As was mentioned in section 2.2, these "levels of granularity" are organized hierarchically in homogenous layers. The projection levels are organized as follows:

disjointness	} no-contact	no projection info
tangency overlap inclusion	} contact	

Other organizations are possible, for example {o,t,d} = non-containement vs. inclusion or {i,c,o} = overlap vs. no-overlap (solids).

The orientation levels, in order of increasingly fine distinctions, are:

0. no orientation information
1. (a) back, front
   (b) left, right
2. back, front, left, right
3. back, front, left, right, left-back, right-back, left-front, right-front

Note that the meaning of `back`, for example, varies depending on the level of granularity it is used: If the only distinction being made at that level is `back/front`, then it means all locations "behind" a given object; if four orientations are available, it means "all locations behind a given object, which are not considered to be `left` nor `right` of that object"; finally, if eight orientations are available, it means "all locations behind a given object, which are not considered to be `left-back` nor `right-back` of that object." For our purposes it is not necessary to make finer orientation distinctions, even though some existing qualitative navigational systems have 16 or even 32 different orientations. Multiple constraints (section 3.1.3) allow an arbitrarily precise localization of objects without increasing the number of basic relations.

Spatial reasoning, both changing the point of view and computing the composition of relations, can be done at coarser or finer levels depending on the kind of information available. In particular, if only coarse information is available, the reasoning process is less involved than if more details are known. This is a very intuitive aspect of our approach, which distinguishes it from other frameworks (e.g., using value ranges or confidence intervals) in which less information means more computation.

4. Experimental System

These representational issues form the basis for an experimental system that we envision as follows: The system gets a 2-D scanned layout plan of a building (reduced to the "relevant" surfaces) and an initial position as input[3]. Note that the layout plan is used as a cheap replacement for an actual perceptual system and is not available as such for internal processing (otherwise we would already have a spatial representation!). Based on the layout plan, the first stage of processing generates the partial views—as seen by a hypothetical agent—thus simulating a robot's "perception" while wandering through the building. Those partial views are used next to create a relative, egocentric representation of the perceived room, a process that includes the recognition of spatial boundaries and exit points and the position of other surfaces relative to those boundaries or exits. This *egocentric* representation, in which all information is relative to an agent, is just the first stage in the process of building a "cognitive map" (Hart and Moore, 1973). In order to perform useful spatial reasoning, the agent must be able to transform this representation into non-egocentric forms: an *allocentric* one, in which spatial information is expressed relative to distinguished reference structures, and a *geocentric* one containing abstract topologic and metric relations in a coordinated system of reference frames. A common characteristic of all stages is the hierarchical clustering of relative information. In our setting, the most important kind of information going beyond single rooms is connectivity, which emphasizes that the way (including temporal order) in which we experience our environment strongly influences the representation we build thereof.

5. Notes

[1] Note, however, that in the projection it is not possible to distinguish which object is above the other one.

[2] The names of the relations were chosen purposely not to coincide with the spatial prepositions to which they are related, because there is more to the use of prepositions in natural language than what a straightforward mapping would suggest.

[3] Since the images consist mostly of straight lines and right angles, we assume the existence of simple recognition mechanisms (classic pattern recognition algorithms or diagrammatic methods such as Funt's retina (Funt, 1980)).

6. Acknowledgements

This research was supported by Deutsche Forschungsgemeinschaft under grant Br 609/4-2 and Siemens AG.

I would like to thank Wilfried Brauer and Christian Freksa for their comments and support. I also received helpful hints from Scott Freundschuh and Hansgeorg Schlichtmann.

7. References

Allen, J.F. (1983) "Maintaining Knowledge about Temporal Intervals", Communications of the ACM 26(11), 832–843.

Davis, E. (1989) "Solutions to A Paradox of Perceptual Acuity", in Proceedings of the First Conference on Theoretical Aspects of Knowledge Representation, Toronto.

Freksa, C. (1990) 'Temporal Reasoning based on Semi-Intervals', Technical Report TR-90-016, International Computer Science Institute, Berkeley, CA.

Freksa, C. (1991) "Qualitative Spatial Reasoning", in D. Mark and A. Frank (eds.), Cognitive and Linguistic Aspects of Geographic Space, NATO Advanced Studies Institute.

Kluwer Academic Publishers, Dordrecht.

Funt, B. V. (1980) 'Problem Solving with Diagrammatic Representations', Artificial Intelligence 13(3), 201–230.

Guesgen, H.W. (1989) 'Spatial Reasoning based on Allen's Temporal Logic', Technical Report TR-89-049, International Computer Science Institute, Berkeley, CA.

Hart, R. and Moore, G. (1973) 'The Development of Spatial Cognition: A Review', In R. Downs and D. Stea (eds.), Image and Environment: Cognitive Mapping and Spatial Behaviour, Aldine Publishing Company, Chicago.

Mesaguer, P. (1989) 'Constraint Satisfaction Problems: An Overview', AI Communications 2(1), 3–17.

Retz-Schmidt, G. (1988) 'Various Views on Spatial Prepositions', AI Magazine 9(2), 95–105.

MATCHING REPRESENTATIONS OF GEOGRAPHIC LOCATIONS

MATTHEW McGRANAGHAN
Department of Geography
University of Hawaii
Honolulu, Hawaii 96822 USA

ABSTRACT. Many references to geographic locations are made linguistically, using words to describe the location in terms of objects in the landscape and the relations among them. In order to attach coordinates to these locations, one must be able to identify them in a database which contains representations of the objects and relations as well as coordinates for these features. This article assesses the correspondence between the features used in describing locations in an herbarium collection and those represented in standard cartographic databases. An alternative representation for geographic data is suggested.

1. Introduction

> "...in general, there is a basic level of specificity of description. With objects, this is Rosch's "basic level"; there seems to be the equivalent of a basic level for spatial relations in scenes, but since scenes and situations are much more difficult to circumscribe than objects, their categorization properties have not been studied." (Herskovits 1986, 113)

This chapter is concerned with automating the conversion of descriptions of geographic locations to systematic coordinates. This facility would speed the integration of a large volume of non-coordinate-referenced spatial data, such as is now held in natural history collections, with geo-referenced data sets. The task is essentially one of matching geographic patterns found in the descriptions with those recorded in a geographic database. While this may seem to involve machines "understanding" descriptions, and indeed, it is convenient to use that metaphor, Winograd and Flores (1986) argue convincingly that computers, are symbol processors, and that any illusion that they "understand" is due to sufficiently manipulable symbols retaining meaning for humans. The importance of representation is clear.

Interpreting a description of a location requires establishing correspondence between two representations of the Earth's surface. One representation is the database of the world, which presumably records all relevant terrain features, including objects, locations, relations, and other attributes. Most typically these are digital recodings of maps; i.e., abstractions from representations which are themselves abstractions. The intent with such data sets has often been to record maps in ways that support subsequent display, perhaps to simply reproduce the original map, or consistency checking to ascertain that all elements of the original map have been encoded.

D. M. Mark and A. U. Frank (eds.), Cognitive and Linguistic Aspects of Geographic Space, 387–402.

This has usually included recording gross feature categorizations and geometric positions from existing map manuscripts.

The other representation is the text describing a specific place, usually from the perspective of someone in the field, by reference to salient landmarks and the visible relations among them. This is based on the less abstract experience of the geography. Establishing correspondence between these representations requires matching patterns of objects and relations. What are the objects and relations that need to be matched, and how should they be represented?

This chapter compares the content of existing cartographic databases with the kind of data required for more human-oriented spatial reasoning. The primary question addressed here is whether current cartographic data bases, such as those distributed by the U. S. Geological Survey, are appropriately rich in detail to facilitate automated comparison with textual descriptions. The approach is to by examine what theory in spatial cognition and practice in herbarium records suggest will be the important features and relations to represent, and compare these to the what is represented in available data sets.

Part of the answer lies in understanding how people think about geographic space. Literature on cognitive categories and linguistic representations of space are particularly relevant. Studies of cognitive categories (Rosch 1973, 1978; Lakoff 1987) reveal the non-Aristotelian nature of mental categories and support the notion of basic-level categories, offering a default level of abstraction for reasoning and discourse about objects and relations. Herskovits (1986) categorized the spatial relations represented by several common prepositions. Her "use types", proffered as characterizations of spatial relations upon which one might base computation, and her recognition that the archetypical meanings of locative phrases are "stretched" for use in any instance are very useful in the present context.

Finally, a style of representation that might be appropriate for interpreting location descriptions is discussed. Many authors (Minsky,1974; Havens and Mackworth, 1983; McGranaghan, 1985; and Shapiro and Rapaport, 1987, for examples) have suggested representation schemes that could support the pattern matching requirement suggested here. Indications are that future cartographic data sets will have richer structures, intended to model geographic more than map reality.

2. Background

The context for this work has been described in detail elsewhere (McGranaghan and Wester, 1988; McGranaghan, 1989a, 1989b). In brief, a set of botanical specimens had been collected at points, the locations of which were recorded in textual descriptions. The descriptions neither included geographic coordinates, nor followed a specified grammar. We were asked to map these locations using standard computer cartographic software. This required that coordinates be derived for the collection site points. Lyndon Wester and Warren Wagner spent considerable effort in manually interpreting the descriptions and in the process we recognized that this process should be automated.

2.1. LABEL CONTENTS

One thousand three hundred forty location descriptions were drawn from specimen labels in the Bishop Museum's Herbarium Pacificum. Each label contains an accession number, identification (genus, species, and sub-specific taxa), collector, preservation method, and a description of where the specimen was collected. This description is usually terse but typically includes: the name of a country (or island, as in Hawaii), a locality (the name of a physical feature or land division), the site's approximate elevation, and, on about 75% of the labels, more detailed locational information. Often it contains directions for reaching the site. Prepositions, proper names of features and places, generic feature names, metric relational terms, and non-metric relational terms figure heavily in such descriptions. Habitat information such as ground cover, soil type and moisture, is also often present. The labels do not contain site coordinates.

2.2. TWO APPROACHES

Two approaches to automation have been examined (McGranaghan, 1989b). The first was a brute force name matching algorithm which relied on matching place names from labels with those in a digital gazetteer (the Geographic Names Information System described below). The second attempted to use grammatical structure in the descriptions to gain meaning beyond the bare place names. Each of these approaches has some merit but neither is completely satisfactory.

2.3. SOME LESSONS

There are a number of difficulties with these approaches. Obviously, if the gazetteer database does not contain names from the label, there is no way to tie the site to a position. The grammar-based approach might use more of the label's information, but the current parser does not differentiate between prepositions. As we will see shortly, identifying a particular preposition points to a family of relations rather than a specific spatial relation; another knowledge base is needed to recognize the relations prepositions encode.

Preposition use involves knowledge of spatial relations and objects. In everyday discourse there is considerable redundancy in using prepositions due to our expectations about the possible relations among the objects. In the examples below, it is easy to fill-in the appropriate leading preposition when we know what the features are:

__?__ Palolo Valley
__?__ Koolau Mountains
__?__ summit ridge
__?__ Nuuanu Pali
__?__ Waialae Iki

but more difficult when the objects are not familiar. Knowing that "Pali" means "very sharp ridge" makes the use of "on" in the fourth example much more likely. The use of "at" in the last example (without knowing what "Iki" means) would indicate that Waialae Iki is generalizable to a point.

We have a sense that in discourse: 1) we communicate about things and the relations among them; 2) often, prepositions code the relations, but knowledge of objects also informs us of the possible spatial relations between things; and 3) there is a need for sufficiently expressive representations for objects and the relations among them. As we shall see, it is not clear that existing databases store the correct knowledge for supporting human-like reasoning about geographic objects and relations.

3. Stores of Geographic Knowledge

The typical location specifying content of an herbarium label was introduced in the previous section. In this section I consider two other stores of locational information. The section opens with a brief view of spatial cognition and mental maps. These are likely to be closely related to the herbarium label contents. Then the remainder of the section introduces digital cartographic data sets currently available from the USGS.

3.1. MENTAL REPRESENTATIONS OF GEOGRAPHIC SPACE

There is considerable evidence that people think about geographic space in terms of the objects in it and the relations among them, rather than in terms of systematic coordinates (Lynch, 1964; Downs and Stea, 1977; Talmy, 1983, 1987, 1989; Herskovits, 1986). Some places have salience as points or areas and serve as landmarks, these are joined by paths and separated by edges. A mental map can be constructed as a network of these objects and relations. A position can be specified by naming a landmark and then adding offsets and constraints to the description. The construction and use of cognitive models of space based on this concept has proved useful in geography, psychology, cognitive science, AI and linguistics (see for examples: Kuipers, 1978; Elliott and Lesk, 1982; and Goldin and Thorndyke, 1982). Whether or not peoples' mental representations of space are ultimately network-like views, there seems to be sufficient reason to include both objective geometric representations and cognitively salient topological relations (or the ability to compute one from the other) in the database against which locations will be matched.

Lakoff (1987) provides an excellent synthesis of recent work on mental categories. The arguments that individuals organize knowledge in structures he terms idealized cognitive models (ICM) rather than in structures like those of classical logic and set theory are compelling. ICMs are complexes of propositional structures, image-schematic structures, metaphoric mappings, and metonymic mappings (Lakoff, 1987, 68). They do not follow the rigid hierarchy of Aristotelian classes, but rather, allow fuzzy and overlapping categories. These structures plausibly support effects like Rosch's "basic level categories." In addition to being psychologically satisfying, ICMs are computationally tractable.

The forgoing indicates that people think about space in a ways that are at once concrete and abstract. The things that we think about, both the objects and the relations among them, tend to be given, and probably seem clearly evident in the environment. But they are not easily amenable to definition by ridged rules into cleanly distinct categories. To give an example, it is very easy to conceive of a “hill” feature or a “mountain” feature, but it is difficult to say exactly what would differentiate the two. How are these ambiguities handled in standard digital cartographic data sets?

3.2. DIGITAL CARTOGRAPHIC DATA BASES

The United States Geological Survey (USGS) has produced four data sets, Digital Line Graphs, Digital Elevation Models, Geographic Names Information System, and the Land Use and Land Cover database, which are distributed for the by the National Cartographic Information Center (NCIC). These are representative of the types of geocartographic databases available today. They record a combination of geometry, topology, classification, attributes, and feature names, intended for the most part to support mapping. They emphasize geometric relations, and exploit topological relations more commonly to check data consistency than to guide spatial search. In this paper we will be concerned with three of these data sets.

3.2.1 *Digital Line Graphs.* The USGS Digital Line Graph (DLG) files are the the first high resolution digital cartographic database to cover the entire US. They record the planimetry and categorization of roads, transportation facilities, hydrography, cultural features and boundaries. The intent has been to encode the information from the maps. Features are conceptualized as points, lines or areas at the fixed scale of the map. Positions are given as geometric coordinates, and the topological relations of the planar graph of the map objects is recorded explicitly. An object, i.e. the abstract geometric entity, may have several uses or functions. For instance, the object may be both the center line of a stream and a boundary between municipalities. In general, each distinct map symbol (not each occurrence of the symbol) will have a unique code.

The DLG feature codes are hierarchical. A two-level hierarchy provides both coarse and very fine categorization. The major code divides features into the twenty broad categories listed in (USGS, 1985). A major code of 030 indicates a stream while 100 indicates a road. Within each of these, minor codes distinguish finer categorical division. Within the 030 category, a minor code of 0201 indicates an intermittent stream and a minor code of 0203 indicates a perennial stream. These feature codes were revised in 1983 (USGS, 1985), and may be changed again for DLG-E, an enhanced version of DLG (Guptill, 1991), but they remain of interest because of current availability and the inertia that will instill.

3.2.2. *Digital Elevation Models.* Digital Elevation Models (DEM) record terrain configuration as surface elevations (spot heights) on a regular grid. Geometric position in three dimensions is explicit and the topological relations between adjacent points are implicit in the regular grid structure. They do not identify, name or label parts of the surface. They only record positions in three dimensions.

Digital Elevation Models (DEM) are typically used to calculate slope, height, aspect, and volume (as for cut & fill operations). They are also used to generate contours, but they represent the terrain in a relatively sterile, geometric form. There are no features per say in a DEM. That is, none are given explicitly in the data base, but there are a number of terrain features (peaks, ridges, mountains, draws, spurs, saddles, etc.) that could potentially be recognized in the elevation grid.

3.2.3. *Geographic Names Information System.* The Geographic Names Information System (GNIS) relates a name, a feature class, latitude longitude coordinate pairs, and other data for places, features, and areas which are identified by a proper name on USGS 1:24,000 scale topographic maps. In later versions of the data, historical names are included.

GNIS place names and coordinates are clearly useful search keys, but the feature class codes are also valuable in abstract knowledge representations. There are sixty three feature classes (see USGS, 1987). Each feature class has a name, a definition, and usually a list of terms that are either synonyms, or members of the named class. Examination reveals that the categories are not classically hierarchical. They are not mutually exclusive, not exhaustive of common terrain features, and not at the same level of categorical abstraction.

The categories differ widely in their number of members. Geyser, arch and oil field are defined very narrowly and have neither synonyms nor class members while "summit" has 26 synonyms and/or class members (including mountain, table, and rock) and valley has eleven (including gulch, gulf, chasm, ravine, cove, and draw).

There are a number of cases in which GNIS categorical hierarchies seem to be distorted. The term "mountain" is in a class called "summit", even though it seems to be the basic-level term and "summit" a subordinate class. Similar situations are found in several other classes. The "beach" category includes "coast, "shore" and "strand". "Bar" includes "ledge", "reef", "sandbar", "shoal", and "spit". "Coulee", "draw", "gully" and "wash" are all members of "arroyo".

Categories overlap. Both the "valley" and the "bay" categories have "gulf" and "cove" as members. "Stream" and "gut" both include "creek". "Rock" is in both "island" and "summit". "Lake" is defined as a class of natural inland water bodies, but also is a member of the "reservoir" class, where it is defined to be manmade.

Several common geographic features are not included in the data set. These include: road or highway, desert, dune, drumlin, esker, kame, kettle, and moraine. The category "area" serves as a catch-all for "any of several areally extensive natural features not included in other categories" and includes "delta". "Desert" is not included elsewhere so presumably would also fall into this category.

4. Representation of Terrain Features

4.1 MENTAL REPRESENTATIONS

The spatial cognition literature suggests the need to represent terrain features that are landmarks, paths and areas, as a minimum. Introspection (admittedly suspect) provides a relatively small set of basic-level terrain features. Perhaps: mountain, valley, ridge, stream, road, shore, ocean, lake, and flat would suffice. Each of these terms refers to a set of objects that admit of a range of differences, but are related by some conceptual structure.

4.2 HERBARIUM LABELS

The Herbarium Pacificum labels refer to a slightly different set of features. The most frequently mentioned of which are (with number of uses): trail (120), mountain (114), range (106), gulch (62), valley (56), ridge (54), stream (54), slope (47), forest (43), head (32), summit (28), fork (21), reserve (21), bottom (20), bank (18), pu (hill) (21), branch (9), cliff (8), pali (8), ravine (5), ditch (5), gully (3), gap (3), gorge (2), declivity (2), and road (2). Several terms (pool, border, flume, peak, tributary, and rim) were used only once in the sampled labels . These features, though nowhere defined in the labels - and only presumed to be understood, should be represented in any database to which one would like to register the descriptions. The following section considers how well represented these features are in standard digital cartographic data bases.

4.3. DIGITAL CARTOGRAPHIC DATA SETS

4.3.1. *Digital Line Graph Features.* The list of base categories in (USGS, 1985) is reasonably close to a set of basic-level landscape elements except for lacking the surface configuration features (e.g., mountains, valleys and slopes). It makes cognitively satisfying distinctions, such as those among hydrographic features, in separating linear water bodies, typically non-linear ones, inland wetlands and coastal features. Similar comments apply to the transportation categories.

The minor codes make finer distinctions than are usually found in the herbarium labels. However, this level of detail could be used to aid reasoning about locations.

4.3.2. *DEM Features*. Representation of terrain features in a DEM depends on the interaction of the grid spacing and its placement with the feature's size and location. If a terrain feature is large enough that is sampled in the elevation grid of a DEM then it is represented. The problem is recognizing that it is there and associating a name with it. There is some progress in this area but, so far, results seem more suited for specific purposes than for general feature extraction. Chen and Guevara (1987), Haralick and Shapiro (1980), Mark and O'Callaghan (1984), and Riazanoff *et al.* (1988) have discussed extracting terrain features from digital elevation models based on configurations of elevations. Typically, these studies have dealt with course lines and ridges. Frank *et al.* (1986) describe attempts to formally define terrain features as patterns in elevation data.

The techniques required to find such patterns are computationally intensive, not involving simple item to item comparison (as with search keys) but rather requiring comparison of assemblages of configurations of values. They are made even less attractive by the difficulty in defining features in terms of such patterns. Defining "mountain" such that: 1) it can be recognized as a pattern of elevations, and 2) people would generally agree that objects selected by the criteria are in fact mountains, is not easy. Determining the spatial limits of a "valley" involves determining where one shifts from being "in the valley" to being "on the ridge". Still, because several terrain features are not represented in any of the other standard data sets but are derivable from DEM data, it is clear that these data and techniques must be included in interpreting locality descriptions.

4.3.3. *GNIS Features*. The categories for terrain features used in GNIS are not as cleanly organized as those in DLG data. The configuration of the terrain is not recorded, but the names of a number of important classes of terrain elements are related to coordinates. This suggests that for symbolic processing, GNIS does provide useful information.

There is fair intersection between the types of features named in the herbarium labels and used as (or in) GNIS categories. Trail, valley, ridge, slope, stream, spring, summit, cliff, falls, woods, forest and reserve are all used in common, though the intended meanings may be different in the two uses. It is interesting to note that "slope" is often present in the labels and appears as a GNIS feature class but that it is used to classify only two objects in the GNIS data file for Hawaii.

5. Representation of Locative Relations

5.1. MENTAL REPRESENTATIONS

The set of relations used to specify relative positions has been studied by a number of authors. Several studies have deduced minimal sets of terms purported to describe all of the spatial relations two objects might possibly share. Freeman (1975) suggested the following as a complete set of spatial relations for objects in an image:

left of	right of
beside	above
below	behind
in front of	near
far	touching
between	inside
outside	

This list was modified for geographic space by Peuquet and Zhan (1987) to include the cardinal directions (north, south, east and west). The problems of defining these relations from coordinates has been discussed by Peuquet (1986, 1987). This approach to identifying a sufficient set of spatial relations has been discussed by Mark and Frank (1989).

Herskovits (1986) indicates that there are relatively few schema, or prototypical relations, to which a locative description might refer. The prototypical relation is extended ("stretched") for use in any instance, and reasoning based on the prototypical situation is tempered by context. She provides analyses of the relations “at,” “in,” and “on,” as well as of a set of projective relations that closely resembles that from Freeman (1975).

Use types (Herskovits 1986, 86 *et seq.*) are the conceptual relations that might be signaled by a preposition. In each of these, the position of some object of discourse is given relative to some other ground object. Herskovits clearly intended that the use types should capture the meanings of relations in a way that was amenable to machine representation and processing.

5.1.1. *At.* The ideal meaning of "at" is for a point to coincide with another, Herskovits (1986, 128) identifies eight use types (see list below) which seem at plausible.

Use Types for "at":

Spatial entity at location
Spatial entity "at sea"
Spatial entity at generic place
Person at institution
Person using artifact
Spatial entity at landmark in highlighted medium
Physical object on line and indexically defined crosspath
Physical object at a distance from point, line, or plane

5.1.2. *On.* Herskovits (1986, 140) gives "on" the ideal meaning:

> for a geometrical construct X to be contiguous with a line or surface Y; if Y is the surface of an object O_y, and X is the space occupied by another object O_X, for O_y to support O_X.

and identifies these use types:

Spatial entity supported by physical object
Accident/object as part of physical object
Physical object attachcd to another
Physical object transported by a large vehicle
Physical object contiguous with another
Physical object contiguous with a wall
Physical object on part of itself
Physical object over another
Spatial entity located on geographical location
Physical or geometrical object contiguous with a line
Physical object contiguous with edge of a geographical area

5.1.3. *In*. "In", denotes "inclusion of a geometric construct in a one-, two-, or three-dimensional geometric construct" Herskovits (1986, 149). Use types are:

Spatial entity in container
Gap/object "embedded" in physical object
Physical object "in the air"
Physical object in outline of another, or of a group of objects
Spatial entity in part of space or environment
Accident/object part of physical or geometric object
Person in clothing
Spatial entity in area
Physical object in a roadway
Person in institution
Participant in institution

5.1.4. *Projective Prepositions*. Herskovits' projective prepositions (Herskovits 1986, 156) capture the sense of how bodies might be related to each other in geographic space.

Horizontal Plane
(at/on/to/by) the (left/right) of
(at/on/in/to/by) the (left/right) {hand} side of
(at/on/in/by/to) the (front/back/side) of
in (front/back) of
before/behind
(right/back/left) of

Vertical Order
above/below
over/under
(at/on/in/by) the (top/bottom) of
on top of

5.2. HERBARIUM REPRESENTATIONS

This section examines the contents of the herbarium labels in light of the "use types" proposed by Herskovits (1986). Recognizing the use of implicit but tacit prepositions (see the discussion and examples above relating to the robustness of our discourse understanding), the following discussion treats explicit uses of these symbols. Do Herskovits' characterizations of "in," "on," "at," and the projective relations correspond to the uses found on the herbarium labels?

5.2.1. *At*. The word "at" was used thirty-five times in the sampled labels. Several uses were not easily assigned to one of the use types and three of the use types ("entity at sea", "person at institution", and "person using artifact" were not used at all.

The most common use type was the "object at a distance (from a plane)" construct. Thirteen of the uses specified an elevation, as in "at an elevation of 1100-1500 ft.," or more succinctly, "at 500-700 feet." This use type is not altogether separable from the "object on line and indexically defined crosspath" uses such as, "Waikane Ditch Trail at 750-900 ft." It seems that some object knowledge is needed to separate them.

Several constructions were difficult to assign to use types. "At intake #24", refers to a landmark (but not in a highlighted medium) or to a location (but not specified in a coordinate system), signalling either the "spatial entity at landmark in highlighted medium" or the "spatial entity at location" construct.

"At head of Punaluu Valley", "at the edge of {trail/stream}", and "at the edge of {an area}" each involve both the "at generic place" and the "landmark in highlighted medium" constructs, but also seem to indicate projective relations which are not in Herskovits' set. The former, because head is generic and not really a landmark in the sense of being a named object in its own right, and the latter because "head" is an identifiable part of the medium (valley).

Examining it in terms of the projective prepositions seems reasonable (it is tacitly "at the head of ...") reveals that "head" in this construction is very similar to the "front/back/top/side" signifying object parts (Herskovits 1986, 175). Herskovits stipulated that object parts be defined geometrically, but in this construction, the object part is a functional or morphological unit.

5.2.2. *On*. "On" was used 77 times in the herbarium sample. Herskovits includes two use types explicitly mentioning geographical locations and geographical areas, but surprisingly few of the herbarium's ostensibly geographic descriptions fit easily into one of these categories. Only one unambiguously specified "on a geographic location" ("Along Castle Trail, on Koolau Range"). Three of the use types, "object as part of physical object", "object transported by a large vehicle", and object on part of itself", were not used at all.

The most common use of "on" was to indicate an object (terrain feature) which literally supported the plant (and the collector), the "supporting object" schema. Most commonly (27 occurrences) this was "on {some kind of} slope". (There were two "on bank", uses that could be lumped with the "on slope" phrases if banks are taken to be a class of slopes.) This was so frequent as to lead one to suspect that "slope" is a basic-level category of terrain features. "On {some kind of} ridge", was used only seven times - surprising given the placement of trails and access in Hawaii. Five specimens came from "on the summit [or crest, or top]," where the feature was named in a separate phrase (for example, "Nuuanu, on the top") making the meaning very similar to a projective use type.

The next most common use was projective. "On {some kind of} side of {something}" describes 19 of the uses of "on". These were most often on the side of a valley, but about a third of the time were "contiguous with the edge of a geographical area." The "contiguous with a line" use type had only ten cases. There were four cases where "contiguous with a wall" constructs were used to indicate being against the (very steep) side or head of a valley.

5.2.3. *In*. "In" was used 81 times in the data set. As with "on" and "at", several of the Herskovits use types are clearly represented and, others are not. The 'in a container' structure is the most common. Taking gulch (18), valley (6), gully (3), draw (1), ravine (1), crevice (2) and ditch (2) together as bowl-like containers, there were thirty-three uses in this category.

The "object in the outline of another" construct was also very common. There were 16 "in" {some kind of} woods, and 13 specimens "in" {some kind of} forest. Four of these being "in" rain forests, two "in" thickets, and one was "in [a] very deep dark forest". Often the community composition (kukui/guava/ohia) was included as a modifier. This construct would also seem to be the best placement of "in rocks", as the plants were almost certainly among rather than inside the rocks.

The "spatial entity in area" captures most of the remaining cases. Four cases were explicitly "in {some kind of} area", four were "in kipuka (remnant patches)", one "in Forest Reserve", one "in pasture", and three "in lava flows." "In shade" (two cases) also seems to fit this construct, although not easily.

One construction resonates between broadening the "object in roadway" construct to become an "object in corridor or conduit" construct, and forcing a stream to be considered as an areal, rather than a linear feature. "In stream" could perhaps be conceptualized as a plant as an obstacle in the path of the flowing water (see Herskovits 1986, 154). It also could be conceptualized as the stream being a container for the plant.

5.2.4. *Projective Relations.* Somewhat surprisingly, there is a real bias for plants to come from the nether regions of areas. Ten specimens came from "back of {a valley, a farm, or a building} but none came from "front of" anything (though, as noted above, several were collected from "head of {usually a valley}"). "Before" and "behind" behave similarly. No plants were "before" anything, while three were from "behind {something}.

Left and right are not used much in this sample. There was one "to right of {a sign post}". Most of the uses of "left" were due to one collector who used both "from left roadside" and "left of palm" in each of the labels for four specimens.

Projective use of cardinal directions was also somewhat less frequent than one might anticipate in such a geographical context. There were 9 "east of"s, 1 "west of", 7 "north of"s, and 1 "south of". Even including the Hawaiian terms for directions in the local polar coordinate system does not capture many more uses; there was one "mauka of" and no uses of "ewa", "makai", or "Diamond Head" as directions.

The exercise of trying to use Herskovits' use types to categorize locative propositions on the herbarium labels was instructive. On reading through them, the use types seem plausible, and indeed almost all of the uses in this database can be fitted into one of the use types. But the exercise also showed that even in this restricted domain, where intent is closely circumscribed, there is ambiguity. Generally, the use types captured the sense I understood in the phrases but occasionally it was difficult to decide which one was being used. A description understanding system would either need some further specification for determining classes or would need to be able to deal with mutually overlapping categories.

5.3. DIGITAL CARTOGRAPHIC DATA SETS

None of the digital cartographic databases explicitly encodes rich inter-feature relations. Only the DLG data sets record inter-feature relations of any kind. There data contain some topological information from which adjacent regions and successive segments along a linear feature can be computed. Further, containment relations could be calculated for the objects which are defined as polygons in this data set.

DLG, DEM, and GNIS all use coordinates to locate features. From these it is possible to compute many distance and direction relationships. Unfortunately, establishing correspondence between these computed values and linguistic categories is not straightforward (Robinson and Wong, 1987; Peuquet, 1986, 1987; Peuquet and Zhan, 1987). From the herbarium record, it seems that getting at cardinal directions, though easy, is not all that useful. The tolerances required to interpret "along," "at," "near" and the like are still mostly unknown. None of these data sets allow easy determination of "the ridges that surround Palolo Valley." Similarly, "which slopes are along Waiomao Stream" is not directly answerable. Yet, it seems that a geographic database should allow very easy answers to these questions.

6. A Different Approach to Geographic Databases

6.1. FRAMES AND NETWORKS OF GEOGRAPHIC OBJECTS

Is there a representation that could make salient objects and relations directly accessible? Minsky is credited (Winston 1977, 179 *et seq*) with the idea of frames, basic structures for representing stereotypical situations which are adapted to represent specific instances of those situations. Frames are sometimes thought of as collections of "slots" and the values which occupy them. The slots may be occupied by other particular values, sets of values, or even other frames. Facts, i.e. knowledge, can be represented in networks of frames.

Mackworth and Havens (1983) describe Mapsee2, a system designed to interpret visual scenes (geographical sketch maps) in terms of hierarchical networks of schemata (frames). This approach, they argue, avoids both representational and computational problems associated with network consistency based approaches. Each schema represents a class of objects which has generic properties and possible relations to other schemata. A scene interpretation is then built as a composition of simpler schemata. A complete representation of a scene might then be composed of (decomposed into) schemata for river systems, road systems and a mountain range. Each of these schemata is in turn composed of (decomposed into) other schemata. Instantiation of a schema can be driven from the bottom up, from the top down or in a hybrid mode (Havens and Mackworth 1983, 95).

Although they do not stress the point, in the spatial context, connectivity and other relations to other schema could be considered and used to build geographic regions. For instance a valley schema could be linked to surrounding ridge schemata and [possible] mouth schemata. Regions built in this way would be consistent with multi-scale phenomena in mental map use, as in route planning/selection where space is chunked at various resolutions for different (parts of) reasoning processes.

McGranaghan (1985) suggested that a list-based data structure might be used to represent any geographic object. This would use slot-value pairs to represent an entity with an arbitrary set of attributes and connections to other entities.

```
(ENTITY name-id
    (TYPE (of_entity (modifier))
    (IN entity)
    (ALONG linear_entity_identifier)
        (BEFORE entity)
        (PAST entity)
        (NEAR entity)
    (CONTAINS list_of_component_entities)
    (NEIGHBORS (entity direction distance)..()..())
    (ORIENTATION (directional_trend_of_entity)
    (MIN_RECT \)
    (COORDS list_of_x,y_coordinates)
    (FUNCTIONS list_of_activities_at_entity)
    (HAPPENED_THERE list_of_events)
    (INTENT goal))
```

A more developed version of this idea has been presented in (Buisson, 1989). There, an object-oriented database was created to support spatial reasoning in a GIS. The slot values could either be given explicitly or could be inferred as needed by pattern matching within the database or by computation.

This type of representation for geographic objects could be modified to include other or variable relations. One advantage of labeled slots over data in which position in a file defines how a value is interpreted (as in GNIS, DLG, and DEM formats) is that it is not necessary to anticipate all possible future slots when designing an object's representation.

6.2. SCHEMA AND OBJECT MATCHING

Given data representations which were structured more around the geographic than the map objects, the problem of finding the coordinates of a described place could be translated into the problem of matching patches of space represented as a net of geographic entities, their properties, and the relations among both. A geometrically constrained, propositionally structured representation might start as a set of all the meaningful terrain units in a region. These would be related by their adjacencies, by functional connections, and by metric relations. Nodes in this net might have terrain unit classifications, coordinates, names and other attributes attached. Each of these attributes would be connected into networks of relations among categories. The result might be thought of as a "hypermap" and should be able to represent a good deal of what is considered the "meaning" of a place, beyond its coordinates.

Interpreting a description would then be recast as finding where the structure of the description's representation matches a piece of the hypermap. Goodness of the match should be allowed to be imperfect. Either the description or the hypermap could have more or less detail than the other in a given case. Software for matching such representations is discussed in Buisson (1989).

7. Conclusions

Existing digital cartographic data sets represent map information, but they do not capture the meaning of the map symbols. Neither, of course, do the maps. Meaning is human. For the data sets to incorporate meaning, it must be present in the representation. Some of it is calculable (at some cost) from the data but other parts are not because human categories for reasoning about terrain are not now adequately defined, and may ultimately be undefinable. Geocartographic data must be made much richer.

People use terrain feature categories that are not easily discriminated by reference to the surface configuration portrayed in DEM or, for that matter in digital contour data. They note relationships among features that are not directly discerned from coordinates via analytic geometry. If computers are to perform tasks that are more like human processing of maps and terrain descriptions, they must be able to recognize and access features and relations that would be meaningful to people. It is not critical that the underlying data be exactly like human representations, only that the machine be able to transform between its internal representations and ones that are meaningful.

Adequate access to objects and relations is not available in the current data sets. Strategies for providing it should be examined. One approach would be to restructure the database(s), processing them into some representations that make salient objects and relations directly accessible. Another approach is to build a database management system that is capable of computing objects and relations from current data as needed. The restructuring option carries a heavy manual and computational cost at the outset and a larger storage requirement, but it seems to allow fine tuning of categories that then may remain stable. The DBMS option may allow smaller storage, but will require heavier computation and will be very dependent on having correct translations between representations. In either case the nature of human categories must be addressed and ranges of machine manipulable symbolic representations mapped to human meaning.

The problem of translating these cartographic data bases into fuller knowledge representations remains one of defining mental categories such that they can be adequately symbolized. The confusion in the categories in GNIS, for instance, reflects both cognitive style and the difficulty in interpreting speech acts. Symbols, i.e. words are used differently in different (linguistic and geographic) contexts. The apparently simple problem of naming terrain features, big, objectively given things in the environment, is not simple: ill-defined terms, and terms with multiple meanings make symbolic processing difficult.

Ultimately, digital databases will better reflect geographic reality. The objects and relations to be represented are well known. A finished data set would be a set of objects that contains every terrain feature that might be recognized and all of the relations among them that are likely to be needed. This clearly entails tighter integration of the information scattered in current data sets. Attaching proper names to features in DLG data and conceiving of the data as representing geographic rather than map objects are steps in the right direction which will be part of the DLG-E standard (Guptill, 1991).

8. Acknowledgements

The author would like to thank the participants and organizers of the NCGIA Research Initiative #2 ("Languages of Spatial Relations") for comments and discussion on some of the ideas presented in this paper, and acknowledge the support of the U.S. National Science Foundation, under NCGIA grant SES 88-10917, during the final preparation of this manuscript.

9. References

Buisson, L. (1989) 'Reasoning on Space with Object-Centered Knowledge Representations', in A. Buchman, O. Gunther, T.R. Smith, and Y.F.Wang (eds.), Design and Implementation of Large Spatial Databases, Springer-Verlag, New York.

Chen, Z. and A. J. Guevara (1987) 'Systematic Selection of Very Important Points (VIP) from Digital Terrain Model for Constructing Triangular Irregular Networks', Proceedings of the International Symposium on Computer-Assisted Cartography (Auto-Carto 8), 50-56.

Downs, R. and Stea, D. (1977) Maps in Minds: Reflections on Cognitive Mapping, Harper & Row, New York.

Elliott, R. and Lesk, M. (1982) 'Route Finding in Street Maps by Computers and People', Proceedings, AAAI-82 Conference (American Association for Artificial Intelligence).

Frank, A., Palmer, B., and Robinson, V. (1986) 'Formal Methods for the Accurate Definition of

Some Fundamental Terms in Physical Geography', Proceedings of the Second International Symposium on Spatial Data Handling, Seattle, WA., 583-599.

Freeman, J. (1975) 'The Modelling of Spatial Relations', Computer Graphics and Image Processing 4, 156-171.

Goldin, S. and Thorndyke, P. (1982) 'Simulating Navigation for Spatial Knowledge Acquisition', Human Factors 24, 457-471.

Guptill, S. (1991) (ed.) An Enhanced Digital Line Graph, U.S. Geological Survey Circular 1048, Denver Colorado.

Grace, G. (1987) Linguistic Construction of Reality, Croom Helm, New York.

Haralick, R. and Shapiro, L. (1980) 'A Spatial Data Structure', in H. Freeman and G. Pieroni (eds.), Map Data Processing, Academic Press, New York.

Havens, W. and A. Mackworth (1983) 'Representing Knowledge of the Visual World', Computer 19(10).

Herskovits, A. (1986) Language and Spatial Cognition, Cambridge University Press.

Kuipers, B. (1978) 'Modeling Spatial Knowledge', Cognitive Science 2, 129-153.

Lakoff, G. (1987) Women, Fire, and Dangerous Things, University of Chicago Press.

Mark, D. and Frank, A. (1989) 'Concepts of Space and Spatial Languages', Proceedings, International Symposium on Computer-Assisted Cartography (Auto-Carto 9), Baltimore, 538-556.

McGranaghan, M. (1985) 'Pattern, Process and a Geographic Language', Ontario Geography 25, 15-27.

McGranaghan, M., and L. Wester (1988) 'Prototyping an Herbarium Collection Mapping System', Technical Papers: 1988 ACSM-ASPRS Annual Convention: GIS 5, 232-238.

McGranaghan, M. (1989a) 'Context-Free Recursive-Descent Parsing of Location Descriptive Text', Proceedings of the International Symposium on Computer-Assisted Cartography (Auto-Carto 9), Baltimore, 580-587.

McGranaghan, M. (1989b) 'Incorporating Bio-Localities in a GIS', Proceedings of GIS/LIS '89, Orlando, 814-823.

Minsky, M. (1974) 'A Framework for Representing Knowledge', in P. Winston (ed.), The Psychology of Computer Vision, McGraw-Hill, New York.

O'Callaghan, J. and D. Mark (1984) 'The Extraction of Drainage Networks from Digital Elevation Data', Computer Vision, Graphics and Image Processing 28, 323-344.

Peuquet, D. (1986) 'The Use of Spatial Relationships to Aid Database Retrieval', Second International Symposium on Spatial Data Handling, Seattle, WA., 459-471.

Peuquet, D. (1988) 'Toward the Definition and Use of Complex Spatial Relationships', Third International Symposium on Spatial Data Handling, Sydney, Australia, 211-224.

Peuquet, D. and Zhan, C. (1987) 'An Algorithm to Determine Direction Between Arbitrarily-Shaped Polygons in a Plane', Pattern Recognition 20, 65-74.

Riazanoff, S., B. Cerevelle, and J. Chorowicz (1988) 'Ridge and valley line extraction from digital elevation models', International Journal of Remote Sensing 9(6), 1175-1183.

Robinson, R., and Wong, R. (1987) 'Acquiring Approximate Representations of Spatial Relations: Theory and Results of Tests', Proceedings of the International Symposium on Computer-Assisted Cartography (Auto-Carto 8), 604-622.

Rosch (Heider), E. (1973) 'Natural Categories', Cognitive Psychology 4, 328-350.

Rosch, E. (1978) 'Principles of Categorization', in E. Rosch and B. B. Lloyd (eds.), Cognition and Categorization, Lawrence Erlbaum Associates, Hillsdale, NJ.

Shapiro, S. and W. Rapaport (1987) 'SNePS Considered as a Fully Intensional Propositional Semantic Network', in Cercone and McCalla (eds.), The Knowledge Frontier, 262-315.

Shapiro, S. and G. Woodmansee (1971) 'A Net Structure Based Relational Question-Answerer', Proceedings of the International Joint Conference on AI, Washington, DC, 325-346.

Talmy, L. (1983) 'How Language Structures Space', in H. Pick and L. Acredolo (eds.), Spatial Orientation: Theory, Research and Application, Plenum Press.

USGS (1985) USGS Digital Cartographic Data Standards: Digital Line Graphs from 1:24,000-Scale Maps, USGS Survey circular 895-C, United States Geological Survey, Alexandria, VA.

USGS, (1987) Geographic Names Information System, (Data Users Guide 6), US Department of the Interior, Reston, VA.

Winograd, T. and F. Flores (1986) Understanding Computers and Cognition, Addison-Wesley, Reading, Mass.

Winston, P. (1977) Artificial Intelligence, Addison-Wesley, Reading, Mass.

THE ROLE OF MODAL LOGICS IN THE DESCRIPTION OF A GEOGRAPHICAL INFORMATION SYSTEM

MICHAEL F. WORBOYS
Midlands Regional Research Laboratory and
Department of Computing Studies
University of Leicester, Leicester LE1 7RH
United Kingdom

ABSTRACT. This paper considers the application of modal logic formalisms to spatio-temporal information systems. It begins with a brief description of modal logic. Then follows a survey of common modal logical systems, especially temporal and dynamic systems. The paper discusses current work on temporal databases and considers the special problems of data with both a spatial and temporal reference.

1. Introduction

Little is yet known about the introduction of the temporal dimension into spatial knowledge bases. Indeed, the study of general temporal databases is still undeveloped. But it is clear that spatio-temporal systems have many practical applications and will certainly be constructed. It is therefore important to establish their theoretical foundations.

Logic has a most important role to play in the representation of knowledge. However, not all types of knowledge can be represented using traditional propositional and predicate logics. Such logics are useful for formalizing the reasoning within static domains, as would occur for example within an unchanging database. But for more dynamic situations, such as occur in spatio-temporal systems, classical logic is not sufficient. Using classical logic, we may be able to reason quite well within a spatial domain which contains information about the state of affairs at particular point in time. We need extra machinery to be able to reason about a spatial domain which is changing over time. A further complication occurs since the knowledge base which holds this information may itself be dynamic, knowledge being inserted, updated and deleted over time. An important step forward is made when we make a clear distinction between time in the knowledge domain (historical or logical time) and time in the knowledge base (registration time).

One approach to the representation of spatio-temporal knowledge is to use modal logic. Specific modal logics concerned with temporal situations are tense or temporal logics. Modal logic applied to general dynamic domains is referred to as dynamic or action logic. We do not wish to make the claim that the formalism of modal logic is sufficient to capture all of our temporal concepts, or all temporal constructions within natural language. However, such formalisms are a first step.

In Section 1 we present a brief description of modal logic. For a fuller treatment, the reader could consult Hughes and Creswell (1972, 1984). Section 2 describes tense and action logics. Section 3 applies these theories to spatio-temporal information systems.

D. M. Mark and A. U. Frank (eds.), Cognitive and Linguistic Aspects of Geographic Space, 403–413.

2. Brief Description of Modal Logic

When studying modal logic we shall confine ourselves to modal propositional logic. The vocabulary of modal propositional logic is formed by adding the modal operators $\Box$ (for "necessarily") and $\Diamond$ (for "possibly") to the vocabulary of propositional logic. In all the systems which we shall study $\Diamond\varphi$ is taken to be an abbreviation for $\neg \Box \neg \varphi$. The intuition we are relying on here is that φ is a necessary truth if, and only if, φ is true in all possible scenarios (or at least all those which may be envisaged from the present scenario). φ is a possible truth if, and only if, $\neg \varphi$ is not a necessary truth.

The formation rules for modal propositional logic are as for propositional logic, with the addition of:

If φ is a well-formed formula then $\Box\varphi$ is also well-formed.

2.1 PROOF THEORY

The proof system we consider first is called the system K (K for Saul Kripke, the inventor of possible world semantics for modal logic). The tradition in modal logic is to treat the proof theory as an extension of a Hilbert style classical logic. That is to say, the axioms are all theorems, and the rules of derivation allow one to derive theorems from theorems.

When a wff φ is provable as a theorem in K, we shall represent this fact by $K \vdash \varphi$, or sometimes, when there is no risk of ambiguity, by $\vdash \varphi$.

Axioms:

$K \vdash \varphi$, where φ is a classical tautology.

$K \vdash \Box(\varphi \to \psi) \to (\Box\varphi \to \Box\psi)$ (distribution axiom)

Rules of Derivation

$$\frac{K \vdash \varphi \to \psi \quad K \to \phi}{K \vdash \psi}$$ (modus ponens)

$$\frac{K \vdash \varphi}{K \vdash \Box\varphi}$$ (necessitation rule)

2.2 SEMANTICS: KRIPKE FRAMES AND MODELS

These semantics are usually attributed to Saul Kripke. In fact, the idea seemed to occur to Kripke, Hintikka, and Kanger independently. The seminal paper that is usually quoted is [Kripke 1963].

A *Kripke frame* $<W, \rho>$ is a set W together with a relation ρ on W. The relation is often called an *accessibility relation* on W. The idea is that W is a set of possible scenarios, or worlds, and ρ is a relationship of accessibility between these worlds. That is $w_1 \rho w_2$ if, and only if, from the scenario of w_1, w_2 is considered a possibility.

A *Kripke model* $\mathcal{M} = <W, \rho, v>$ of a set of wffs $\mathcal{F}$, is a Kripke frame together with a *valuation function* $\mathcal{V}$, such that:

$$\mathcal{V}: \mathcal{F} \rightarrow \mathcal{P}(W)$$

and:

$$\mathcal{V}(p \wedge q) = \mathcal{V}(p) \cap \mathcal{V}(q)$$

$$\mathcal{V}(\neg p) = W - \mathcal{V}(p)$$

$$\mathcal{V}(\Box p) = \{w: \forall x \in W \text{ (If } w\rho x\text{, then } x \in \mathcal{V}(p))\}$$

The intuition here is that $\mathcal{V}(p)$ consists of all worlds in which p is true. The last statement may be interpreted as stating that $\Box p$ is to be true in world w if, and only if, p is true in every world accessible to world w.

Let $\mathcal{M}$ be a Kripke model $<W, \rho, \mathcal{V}>$ of a set of wffs $\mathcal{F}$. Let $w \in W$ and $p \in \mathcal{F}$. Then:

$<\mathcal{M}, w> \models p$ if, and only if, $w \in \mathcal{V}(p)$.

p is *valid in the Kripke model* $\mathcal{M}$ if, and only if, $<\mathcal{M}, w> \models p$, for all $w \in W$. (In this case we write $\mathcal{M} \models p$, or often just $\models p$ if no ambiguity arises.)

p is *valid in a Kripke frame* $<W, \rho>$ if, and only if, p is valid in $\mathcal{M}$, for all Kripke models $\mathcal{M}$ which are based upon $<W, \rho>$. (We write $<W, \rho> \models p$, or often just $\models p$ if no ambiguity arises.)

Soundness and completeness theorem

φ is provable in K if, and only if, φ is valid in all Kripke frames

3. Survey of Common Modal Logical Systems, especially Temporal and Dynamic Systems

This section introduces two modal systems which are particularly relevant to spatio-temporal information systems. The first, temporal or tense logic, allows the possibility of formulating tensed propositions. The second, dynamic logic, enables a formalization of knowledge held within a dynamically changing database system.

3.1 TENSE LOGIC

Tense logic is a modal logic formed by the introduction of some operators which express tenses. The basic vocabulary is as follows:

$G\varphi$: φ will always be the case.
$H\varphi$: φ has always been the case.
$P\varphi$: φ has been the case (at some time in the past).
$F\varphi$: φ will be the case (at some time in the future).

G and H correspond, roughly, to two different versions of $\Box$, and P and F to two different versions of $\Diamond$. P is $\neg H \neg$, F is $\neg G \neg$.

3.1.1 *The minimal system - temporal K.* The system below formalizes relationships between the temporal operators G, H, P, and F. This system, which is a development of Kripke's work, described in Section 1, is attributed to E. J. Lemmon and D. Scott (notes not fully published).

Axioms:

All propositional tautologies.
$G(\varphi \rightarrow \psi) \rightarrow (G\varphi \rightarrow G\psi)$
$H(\varphi \rightarrow \psi) \rightarrow (H\varphi \rightarrow H\psi)$
$\varphi \rightarrow GP\varphi$
$\varphi \rightarrow HF\varphi$

Rules of Derivation:

$$\frac{\vdash \varphi}{\vdash G\varphi} \qquad \frac{\vdash \varphi}{\vdash H\varphi}$$

and modus ponens.

3.1.2 *Semantics for Temporal K.* Frames and models are defined as before, except that we have to extend our account of the valuation function to take account of G, H, P, and F.

$$\mathcal{V}(G\varphi) = \{w \mid \forall w'(w\rho w' \rightarrow w' \in \mathcal{V}(\varphi)\}$$
$$\mathcal{V}(F\varphi) = \{w \mid \exists w'(w\rho w' \wedge w' \in \mathcal{V}(\varphi))\}$$
$$\mathcal{V}(H\varphi) = \{w \mid \forall w'(w'\rho w \rightarrow w' \in \mathcal{V}(\varphi))\}$$
$$\mathcal{V}(P\varphi) = \{w \mid \exists w'(w'\rho w \wedge w' \in \mathcal{V}(\varphi))\}$$

Our intuition here is that the possible worlds in a temporal frame are instants of time, and that wrw' if, and only if, w is earlier than w'.

3.1.3 Soundness and Completeness Theorem.

φ is provable in Temporal K if, and only if, φ is valid in all Kripke temporal frames.

3.1.4 Branching and Linear Temporal Systems.

Linear time

In 3.1.1, there is no restriction on the structure of the temporal frames. It is possible to quite easily impose restrictions which may more closely mirror our view of the structure of time for particular applications. For example, we may view time as a linear stream. Absolute time in Newtonian physics has this structure. The instants of time may be imagined to be arranged along a line, as shown below.

This restriction on the temporal frame structure is mirrored by adding a further axiom to Temporal K:

$$PF\varphi \rightarrow (P\varphi \vee \varphi \vee F\varphi)$$

There are appropriate soundness and completeness results.

Branching time

Branching time is a weaker system, in that every linear time frame is branching. The intuition that we intend to capture here is that at any present moment there is the possibility of a number of different futures, but only one past. The diagram below shows the kinds of frames possible.

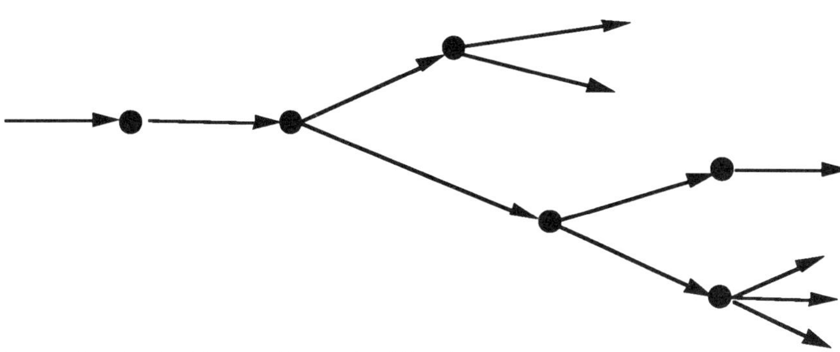

The axioms to be added to Temporal K are the following:

$$G\varphi \rightarrow GG\varphi$$
$$H\varphi \rightarrow HH\varphi$$

$$(H(\varphi \vee \psi) \wedge H(\varphi \vee H\psi) \wedge H(H\varphi \vee \psi)) \rightarrow (H\varphi \vee H\psi)$$

Again, there are appropriate soundness and completeness results.

3.2 DYNAMIC LOGIC

To motivate the introduction of a further extension of modal logic, called *dynamic*, or *modal action* logic, we begin with the semantics, and take the set W of possible scenarios to be the set of possible states of a knowledge base, and allow one world to be accessible from another if there is a transition from the former to the latter by means of a sequence of transactions (e.g. updates) on the knowledge base. There will be a set of accessibility relations, each relation corresponding to a transaction sequence (or *action*). More formally:

$w \rho_a w'$ if, and only if, we can get from state w to state w' via action a.

Suppose that a is some action. Then we can understand $[a]\varphi$ to mean that after action a has been performed upon the knowledge base, then φ must be the case.

[a] can be treated as a modal operator. A little reflection will quickly show that the necessitation rule and the distributive axiom are entirely reasonable when [a] is understood in this way. We also have an extra set of axioms concerning negation which can be asserted if each state of the knowledge.base is a consistent state. As before, <a> is an abbreviation for $\neg [a] \neg$.

3.2.1 Proof system.

Axioms:

All propositional tautologies

$[a](\varphi \rightarrow \psi) \rightarrow ([a]\varphi \rightarrow [a]\psi)$ for all actions a.

$[a]\neg\varphi \rightarrow \neg[a]\varphi$ for all actions a.

Rules of Derivation:

$$\frac{\vdash \varphi}{\vdash [a]\varphi}$$

and modus ponens.

3.2.2 *Semantics.* An *extended Kripke frame* $\langle W, \{\rho_a : a \text{ an action}\}\rangle$ is a set W together with a set of relations ρ_a on W (one for each action a). Models are defined as before, except that the result of the valuation function $\mathcal{V}$ acting on $[a]\varphi$ is given by:

$$\mathcal{V}([a]\varphi) = \{w \mid \forall w'(w \rho_a w' \rightarrow w' \in \mathcal{V}(\varphi)\}.$$

There are soundness and completeness results, as for the earlier system. The topic of dynamic logic is developed fully in (Harel, 1979), and its application to knowledge base specification is discussed in (Khosla *et al.*, 1986).

4. Current Work on Temporal Issues in Databases and GIS

For a database management system to be labelled temporal, it should be capable of supporting a model of a time-varying world. There have been three main areas of research here: the development of temporal query languages, the development of an underlying logical model and theory (cf relational theory for relational databases), and of course the question of physical implementation.

Characteristic problem areas for temporal databases are storage, object identity, and representation of temporal knowledge. Regarding storage, it is clearly inefficient to store the contents of the entire database of information recorded at different times. Much of the data may be time-invariant, and this method would lead to redundant duplication of information. However, it is not clear at this time what is the best approach to storage for temporal databases. (Langran, 1989) lists various approaches taken in the literature.

Object identity is required if the system is to recognize differing versions of the same time-varying object. The principle here is that an object has an identity independent of its attributes. Thus if the attributes change, the object is still intact. Object identity is not supported by the pure relational model, although it can be argued that it is supported by Codd's extended model RM/T (Codd, 1979). However, RM/T is not implemented in any currently marketed system known to the author. Object-oriented database systems have addressed this problem of object identity.

4.1. TEMPORAL KNOWLEDGE REPRESENTATION

Temporal knowledge may be represented using the modal formalism outlined in section 3. There is a distinction between *registration* time (also called transactional, recording history, or system time), when a transaction takes place and the *logical* time (also called valid, historical, extrinsic, or object level time) at which the events recorded actually happened. At one extreme, a database may be *dynamic* in that the full transaction history is recorded, and yet contain information with no temporal reference. At the other, a database may be *static* with no continuous stream of transactions, (maybe with no update facility), but contained temporally referenced information.

For a dynamic database holding temporally referenced data, transactions may be *retroactive* (i.e. coming into effect after the time to which the data was referenced), or *proactive* (i.e. coming into effect before the time to which the data was referenced).

There may be more than one type of logical time to be recorded in the same database. For example, if written forms are used, the time of the document which contains the observation may be important. Thus, in the harbor information system described below, notification of the placement of a buoy in the harbor at coordinates (45,46) may have the following times associated with it::

office historic time	notifying document dated	28 August
registration time	data entered into GIS	1 September
harbor historic time	buoy placed in harbor	8 September.

4.2. DATA WITH A SPATIO-TEMPORAL REFERENCE

Future GIS should be capable of managing data referenced to both space and time. Langran (1989) proposes that a temporal GIS should be able to provide answers to such questions as:

> Where was this feature located two years ago?
> How has this feature changed over the last five years?
> Has this feature moved?
> Have these two features ever been juxtaposed?

4.2.1 *Extended HYDRO*. We use the example HYDRO given in (Langran, 1989). HYDRO contains marine information. Each piece of information relates to a feature, gives an (x,y)-position, and gives the date of the source document which refers to it.

On 30 August, HYDRO contains information:

Feature	X	Y	Effective
Beacon	38	38	8 March
Shoal	23	32	2 May
Hazard	56	34	3 March
Bell	44	28	8 August.

On 1 September, the following transactions occurred:

Buoy	added proactively; coordinates (45,46); taking effect from 8 September. (Plans are made to place it in the harbor the following week.)
Shoal	coordinates altered to (15,21); new survey information; document dated 18 August.
Light	added; coordinates (12,18); document dated 18 August.
Hazard	deleted.
Bell	deleted.

On 5 September, the following transactions occurred:

Hazard	restored.
Light	coordinates corrected postactively to (18,18); document dated 18 August.

Example constraints:

Below is given a subset of the possible integrity constraints which the harbor system must satisfy. A distinction is made between static constraints, which relate to individual states of the system, and dynamic constraints, which operate on temporal sequences (in both historical and registration times) of states of the system.

Static:	Each feature has a unique coordinate set.
Temporal (historical):	Lighthouses do not move.
Temporal (registration):	The date of a document specifying an observation must be earlier than the date of insertion of the corresponding data into the database.

Example queries:

Next is given a pair of typical retrieval requests to the system. Both queries include time as an integral component.

Retrieve all features currently active within a spatio-temporal region.
Retrieve all features that may never vary temporally.

Formalization

We illustrate the use of a typed tense first order predicate logic in this context by formalizing two of the above constraints.

That each feature has a unique coordinate reference may be expressed as:

$\forall x, x', y, y' : \mathbb{R}$ (the class of real numbers)
$\forall f : \text{Feature}$
$(\text{coordinates}(f, x, y) \wedge \text{coordinates}(f, x', y')) \rightarrow (x = x' \wedge y = y')$

That lighthouses do not move may be expressed as:

$\forall x, y : \mathbb{R}$
$\forall l : \text{Lighthouse}$
$\text{coordinates}(l, x, y) \rightarrow G(\text{coordinates}(l, x, y))$

Notice that only the second constraint requires a temporal operator.

4.2.2 *Boundary change example.* The next example shows up some of the special problems of a database of spatio-temporally referenced data. The difficulties of transforming data from one spatial reference unit to another are well known. These difficulties become compounded if the spatial units are time-variable. A brief discussion of these matters is given in (Hearnshaw *et al.*, 1989). We illustrate with a local example. The city of Leicester, UK, is partitioned into a number of district electoral wards for purposes of the census and local administration. The partition is regularly modified as boundaries are changed and wards are inserted or deleted. Attribute data is collected, which is referenced to the partition at a particular time (or set of times).

Imagine that at time t_1, Leicester was divided into two wards, A and B. At time t_2, in order to adjust for population movements, the boundary between A and B was altered. At time t_3, ward B was divided into wards C and D. Throughout this time period, feature f is always in ward A. We illustrate this in the diagram below.

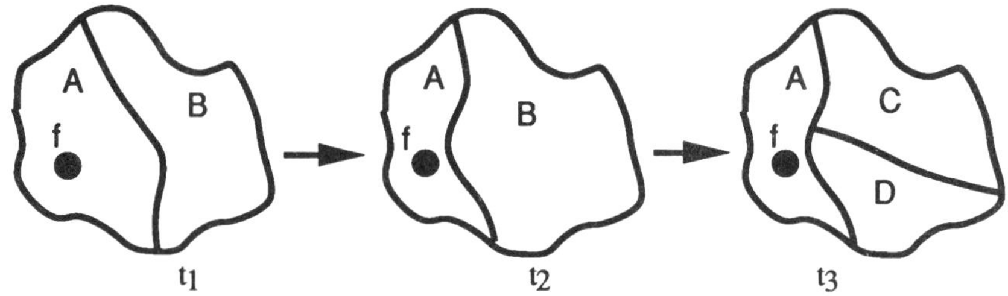

The knowledge representation problem is how to formalize this information. To illustrate the approach using tense logic, the knowledge that at time t_1 feature f is and will always be inside region A may be expressed as:

$$G\ (\mathrm{inside}\ (f, A))$$

We can also imagine a query language which is a modal extension of the relational calculus which could be used to retrieve information stored in a database about such a system. In (Worboys *et al.*, 1990), we develop an object retrieval language (ORC) which interacts with an object oriented database. ORC is an object version of the domain relational calculus, considered in (Lacroix and Pirotte, 1977). It is a straightforward matter to furnish ORC with a modal component to handle spatio-temporal object systems.

5. Conclusions

This paper has raised more questions than it has answered. The plan has been to show that there is a well developed formalism for representing knowledge about spatio-temporal systems, namely modal logic. We have examined two special forms of modal logic, temporal and dynamic logic, and considered the role of each for working with information in a temporal GIS. We concluded with some examples.

Much further work remains. The details of the representation of spatio-temporal knowledge, and in particular the use of modal logic to represent topological/geometrical knowledge in a temporal setting, are still to be worked out. The interaction between the representation of historical and registration time needs consideration. Storage problems are unsolved.

Regarding implementation, the basic problem is how to store the information efficiently and effectively. At one extreme, we could store all the geometry/topology of the partition at each change time, even if only a single boundary is altered. This uses a great deal of storage, but the model is simple and retrieval will be fast. At the other extreme, store a start state of the partition and then only changes (*deltas*) to the partition. This *temporal spaghetti* will be low on storage space but the model is messy and retrievals will be inefficient. A compromise would be to store the whole partition, store the deltas, and after a specified number of deltas store the whole partition again, more deltas, and so on. The question of temporal database storage is still unanswered in the basic relational case, and so *a fortiori* for spatial data.

We certainly do not claim that modal logic holds all the answers to temporal knowledge representation. We hope to have acquainted the reader with some of the issues in this fascinating area, and indicated some of the directions for future research.

6. Acknowledgements

The author thanks Nick Measor, University of Leicester, for introducing him to modal logic and for interesting discussions.

7. References

Clifford, J., Warren, D. S., (1983) 'Formal semantics for time in databases', ACM TODS 8, 214-254.

Codd, E. F., (1979) 'Extending the database relational model to capture more meaning', ACM TODS 4(4), 397-434.

Harel, D., (1979) First-Order Dynamic Logic, Springer-Verlag, Berlin.

Hearnshaw, H. M., Maguire, D. J., Worboys, M. F., (1989) 'An Introduction to Area-Based Spatial Units: A Case Study of Leicestershire', Midlands Regional Research Laboratory, Research Report No. 1.

Hughs, G. E., Cresswell, M. J., (1968) An Introduction to Modal Logic, Methuen, London (reprinted with corrections 1972).

Hughs, G. E., Cresswell, M. J., (1984) A Companion to Modal Logic, Methuen, London

Kholsa, S., Maibaum, T. S. E., Sadler, M., (1986) 'Database Specification', in Steel, T. B., Meersman, R. (eds.), Database Semantics, Elsevier, Netherlands.

Kripke, S., (1963) 'Semantical Analysis of Modal Logic I: Normal Propositional Calculi', Zeit. Math. Logik Grund 9, 67-96.

Lacroix, M. and Pirotte, A., (1977) 'Domain-Oriented Relational Languages', Proceedings, 3rd International Conference on Very Large Data Bases.

Langran, G., (1989) 'A review of temporal database research and its use in GIS applications', International Journal of Geographical Information Systems 3(3), 215-232.

Reschner, N. and Urquehart, A., (1971) Temporal Logic , Springer-Verlag.

Worboys, M. F., Hearnshaw, H. M., and Maguire, D. J., (1990) 'Object-oriented data and query modelling for geographical information systems', Proceedings, 4th International Symposium on Spatial Data Handling, Zurich.

SECTION 6:
USER INTERFACES AND HUMAN-COMPUTER INTERACTION

User interfaces for Geographic Information Systems represent an area in which results of the cognitive research on spatial languages can be applied directly. Users need to communicate with systems, both to issue commands and queries and to understand the responses from the systems. In this section, most of the discussion focuses on the communication between the system and the user about the system actions—a "meta" level, compared to the communication between system and user about the real world objects represented in the system. Many current systems require the user to learn a command language in order to interact with the system. This language is based on a set of concepts that the system designers—mostly English speaking programmers—built into the programs. Command and concept names were selected by them according to their concepts of space and spatial operations, in their natural language, and without human-subjects testing as to optimal command names. Indeed the terminology is often related to the internal structures of the program and not to meaningful concepts in the user domain.

The first two papers in this section are concerned with this conceptual level, arguing that the cognitive sciences can contribute to the design of GIS user interfaces. They both start with the assumption that systems which mimic human cognitive structure and behavior are more likely to be useful and effective. Thus they propose to use the results from cognitive research and other cognate disciplines, specifically geography, for the design of GIS user interfaces. In the first paper **Werner Kuhn** and **Andrew Frank** discuss the user interface from a very abstract point of view. They begin by noting Lakoff's theme that metaphorical thought pervades communication, and by observing the success of the "desktop" metaphor popularized by Apple with the Macintosh personal computer. Based on this, Kuhn and Frank recommend that user interface designers apply metaphorical mappings from a familiar domain to the unfamiliar world of computer operations. They describe a method (based on algebraic specifications) for formalizing the user interface objects and operations and show that the metaphorical mapping can be formalized as a morphism between algebras. In their example they show the similarity of the operations that apply to real world office objects like file-folders, etc., and their electronic counterparts.

The use of metaphors, however, does potentially increase the cultural dependency of an interface. The "trash" can on the Apple desktop had to be renamed for other countries, and the icon used should also be changed—not every office in the world contains this object so familiar to most North Americans or Europeans. (Actually, few North American offices contain things that look like a standard *outdoor* household garbage can, and refuse receptacles seldom sit on desk tops!) Using these formalizations, Kuhn and Frank quickly show that the electronic clipboard included in the Apple desktop is slightly different from a regular clipboard, the difference being mostly an improvement, but sometimes causing problems—for example, adding something to a real clipboard does not destroy the previous contents. This method of formalization and comparison is important, as it helps to evaluate a user interface metaphor objectively. The "spatialization" of human thought processes automatically leads to visualization of non-spatial information, a topic that will be picked up in the last paper in this chapter.

The second paper in the section, by **Michael Gould**, describes research that addresses the role of a user's natural language in the design of user interfaces. His research proposes to interview GIS users in the United States, Spain, and Ecuador to observe what language they use to describe spatial situations and spatial problem-solving. It has been observed that the translation of English terms in the user interface can cause problems, as pointed out initially in this volume by Campari. One finds Spanish or Italian users grappling with the English terms "digitize" or "buffer", for which no original local words exist. But this is only a superficial difference; more important would be differences in the conceptualization of space and the management of space. We are reminded of a fundamental, but eminently practical problem faced by those attempting to provide technical assistance in some developing nations: the concept of "time is money" and all its implied notions related to time management, planning etc., is conspicuously absent. The proverbial *mañana* is not always exactly "tomorrow", but may refer to an indeterminate point in the future. We may encounter similar linguistic and conceptual differences when it comes to the spatial concepts embedded in current geographic information systems.

The next two papers deal with today's design problems of user interfaces to GIS, especially the issues associated with query languages. **Jonathan Raper** and **Michael Bundock** give a blueprint for the design of a general-purpose GIS user interface. One of the immediate problems is the enormous variability between different user groups: user groups differ in the tasks they want to perform, their levels of spatial, application related knowledge and computer skills, and their interest in learning about the GIS technology. Raper and Bundock draw on their experience with the successful *GIS Tutor* system and stress that a number of different solutions are necessary, supporting different user needs. In the second part of their paper, they concentrate on the database language SQL, which is widely used for access to relational databases both in general and in GISs. SQL has been standardized and is available from many commercial vendors, and can in principle be extended to serve as a GIS query language. Raper and Bundock compare the advantages of such extension with the disadvantages and then detail their proposal for SQL-SX, a spatial extension of SQL, based on the latest standard and proposed future standards for SQL. They found it necessary to add several new data types, and a long list of operations and relations is given.

The subsequent paper by **Max Egenhofer** continues this topic, but with a much narrower focus: he investigates the suitability of SQL as an *ad hoc* query languages for GIS. Some users need very flexible access to the data in a GIS, and their needs cannot be anticipated and programmed for ahead of time. Interactive query languages are the accepted tool for this user group when using standard databases, and a similar tool is required for GIS. Egenhofer starts with a short list of questions that a GIS user may have, all taken from a navigation and tourism environment. He then extensively surveys proposals for SQL extensions and determines how they would resolve the various issues. He finds a list of difficulties, in which the basic design of SQL, which was developed for non-geometric data arranged in tabular format, cannot cope with the requirements posed within a GIS. Despite the fact that most of the immediate problems, e.g., spatial data types with appropriate relations, can be patched into SQL, the overall structure of that language is not appropriate for the GIS requirements. A major shortcoming remains the complete lack of a method to direct the format of the display. Another is the lack of expressive power—SQL is not computationally complete and is first-order and not reflective, i.e., it cannot treat data and meta-data in the same command. Egenhofer concludes that extensions to SQL may be a short term solution for access to GIS data, but that it is not the appropriate method for spatial data handling, and a new design is necessary.

The next paper, by **Elsa Maria João,** concentrates on the user interface aspects of cartographic generalization. She observes that current GISs have only very limited facilities to generalize data—and that the facilities differ for vector- and for raster-based systems. She advocates that generalization methods should be useful both for display and for analysis, and must apprise the user of their effect on data quality. João gives an over-all design for a system to assist the user, not to supplement user judgement by automatic processing. The implementation of the concept requires a number of modules, some of which must be based on expert systems. It is therefore desirable to collect cartographic generalization rules in a format that can be used in a knowledge base.

As the final paper in this book, we choose **Robert Jacobson**'s look into the future of "Virtual Reality". Jacobson critiques the technical media, from books to television to computer-based systems, as providing information to the human user in a format that does not appeal to our deep understanding of being at a place. The modern media use only a very limited proportion of our senses. They can inform us about places we may never experience personally, thus enriching our intellectual understanding, but do not feed our senses of place. Jacobson uses the term "virtuality" to describe this more complete experience of space, which we can now create artificially in the laboratory. This "virtual reality" allows people to experience spatial arrangements at another location, or even situations that do not exist outside of the computed environment. Virtual reality involves more senses than vision and verbal communication. It becomes feasible to organize and experience spatially different environments, which are normally not accessible to humans: we may travel into the extremely small—to places inside the molecules—or to the extremely large—to traverse the universe at speeds faster than speed of light (physical laws can be suspended in virtual reality). Perhaps even more significantly, we may also arrange other types of data spatially, and use our spatial senses to understand the data. Some hypermedia systems already permit navigation in multidimensional knowledge spaces. In a virtual world, one may traverse knowledge using the full complement of experiences that we use to navigate in the natural environment. Using this technology wisely and effectively will require us to understand more about how people structure space, how they accumulate information about an environment and find their way, what makes the essence of a place, etc. And these issues, all discussed at various places in this volume, will directly contribute to the structuring of the virtual realities that future users will experience.

A FORMALIZATION OF METAPHORS AND IMAGE-SCHEMAS IN USER INTERFACES

WERNER KUHN and ANDREW U. FRANK
National Center for Geographic Information and Analysis
Department of Surveying Engineering
University of Maine
Orono, Maine 04469 USA

ABSTRACT. Sound engineering approaches to user interface design require the formalization of key interaction concepts, one of them being metaphor. Work on interface metaphors has, however, been largely non-formal so far. The few existing formal theories of metaphor have been developed in the context of natural language understanding, learning, or reasoning. We propose to formalize interface metaphors by algebraic specifications. This approach provides a comprehensive formalization for the essential aspects of metaphorical user interfaces. Specifically, metaphor domains are being formalized by algebras, metaphorical mappings by morphisms, and image-schemas by categories. The paper explains these concepts and the approach, using examples of spatial and spatializing metaphors.

1. Introduction

Metaphor pervades communication. Metaphorical thought, action, and language are not only essential to interpersonal communication (Lakoff and Johnson, 1980), but to human-computer communication as well. Since communication with computers is an outcome of design, building effective user interfaces requires a formal theory of metaphor (Carroll, *et al.*, 1988).

This paper presents a methodological foundation for such a theory. It proposes a formalization of interface metaphors based on abstract algebra (MacLane and Birkhoff, 1967). The algebraic approach allows designers to specify not only metaphor domains, but metaphorical mappings and image-schemas as well. The paper describes the basic elements of the approach.

After explaining the underlying understanding of interface metaphors in Section 2, important related work is discussed in Section 3. The formalization method is presented with a running example in Section 4 and applied to spatial zooming operations in Section 5. The last section contains conclusions and suggestions for future work.

2. Interface Metaphors and Image-Schemas

Systems with user interfaces based on the DESKTOP metaphor and its variants (Apple Computer, 1987, Smith *et al.*, 1982) have established metaphor as a powerful means to control complexity in human-computer interaction (Carroll, Mack, and Kellogg, 1988). The structure and role of interface

D. M. Mark and A. U. Frank (eds.), Cognitive and Linguistic Aspects of Geographic Space, 419–434.

metaphors are, however, still understood in many different ways (Blumenthal, 1990; Carroll, Mack, and Kellogg, 1988; Erickson, 1990a; Smith *et al.*, 1982, Wozny, 1989).

The understanding of interface metaphors underlying our formalization derives from recent work in cognitive linguistics (Johnson 1987b; Lakoff, 1987; Lakoff and Johnson, 1980). Johnson (1987b, XIV) has characterized metaphor as,

> "...a pervasive mode of understanding by which we project patterns from one domain of experience in order to structure another domain of a different kind."

The two domains are commonly called the *source* and *target* domains of a metaphor and the metaphorical projection is a *mapping* from source to target.

Interface metaphors, in this projective view, go beyond explaining unfamiliar domains to novices. They determine how labor is distributed between a user and a system, what concepts a user has to deal with, and in what terms user and system communicate. In short, they structure application domains and organize tasks.

It is not clear how much metaphor-independent structure there can be in a target domain. A recent case study of metaphors in interface design (Erickson, 1990a) indicates that the structure available to the designer before a metaphor is consciously chosen is either quite abstract or already metaphorical. In Erickson's example of a DESKTOP metaphor extension, that structure consists of links between documents. Interestingly, the link structure is a so-called *image-schema.* Image-schemas are image-like reasoning patterns, consisting of a small number of parts and relations, made meaningful by sensorimotor experience (Johnson, 1987b; Lakoff and Johnson, 1980). Examples are the CONTAINER, SURFACE, LINK, PATH, NEAR-FAR, PART-WHOLE, and CENTER-PERIPHERY schemas.

Image-schemas may in fact be the kind of structure which is preserved by interface metaphors. This assumption agrees with Lakoff's invariance hypothesis (Lakoff, 1990) which claims that image-schemas remain invariant under metaphorical mappings. It also generalizes to user interfaces of any kind the suggestion that image-schemas play a fundamental role in interfaces of Geographic Information Systems (Mark, 1989). The design of interfaces for international use (Nielsen, 1990) is one area where image-schemas, due to their presumed culture-independence, are likely to be of particular importance.

3. Existing Metaphor Formalizations

While there is a vast literature on metaphor and analogy (e.g., Johnson 1981; Lakoff and Johnson, 1980; Mac Cormac, 1985; Ortony 1979; Vosniadou and Ortony, 1989), formal approaches are rare (see Hall, 1989, for a survey, and Indurkhya, 1989; Martin, 1990, for recent treatments). Most existing formalizations employ traditional knowledge representation formalisms, such as semantic networks, to specify metaphorical domains, but not mappings or image-schemas. Furthermore, the formalizations have been developed for natural language understanding, machine learning or automated reasoning, while discussions of metaphor in human-computer interaction have been largely non-formal so far.

The best known formal theory of metaphor and analogy is Gentner's structure mapping theory (Gentner, 1983, 1989). It describes analogies as mappings between source and target domains, each represented by semantic networks. It does not formalize the mappings themselves, however, and rests on a syntactical distinction of different kinds of relations. While Gentner's theory deals with structural aspects, it neglects the role of tasks in metaphor use (Carroll, Mack, and Kellogg, 1988).

Our formalization addresses these problems by formalizing mappings as morphisms and expressing tasks and actions through algebraic operators and their effects. Describing domains algebraically rather than relationally may only be a syntactic difference; it does, however, allow for relating metaphors to tasks. For example, the informal task description of parts of the DESKTOP metaphor given in (Carroll, Mack, and Kellogg, 1988) can directly be expressed by an algebraic formalism (see next section).

In the terminology of Carroll *et al.*, our approach is primarily *structural*, addressing the representation of metaphor domains and mappings, but incorporates the *pragmatic* aspects of actual user tasks. It does not deal with operational analyses of the effects of interface metaphors on learning or using a system.

Most theories of metaphor stress the importance of mapping source structure into the target domain. Surprisingly, the mathematical devices for describing structure mappings, i.e., morphisms, are rarely applied. An exception is a forthcoming treatment (Indurkhya, 1989) which describes cognitive models by algebras and metaphorical projections by morphisms.
Our independently developed algebraic approach differs from Indurkhya's theory mainly by:

- describing metaphorical domains as elements of algebra *classes*, thereby dealing with the relation of metaphors to image-schemas;
- specifically addressing *interface* metaphors;
- using the established software design technique of algebraic specifications.

Existing metaphor formalizations provide little support for interface designers because they are either purely theoretical or rely on specialized software in prototype form. The advantage of using the technique of algebraic specifications is that it allows designers to apply the method immediately, even with the possibility of using computer supported design tools such as Larch (Guttag, Horning, and Wing, 1985) or CLEAR (Burstall and Goguen, 1981).

4. An Algebraic Approach

4.1. ALGEBRAIC SPECIFICATIONS

Algebraic specifications describe objects in terms of operations. The associated theory has been developed and used for specifying abstract data types (Goguen, Thatcher, and Wagner, 1978; Guttag and Horning, 1978); practical introductions can be found in (Ehrig and Mahr, 1985; Liskov and Gutta, 1986; Woodcock and Loomes, 1989). An algebraic specification consists of three parts:

- a list of *sorts*, naming the objects involved;
- a list of *operators* (Ops), giving the sorts of arguments and results;
- a list of *axioms* (Eqs), defining equivalences for different sequences of operations.

Sorts and operators together are called the *signature* of an algebraic specification. *Algebras* are interpretations of signatures, assigning a set to each sort and an operation on these sets to each operator. An algebraic specification, i.e., sorts, operators, and axioms together, describes a

category, i.e. the class of algebras satisfying the specification (Ehrich, Gogolla, and Lipeck, 1989).
The technique of algebraic specifications is best demonstrated by applying it immediately to the formalization of source and target domains from Apple's composite DESKTOP metaphor (Apple Computer, 1987).

4.2. SOURCE AND TARGET DOMAINS

The following is an algebraic specification of physical office desktops. It lists three sorts: Desktops, Folders, and Bools (truth values). Desktops are the sort actually being specified, Folders will be specified below, and Bools are assumed to have been specified previously. There are operators to create a new desktop, put a folder on a desktop, get a folder from a desktop, and check whether a folder is on a desktop. The meaning of these operators is defined by the equations and does not rely on an interpretation of their names. The equations make free use of universally quantified sort variables (dt for desktops, f for folders, etc.).

```
Desktop
Sorts                 Desktop, Folder, Bool
Ops                   new:   →Desktop
  put:                Desktop x Folder → Desktop
  get:                Desktop x Folder → Desktop
  on:                 Desktop x Folder → Bool
Eqs                   on(new,f) = false
  on(put(dt,f1),f2) = if f1=f2 then true else on(dt,f2)
  get(put(dt,f1),f2) = if f1=f2 then dt else put(get(dt,f2),f1)
```

Now, the technique is applied to specify the corresponding target domain in Apple's DESKTOP metaphor. The sort and variable names for the electronic analogues of office objects are prefixed by "El" and "el", respectively.

```
ElDesktop
Sorts                 ElDesktop, ElFolder, Bool
Ops                   new:   → ElDesktop
  put:                ElDesktop x ElFolder → ElDesktop
  get:                ElDesktop x ElFolder → ElDesktop
  on:                 ElDesktop x ElFolder → Bool
Eqs                   on(new,elf) = false
  on(put(eldt,elf1),elf2) = if elf1=elf2 then true else on(eldt,elf2)
  get(put(eldt,elf1),elf2) = if elf1=elf2 then eldt else
                      put(get(eldt,elf2),elf1)
```

Obviously, these two specifications are identical up to sort and variable names. In practice, this means that Macintosh desktops behave like real desktops with respect to the operations defined. For the sake of simplicity, only the most salient features have been specified. By including additional operators, dissimilarities could be shown, such as the fact that things can fall from physical, but not from electronic desktops. In an actual interface design process, the designer has to decide which features of a source domain are to be considered salient and which are not.

Next, clipboards are specified. First the physical variety, with the assumption that clips are added to and removed from the top. Note that this specification resembles that of desktops. We will come back to this parallel when discussing image-schemas in 4.4.

Clipboard	
Sorts	Board, Clip, Bool
Ops	new: $\to$ Board
put:	Board x Clip $\to$ Board
get:	Board $\to$ Board
on:	Board x Clip $\to$ Bool
Eqs	on(new,c) = false

on(put(b,c1),c2) = **if** c1=c2 **then** true **else** on(b,c2)
get(put(b,c)) = b

Secondly, the electronic variety of clipboards, as provided by the Macintosh to move data between documents, is specified:

ElClipboard	
Sorts	ElBoard, ElClip, Bool
Ops	new: $\to$ ElBoard
put:	ElBoard x ElClip $\to$ ElBoard
get:	ElBoard $\to$ ElBoard
on:	ElBoard x ElClip $\to$ Bool
Eqs	on(new,elc) = false

on(put(elb,elc1),elc2) = **if** elc1=elc2 **then** true **else** false
get(put(elb,elc)) = put(elb,elc)

The axioms show two differences between physical and electronic clipboards:

- a clip remains on the Macintosh clipboard only until the next clip is put on - then it is lost;
- getting back a clip does not change the electronic clipboard and the same clip can be retrieved multiple times.

While the first mismatch may be harmful, the second is rather an advantage. The point is, however, that the specification *reveals* these mismatches. In an actual interface design process, it is the designer's task to decide about their desirability (Halasz and Moran, 1981; Johnson 1987a).
Folders are another part of the DESKTOP metaphor. Here is a specification of the salient features of physical office folders:

```
Folder
Sorts                 Folder, Document, Name, Bool
Ops                   new:    → Folder
  label:              Folder x Name → Folder
  insert:             Folder x Document → Folder
  remove:             Folder x Document → Folder
  empty:              Folder → Folder
  in:                 Folder x Document → Bool
Eqs                   remove(label(f,n),d)  =  label(remove(f,d),n)
  remove(insert(f,d1),d2)  =  if d1=d2 then f else
  insert(remove(f,d2),d1)
  empty(new) = new
  empty(label(f,n))  =  label(empty(f),n)
  empty(insert(f,d))  =  empty(f)
  in(new,d)  =  false
  in(label(f,n),d)  =  in(f,d)
  in(insert(f,d1),d2)  =  if d1=d2 then true else in(f,d2)
```

The following specification of electronic folders is very similar. The only difference is that the empty operator is missing because the Macintosh does not offer such an operation. (It does, of course, allow users to select all documents in a folder and remove them or to discard an entire folder, but these are different operations.)

```
ElFolder
Sorts                 ElFolder, ElDocument, Name, Bool
Ops                   new:    → ElFolder
  label:              ElFolder x Name → ElFolder
  insert:             ElFolder x ElDocument → ElFolder
  remove:             ElFolder x ElDocument → ElFolder
  in:                 ElFolder x ElDocument → Bool
Eqs                   remove(label(elf,n),eld)  =
label(remove(elf,eld),n)
  remove(insert(elf,eld1),eld2)  =  if eld1=eld2 then elf else
                      insert(remove(elf,eld2),eld1)
  in(new,eld)  =  false
  in(label(elf,n),eld)  =  in(elf,eld)
  in(insert(elf,eld1),eld2)  =  if eld1=eld2 then true else in(elf,eld2)
```

The final pair of DESKTOP metaphor domains to be specified here is that of trash cans. Some operators (e.g., the possibility to discard folders) have been omitted for brevity:

Trash Can

```
Sorts            Trash, Document, Bool
Ops              new:    → Trash
  discard:       Trash x Document → Trash
  remove:        Trash x Document → Trash
  empty:         Trash → Trash
  in:            Trash x Document → Bool
Eqs              remove(discard(t,d1),d2) = if d1=d2 then t
else             discard(remove(t,d2),d1)
  empty(new) = new
  empty(discard(t,d)) = new
  in(new,d) = false
  in(discard(t,d1),d2) = if d1=d2 then true else in(t,d2)
```

The specification of the target domain is similar to that of the source domain; however, it also reveals a mismatch:

ElTrash

```
Sorts            ElTrash, ElDoc, Bool
Ops              new:    → ElTrash
  discard:       ElTrash x ElDoc → ElTrash
  remove:        ElTrash x ElDoc → ElTrash
  empty:         ElTrash → ElTrash
  in:            ElTrash x ElDoc → Bool
Eqs              remove(discard(et,ed1),ed2) = if ed1=ed2 then
et else                   discard(remove(et,ed2),ed1)
  empty(new) = new
  empty(discard(et,ed)) = new
  in(new,ed) = false
  in(discard(et,ed1),ed2) = true or false
```

The difference in the last equation captures the fact that a Macintosh user does not have complete control over the Trash can contents. A document can be retrieved immediately after discarding it, but may be lost after another operation, such as opening a document or inserting a disk.

Each specification of a source or target domain given in this section describes a class of algebras. The *designer's* conceptual model of a domain is seen as an algebra in this class, i.e., an interpretation of the corresponding specification which assigns symbols to the sorts and operators such that the axioms hold. An important challenge in implementing an interface metaphor is to make sure that the *user's* conceptual model belongs to the same class.

4.3. METAPHOR MAPPINGS

Metaphors are mappings from source to target domains. When the domains are specified algebraically, it is natural to use morphisms, which are mappings between algebras, to define metaphorical mappings.

There are different kinds of morphisms, reflecting different degrees of similarities between algebras. The weakest kind of similarity is a *signature-morphism* which is a correspondence between the signatures defining two algebras (Ehrich, Gogolla, and Lipeck, 1989). A stronger kind is a *homomorphism* between two algebras A and B. It is a family of mappings from the sets of A to those of B, such that it does not matter whether an operator is first applied in A and the result mapped to B, or the arguments are first mapped to B and the corresponding operator in B is applied. Thus, homomorphisms are structure-preserving mappings. A one-to-one homomorphism is called an *isomorphism.*

If the source and target domain are both specified, as in the above examples, the morphism is determined. It indicates the degree of similarity between source and target for the specified (and, in this case, implemented) interface metaphors:

- The mapping between office and electronic *desktops* is an isomorphism, since the effects of the specified operators on office desktops correspond to the effects of the analogous operators on electronic desktops, and vice versa.
- The mapping from physical to electronic *clipboards* is a signature-morphism. The signatures are the same, but the effects of some of the operators are different. For example, the signatures of the physical put and get operators map to those of the electronic ones, but the effects of applying get after put are different, as can be seen from the third axioms in the two specifications.
- The *folder* specifications exemplify another isomorphism, but only with respect to the new, label, insert, remove, and in operators; the empty operator has no equivalent in the target domain.
- The *trash can* specifications are linked by another signature-morphism. The signatures correspond, but the result of the in operator after discarding a document can be different for the two domains.

In the normal course of an interface design, the target domains are not predetermined. Source domains are specified first and appropriate morphisms are chosen to achieve the intended matches. These morphisms then determine parts of the target domain by projecting the desired elements of the source structure.

4.4. IMAGE-SCHEMAS

The algebraic specifications of source and target domains from the DESKTOP metaphor exhibit two image-schemas: Desktops and clipboards are instantiations of the SURFACE schema; folders and trash cans represent the CONTAINER schema. The SURFACE schema embodies the logic of a surface on which items can be put. The CONTAINER schema consists of an interior, an exterior, and a boundary between them, so that items can be in a container or outside of it (Lakoff, 1987).

Thus, not only are image-schematic structures believed to be invariant in metaphorical mappings, they can also be common to different parts of a composite metaphor. Composite metaphors whose domains are related to common image-schemas are more likely to be perceived as *coherent* than those where this is not the case. It can be argued (Erickson, 1990b) that the success of the Macintosh DESKTOP metaphor is in part due to a largely consistent use of the CONTAINER image-schema.

If an image-schema is invariant in a metaphorical mapping, it must be a common part of the source and target domains. It is therefore possible to obtain the algebraic specifications of these domains by extending a common core specification which formalizes an image-schema. Such a process of adding operators to algebraic specifications is called an *enrichment* (Ehrich, Gogolla, and Lipeck, 1989). Thus, a formal version of Lakoff's invariance hypothesis is:

> For any metaphor, there is an algebraic specification which describes an image-schema or a combination of image-schemas and which can be enriched toward specifications of the source and target domains.

Since an algebraic specification describes a class of algebras or category, image-schemas are formalized as categories. These categories contain the algebras of the source and target domains as well as the morphisms between them.

The following is an algebraic specification of the SURFACE image-schema. It can easily be enriched toward the above specifications of the physical and electronic desktops by renaming the sorts (Surface → Desktop, Item → Folder, etc.) and adding the get operator with its axiom. The same can be said for the physical, but not for the electronic clipboard: The fact that a previously clipped item is lost when a new item is put on is not only a deviation from the behavior of physical clipboards, but also from the logic of the SURFACE schema.

```
Surface
Sorts                  Surface, Item, Bool
Ops                    new:     → Surface
  put:                 Surface x Item → Surface
  on:                  Surface x Item → Bool
Eqs                    on(new,i) = false
  on(put(s,i1),i2) = if i1=i2 then true else on(s,i2)
```

The other image-schema dominating the DESKTOP metaphor is the CONTAINER. Its specification happens to be isomorphic to that of the SURFACE schema. This correspondence may, in fact, contribute to the coherence of the DESKTOP metaphor. From a cognitive perspective, however, two different kinds of experiences are described. The CONTAINER schema can be specified as follows (see also Guttag, Horning, and Wing, 1985):

Container

Sorts	Container, Item, Bool
Ops	new: → Container
insert:	Container x Item → Container
in:	Container x Item → Bool
Eqs	in(new,i) = false
in(insert(c,i1),i2) =	**if** i1=i2 **then** true **else** in(c,i2)

This specification can again be enriched toward the previously given specifications of physical and electronic folders as well as physical trash cans. Thus, all these objects behave the way one intuitively expects from containers. No enrichment is possible, however, for electronic trash cans: The second axiom of the CONTAINER specification is violated by those cases where a discarded document is not in the trash without having been removed. Given how deeply rooted image-schemas are in human cognition, such mismatches with image-schemas are more likely to be harmful than source-target differences which are above the image-schematic level.

5. A Spatial Application

5.1. SPATIALIZED AND SPATIAL DOMAINS

The DESKTOP metaphor affords a spatialization for an abstract, non-spatial application domain: The source concepts of the physical office impose a spatial structure on the abstract target domain of information processing. A wide range of other spatializing interface metaphors provides evidence for the crucial role of spatialization in human-computer interaction. Examples include the NAVIGATION (Conklin, 1987), TOURIST (Fairchild, Meredith, and Wexelblatt, 1989), ROOM (Henderson and Card, 1986), and MUSEUM (Travers, 1989) metaphors. Indeed, the two- or three-dimensional nature of visual interaction itself implies a spatialization for visual interfaces (Tauber, 1987) in general.

At the same time, cognitive science has established that "most of our fundamental concepts are organized in terms of one or more spatialization metaphors" (Lakoff and Johnson, 1980) and that many image-schemas are spatial in nature. These findings suggest, too, that spatialization is probably the rule rather than the exception for interface metaphors.

When dealing with geographic space, however, interface metaphors for inherently spatial applications are of primary interest. An example of such an interface metaphor is furnished by zoom and pan operations in Geographic Information Systems (GIS) and other graphics applications.

The terms "zoom" and "pan" are used in computer graphics in analogy to manipulations of cameras or other optical instruments. An investigation of zoom and pan interface designs for GIS concluded, however, that intuitive and effective interface tools require a deeper understanding of the metaphors involved (Jackson, 1990). The remainder of this section applies the formal tools developed in the previous section to an analysis of the metaphorical and image-schematic structure of zooming.

5.2. ZOOMING AND THE "DISPLAYS ARE VIEWS" METAPHOR

The Oxford English Dictionary (second edition, 1989) defines the original meaning of "zoom" as follows:

> "To make a continuous low-pitched humming or buzzing sound;
> to travel or move (as if) with a 'zooming' sound;
> to move at speed, to hurry."

The use of the term in photography and cinematography is, thus, already doubly metaphorical: It explains the variation of the focal length by a (fictive) motion of rapidly closing up on a subject which, in turn, is metaphorically related to the corresponding sound effect. (It is an interesting twist that one of the key metaphors in visual interfaces is originally echoic, i.e. rooted in auditory perception.)

Combined with our deeply rooted visual experience that the viewing distance determines *what* we see, zooming acquires the interpretation of changing the level of detail at which we perceive and conceptualize the world or a computer model.

This understanding of zooming suggests the more general metaphor DISPLAYS ARE VIEWS, where views are understood as visual fields, containing what humans see in a given situation (Kuhn, 1990). This metaphor accommodates various transformations of the visual field in a common framework: Zooming is the transformation caused by changing the viewing distance; panning is moving the view to another part of a "panorama" without changing the level of perceived detail. Since transformations of the visual field correspond to "basic cognitive processes such as focusing, scanning, superimposition, figure-ground shifting, vantage-point shifting" (Lakoff, 1988), they are ideal candidates for metaphor sources.

5.3. IMAGE-SCHEMAS IN "DISPLAYS ARE VIEWS"

The basic image-schema involved in the visual field is the CONTAINER schema. It structures the visual field as a bounded space, consisting of a boundary, an interior, and an exterior: Things are either in or out of sight and they come into or go out of it (Lakoff, 1987). The visual field has also a center of attention and a surrounding region. Thus, a CENTER-PERIPHERY schema is combined with the CONTAINER schema, providing for the distinction of foveal and peripheral vision.

Enriching the previously specified CONTAINER schema by a center-periphery distinction produces the following specification (at this point, no axioms are given to constrain the inCenter operator):

```
Container with Center
Sorts                  CentCont, Item, Bool
Ops                    new:     → CentCont
  insert:              CentCont x Item → CentCont
  in:                  CentCont x Item → Bool
  inCenter:            CentCont x Item → Bool
Eqs                    in(new,i) = false
  in(insert(cc,i1),i2) = if i1=i2 then true else in(cc,i2)
```

The visual field is further characterized by an interaction of the PART-WHOLE with the NEAR-FAR schema. We experience objects in the world as configurations of parts, forming wholes. Our perception has evolved so that it can distinguish parts and wholes. Unaided visual perception, in particular, relies on motion of the body or of the objects to extend this distinction beyond the limited range of configurations present in one view: Getting parts into view involves moving nearer and vice versa. This connection is the essence of zooming.

To formalize the superimposition of the PART-WHOLE and NEAR-FAR schemas on top of the centered CONTAINER, the above specification is further enriched by a closeUp operator and a part-whole relation (which actually belongs to a separate Item specification). Let us call this multiply enriched container a View Container. With these additional operators, it becomes possible to capture the meaning of zooming in one axiom for the closeUp operator:

```
View Container
Sorts                  ViewCont, Item, Bool
Ops                    new:     → ViewCont
  insert:              ViewCont x Item → ViewCont
  in:                  ViewCont x Item → Bool
  inCenter:            ViewCont x Item → Bool
  partOf:              Item x Item → Bool
  closeUp:             ViewCont → ViewCont
Eqs                    in(new,i) = false
  in(insert(vc,i1),i2) = if i1=i2 then true else in(vc,i2)
  in(closeUp(vc),i) = inCenter(vc,i) or (inCenter(vc,i1) and partOf(i,i1))
```

Note that the closeUp operator is specified quite generally: It is only constrained to keep center items in view and get their parts into view; no details about the degree of zooming or the extent of the center region are given. More specific axioms would require metric concepts, while an image-schematic analysis is only concerned with topological properties.

Further enrichments with metric concepts would lead to complete source (view) and target (display) specifications for a particular user interface. They could involve, for example, a viewing distance in the source domain and a zoom factor or view extent in the target domain.

It should be emphasized that DISPLAYS ARE VIEWS is a metaphor and not a literal equivalence. The matching between the two domains of a metaphor is by definition partial. Aspects which are not image-schematic, such as the shape of the container boundary (elliptic, rectangular) or the type of optical projection involved (central, parallel), need not be equal or even comparable.

6. Conclusions

An algebraic approach to the formalization of interface metaphors has been presented. Algebraic specifications define metaphor domains as algebras, metaphorical mappings as morphisms, and invariant image-schemas as categories. The gist of the approach should be captured in the examples from a spatialization metaphor (DESKTOP). The sketched application to zoom operations should demonstrate the relevance to the design of user interfaces for systems dealing with spatial information.

The proposed formalization of interface metaphors is not a formal method for the design of metaphorical interfaces, but we believe it is a useful and immediately applicable step toward it. The principal strengths of the proposal are that:

- the underlying understanding of metaphors is in agreement with recent findings in cognitive science;
- the algebraic approach provides a comprehensive formalization of all essential aspects of metaphors: conceptual domains, structure-mappings, and image-schemas;
- the technique of algebraic specifications is well established in software design and supported by design tools.

The proposed formalization has yet to be validated in actual user interface designs. Our initial experiences from applications to Geographic Information Systems (Jackson, 1990) and to an interface based on book metaphors (Volta, Cicogna, and Kuhn, 1990) have been positive. The rigor imposed and the questions raised by the approach have clearly been beneficial in the design process.

Formal approaches to design are mostly motivated by the need to evaluate designs. Distinguishing different kinds of morphisms and separating the image-schematic structure from surface-level structure are solid bases for such evaluations, but much more theoretical and experimental work needs to be done in this direction. Evaluating interface metaphors also involves dealing with the coherence of composite metaphors. Since image-schemas are believed to play an important role for coherence, their formalization is a key feature of this approach.

Some aspects of our work may have implications beyond human-computer interaction. The description of metaphorical domains by algebraic specifications, for instance, is an attempt at formalizing the notion of Idealized Cognitive Models (ICM) (Lakoff, 1987). It accommodates propositional and image-schematic structure as well as metaphoric mappings. Another general result is the formal version of the invariance hypothesis.

7. Acknowledgements

Support from the NCGIA under U.S. National Science Foundation grant SES 88-10917, from Digital Equipment Corporation under grant TP-765536, and from Intergraph Corporation is gratefully acknowledged. We also thank Max Egenhofer and Jeffrey Jackson for discussions on the topic.

8. References

Apple Computer, Inc. (1987) Human Interface Guidelines: The Apple Desktop Interface, Reading, MA., Addison-Wesley.

Blumenthal, B. (1990) 'Incorporating Metaphor in Automated Interface Design', Proceedings Interact '90, 659-664.

Burstall, R.M., and J.A. Goguen (1981) 'An Informal Introduction to Formal Specifications Using CLEAR', in R. S. Boyer and J. Moore (eds.), The Correctness Problem in Computer Science, New York: Academic Press, 185-213.

Carroll, J.M., R.L. Mack, and W.A. Kellogg (1988) Interface Metaphors and User Interface Design, in M. Helander (ed.), Handbook of Human-Computer Interaction, North-Holland: Elsevier Science Publishers, 67-85.

Conklin, J. (1987) 'Hypertext: An Introduction and Survey', IEEE Computer 20(9), 17-41.

Ehrich, H.-D., M. Gogolla, and U.W. Lipeck (1989) Algebraische Spezifikation abstrakter Datentypen, Leitfäden und Monographien der Informatik, Stuttgart: B.G. Teubner.

Ehrig, H., and B. Mahr (1985) Fundamentals of Algebraic Specification, Berlin: Springer-Verlag.

Erickson, T.D. (1990a) Working with Interface Metaphors, in B. Laurel (ed.), The Art of Human-Computer Interface Design, Addison-Wesley, 65-73.

Erickson, T.D. (1990b) Personal communication.

Fairchild, K., G. Meredith, and A. Wexelblatt (1989) 'The Tourist Artificial Reality', Proceedings ACM CHI'89, 299-304.

Gentner, D. (1983) 'Structure-Mapping: A Theoretical Framework for Analogy', Cognitive Science 7, 155-170.

Gentner, D. (1989) The mechanisms of analogical learning, in S. Vosniadou and A. Ortony (eds.), Similarity and Analogical Reasoning, Cambridge, UK, Cambridge University Press, 199-241.

Goguen, J. A., J. W. Thatcher, and E. G. Wagner (1978) An Initial Algebraic Approach to the Specification, Correctness, and Implementation of Abstract Data Types, in R. Yeh (ed.), Current Trends in Programming Methodology, Englewood Cliffs, N.J., Prentice-Hall. 80-149.

Guttag, J.V., and J.J. Horning (1978) 'The Algebraic Specification of Abstract Data Types', Acta Informatica 10(1), 27-52.

Guttag, J.V., J.J. Horning, and J.M. Wing (1985) Larch in Five Easy Pieces, Digital Equipment Corporation, Systems Research Center.

Halasz, F., and T.P. Moran (1981) 'Analogy Considered Harmful', International Journal of Man-Machine Studies 14, 383-386.

Hall, R.P. (1989) 'Computational Approaches to Analogical Reasoning: A Comparative Analysis', Artificial Intelligence 39, 39-120.

Henderson, D.A., Jr., and S.K. Card (1986) 'Rooms: The Use of Multiple Virtual Workspaces to Reduce Space Contention in a Window-Based Graphical User Interface', ACM Transactions on Graphics 5(3), 211-243.

Indurkhya, B. (1989) Metaphor and Cognition, Boston University, unpublished manuscript.

Jackson, J.P. (1990) Visualization of Metaphors for Interaction With Geographic Information Systems, Masters of Science Thesis, University of Maine, Department of Surveying Engineering.

Johnson, J. (1987) 'How faithfully should the electronic office simulate the real one?', ACM SIGCHI Bulletin 19(2), 21-25.

Johnson, M. (1981) 'Metaphor in the Philosophical Tradition', in M. Johnson (ed.), Philosophical Perspectives on Metaphor, Minneapolis, University of Minnesota Press, 3-47.

Johnson, M. (1987) The Body in the Mind, Chicago, The University of Chicago Press.

Kuhn, W. (1990) 'Are Displays Maps or Views?', Technical Report, Department of Surveying Engineering, University of Maine.

Lakoff, G. (1987) Women, Fire, and Dangerous Things, Chicago, The University of Chicago Press.

Lakoff, G. (1988) Cognitive Semantics. In Meaning and Mental Representations, in U. Eco, M. Santambrogio and P. Violi (eds.), Bloomington, Indiana University Press, 119-154.

Lakoff, G. (1990) 'The Invariance Hypothesis: is abstract reason based on image-schemas?', Cognitive Linguistics 1(1), 39-74.

Lakoff, G., and M. Johnson (1980) Metaphors We Live By, Chicago, The University of Chicago Press.

Liskov, B., and J. Guttag (1986) Abstraction and Specification in Program Development, The MIT Electrical Engineering and Computer Science Series, Cambridge, MA., The MIT Press (MacGraw-Hill).

Mac Cormac, E.R. (1985) A Cognitive Theory of Metaphor, Cambridge, MA., The MIT Press.

MacLane, S., and G. Birkhoff (1967) Algebra, New York. Macmillan.

Mark, D. (1989) 'Cognitive Image-Schemata for Geographic Information: Relations to User Views and GIS Interfaces', Proceedings GIS/LIS'89 2, 551-560.

Martin, J.H. (1990) A Computational Model of Metaphor Interpretation, New York: Academic Press.

Nielsen, J. (1990) Designing User Interfaces for International Use, Amsterdam: Elsevier Science Publishers.

Ortony, A. (1979) Metaphor and Thought, Cambridge, UK, Cambridge University Press.

Smith, D.C.S., C. Irby, R. Kimball, B. Verplank, and E. Harslam (1982) 'Designing the Star User Interface', Byte 7(4), April 1982, 242-282.

Tauber, M.J. (1987) On Visual Interfaces and their Conceptual Analysis, in P. Gorny and M.J.Tauber (eds.), Visualization in Programming, Springer-Verlag, 106-123.

Travers, M. (1989) 'A Visual Representation for Knowledge Structures', Proceedings Hypertext '89, 147-158.

Volta, G.S., R. Cicogna, and W. Kuhn (1990) The MaineBook, Technical Report, Department of Surveying Engineering, University of Maine.

Vosniadou, S., and A. Ortony (1989) Similarity and Analogical Reasoning, Cambridge, UK, University Press.

Woodcock, J., and M. Loomes (1989) Software Engineering Mathematics, SEI Series in Software Engineering, Reading, MA., Addison-Wesley.

Wozny, L.A. (1989) 'The application of metaphor, analogy, and conceptual models in computer science', Interacting with Computers 1(3), 273-283.

ELICITATION OF SPATIAL LANGUAGE TO SUPPORT CROSS-CULTURAL GEOGRAPHIC INFORMATION SYSTEMS

MICHAEL D. GOULD
National Center for Geographic Information and Analysis
Department of Geography
State University of New York at Buffalo
Buffalo, New York 14261 USA

ABSTRACT. There exists a conceptual gap between the methods people naturally use to structure spatial relations and the functionality of user interfaces for geographic information systems to support these methods. Spatial cognitive methods may be inferred by eliciting subjects' verbal description of "real world", everyday spatial relations. The resulting collection of natural spatial language may then be used as a realistic goal in the GIS user interface design process, especially in query language formulation. Furthermore, because many potential GIS users are non-English speakers it is also helpful to survey the cross-cultural extent of spatial-linguistic terminology; to do so foreign subjects should be tested in their native environments. Verbal descriptions of geographic space may then be empirically compared to performance in executing simple geographical analyses using GIS-like testing software, or may directly support theories of spatial cognition. It is suggested that instead of teaching users to perform better using current GIS technology, we may support the design of more intuitive GISs by studying the way people naturally think about, and describe, geographic space. A summary of such a study, a related pilot study, and a methodological template for testing Spanish and English-speakers are each described here.

1. Introduction

The geographic information system (GIS) has been a valuable tool to geographers, foresters, ecologists, survey engineers and other geographic analysts since the mid-1960s (see Tomlinson, 1984, for a brief history). Traditionally, GISs have been developed with the expert technician (i.e., a system programmer) as target user. Because of this, the user interface portion of GISs has been arcane and cryptic at best. GISs were most commonly designed using human-computer interaction concepts inherited from more general computing systems, primarily from business (MIS) and engineering (CAD/CAM). Although GISs have successfully adopted alphanumeric data management and graphic display concepts, they remain fundamentally different from their 'parent' computing systems in that spatial relationships must be addressed explicitly. Problematic is that these spatial relationships must be addressed *simultaneously* by the system (internally, digitally) and by the user (externally, in his/her real world). Thus far, the morphism between the internal and external representation of spatial relationships in GIS has been unnatural and incomplete. Rather than GIS design being centered upon spatial relations that are natural and as people commonly structure them, GIS users are expected to acquire a set of data structure-based spatial concepts and the accompanying vocabulary. Users are conditioned to think about space as being organized by gridded fields (the raster GIS model) or by nodes, arcs, and polygons (the vector

D. M. Mark and A. U. Frank (eds.), Cognitive and Linguistic Aspects of Geographic Space, 435–448.

model). While it is one opinion that "users must be familiar with topology and data structures" (Ashworth, 1989, 20), we suggest that users need only to become more familiar with their cognate spatial science and that GISs need to be built so as to better accommodate human spatial cognition. The weak human-computer morphism that exists in today's GIS does not serve to "best liberate users' intellectual capacities" (Bjorn-Andersen, 1988, 390), to allow them to better solve social and environmental problems.

The relationship between spatial language and GIS may appear tenuous: one may be considered the domain of linguistics or psychology, the other a computerized exploratory tool. It is the union of the two, however, that is here the focus of attention. In order for GIS to be optimally applied to environmental and social problems, many more of the world's "spatially aware professionals" (Raper and Bundock, 1991) must be given access to it. Furthermore, for GIS to become publicly accessible, in libraries, town halls, and taxicabs, verbal interaction will be a necessary option. To allow such widespread access, however, we must first learn more about the potential users' conception of space and how they choose to describe it.

The spatial language—GIS relationship may be studied at several levels of resolution (e.g., the individual, the disciplinary group, the society, the culture). Here the main interest is at the cultural level, but primarily that portion of cultural differences revealed in language. Recently, this topic—cross-linguistic differences in spatial language for GIS—has been a focal point of Research Initiative 2, *Languages of Spatial Relations*, of the U. S. National Center for Geographic Information and Analysis (NCGIA, 1989). What direction has the research taken so far? One direction, among many, has been in making explicit an obvious yet oft-ignored fact: that most geographic information systems have been designed in English. This is true even of some systems designed by non-anglophones (i.e., MICROM España, 1989). Thus, query languages, graphical user interfaces and GIS data models (e.g., arc/node, object-oriented) contain artifacts not only of the few available data structures, but also of the spatial relations which can be described in English (see Mark, Gould, and Nunes, 1989, for examples of relations not allowed in English). Natural language studies, particularly those based on cognitive linguistics (Talmy, 1983; Lakoff, 1987, 1990; Herskovits, 1989), are important for GIS designers to consider because they can help pinpoint situations where users might prefer to (or need to) "vicariously interact with real or possible phenomena of the world" (Mark, 1989) using natural language. These situations may involve domain-specific language for disciplinary groups of GIS users (e.g., foresters, wildlife ecologists); input of geographic information in text form, such as biological localities of collected specimens (McGranaghan, 1989); and, at a more abstract level, spatial object definition and data-structure design based on human cognition, in order to support natural 'user views' of spatial data.

To summarize, it is important that GIS designers look beyond the bounds of the English speaking world of computer experts in pursuit of interfaces to support a highly diverse potential user population. Important long-term gains in GIS human-computer interaction will likely come from cross-cultural studies in cognitive science and not necessarily from the development of a better computer peripheral or database management system. It is suggested that systems which mimic human cognitive structure and behavior are more likely to be useful and effective than systems which are designed based primarily upon current hardware and software capabilities (Carroll, Mack, and Kellogg, 1988; Mark, Gould, and Nunes, 1989; Gould, 1989; Mark, 1989; Gould and McGranaghan, 1990). GIS designers cannot rely solely upon introspective research and personal assumptions regarding the cognitive structure of spatial relations in the real worlds of perspective users. Rather, they must test potential (and actual) users in their natural environments, in search of core commonalities.

2. Two Relevant Approaches to Human-Computer Interaction

Two multidisciplinary research approaches are applicable to the search for improved spatial human-computer interaction for GIS; they are described in sections 2.1 and 2.2. To date, the human-computer interaction problem (i.e., the conceptual gap between human and automated structuring of the "real world") has not fully been recognized by GIS researchers. Only a dozen or so articles discussing human-computer interaction have entered the GIS literature during the past few years (an early example is Porter, 1985). Many of these articles focus upon previously implemented interface concepts from nonspatial disciplines; attempts to link these concepts to spatial data handling have been weak. This is partly due to the fact that we know very little about space, or at least have not agreed on which interpretation to accept (Harvey, 1969; Mark and Frank, 1989). This weak link to spatial data handling is also partly due to adherence to the engineering philosophy of system design, which often relies on teaching users to cope with current technology rather than examining the strengths and weaknesses of the human mind. This is not to suggest that engineers have not contributed to the human-computer interaction problem, only that blind adherence to the engineering philosophy often neglects the end user.

An example of possibly misguided engineering philosophy is the current focus on extending structured query language (SQL) to form a general spatial query language (Egenhofer, 1987; Ashworth, 1989; Raper and Bundock, 1991). Proponents extoll the current popularity and emerging standard of SQL as validating its incorporation by the next generation of GIS. An opposing view, however, sees SQL extension as a myopic solution to the real problem of spatial query language interaction: representation of the user's view of spatial phenomena. In this view SQL is an unnatural abstraction, a cognitive hurdle in the path of the potential user. He or she must rifle through a textbook to learn SQL—a language optimized for the relational DBMS model and mainframe computers, not for humans— in order just to communicate with the GIS. Many people, including engineers and computer scientists (see Egenhofer, 1991), recognize the danger of adhering to the SQL *de facto* standard solely for the sake of homogeneity. Worboys and others (1990) concur, and explain the options:

> "There are two main approaches to the construction of a spatial query language. Either one starts from fundamentals or one builds on top of an existing language such as SQL. At present, the latter is the favoured approach. However, it is well recognized within the database community [see, for example, Whittington 1988] that SQL itself has shortcomings. The authors are engaged in the construction of a spatial query language using the former approach." (Worboys *et al.*, 1990, 382)

These two views of SQL highlight the philosophical polarity within the human-computer interaction community. The two philosophies can, however, peacefully coexist. The engineering philosophy pragmatically ensures that new ideas get 'up-and-running' as soon as possible, or that they get implemented at all! The cognitive (scientific) philosophy has a less pragmatic research agenda, concerned not with products but with answers to fundamental questions. Both philosophical approaches will make contributions, as explained in the following two sections, to the human-computer interaction problem. In the worst case they will serve as checkstops and will guide each other's progress.

2.1 THE ENGINEERING APPROACH: HUMAN FACTORS

Human factors (often called *ergonomics*) is a growing subdiscipline of engineering that deals with human response to existing technology in order to aid the design of future technology. Thus, human factors can be thought of as an application of the engineering design philosophy. Although its roots are in the design of mechanical devices to support pilots during World War II, a significant part of human factors is now dedicated to human-computer interaction (HCI). It is of interest to this book that many of the difficulties associated with human-computer interaction are linguistic; for example, some languages—such as German— are unable to condense complex, multiple word concepts, which makes precise on-screen menu interaction difficult (Sukaviriya and Moran, 1990). Human factors studies have focussed some attention on cross-linguistic and cross-cultural aspects of software design during the last several years. Exemplary of this attention are foreign language versions of American-designed word processors, which allow for vertical (Asian languages) or right-to-left (Hebrew) text formatting, as well as differing hyphenation methods (needed for Spanish). Design of cross-linguistic menus, keyboard layouts, query languages, etc. is commonly in the domain of human factors engineers (see Helander, 1988, for a full review).

The human factors approach proceeds top-down in search of better human-computer interaction solutions. This involves testing existing (fully implemented or prototype) systems and recording user performance. A critical view of this approach may describe it as follows: "I'll give the user the system; he'll break it; I'll tell him what went wrong." This may prove the quickest path to discovering optimal interfaces for GIS, but it may also lead to dead-ends, due to the lack of a heterogeneous sample of GIS solutions for testing.

The discipline of human factors is of considerable value in that it offers interesting points-of-view from non-traditional spatial analysts—people who at times deal with GIS topics without labelling them as such. The human factors literature is rich in analogs to GIS (e.g., facility management systems, military command systems, navigation systems) which may be exploited in support of GIS design (Gould, 1989). The role to be played by human factors (relative to other disciplines) in bridging the user-GIS gap is indicated in figure 1.

2.2 THE COGNITIVE APPROACH

The cognitive approach (termed the "behavioral approach" by Bjorn-Andersen, 1988) to human-computer interaction has tended to begin by addressing models of cognition and behavior without regard to the state of current technology. Rather than teaching users to adapt in a manner that will subjectively optimize system performance, the cognitive approach encourages the design of systems which may be used efficiently with minimal modification of cognitive style and behavior. This approach stresses constant user involvement in the iterative design process (that is, *before* the product is released to the public). This is the user interface design approach advocated by John Carroll (1987; Carroll and Olson, 1988; Carroll, Mack, and Kellogg, 1988) and other applied cognitive psychologists.

Looking beyond the roster of full-time, cognitively-inclined HCI researchers, we suggest that many behavioral geographers (Golledge, Rushton, Lloyd), planners (Lynch, Appleyard), cognitive psychologists (Spencer, Blades, Thorndyke, Hayes-Roth), and computer scientists (Kuipers, McDermott) should be recognized as well (citations for many of these researchers are found in Freundschuh, 1991). Most of these researchers have no connection with, and perhaps no interest in GIS, yet they provide necessary building blocks to the understanding of spatial cognition, suggested here as a critical element of human-computer interaction for GIS.

The role to be played by cognitive scientists (under the general category 'cognitive studies') is indicated in figure 1. It occupies the region along the human-computer bridge closest to the human, and it remains near to the social sciences (i.e., geography).

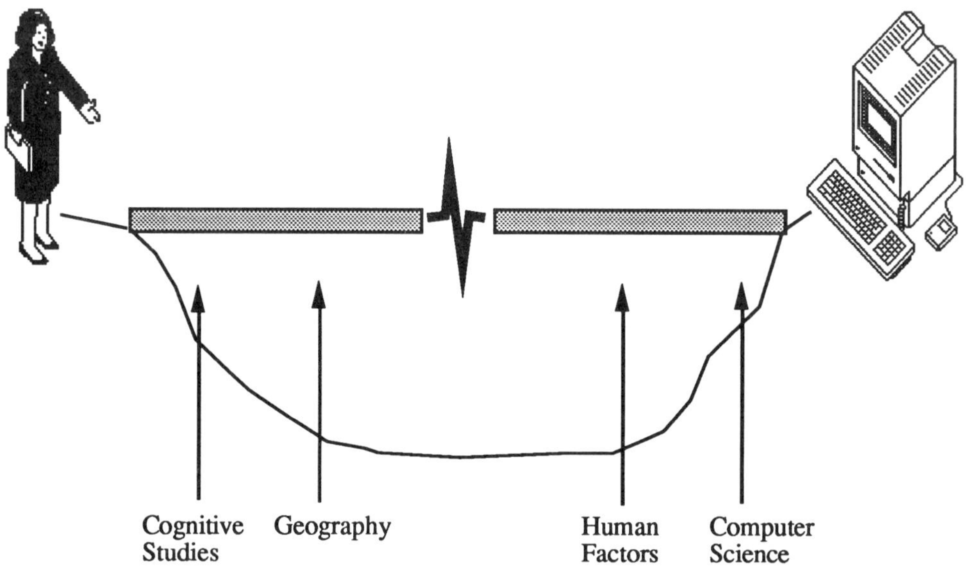

Figure 1. The GIS human-computer interaction bridge is a broken continuum; related disciplines may contribute at appropriate levels along this continuum in order to complete the bridge.

2.2.1. *Cognitive Engineering?* An integrative perspective is taken by Don Norman (1986, 1987), who posits that while cognitive science affects the applied field of human-computer interaction, the converse can happen as well. He emphasizes that which may be learned about cognition based upon analysis of people's performance using computers. Norman's "cognitive engineering" may serve to exploit the best of both the cognitive and engineering design philosophies (see also Rasmussen and Vicente, 1989, on this topic). It may also serve as a mere compromise to both positions. Gould and Helander (in preparation) point out that while this systems analysis approach to the HCI problem is functionally adequate for a closed environment, such as the control room of a power plant, it fails to consider the external environmental aspects necessary to support the system—>real world linkage needed for GIS. More investigation into this curious marriage of engineering and scientific philosophies is needed.

3. Collecting Natural Spatial Language

To support the design of cognitively-attuned natural language interfaces for GIS, we must sample true natural spatial language, taken from within the user's environment. This section describes an on-going research project for doing just that—a project which exploits conceptually and methodologically both approaches described above. The project is designed to elicit common methods of structuring and expressing geographic spatial relations, in both English and Spanish speaking cultures and to then offer GIS user interface design strategies for supporting these methods. By "geographic spatial relations" is meant those spatial relations not visible in their entirety from a single vantage point and, thus, often necessitating the use of maps. Thus, egocentric spatial relations, such as in "my hand is on the table," and extraterrestrial spatial relations, such as in "the moon revolves around the earth, " are not of direct interest here. After describing, in sections 3.1 and 3.2, the significance of the project, section 3.3 describes a pilot study conducted during the summer of 1990.

This research project seeks answers to three basic questions:

- What are common methods, across English and Spanish-speaking cultures, of structuring and describing geographic space?
- How do the methods by which people structure geographic space relate to the design of GIS?
- What new, uniquely spatial, user interfaces might be designed to allow more effective use of GISs?

In order to answer these questions, we must disregard the assumptions and hearsay of GIS researchers (including ourselves), and we must instead test potential and actual GIS users in the field. In this project, college students and professionals from various disciplines are being tested in Buffalo, New York and in Quito, Ecuador (cross-linguistic implications of this research are discussed in section 3.2). The project relies on two coordinated methods of data collection and analysis. The first method takes the form of an open-ended, oral questionnaire (see below), the results of which are tape-recorded and analyzed using automated verbal protocol analysis (Ericsson and Simon, 1984; see Lundberg, 1984, for application to spatial behavior). The second method, not discussed here, takes the form of observed GIS user interface experimentation: a simulated GIS, implemented on an Apple Macintosh using rapid-prototyping tools (described in Gould and Love, 1988), is made available for each subject to experiment with. Several types of user actions are recorded automatically and observations are made by the experimentors.

3.1 SIGNIFICANCE OF STUDYING SPATIAL HUMAN-COMPUTER INTERACTION

What contribution to knowledge and to science can be made by studying user interfaces for geographic information systems? A clue is offered by the executive director of the Association of American Geographers, Ronald Abler. He compares the GIS to the microscope in its potential contribution to the sciences (others argue the macroscopic, rather than microscopic, value of GIS). Abler states that GIS provides geographers with the increased "opportunity to address the themes that will dominate science in the 1990s: global environmental problems and globalization of socioeconomic affairs" (Abler, 1989, 2). In order to do so, however, GIS must be designed to be more easily used in both novice/expert and cross-cultural contexts. In a sense, it is necessary that GIS attain at least the level of acceptability and ease-of-use as did the word-processor during the previous decade. Waldo Tobler (1987) wonders what Torsten Hägerstrand or other "world class" geographers might have accomplished having had early access to GIS. Extending that idea, Gould (1989) wonders what novices (realtors, ambulance drivers, history teachers), of any ethno-linguistic group, might be able to accomplish if they were able to *understand and comfortably use* today's GIS. This question now applies directly to environmental and social scientists of the future. What is needed is a fresh look at the linkage between the human and spatial information through the GIS. The key to this linkage is the human-computer interface. In order for geographers and other spatial scientists to maintain any influence over such a valuable tool as GIS, we must work diligently to define how humans manipulate geography in their "real worlds", so that we can influence GIS developers (often computer scientists) to support these manipulations more directly.

3.2 SIGNIFICANCE OF CROSS-LINGUISTIC ISSUES

While this paper mentions often the term cross-cultural, the aspects of culture revealed through language are the main concern here. Nations that speak languages other than English can be expected to make up a large proportion of the GIS market in the future. Therefore, design of GIS should consider language-based schemata and primitives derived from cross-linguistic studies of geographic concepts and relations. Otherwise, transferring many GIS concepts from English-designed systems to other languages will potentially offer nothing more than ill-formed word-for-word translations. This can be a serious impediment, not only to basic system operation but to geographic problem solving as well. These two levels of GIS use must be distinguished. The former is related primarily to the ability to set-up, turn-on, and manipulate the system, whereas the latter involves problem solving and theory development which may otherwise not have happened (as quickly, or at all) without the GIS. The first type of usage demands, at a simplistic level, salient command names and well-written documentation, while the second type demands that the system take on a familiar appearance and incorporate natural thought processes. This means, among other things, that the system accommodate the intentions of the user, often expressed in language.

For example, suppose that a GIS procedure has been successfully written to determine the geographical meaning of the English word "on", as in "hospitals on the Amazon River". Translation to Spanish would present possible ambiguity. Most geographical uses of "on" would become "en" in Spanish, but "en" also includes many cases where English would use "in." Furthermore, "sobre" and "encima" may occasionally be used for situations represented by the German "auf," but never for German "an" situations meaning "on" (Mark, Gould, and Nunes, 1989). Some of these spatial preposition-based relationships have been addressed by Herskovits (1987, 1989) and Lakoff (1987) but only in English and not exhaustively. Other languages, especially those non-indoeuropean, cause even more pronounced ambiguity or misunderstanding.

As problematic, perhaps, is that some computer system commands are formed as regional idiomatics or in technical jargon. An example was discovered on a visit to Spain during the summer of 1989: a cadastral management system, designed in a small city in Cataluña, contained a menu command titled "digitalización nube de puntos," literally meaning "digitize a cloud of points." What type of system function or manipulation is evoked by this command? This may seem as strange to English speakers as GIS commands such as "buffer" and "spread" are to many native

Spanish speakers. Buffer, for example, does not have a direct Spanish counterpart (especially when used as a verb by GISs). In the GIS context, it usually means to delineate a zone around a punctiform or linear feature (such as a house or a river). The selected zone (buffer) is then used to constrain future spatial database queries, as in the case, "show all schools within [a buffer of] 500 meters of a major river". This use of buffer creates a continuous zone, whereas the closest Spanish verb, escampar, is best used to specify a scattering (of discrete objects) rather than the spreading assumed by the buffer command. The result of this lack of direct translation is that the GIS command in Spanish is now simply called "buffer"; the user must either learn the concept in English or issue the command by rote as a pattern-matching exercise devoid of semantics. Criticism of this single command may seem trivial, but many of today's most popular GISs (such as Arc/Info and TIGRIS), have 500-1500 commands. The interface complexity of these systems is a major threat to their cross-cultural adoption

3.3 A PILOT STUDY

As a pilot to the research project described above, Spanish subjects were tested during a visit to Barcelona, Spain (June 1990). Logistical support was provided by the Department of Geography at the Autonomous University of Barcelona. Following is a synopsis of the pilot study and a brief discussion of its results.

3.3.1. *Study Summary.* During three days of testing, an oral questionnaire was used to elicit qualitative information about Spanish-speakers' manipulation (using maps) and description of particular geographic spatial relations. Results were tape-recorded and later analyzed. The second part of the project described above, GIS user interface testing, was not included in this pilot study due to time constraints.

Why Spanish? Spanish was recently identified at the NCGIA Research Initiative 2 (Languages of Spatial Relations) Specialist Meeting as being a high priority language for cross-linguistic research associated with GIS (NCGIA, 1989). This is party due to strong interest in applications of GIS to the growing environmental and economic problems of Central and South America. This interest is evident by the high attendance at the Second Latin American Conference on Geographic Information System Technology, in Mérida, Venezuela in September 1989, and by the recent emergence of large GIS projects in Brazil, Chile, Colombia, Costa Rica, and Ecuador among other nations.

3.3.2. *Human Subjects.* The 14 experimental subjects in Barcelona were undergraduates (3), graduate students (9), and professors (2) from various disciplines within the social sciences. These subjects were not remunerated for their participation but were instead each asked to volunteer 20-30 minutes of their time. This experiment was earlier approved by the University at Buffalo Human Subjects Review Board.

3.3.3. *Questionnaire Administration.* An oral questionnaire was administered in Castilian Spanish, individually, to each subject in the pretest. It should be noted that while in Barcelona Castilian Spanish is considered a language secondary to Catalan, it is still widely in use and was in may cases the primary language for many of the subjects during their formative years.

The questionnaire consisted of 21 questions related to personal background, geographic-scale spatial relations, and the subject's structuring and manipulation of these relations. Questions were the same for each subject and were designed to elicit qualitative rather than quantitative responses;

most were suitably open-ended. Subjects were allowed (neither encouraged nor discouraged) to draw on a pad of white paper while answering. Subjects were asked to consult several maps of Puerto Rico (standard US Geological Survey maps with scales of 1:20,000 and 1:240,000) for the specific spatial-relation questions. These maps include both English and Spanish text. Puerto Rico was chosen as the study area for the tests because of its relative geographic neutrality and linguistic duality, given that the subjects were from Spain and later would be from Quito and New York. The oral questioning was tape recorded and written notes were recorded when necessary. A graduate student from the geography department in Barcelona assisted logistically, although the questionnaire was administered solely by this author, a near-fluent Spanish speaker. The administrators answered questions and prompted the subjects only when absolutely necessary; for the most part, they remained silent and allowed the monologues to stray tangentially if the subjects wished so.

Examples of the questions are as follow.

Background

- What is your area of specialty (academically)?
- What is your current position? (e.g., student, professor, secretary)
- What is your level of competency in English? What formal language training do you have? What other languages do you know?
- Where were you born? Where do you now live?
- Have you ever travelled to Puerto Rico? The Caribbean? The USA?
- Have you visited any island? If so, which ones?

General spatial relations

- Describe the relationship between your house and your school campus.
- Describe the relationship between the Barcelona train network and your school campus.

Specific spatial relations

- Given several maps of the city (San Juan, PR), how would you go about determining the best location to build a new public school? Where would you locate one?
- Given various maps of Puerto Rico, explain where you would locate a (vacation) home. Why would you choose this location over all others?
- Given various maps of Puerto Rico, explain how and where you would locate a disposal site for dangerous chemicals or explosives. Why there?
- Given the same maps, what would be the best route for transporting dangerous materials from San Juan to the site chosen in the previous question? What mode of transportation is best?

This questionnaire may, with slight modification perhaps, serve as a template for use by other researchers in the elicitation of comparative Spanish and English spatial language. The mapped study area would need to be substituted in order to accommodate the comparison of other linguistic pairs. While this questionnaire may seem to lack specificity and depth it should be noted that it is relatively difficult to convince subjects to participate in any study of duration longer than 10 or 15 minutes, and a 15 minute oral questionnaire may potentially generate 5 or 6 pages of dialog when transcribed. Therefore, researchers are advised to consider potential diminishing returns of administering oral questionnaires significantly longer than this one.

3.4 GENERAL PILOT RESULTS

Responses to a similar oral questionnaire in Quito (the main study) were tape-recorded and are being analyzed using automated verbal protocol analysis. This will allow the unstructured dialog to be properly categorized for comparison with that of the English-speaking subjects. The Barcelona (pilot) responses, however, were not analyzed using this procedure but are here summarized based upon a review of the tapes.

Like previous elicitation of spatial language (i.e., collection of driving directions in Gould, 1990a), the result of this pilot study is a collection of unstructured dialog recorded on tape. While this is not the inferential statistician's preferred data format, it does provide the advantage of considerable grounding in reality, or "ecological validity" as it is called in psychology (see Ericsson and Simon, 1984, for more on "verbal reports as data"). Ecological validity is strongest, perhaps, when the data are gathered covertly as in Gould (1990)— that is, subjects behave less naturally when they are aware of being recorded or observed. On the other hand, methodological purists will argue that covert experiments cannot possibly account for the variability in testing conditions, human interaction, experimenter bias, sample selection, etc. Therefore, a compromise must be made when designing a spatial-cognitive experiment, between a natural yet uncontrolled environment or a rigorously controlled environment (the laboratory). This pilot study attempted to include positive attributes of both environments: the naturalness associated with the withholding of the actual testing goal (i.e., the linguistic medium used to convey answers, not the answers themselves) and the control of a consistent testing environment (i.e., same office setting, single experimenter / single subject, scripted questionnaire, map-based study area).

3.4.1. *Descriptions of General Spatial Relations.* Questions regarding general spatial relations (those related to Barcelona rather than the Puerto Rico study area) were included both as a subject warm-up and to elicit a small amount of information about the subject's familiar environment. The questions were worded in such a way as not to imply that a spatially-oriented response was being solicited. This was intentional, to reveal whether or not the subjects would assume that the "relation" between two entities (i.e., home and school) would be considered spatially, as locations, or in another context. The responses show a distinct tendency toward a spatial, linear or procedural view of the environment. When asked to describe the relationship between their house and the campus, nearly all subjects (12 of 14) described this relationship in terms of driving time or distance. The most common form of description was a route simulation, in the present or conditional tense—the Spanish equivalent of "I leave my house, then I drive for 15 minutes on the highway, then I take exit 999" for example. The same spatial, procedural view was evident for the question (again, open-ended) of the "relation" between the "Catalanes" local train network to their campus. To this question, two subjects responded they did not know anything of this relationship, due to having never used the train to commute to school. Because these two subjects had been students on the campus for 3 and 4 years, respectively, and thus are not newcomers to the environment, it may be assumed that they do not hold "map view" or configurational knowledge of the campus environs (see Freundschuh, 1991; Blades, 1991, for more on these types of spatial knowledge).

It will be interesting to compare these spatial, procedural views of spatial relations with the results of this questionnaire from Quito and Buffalo, and from other locations hopefully to be studied by other researchers in the near future.

3.4.2. *Descriptions of Map-based Spatial Relations*. The questions regarding the study area of Puerto Rico also produced interesting, yet not surprising, responses. For example, when asked to locate a summer house on the island, most subjects chose a location on the outskirts of a major city (either San Juan or Ponce). Several comments were made regarding the need to remove oneself from the pollution and traffic of a city the size of San Juan, yet to retain access to the city's amenities. Moreover, 4 of the 14 subjects chose to locate their hypothetical summer house in such a region in disregard or ignorance of the terrain as indicated by the topographic maps supplied. Each located his house in the midst of marsh/swampland, perhaps due to cartographic illiteracy or to ignorance of physical factors in favor of optimizing Euclidean distance from their service center. This may also be explained by the fact that, these days, we in Western cultures often simply fill-in marshland in order to build to suit our needs.

In responding to another question regarding the location of a dump for dangerous materials (such as explosives), several subjects chose off-shore locations, on smaller islands. Another subset of subjects chose to transport the materials by rail or truck into the central, mountainous "Cordillera" region of Puerto Rico. Only 1 of the 14 subjects (a geologist) mentioned geologic or pedologic considerations. Most subjects were concerned only with minimizing perceived local population near the dump site.

Analysis of this pilot data set is still in progress; further summary of these oral responses including in-depth scrutiny of the spatial language used will appear in a later publication, at which time they will be compared to the results obtained in Buffalo and in Quito.

4. Summary and Future Research Plans

All too often, design decisions are made on the advice, consent, and assumptions of researchers operating introspectively, in isolation from the "real worlds" of the potential users. In the case of GIS, too often the system designer (often from a nonspatial discipline) decides what should appear within the system based upon his/her desires as a user. Not only are potential users from diverse subdisciplines such as architecture and wildlife biology ignored, but users in cultures other than Anglo-American suffer from even greater ignorance.

This research project seeks to support both English and Spanish-speaking GIS users and potential users. Some preliminary user testing in Barcelona was described. Testing is now underway in Buffalo, New York and in Quito, Ecuador. This includes administration of the oral questionnaire, described above, followed by the introduction to several simulated Macintosh-based GISs about which the subject is asked to comment. As in Barcelona, all instructions, questions, and further assistance in Quito are presented in Spanish with the help of a technically-oriented Spanish translator. This research project is designed to survey the potential needs of a small set of Spanish and English-speakers with regard to geographic analyses with and without the benefit of GIS. It is suggested here that only through the systematic collection of natural spatial language are we able to generalize (a primary goal of science) across disciplinary and cultural groups with regard to the possible needs of future GIS users.

5. Acknowledgements

This paper represents part of Research Initiative 13, *User Interfaces for Geographic Information Systems*, of the National Center for Geographic Information and Analysis, supported by the U.S. National Science Foundation. Field work in Barcelona was funded by NSF Grant SES 90-01229; the National Science Foundation is gratefully acknowledged. Thanks to Joan Nunes and Joan Carles Llurdei for useful suggestions and logistical assistance in Barcelona, and to David Mark for travel support to the NATO ASI in Las Navas del Marqués.

6. References

Abler, R. (1989) Executive Director's Column, AAG Newsletter 24(8), Association of American Geographers, 2.

Ashworth, M. (1989) 'Discussion Regarding Extensions to SQL', in D. M. Mark, A. U. Frank, M. Egenhofer, S. Freundschuh, M. McGranaghan, and M. White (eds.), Languages of Spatial Relations: Initiative Two Specialist Meeting Report, NCGIA Technical Paper 89-2, University of California, Santa Barbara, 22.

Bjorn-Andersen, N. (1988) 'Are 'Human Factors' Human?', The Computer Journal 31, 386-390.

Blades, M. (1991) 'The Development of the Abilities Required to Understand Spatial Representations', in D. M. Mark and A. U. Frank (eds.), Cognitive and Linguistic Aspects of Geographic Space, Kluwer, Dordrecht.

Carroll, J. M. (ed.) (1987) Interfacing Thought: Cognitive aspects of human-computer interaction, MIT Press: Cambridge, Massachusetts.

Carroll, J. M. and Olson, J. R. (1988) 'Mental Models in Human-Computer Interaction', in M. Helander (ed.), Handbook of Human-Computer Interaction, North-Holland.

Carroll, J. M., Mack, R., and Kellogg, W. A. (1988) 'Interface Metaphors and User Interface Design', in M. Helander (ed.), Handbook of Human-Computer Interaction, North-Holland.

Egenhofer, M. J. (1987) 'An Extended SQL Syntax to Treat Spatial Objects', in Y. C. Lee (ed.), 2nd International Seminar on Trends and Concerns of Spatial Sciences, New Brunswick, Canada, 83-95.

Egenhofer, M. J. (1991) 'Deficiencies of SQL as a GIS Query Language', in D. M. Mark and A. U. Frank (eds.), Cognitive and Linguistic Aspects of Geographic Space, Kluwer, Dordrecht.

Ericsson, K. A. and Simon, H. A. (1984) Protocol Analysis, The MIT Press, Cambridge, MA.

Freundschuh, S. F. (1991) 'The Effect of the Pattern of the Environment on Spatial Knowledge Acquisition', in D. M. Mark and A. U. Frank (eds.), Cognitive and Linguistic Aspects of Geographic Space, Kluwer, Dordrecht.

Goodchild, M. F., Kemp, K., Mark, D. M., and Frank. A. U. (1989) 'The U.S. National Center for Geographic Information and Analysis: An Overview of the Agendas for Research and Education', Proceedings, Challenge for the 1990s: GIS, Ottawa, Canada, 918-925.

Gould, M. D. (1989) 'Human Factors Research and its Value to GIS User Interface Design', Proceedings, GIS/LIS '89, American Congress on Surveying and Mapping (ACSM),Orlando, Florida, 542-550.

Gould, M. D. (1990) 'Comparing driving directions in English and Spanish-speaking cultures', Presented at the Annual Meeting, Association of American Geographers, Toronto, Ontario, April 21, 1990.

Gould, M. D. and Helander, M. (in prep.) 'Rasmussen's Cognitive Engineering: Implications for geographic information systems', State University of New York at Buffalo, Buffalo, NY.

Gould, M. D. and Love, S. R. (1988) 'An Easily-customized User Interface for GIS and Map Display Systems', Technical Papers, American Congress on Surveying and Mapping (ACSM) Fall Meeting, Virginia Beach, 255-264.

Gould, M. D. and McGranaghan, M. (1990) 'Metaphor in Geographic Information Systems', Proceedings, 4th International Symposium on Spatial Data Handling, Zurich, Switzerland, 433-442.

Harvey, D. (1969) Explanation in Geography, Edward Arnold, London.

Helander, M. (ed.) (1988) Handbook of Human-Computer Interaction, North-Holland.

Herskovits, A. (1987) Spatial Prepositions in English, Cambridge University Press, Cambridge, Massachusetts.

Herskovits, A. (1989) Language and Spatial Cognition, Cambridge University Press, Cambridge, Massachusetts.

Lakoff, G. (1987) Women, Fire, and Dangerous Things: What categories reveal about the mind, University of Chicago, Chicago.

Lakoff, G. (1990) 'The Invariance Hypothesis: Is abstract reason based on image-schemas?', Cognitive Linguistics 1(1), 39-74.

Lundberg, G. (1984) 'Protocol Analysis and Spatial Behavior', Geografiska Annaler 66B, 91-97.

Mark, D. M. (1989) 'Cognitive Image-schemata for Geographic Information: Relations to user views and GIS interfaces', Proceedings, GIS/LIS '89, American Congress on Surveying and Mapping (ACSM), Orlando, Florida, 551-560.

Mark, D. M. and Frank, A. U. (1989) 'Concepts of Space and Spatial Language', Proceedings, International Symposium on Computer-Assisted Cartography (Auto-Carto 9), American Congress on Surveying and Mapping (ACSM), Baltimore, 538-556.

Mark, D. M., Gould, M. D., and Nunes, J. (1989) 'Spatial Language and Geographic Information Systems: Cross-linguistic issues', Proceedings, II Conferencia Latinoamericana sobre la Tecnologia de los Sistemas de Información Geográficos, Mérida, Venezuela, 105-130.

McGranaghan, M. (1989) 'Context-free, Recursive-decent Parsing of Location-description Text', Proceedings, International Symposium on Computer-Assisted Cartography (Auto-Carto 9), American Congress on Surveying and Mapping (ACSM), Baltimore, 580-587.

MICROM España (1989) MICROM Image Processing System: user documentation, Barcelona, Spain.

NCGIA (1989) 'Languages of Spatial Relations: Researchable questions', Technical Paper 89-2a, National Center for Geographic Information and Analysis, University of California, Santa Barbara.

Norman, D. A. and Draper, W. S. (eds.) (1986) User-centered Design: New perspectives in human-machine interaction, Erlbaum Associates, Hillsdale, NJ.

Norman, D. A. (1987) 'Cognitive Engineering—Cognitive Science', in J. M. Carroll (ed.), Interfacing Thought: Cognitive aspects of human-computer interaction, The MIT Press, Cambridge, MA., 325-336.

Porter, E. M. (1986) 'Human Factors Design Considerations for the Next Generation of Geographic Information Systems', in B. Opitz (ed.), Geographic Information Systems in Government, A. Deepak Publishing, Hampton, VA., 455-463.

Raper, J. F. and Bundock, M. S. (1991) 'UGIX: A layer based model for a GIS user interface', in D. M. Mark and A. U. Frank (eds.), Cognitive and Linguistic Aspects of Geographic Space, Kluwer, Dordrecht.

Rasmussen, J. and Vicente, K. (1989) 'Cognitive Engineering: An ecological frontier', Sixth International Conference on Event Perception and Action, July 1989, Miami Univ. of Ohio (available from Riso National Laboratory, Roskilde, Denmark).

Sukaviriya, P. and Moran, L. (1990) 'User Interfaces for Asia', in J. Nielson (ed.), Designing Interfaces for International Use, Elsevier Science Publishers, 189-218.

Talmy, L. (1983) 'How Language Structures Space', in H. Pick and L. Acredolo (eds.), Spatial Orientation: Theory, research, and application, Plenum Press.

Tobler, W. (1987) Discussant Paper, Proceedings, International Geographic Information Systems Symposium (1), NASA, Crystal City, VA., 49-52.

Tomlinson, R. (1984) 'Geographic Information Systems: A new frontier', Proceedings, First International Symposium on Spatial Data Handling, Zurich, Switzerland, 1-14.

Whittington, R. P. (1988) Database Systems Engineering, Oxford University Press, Oxford.

Worboys, M. F., Hearnshaw, H. M. and Maguire, D. J. (1990) 'Object-oriented Modelling for Spatial Databases', International Journal of Geographical Information Systems 4(4), 369-383.

UGIX: A LAYER BASED MODEL FOR A GIS USER INTERFACE

J. F. RAPER and M. S. BUNDOCK
Department of Geography
Birkbeck College
7-15 Gresse Street
London W1P 1PA United Kingdom

ABSTRACT. Work is under way in the UGIX project to define a layer based model of user interaction with a GIS. The key layer is a command language which can be assembled into procedures reflecting certain tasks required of a GIS. The highest level is a graphical user interface (GUI) customized to satisfy specific functional requirements. This interface model is accessed by distinct types of users at the appropriate level i.e., the inexperienced, infrequent and/or job-limited user at GUI level; the spatially aware professional (analyst/engineer) at procedure/GUI level; the system customizer and database administrator at procedure level; the system developer at command language or application program interface (API) level. This paper reviews the layered structure of UGIX and then looks more closely at two of the layers - in particular a specification for a spatially extended version of SQL and a functionally equivalent iconic query language.

1. Introduction

The user interface is a vital element of any GIS. Long ignored as an esoteric aspect of GIS design whilst GIS development was driven by the need to extend functionality, the user interface is now beginning to attract its due attention. However, the implementation of a GIS user interface involves considerably more than the improvement of the human-computer interaction (HCI) process. Since GIS are conceptually complex and involve diverse operations ranging from data modelling to geometric transformations, improving the HCI cannot be a complete solution to the improvement of GIS use. Hence the shortcomings of the use environment for current GIS can be characterized as, firstly, task complexity— since the system functions implementing spatial and database operations are often complex or obscure, secondly, the inadequate nature of most database view tools- since few systems make available database views meaningful to the user, and finally the restricted form of the help system available.

2. Recent Research Into GIS User Interfaces

Research into the form and content of user interfaces for Geographic Information Systems (GIS) has accelerated rapidly in the last couple of years. Initially this work has followed from the general acceptance of the window-icon-mouse-pop-up menu (WIMP) system as an effective form of

D. M. Mark and A. U. Frank (eds.), Cognitive and Linguistic Aspects of Geographic Space, 449–475.

human-computer interface. Within GIS the unique appeal of the GUI has been the desire to improve the ease of use for systems involving complex graphical display and spatial manipulation.

Initial work in the GIS field on user interfaces has concentrated on the screen designs appropriate for spatial data handling. Thus, Egenhofer and Frank (1988) suggested a way in which the WYSIWYG principle of 'what-you-see-is-what-you-get' could be extended to GIS and outlined the components of their own system which included a study of the selection of objects and areas, legends and query specification. Subsequent work by Cassel and Parker (1989) carried out tests on the effectiveness of different designs of GUI, focussing especially on the use of icons to represent classes of spatial data. This work demonstrated some of the special problems which have to be solved to design good GUIs for GIS: thus, an icon of a single pine tree caused some confusion when presented to a forester since he could not identify its species! This indicates how legends and graphical cues must generalize a complex reality to provide appropriate 'cues for action' (Mark, 1989).

Another key aspect of GIS user interface design is that it can structure and filter human perception of space and spatial relationships when expressing an operation and passing it to a GIS. Research into general aspects of the perception of spatial relationships began long before the special problems presented by GIS existed, and is exemplified by the classic work of Gould and White (1974) on "Mental maps". In the GIS context such issues are becoming increasingly important as GIS use grows rapidly and reaches a wider audience: this concern has created a new sub-discipline of 'spatial language' research. Most current GIS user interface developments are based around the assumption that users adjust their conceptual model of the spatial phenomena under study to that provided by the GIS, and not vice versa.

The importance of the cognitive structuring of space is expressed by Mark and Frank (1989), who set out the research agenda for the "Spatial languages" Research Initiative of the U.S. National Center for Geographic Information and Analysis (NCGIA). Cognitive aspects of spatial perception place constraints on the definition of a spatial language intended to be used as part of a user dialogue with the computer. Thus, Mark (1989) suggests that 'image- schemata' (large scale spatial metaphors) can be used to structure perceptions of space, and provides examples drawn from Johnson (1987) such as CONTAINER (associated with preposition "IN") and PLATFORM (associated with the preposition of "ON"). This consideration can affect the process of data modelling since the cognitive models available define the 'entities thought to be relevant to the task in hand'. Mark (1989) applies this consideration to the Digital Line Graph, Extended (DLG-E) model of cartographic representation used by the U.S. Geological Survey mapping programme and comments that the rules and feature classes adopted do not relate well to an image-schemata, and hence will be difficult to generalize to different scales.

Similarly, different linguistic traditions structure space in distinct ways (Mark, 1989). Awareness of these characteristics of language can help in the design of a generic spatial language with which a wide variety of users can interact with a spatial database. A clear objective in this field must surely be the evaluation of the language of groups of users. However, attempts to translate the Geographic Information System Tutor (GISTutor) (Raper and Green 1989) into a number of European languages have encountered two main difficulties: firstly, the local adoption of English for spatial terms (and therefore concepts?) due to the English lexicon of most commercially available GIS, and secondly the difficulty of providing for translation of terms with no direct counterpart., especially in the field of spatial prepositions.

The spatial query language tools which a GIS uses for general interaction with spatial data and for the elucidation of spatial relationships has also been the subject of recent research (Egenhofer, 1989). However, spatial query tools are frequently linked to the data model employed by the GIS:

hence the use of multiple spatial data structures has led to the proliferation of query processors. The result has been that in many GISs spatial queries initiated by the user must be formulated in terms of a software-specific command language. However, the fact that many new systems use relational database storage for spatial and non-spatial data has led to a wide interest in the adaptation of Structured Query Language (SQL) for spatially related query operations in GIS. Thus, SQL is now seen as one possible means of establishing spatial queries across many different spatial data models and data structures. Thus, Herring, Larsen, and Shivakumar (1988) described the design of SQL extensions to a new object-oriented GIS; Abel (1988) has reported work on the creation of SQL based spatial extensions to a relational database in the SIRO-DBMS project; and Ingram and Phillips (1987) have designed spatial extensions to a hybrid GIS data model.

Egenhofer (1989) also reported the test development of a prototype extended SQL (XSQL), adding concepts of spatial relations, spatial operations, graph representation and context. However, when they tested the system they found that the queries became far too long and complex to parse and execute for user initiated queries. This may indicate either that user interfaces to query structures can not yet replicate human processes of spatial reasoning in a reasonable time, or that current extensions to SQL are inefficient.

Hence, the link between the user interface and the spatial query processor is a problem of key importance, and the subject of considerable debate. Egenhofer (1989) in work on the testing of query processors suggests that it is not possible to hide the implementation structure of the query at the interface. However, Rhind, Raper and Green (1989) suggest that a Universal Geographic Information eXecutive (UGIX) can be created to interface between the user and a wide variety of different systems. However, the UGIX concept is based on the key assumption that current GIS users are motivated and capable of exercising spatial reasoning which is adapted to implemented systems. These users are demanding access to tools which extend the spatial database as well as to make spatial analyses. This may indicate that a different approach is required for "public" spatial information systems used by people who are not 'spatial professionals'. These systems may have to be restricted to passive use based on the extraction of information in well defined forms perhaps related to 'image schemata'.

The design of help systems and tutors for geographical information systems also presents some significant challenges to the developers of GIS user interfaces However, Rhind, Raper and Green (1989) identify that the availability of such systems may be a crucial component of the user interface. Difficulties include the need to show complex graphic operations, the explanation of algorithms and file structures, and the application of GIS techniques to decision making in a variety of application areas. One example of a solution to this problem is the creation of the GISTutor described by Raper and Green (1989).

However, whilst much of this research addresses long term inadequacies in the GIS use environment three specific problems are encountered in the interaction with existing commercial GIS systems which need to be tackled in the short term. Firstly, the design of graphical user interfaces (GUIs) is generally inappropriate for GIS purposes, especially spatial task definition; secondly, the language specifications for interaction with spatial databases using spatial operations and queries are too complex and system specific; and, thirdly, standard help systems have a limited content and rigid structure. It is therefore suggested that the design of next generation user interface must be rooted in the creation of a user environment which integrates the data modelling process, the availability of spatial operations and queries, and the system of assistance and concept support within the interface.

3. User Interface Development: A Case Study of UGIX

The UGIX system design as described by Rhind, Raper and Green (1989) contains 3 main modules, viz. (A) containing the screen interfaces, dialogues and spatial command processor; (B) containing a help and information system for a GIS; and (C) an expert system shell or high level system access module. The structure of UGIX is illustrated in figure 1. This section describes the approach taken in the UGIX project, through first and second generation implementations.

The first generation approach to interface design within the UGIX project has been to prototype using Hypercard for the Apple Macintosh, where the Hypercard application (complete with in-built communications software) acts as a client to a host processor running the GIS application software. This approach is similar to the one used by Cowen and Love (1988) to create an interface to the South Carolina Historic Preservation Office GIS database. Hypercard with its standard set of buttons, scrolling boxes and cards makes use of the GIS less daunting for the less technically-aware user. In addition, with the rich graphics environment available in Hypercard it is possible to show a graphic to illustrate the effect of various options available at any one point. It is also desirable to display all the commands available to the user in one place, with a pop-up explanation for each option.

Screen design has involved the standardization of button and text field formats as well as card and background layout for different areas of activity such as:-

- Introduction and explanation (using a map guide);
- Map and file selection (using standard Macintosh file selection dialogue);
- Session screens for command processing;
- Help environment (UGIX (B) based on GISTutor version 2);
- A Gallery for maps and images generated in the GIS (along with button to redraw them).

Screen metaphors have been developed for each of these areas to make location in the system a graphical attribute. Currently this system interface 'shell' is being implemented for the GIS ARC/INFO, and is known as 'HyperArc'.

The first generation system, HyperArc, is currently under test with users at 'novice' and 'competent' levels of expertise: in addition to feedback on the use of the system, the aim of the evaluation phase is to define a core area of functionality in common use to help optimize the UGIX system structure. A key consideration for the development of a spatial language interface to GIS is that many different groups of users exist within the spatial data handling community. It is recognized within the UGIX design that to some high level users skilled in the manipulation of spatial data a command line interface to a GIS offers the maximum flexibility and power available. However, it is considered that during the first stage of the expansion of GIS use the largest group of new users are non-computer literate professionals who frequently have a highly developed insight into spatial data. It should therefore be emphasized that a major objective of the UGIX project is to improve access to existing GIS by converting the current function-orientation of the native system interface (primitive and implementation specific operations) to a task-oriented interface (sequences of high level spatial operations) usable by a spatially aware user (see figure 2 for a taxonomy of interfaces for ARC/INFO). The second generation of the UGIX project aims to build on the experience of constructing such task-oriented interfaces to create generic interfaces capable of communication with any GIS.

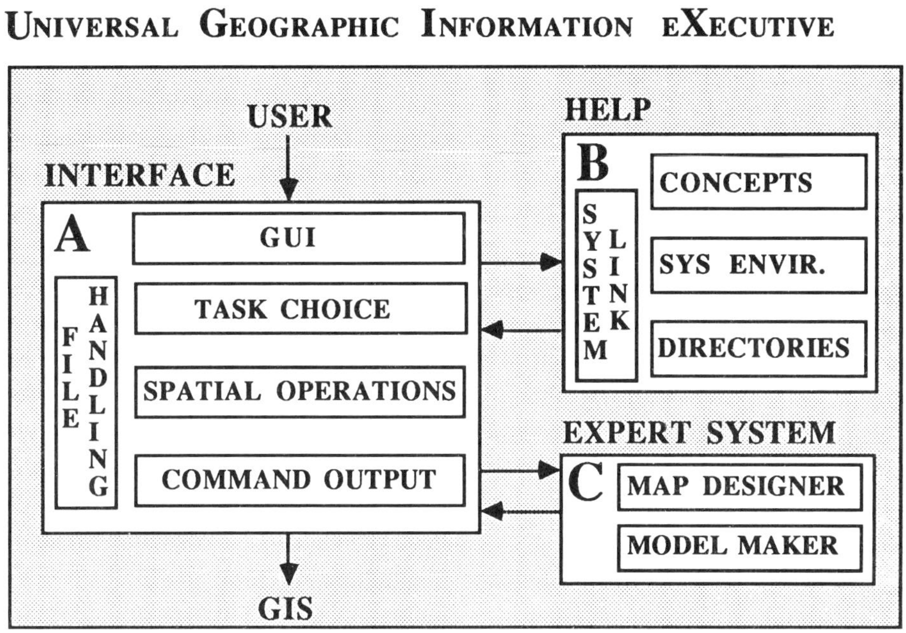

Figure 1. The three primary modules of the UGIX architecture.

Another key objective in the development of HyperArc has been the creation of file handling procedures to harmonize the user's concept of maps with that of ARC/INFO. This establishes that maps are both 'views' of spatial data and sheets within a series i.e., spatial tiles. Thus, search routines to find maps with particular names, to sort maps by type (e.g. point or polygon based), to access the map tiling system and to select the part of the sheet to view have all been created. In the first generation of this project the user specifies spatial queries using system implementation concepts which are made comprehensible to the user diagrammatically (user testing is helping to refine this aspect). Hence HyperArc forces the user to work with ARC/INFO concepts, but tries to connect them with the user's view of the problem under study. This is ultimately restrictive to the user since the data structure is fixed, and maps are files which the user needs to manipulate in some way.

Hence a basic principle of the UGIX design is that in order to make a GIS easy to use the process of making a database enquiry, displaying a map or carrying out spatial analysis must be broken down into a series of logical parts, linked by a pathway for the user to follow. Following such a path and gaining experience with the alternative options is an excellent way to improve a user's end-to-end understanding of the components of spatial data processing. Appropriate information needed for a user to make a decision is also retrieved before indicating the command options, for example only maps with the correct specifications are presented (e.g. with topological relations already created), when this is necessary for the operation.

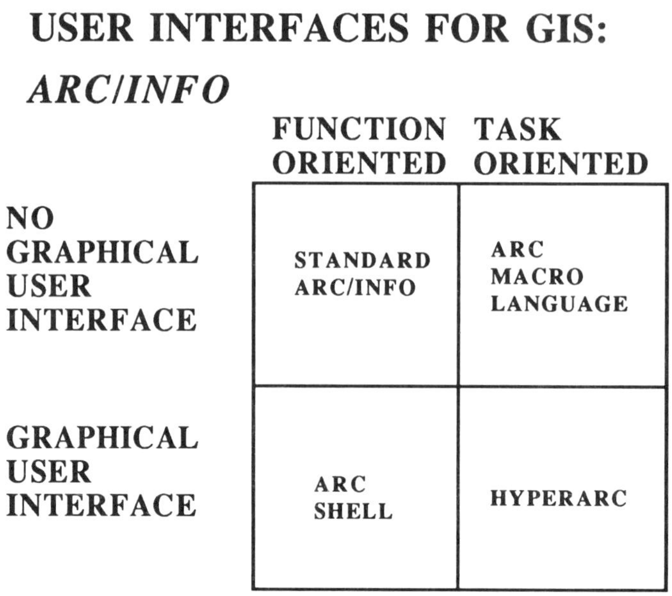

Figure 2. Taxonomy of interfaces for ARC/INFO.

However, in order to implement such an interface in a generic way requires a new form of software architecture which is independent of specific implementations, does not enforce a particular data model, and adheres to the standards in the user community which are most crucial to the success of GIS in heterogeneous computing environments.

4. The Layer Based Model

4.1. UGIX (A) - A LAYERED USER INTERFACE MODEL

One of the objectives of the UGIX development is to produce a generic GIS user interface that is capable of driving a number of different GIS implementations as opposed to HyperArc. The common GIS functions supported by most commercially available systems must be accessible from within UGIX. Use of UGIX should, as much as possible, protect the user from variations in implementation strategy adopted by the various GIS vendors. The mapping between the UGIX user interface elements and the underlying system-dependant commands, procedures etc. must be transparent and readily specified. It is also highly desirable that the mapping be supported at both the third Generation Language (3GL) level and the command or 4GL level.

An important aspect of the UGIX design is that it will incorporate a powerful query language based on SQL with spatial extensions. This will allow the user to interact with data from spatial databases based on a wide variety of data models since the query language SQL-SX described below will interact directly with implemented data types in the host GIS. A number of generic spatial functions will be accessed from within UGIX alongside the query language by mapping the commands into the host GIS. These are derived from experience in testing HyperArc and from an analysis of generic GIS functions: typically they will be expressed as an icon set. Higher level spatial tasks will be supported as in HyperArc by linking sets of lower level functions together. UGIX will also access additional GIS-specific functionality by extending the icon set using the standard customization tools developed within UGIX.

To achieve this a layered model is suggested that protects the user interface from the actual implementation mechanisms provided by each GIS vendor. Each layer within the model will perform a particular task and have a well defined interface to the layers both above and below. Some of the layers within UGIX will be able to communicate directly with the underlying GIS at a matching level.

4.2. USER REQUIREMENTS

The user requirements to be addressed by UGIX are diverse. In particular, the design of UGIX attempts to address the needs of multiple groups of users, including:

- the inexperienced, infrequent and/or job-limited user
- the spatially aware professional (analyst/engineer)
- the system customizer and database administrator
- the system developer.

Each has a different, although sometimes overlapping, set of requirements. For example, the inexperienced user requires a task-oriented interface to guide him/her through a particular application-specific task. Ease of use and an intuitive user interface are primary requirements. Such users would generally have a limited set of jobs to perform. These would utilize a limited set of lower level functions and operate on a restricted section of the database. The customized environment must restrict access to only the data and functions required to perform the jobs.

The analyst or engineer requires a much less restricting interface. Typical system usage might be one-off analysis (e.g. in a utility environment, mains fault/burst analysis - number of bursts per 100 meters per year). Read-only access to much of the database is required plus the ability to create test data sets to perform complex modelling and "what-if" analysis. Virtually unrestricted access to the functionality is required, to allow experimentation and analysis via a number of alternative techniques. The user interface must again be easy to use, but direct access to the functions is required, often in a command-line/4GL/procedural manner. Furthermore, the ability to link a number of functions together to perform a complex query repetitively is required.

The customizer and DBA must be able to synthesize the end-user environments for the other users. The creation of a task-oriented GUI for the infrequent user must be achievable in a short time frame. The integration of Computer Aided Software Engineering (CASE) tools, object oriented analysis/design tools and a graphical customization environment within the GIS would assist the customizer and DBA in performing their functions. If the CASE tool was already primed with knowledge of the spatial data types and spatial operators supported by the GIS, significant simplification of the functional analysis, data analysis, data modelling and database design stages of customization could be achieved.

4.3. POSSIBLE USER INTERFACE STYLES

The style of interface offered must satisfy the user's requirements for ease of use, in particular, access to functionality. The use of GUIs is becoming more common not only for end-user applications, but also within customization and development environments. Consequently, UGIX will offer a GUI interface for task- oriented applications, common GIS functions (e.g. spatial and aspatial interrogation and analysis), access to the help system, customization functions (such as task specification or user interface design) and DBA functions (e.g. access control specifications, DB schema definitions and backup). In addition, a separate layer within UGIX will offer a command line facility to perform the same functions. This will be based on SQL-SX, a spatially extended version of SQL. SQL-SX is defined for the interactive command-line/4GL environment and as a set of extensions for embedded SQL within 3GL environments.

4.4. THE STANDARDS

A number of "standards" for User Interface Management Systems (UIMS) have recently developed. In particular the X.Windows system based on the MIT X Consortium specifications provides a standard for a low level toolkit of functions called "Xlib". However, to facilitate a more productive development process, and provide a consistent look-and-feel, most development is now being performed using one of the higher level "standard" user interface toolkits such as OSF-MOTIF, OpenWindows, etc. It is proposed that UGIX be based on one particular high level standard and be migrated to other environments when UIMS conversion utilities become available (e.g. in the PC world, MS-Windows applications may now be automatically converted to operate under the OS-2 Presentation Manager).

Standards for data management systems have evolved rapidly in recent times. In particular, the RDBMS (Relational Database Management Systems) has become a *de facto* standard for the management of corporate data, and SQL has been virtually unanimously adopted as the mechanism for defining, maintaining and interacting with the database. It would appear that RDBMS will be the means of managing large corporate databases for at least the next 5-10 years. Codd (1985) and Date (1986-b) have provided definitions for what constitutes a fully relational system, and the vendors have responded by producing systems that satisfy the definition. Furthermore, SQL has become the standard mechanism for defining, accessing and managing the database. SQL is now an internationally accepted standard ratified by both ANSI (ANSI 1986, 1989) and ISO (ISO 1987, 1989). Versions of SQL are offered as part of virtually all commercially available RDBMS, for PC, workstation and host machines.

5. SQL-SX: A Spatially Extended Version of SQL

5.1. WHY SQL BASED

5.1.1. *The Case For:* The relational language SQL forms a suitable base on which to develop spatial extensions due to its widespread acceptance, its availability within a large number of commercial DBMS products, its standardization in terms of internationally accepted standards and its ongoing evolution and development thus ensuring a long term future for the language. User organizations have developed significant expertise in its use by investment in training and employment of SQL knowledgeable staff.

SQL was originally developed by IBM within the System R (Astrahan et. al 1976) project and subsequently adopted by IBM (Chamberlain 1981) and other RDBMS vendors as the basic interrogation and manipulation tool for commercial products. A number of other relational languages were also developed, e.g. QUEL, QBE, etc. although none gained the mass acceptance of SQL.

SQL has a well defined specification that defines the language in terms of statements and clauses (see for example Date 1989). Statements are effectively the equivalent of English sentences while the clauses form subcomponents of the sentences. For example, the SELECT, DELETE and UPDATE statements may each contain a WHERE clause, the format of which is consistent in each case.

SQL had a significant number of advantages over previous database languages including:

- simple data structure
- powerful operators
- short learning period
- data independence
- single language integrating data definition, data manipulation and database administration
- integrated data dictionary.

Furthermore it is made available both as an interactive language for ad hoc queries and embedded within a number of 3GLs.

SQL has been an internationally accepted standard since 1987. In fact multiple organizations now provide definitions of the SQL standard including ANSI, ISO, X/Open and the SQL/Access Group. The original American (ANSI) standard has been adopted as an international (ISO) standard, and in the USA as a Federal Information Processing Standard (FIPS). This standard defined two language levels, the first being basically a lowest common denominator of existing SQL implementations, while the second provided a more comprehensive definition. Subsequently, an Integrity Enhancement Feature (IEF) has been defined as a major extension to original ANSI/ISO standard. In 1987, the X/Open consortium adopted a version of SQL for the UNIX environment, with the aim of producing a specification for a portable SQL. More recently, a group of computer vendors and DBMS suppliers, dissatisfied with the slow pace of the standards initiatives got together to form the SQL/Access group. The objective was to speed up specification of a standard that would allow future SQL application portability and interoperability. This group has adopted the X/Open standard as a basis for its efforts. IBM is currently not an active member of any of these standards committees and has its own specification for DB2 within its Systems Applications Architecture. The DB2 dialect of SQL has some significant differences to the ANSI/ISO standard, although it is likely to become a super-set of the standard at some point.

The vendors of RDBMS support versions of SQL across all platforms, from PCs to mainframes. This availability has lead to the widespread adoption of SQL as the standard mechanism for interactive ad-hoc interrogation and manipulation of RDBMS. Vendors of non-relational and pseudo-relational DBMS have also developed SQL interfaces to their systems. Admittedly, strict compliance to the ANSI/ISO SQL standards are often difficult to achieve.

The availability of SQL as an interface on a wide variety of platforms and operating systems, its acceptance as an international standard and the acceptance of the relational model in general has lead to many (most?) large organizations moving towards adoption of RDBMS and SQL as the corporate standards for the management of their organization's data. It is safe to assume that RDBMS and SQL will be the underlying technology behind the management of the world's Corporate Databases for at least the 1990s. GIS technology is beginning to be viewed as a data integration technology, providing spatial/aspatial data integration and potentially bringing benefit to corporate applications traditionally bound to purely aspatial/textual environments. As GIS becomes more heavily integrated into the corporate database environment, it is only natural that the database interrogation and manipulation language used for other corporate applications will be requested as an integral part of the GIS. The investment made by large organizations in training their personnel on RDBMS and SQL, and the availability of SQL literate computer professionals generally will be a further stimulus for those organizations attempting to acquire and integrate GIS technology into the corporate computing environment to specify SQL compliance as a mandatory requirement.

The number of organizations actively promoting SQL, and the activities of the standards organizations in further developing the standard ensures that SQL will be here for some time. The work being put into development of extensions for the ANSI/ISO SQL2 and SQL3 standards is being undertaken world-wide. Working groups in many countries are putting forward proposals for adoption within SQL3 at present. Many of the enhancements involve incorporation of object-oriented concepts and extensibility. In particular, the ability to support the definition and use of domains, abstract (user-defined) data types (ADTs), their associated operators, extensible indexing mechanisms, triggers, inheritance and nested relational constructs all make SQL appear much more viable as a mechanism for interrogation and manipulation of spatial data.

5.1.2. *The Case Against:* SQL is by no means a perfect language. It contains many inconsistencies, weaknesses, missing functionality, and in some ways fails to realize the full potential of the relational model (Codd 1988), (Date 1986a). It was designed to manipulate data types that could be viewed textually, and intended initially as a general purpose language for ad hoc queries and database definition. It contains no constructs for controlling program logic - rather any such facilities must be supplied by the programming language that SQL may be embedded within. With time, the use of SQL and the extensions that support those uses have become extremely diversified, so that now we see attempts to use SQL within information systems managing both spatial and temporal data.

Date (1986a) discusses the weaknesses of SQL at some length, although they may be categorized as:

- lack of orthogonality: expressions
- lack of orthogonality: built-in functions
- lack of orthogonality: miscellaneous items, inconsistencies
- formal definition - incomplete, based on implementations
- mismatch with host languages
- missing functions, especially related to cursors
- mistakes
- aspects of the relational model not supported.

A number of these areas are being addressed in the proposals for SQL2 and SQL3.

The idea that language may influence spatial perception is a potential problem for SQL extensions. Its rigid syntactical constructs and tight definition of keywords (operators) may result in slightly different perceptions of the spatial features being modelled and in the general usage of the spatial operators.

A number of other problems are also relevant to spatial applications (in addition to lack of spatial data types, operators etc.):

- the SELECT statement performs 2 functions, selection and output
- non-elementary selections require nested select structures which are not easy even for experienced users
- no recursive facilities. These are required for network trace, parts explosion and bill-of-materials applications
- multiple ways to produce the same result, not all of which are evaluated in the same way. Consequently it is easy to write inefficient SQL.
- no inbuilt facility to advise a database designer of the expected response times of a query, nor to recommend use of indexes, denormalization etc.

The existing interactive SQL SELECT statement is intended to be the primary mechanism by which ad hoc enquiries are performed on an alphanumeric database. Its characteristics may be briefly summarized as:

- it identifies which columns are to be selected from which tables, under which aspatial criteria
- it specifies any functions that are to be applied to the retrieved values, and
- it specifies whether the result is to be displayed either directly on the screen in the form of a report or be saved in a temporary table.

For spatial applications it would be advantageous to separate the selection function from the display function, since the display requirements are much more complex than for alphanumeric applications- e.g. combined graphical and textual reports where the results may be displayed as a group or one at a time.

Simple queries in SQL involving a small number of tables and columns are moderately easy to develop. However, as soon as subqueries (nested selects) are required, query development is not simple, even for experienced users. Furthermore, simple operations (e.g. difference) in relational algebra are complex in SQL.

Applications requiring a generalized recursive capability are not possible to implement using interactive SQL. Consequently, network trace, bill-of-materials and other parts explosion problems must be tackled in a lower level language.

A further problem occurs largely due to the general lack of orthogonality within SQL. SQL provides multiple methods to generate the same result. Due to the difficulty in developing efficient query optimizers, the same problem stated in multiple ways may result in extreme performance differences and in a few extreme cases different (or unpredictable) results. The result is that developers using SQL must be aware of the strengths and weaknesses of the query optimizer supplied and the physical database design of the tables referenced by the application. The effect of poor query formulation can be extreme (e.g. response time differences of 10 hours reduced to 1 minute by a simple restructuring of predicates).

5.2. PREVIOUS SQL EXTENSIONS

A number of authors have presented papers on the integration of spatial data within relational DBMS including Abel (1988, 1989), Mainguenaud (1990), Egenhofer (1987) and Bundock (1987). Furthermore a number of GIS vendors have in the past five years attempted to provide some form of spatial extensions to SQL. In particular, vendors such as Kork (Ingram & Phillips, 1987), Intergraph (Herring, Larsen & Shivakumar, 1988), Wild Heerbrugg (now Prime Wild GIS) (Charlwood, Moon & Tulip, 1987) and GeoVision (Westwood, 1989) have all reported efforts to provide spatial predicates and other limited spatial data manipulation facilities within SQL (or SQL-like query languages). Unfortunately, the basic query language in each case was a very incomplete implementation of the ANSI/ISO SQL standard, and the spatial extensions were fairly minimal and elementary. To make matters worse, the extensions in general did not maintain consistent syntactic and semantic constructs with the rest of SQL. For example, spatial predicates are not in general supported within the WHERE clause, but rather within a separate clause.

Pong-Chai Goh (1989) reported on the design of a graphic query language based on SQL that supported graphic representations for attributes. However, no operators or functions were provided to select or manipulate the data based on spatial criteria. Earlier work by Sacks-Davis, McDonell and Ooi (1987) described a spatially extended version of SQL called GEOQL, which incorporated spatial data types and operators. It also included the concept of virtual objects derived by applying spatial operators to "real" database objects (e.g. the line joining two spatial objects). However, no description is given for how the spatial data types may be graphically displayed nor interacted with in terms of a graphics screen and cursor/mouse.

5.3. THE SPATIAL EXTENSIONS TO SQL - SQL-SX

5.3.1. *Summary of Extensions*. The extensions necessary to make SQL usable within GIS applications for query of both spatial and aspatial data include:

- spatial data type(s)
- spatial operators (predicates)
- spatial functions
- long transaction management statements
- report specification facilities - both textual and graphical

In addition, functional requirements demand that

- the spatial data types be able to be displayed graphically as a result of being SELECTed
- the data dictionary and DDL support the extensions
- spatial access control (protection) may be provided by inclusion of spatial data types, predicates and functions within VIEW definitions
- spatial integrity maintenance may be provided by support of spatial data types, predicates and functions within the CONSTRAINT clause.

Each of the above items is discussed in more detail below. It should be noted that the extensions suggested here are not constrained to those items that we think can be supported now (within the current standards and implementations) or that we expect to be supported in the short term - rather we are attempting to identify the key requirements, suggest syntax, and identify which extensions being proposed for SQL2 and SQL3 (ISO-ANSI 1989) must be supported for spatial data handling.

5.3.2. *Design Criteria.* The underlying design criteria for SQL-SX have been:

- simplicity
- consistency
- concept independence (orthogonality)
- formal definition
- separation of the language syntax and behavior from the implemented spatial data models and data structures

The first four criteria are essentially the same as those described more fully in Date (1986a). It should be noted that we are not attempting to define a complete GIS manipulation language in our spatial extensions to SQL - rather we are attempting to provide a standardized, general purpose query and manipulation facility that may solve some of the functional requirements for many GIS applications. Furthermore, we are not proposing that SQL-SX be a map specification language—rather that it offer limited facilities for formatting soft-copy graphical output at the workstation.

The spatial extensions in SQL-SX must be easy to remember and use. In support of this requirement, the number of extensions should be kept to a minimum, since the availability of a large number of data types and functions makes each individual component more difficult to remember and use effectively. The abstraction of general characteristics for spatial data types and operators provides a key to the success of the extensions. "Power through simplicity, simplicity through generality, should be the guiding principle" (Morrison, 1981).

The spatial extensions must fit into SQL as an integral part. They must not impact the use of the standard SQL capabilities so that if the spatial extensions are not used, the language IS the ANSI/ISO (X/Open etc.) standard. Furthermore the extensions must utilize the same syntactic and semantic constructs as the rest of the language. Consequently, for extensions to the SELECT statement, spatial predicates must be confined to the WHERE clause, items to be displayed (textually or graphically) must be identified in the selection list as column or function references.

The principle of orthogonality, or concept independence, refers to the principle that distinct concepts should be cleanly separated —never bundled together. In terms of the spatial extensions, each new construct (data type, operator etc.) should have a distinct meaning and not duplicate any other data type or function. The advantage of orthogonality is that it leads to a coherent language - i.e., one that possesses a clean and simple structure (both semantic and syntactic) and that is easy for the user to grasp. One of the mechanisms by which orthogonality is ensured is to keep the number of concepts small (though not necessarily minimal). Consistency of both syntactic and semantic constructs ensures that the same syntax and meaning is applied throughout the language.

The extensions should have a formal definition created and accepted prior to the widespread availability of particular implementations. This was not the case with the original versions of SQL - the strict definitions and the standard came after widespread use had already started. It lead to diverse variations and inconsistencies in the language which will continue to be present for some time (probably until it dies in favour of another query language!). The definition must be precise, providing definitive answers to technical questions. However, the definition must not dictate how to implement the language - rather it must define what its syntax is and how it should behave.

Many existing GIS implementations and the interrogation and manipulation languages they use, require that the user understands the particular implementation technique. In particular the user is often required to manipulate the geometric representation of features (grid cells, points, nodes, arcs, polygons etc.) rather than the features themselves. Furthermore, the user must understand the mechanisms by which spatial relationships are managed within the system, when they are available and when they must be (re)created. This requirement to understand the underlying mechanics behind the particular implementation of spatial/topological relationships is a severe impediment to the more widespread use of GIS technology. A key design criterion for the development of SQL-SX is that the user need not know how the spatial relationships are managed, and instead be able to concentrate on the application requirements.

The specification of SQL-SX specifically does not define the data model to be used for management of the spatial (geometric & topological) data components. Rather it only defines the spatial data types and how they must behave - not how to implement them. This allows the vendors to develop their own data structures, indexing mechanisms etc. to support the requirements - i.e., syntactical correctness, performance, and behavior. It also, in the short term, allows vendors to implement hybrid solutions where the spatial data is not necessarily managed directly within the corporate RDBMS. It is the authors impression that it is highly advantageous to be able to manage all data in a single, homogeneous DBMS, for reasons of simplicity, recovery, integrity and multi-user concurrency. However, it is likely that GIS vendors will still require some time to discover mechanisms for physical database design with RDBMS that provide the necessary performance.

5.3.3. *Spatial Data types.* The approach taken here is to consider geographic phenomena as having a set of attributes, one (or possibly more?) of which may be a spatial attribute. No significant separation is made between spatial and aspatial attributes. For example, if an entity has a spatial characteristic of a polygon, then we shall say that the entity possesses an attribute of type polygon.

The data types supported within ANSI/ISO standard SQL are character strings, fixed and floating point decimal numbers. A number of other data types are available in some of the RDBMS products outside of the standard - e.g. date, date-time, money, binary, boolean and bit strings. We propose a set of new spatial data types to support spatial applications.

The definition for data type used here is as per Cleaveland (1986) "A data type is a set of values and a set of operations on the set of values", i.e., a domain and a set of operators on the domain. Consequently to specify a data type we define the domain of values it may take and the behavior of the data type under the application of a set of operators.

The data types do not have to be elemental - for example a CHARACTER string is a data type - yet it is a list of characters, and has its own set of operators e.g. equality, sub-string, upper/lower case conversion etc. Furthermore, definition of the domain and the operators on the domain does not define the implementation technique that must be used to satisfy the requirements of the data type - for example both character strings and numeric values are implemented quite differently within DB2 and Oracle. Consequently we can separate the definition of the data type from its implementation.

A large body of research literature is available describing spatial data types including Güting (1988), Bundock (1987) and Moellering (1986). The different approaches yield variations in the operations that may be applied, or to the ease with which the operators may be applied.

There are clearly only two alternative approaches to the incorporation of spatial data types:

- creation of a single SPATIAL data type that may be used to represent any type of spatial characteristic, or
- creation of multiple spatial data types such as POINT, NODE, LINK, POLYGON etc.

The first approach potentially has the advantage of simplicity for definition and use. However, some of the operators may be meaningless for some of the domain values - e.g. area of a point (or is this just zero or null?). Consequently, at this stage we are adopting the latter approach of defining multiple spatial data types.

It should be noted that the names used here are fairly arbitrary and hence debatable and the definition of the data types is supplied partly by narrative and partly by the definition of the operators which are also open to debate.

Based on the approach of defining multiple spatial data types we define the following:

• POINT	A position in 3D space
• NODE	A POINT that provides (potential?) topological connectivity
• LINE	A linear feature terminated by 2 NODES and containing any number of internal POINTS that define its geometry. The LINE may be any type of generalized arc, spline etc.
• POLYGON	A closed area delimited by an ordered set of LINEs. POLYGONs may contain holes (other POLYGONs) which may contain holes.
• SURFACE	A topologically connected set of POLYGONs. SURFACES may contain holes etc..
• VOLUME	A closed volume in space delimited by a set of SURFACEs. VOLUMEs may contain holes etc..
• AMALGAMATION	A complex set of other spatial components
• FIELD	An n dimensional field. At each location in space covered by the field there exists a set of n attributes.

The external view of the data types is purely by graphical means. Each of the data types is only viewable as a graphic - e.g. a POINT is viewable only as a possibly symbolized dot, a LINE is viewable only as a line on the map/screen. Specification of how to display the data types is left to report format specification (see Section 5.3.7.). Retrieval of the geometric properties (e.g. area, perimeter, coordinates etc.) is performed by functions operating on the data types (see Section 5.3.5.). Note that each of the data types must have a well behaved NULL value independent of the ZERO value (i.e., a line of zero length is a valid line).

5.3.4. *Spatial Predicates and Operators*. This paper differentiates groups of operators into those that return a boolean value and those that do not. Operators that return a boolean value retain the name *operator* and may be used within predicates, while those that return other values are referred to as *functions*. This is a convention used within some descriptions of SQL - e.g. Date (1989). A predicate is a clause that returns a boolean value (TRUE or FALSE). A spatial predicate is a predicate (i.e., returns a boolean value) containing spatial operators and spatial data types. Spatial operators accept one or more spatial data types as operands.

Freeman (1975) proposed a set of primitive spatial relations for elements within a 2D picture (potentially a view of a 3D small scale space). These are:

• left of	• right of
• above (over, on top of, higher)	• below (under, underneath, lower)
• in front of	• behind (at back of)
• near (close to, next to)	• far
• touching	
• between	
• inside (within)	• outside (without)

Other authors have also defined classifications of spatial relationships, for example Feuchtwanger (1989), Pullar & Egenhofer (1988), Herring (this volume) and Egenhofer & Frank (1987). Peuquet (1988) described a classification of topological relationships between spatial objects, and describes spatial queries that appear complex and not easily solved by the existing classifications.

Mark (1989) reported on Johnson's (1987) work describing an image-schema-based model of cognition, and identified a set of spatial image-schemata, including : CONTAINER, BLOCKAGE, PATH, SURFACE, LINK, NEAR-FAR, CONTACT, CENTER-PERIPHERY and SCALE. These may be an alternative perspective on the problem of identifying and describing spatial cognition and consequently spatial operators.

The following are a suggested set of spatial operators that support interrogation of spatial relationships between point, linear, areal and network topology. Many are reproduced as per the work of the previously mentioned authors. This is neither a complete nor a minimal set, but is based on the authors' experience with GIS requirements in utility, cadastral, environmental and planning applications.

Operator	Symmetric Operator
AT	
NEAR (close_to)	FAR
IN_BETWEEN	
CROSSING	
CONNECTED_TO	DISJOINT_FROM
ADJACENT_TO	
INSIDE (enclosed_by, in)	ENCLOSES
IN_OR_CROSSING	OUTSIDE
WITHIN numeric units OF	
ON (on surface of)	
ABOVE (over, on top of)	BELOW (under, underneath)
IN_FRONT_OF (closer to)	BEHIND (beyond)
LEFT_OF	RIGHT_OF
WEST_OF	EAST_OF
NORTH_OF	SOUTH_OF
BEGINS_AT	ENDS_AT

Examples of the use of the spatial data types (in italic) and operators (in bold) are given below.

(a) Report the names of all lakes within 35 miles of London that have windsurfing facilities.

```
SELECT    LAKE.NAME
FROM      LAKE, CITY
WHERE     LAKE.BOUNDARY WITHIN 50 MILES OF CITY.CENTER
    AND   LAKE.WINDSURFING = 'TRUE'
    AND   CITY.NAME = 'LONDON'
```

(b) Display graphically the longest tributary of the river Mondego

```
SELECT    RT.LINE, MAX(RT.LENGTH)
FROM      RIVER RT, RIVER RM
WHERE     RT.LINE ENDS_AT RM.LINE
AND       RM.NAME = 'Mondego'
```

The following descriptions define more clearly the function of each of the spatial operators and the data types to which they may be applied. If the operator is supported for all spatial data types, the data types will not be listed but rather will be abbreviated "spatial".

Proposed Spatial Operators

AT	Is a spatial value located at a defined location? **spatial AT point -> BOOLEAN**
NEAR	Is a spatial value located near another spatial value? **spatial NEAR spatial -> BOOLEAN**
FAR	Is a spatial value located far from another spatial value? **spatial FAR spatial -> BOOLEAN**
IN_BETWEEN	Is a spatial value in the area between one spatial value and another? **spatial IN_BETWEEN spatial AND spatial-> BOOLEAN**
CROSSING	Is a spatial value crossing another? **spatial CROSSING spatial -> BOOLEAN**
CONNECTED_TO	Is a spatial value topologically connected to another spatial value? **spatial CONNECTED_TO spatial -> BOOLEAN**
DISJOINT_FROM	Is a spatial value disjoint from another spatial value? **spatial DISJOINT_FROM spatial -> BOOLEAN**
ADJACENT_TO	Is a spatial value adjacent to another spatial value? **spatial ADJACENT_TO spatial -> BOOLEAN**
INSIDE	Does a spatial value fall totally inside a polygon value? **spatial INSIDE polygon -> BOOLEAN**
ENCLOSES	Does a polygon value totally enclose another spatial value? **polygon ENCLOSES spatial -> BOOLEAN**
IN_OR_CROSSING	Is a spatial within or overlapping with a polygon value? **spatial IN_OR_CROSSING polygon -> BOOLEAN**
OUTSIDE	Is a spatial value totally outside of another spatial value? **spatial OUTSIDE polygon -> BOOLEAN**
WITHIN num. OF	Is a spatial value within a specified distance of another spatial value? **spatial WITHIN numeric units OF spatial-> BOOLEAN**

ON	Is a spatial value on the surface of another spatial value? **spatial ON spatial -> BOOLEAN**
ABOVE	Is a spatial value over or on top of another spatial value? **spatial ABOVE spatial -> BOOLEAN**
BELOW	Is a spatial value under or underneath another spatial value? **spatial BELOW spatial -> BOOLEAN**
IN_FRONT_OF	Is a spatial value closer to the users point of reference than a second spatial value and in line with the reference? **spatial IN_FRONT_OF spatial -> BOOLEAN**
BEHIND	Is a spatial value further from the users point of reference than a second spatial value and in line with the reference? **spatial BEHIND spatial -> BOOLEAN**
BEGINS_AT	Does a spatial value begin at a specified location? **spatial BEGINS_AT point -> BOOLEAN**
ENDS_AT	Does a spatial value end at a specified location? **spatial ENDS_AT point -> BOOLEAN**
LEFT_OF	Is a spatial value to the left of a second spatial value? **spatial LEFT_OF spatial -> BOOLEAN**
RIGHT_OF	Is a spatial value to the right of a second spatial value? **spatial RIGHT_OF spatial -> BOOLEAN**
WEST_OF	Is a spatial value to the west of a second spatial value? **spatial WEST_OF spatial -> BOOLEAN**
EAST_OF	Is a spatial value to the east of a second spatial value? **spatial EAST_OF spatial -> BOOLEAN**
NORTH_OF	Is a spatial value to the north of a second spatial value? **spatial NORTH_OF spatial -> BOOLEAN**
SOUTH_OF	Is a spatial value to the south of a second spatial value? **spatial SOUTH_OF spatial -> BOOLEAN**

5.3.5. *Spatial Functions.* Functions in SQL are operators that return non-boolean values. They are normally used to derive aggregate values, such as sum, minimum, maximum, average etc. Within SQL, they are subdivided into two distinct categories:

- column functions (SUM, MIN, MAX, AVG, COUNT), and
- table functions (COUNT(*), EXISTS, DISTINCT, UNION).

Column functions operate on an entire column and return a single scalar value. The table functions COUNT(*) and EXISTS each return a scalar value, while DISTINCT and UNION each return a new relation (table). In addition to these "standard" functions, various database implementations (e.g. DB2) include additional functions such as SUBSTR and LENGTH for character string manipulations. We propose similar types of functional extensions that derive spatial and numeric values.

The first group return a single scalar value and are fairly self explanatory. Note that some conventions may be defined so that the operator may be generalized to apply to all spatial data types

e.g. LENGTH_OF(point) -> 0,
PERIMETER_OF(line) = 2.0*LENGTH_OF(line) etc.

```
LENGTH_OF(spatial)                     -> numeric
DISTANCE_BETWEEN(spatial, spatial)     -> numeric
AREA_OF(spatial)                       -> numeric
PERIMETER_OF(spatial)                  -> numeric
XCOORD(point)                          -> numeric
YCOORD(point)                          -> numeric
ZCOORD(point)                          -> numeric
DIRECTION(spatial, spatial)            -> numeric (bearing?)
```

The following functions return a single spatial value:

```
CURSOR()                               -> point
CENTER_OF(spatial)                     -> point
LINE_BETWEEN(spatial, spatial)         -> line
BOUNDS_OF(spatial)                     -> polygon
BUFFER(spatial, numeric)               -> polygon
CLIP(spatial, polygon)                 -> spatial
```

The following functions return a complete relation (table) and consequently may only be used where a relation is expected. For example, in a selection clause, no other expressions may exist within the same selection clause and only a single spatial value may be used as the argument.

```
COORDINATES_OF(spatial)                -> relation (set of x,y,z triples)
POINTS_OF(spatial)                     -> relation (a set of points)
LINES_OF(spatial)                      -> relation (a set of lines)
NODES_OF(spatial)                      -> relation (a set of nodes)
```

The following functions operate on entire relations. More thinking is required on both the input relations and the resulting relations.

```
SPATIAL_UNION(polygon_relation, polygon_relation)          -> polygon_relation
SPATIAL_INTERSECTION(polygon_relation, polygon_relation)   -> polygon_relation
SPATIAL_DIFFERENCE(polygon_relation, polygon_relation)     -> polygon_relation
```

Some examples of the use of the spatial functions is given below.

(a) Report the name, area and perimeter, and display graphically the center and boundary of all nature reserves containing endangered species that are adjacent to major population centers.

```
SELECT    R.NAME, AREA_OF(R.BOUNDARY), PERIMETER_OF(R.BOUNDARY)
          CENTER_OF(R.BOUNDARY), R.BOUNDARY
FROM      RESERVE R, CITY C
WHERE     R.ENDANGERED = 'TRUE'
    AND   R.BOUNDARY ADJACENT_TO C.BOUNDARY
    AND   C.POPULATION > 100000
```

(b) Peuquet (1988) identifies the following example as problematic. "Find the locations of all nuclear power plants within 50 miles and upwind of any urbanized area within the US"

```
SELECT   U.NAME, U.BOUNDARY, P.SYMBOL
FROM     URBAN U, POWER_STATION P, COUNTRY
WHERE    P.TYPE = 'Nuclear'
   AND   COUNTRY.NAME = 'US'
   AND   U.BOUNDARY INSIDE COUNTRY.BOUNDARY
   AND   BUFFER(U.BOUNDARY, 50) ENCLOSES P.SYMBOL
   AND   ABS (DIRECTION (CENTER_OF(U.BOUNDARY), P.SYMBOL)
             - U.WIND_DIRECTION) < 50
```

5.3.6. *Report Format Specification: Alphanumeric Reports.* Current SQL standards do not provide a specification of how alphanumeric reports are to be formatted. In particular, specification details such as page size, character fonts, column widths, column names, column formats, page numbering etc. are all left to the vendors discretion. Consequently, each vendor has implemented their own syntax and facilities for providing such support. It is felt that some standardization may be advantageous to permit greater application portability.

5.3.7. *Report Format Specification: Graphical Reports - Maps and Displays.* It is proposed that a standard syntax be provided for graphical report formatting. Details such as color, fonting, symbolization, shape fill, prioritizing etc. need to be specified to get visually acceptable displays. Note however that each is related to the user entity type, purpose, scale etc., not just to the spatial data type involved.

5.3.8. *Management of the Long Transaction.* We define a transaction to be a logical piece of work within which the database may be in an inconsistent state, and not satisfying the predefined integrity rules. A transaction will normally have three well defined stages:- a beginning, a middle and an end (commit). The manual equivalent of a transaction in terms of a cartographic process might be:

- the transaction start is indicated by removing the master version of a map from the cabinet
- the cartographer/engineer/... performs the updates to the map. This may take minutes, days or months. During this time, other users are not generally allowed to use this master map for other work, but may reference copies of the map made before the beginning of this transaction. Multiple alternative proposals may be made during this period resulting in potentially different master maps.
- the cartographer/engineer gets one set of changes approved by some higher authority, commits these changes in ink onto the map and places the map back on the rack, to allow general reference. This signals end of transaction.

Most existing DBMS transaction management strategies make the assumption that the transaction will take a relatively short time, and that all potential conflicts must be avoided until the transaction is completed. Consequently, locking mechanisms are employed to eliminate the possibility of other users viewing the same data or attempting to update the same data. In a GIS environment, where transactions may be extremely long, this approach is not reasonable. Alternative strategies have recently been proposed by Newell & Easterfield (1990),that support long transactions by the management of versions (see also Easterfield, Newell & Theriault 1990).

SQL-SX must be extended to provide a new type of transaction management that allows definition of the start and commit stages of long transactions. Furthermore, we wish to protect the user interface (SQL-SX syntax and iconic equivalent) from the actual mechanism used within the GIS and DBMS. Note that use of the normal DBMS transaction management facilities are still recommended for short updates within the overall process, since these provide short term rollback and roll-forward recovery protection and short term integrity enforcement.

We suggest a syntax for long transaction management as:

```
BEGIN LONG TRANSACTION transaction-name IN polygon-exp [ ON entity/feature
specification]
COMMIT LONG TRANSACTION transaction-name
ROLLBACK LONG TRANSACTION transaction-name
```

5.4. OTHER ASPECTS OF THE SQL STANDARD

5.4.1 *Relationship with Relational Algebra and Relational Calculus.* We believe that the extensions proposed do not invalidate the operations of relational algebra and relational calculus.

5.4.2 *Schema Definition Language.* No explicit extensions are required to the schema definition language statements in order to support the use of spatial data types. Clearly, however, the definition of the spatial data types must be achieved first. Current proposals for SQL3 support the definition and use of user defined data types. We envisage abstract data types as the method by which spatial data types will be declared. (See Section 5.3.3. and Kerridge (1989) for more details). This technique will allow spatial data types, operators and functions to be used wherever standard data types, operators and functions may be used.

The following examples illustrates the use of spatial data types when defining tables and views.

```
CREATE TABLE  CITY
   (   NAME          CHAR(20)      NOT NULL,
       POPULATION    INTEGER,
       BOUNDARY      POLYGON)

CREATE VIEW   MAJOR_CITY ( NAME, POPULATION, BOUNDARY, CENTER )
   AS  SELECT    C.NAME, C.POPULATION, C.BOUNDARY,
                 CENTER_OF(C.BOUNDARY)
       FROM      CITY C
       WHERE     C.POPULATION > 100000
```

5.4.3 *Spatial Security and Access Control.* There is occasionally a need to provide access control based on spatial criteria (e.g. deny access data describing the area within a military complex). It is proposed that the normal view definition constructs be used to provide such a facility. No extensions are required to the current standard other than to allow the use of spatial data types, operators and functions wherever existing data types, operators and functions may be used.

5.4.4 *Database Integrity - The Constraint Clause.* The constraint (CHECK) clause within SQL may be utilized with spatial data types to enforce correct topology and other spatial relationships between entities. No extensions are required to the standard to support such facilities. The ability to automatically validate that spatial relationships are being correctly maintained at time of data capture and update may be a very important capability. Furthermore, definition of such constraints once within the database removes the need to replicate the checking throughout the application.

The following example illustrates the use of spatial constraints to limit access to utility information within a military compound.

```
CREATE VIEW  PUBLIC_MAIN
             (ID, DIAMETER, MATERIAL, DATE_INSTALLED, GEOMETRY)
    AS  SELECT    ID, DIAMETER, MATERIAL, DATE_INSTALLED, GEOMETRY
        FROM      MAIN, LAND_PARCEL
        WHERE     MAIN.GEOMETRY  OUTSIDE  LAND_PARCEL.BOUNDARY
            AND   LAND_PARCEL.LAND_USE  =  'Military'
```

5.5. COMPATIBILITY WITH EXISTING SQL2 & SQL3 PROPOSALS

Changes to the ISO 9075 standard for SQL2 (ISO-ANSI 1989) are primarily concerned with corrections, replacement of deprecated features, upgrading the standard to support functionality already implemented, and other easily achievable extensions. Consequently, the standard will be extended to support a wider range of data types (e.g. date, time, timestamp, interval, binary etc.), enhancements to the integrity control, domain definition, referential integrity and string functions (SUBSTR, UPPER, LOWER, CONCATENATE, LENGTH etc.), and relaxation of restrictions (e.g. ability to embed a relation in a from clause rather than just a list of table names). These enhancements, although significant, are not sufficient to allow the development of the spatial extensions within the bounds of the proposed SQL2 standard.

However, the extensions proposed for SQL3 (ISO-ANSI 1989) are somewhat more far reaching. They include concepts from the object oriented paradigm, including subclasses (subtypes) by means of an *is-a* operator, user defined (abstract) data types, triggers, explicit join operators, recursive constructs, multi-valued nulls and external procedures. Kerridge (1989) has proposed to the ISO SQL standards committee extensions for SQL3 that support the definition and use of abstract data types (ADTs). He describes the functionality and syntax required to support ADTs that may be used in a manner compatible with the data types currently supported. We envisage that this is the mechanism by which spatial data types, spatial operators and spatial functions will be developed.

The use of a visualization operator to display a spatial data type graphically (rather than as text) currently appears to be the only missing concept. Selecting a column containing spatial values must result in the graphical display of those spatial values. Consequently, there must be a default operator initiated by the selections clause. Clearly, there is already a default operator being used to display values as text - this needs to be rebound to the appropriate operator (operator overloading) for spatial data types.

6. The Iconic Query Language

6.1. QUERY FUNCTIONALITY

It is intended that the query functionality supported by SQL-SX will be translated into a group of icons. Icons will be provided for each of the functions, operators and statements within SQL-SX. Selection of entity types and data types on which to operate will be available either via selection of an icon, or by selection of an exemplar. At present, there is some debate concerning the consistency of syntax for this language - for example, should there be a consistent ordering of tokens, such that feature selection is performed prior to the operator selection. Clearly, this may be invalid for operations requiring specific entity types or data types as input, since the knowledge of which entity types and data types to select could have been used during feature selection if the operator was specified first. A preliminary study of existing GUIs suggests that some inconsistency in token ordering is not a severe problem, so long as a systematic approach is used wherever feasible.

6.2. QUERY THE DATA MODEL

A difficulty reported by a number of GIS end-users is the inability to quickly determine the logical data model provided by the database. The ability to rapidly view a symbolic representation of the application data model should be provided to allow identification of the relevant entity types (object classes) and their relationships to other entity types within the database. This is an important facility missing from many of todays GIS implementations. Selection of entity types directly from the model should also be supported as a mechanism for building the query or function.

6.3. GIS ANALYSIS FUNCTIONALITY

Access to more complex analytical and implementation-specific GIS functionality must also be provided from within UGIX. A consistent set of data selection facilities must be available that allows the creation of named sets of features. It should be possible to add features to and remove features from a named set. Both generic functionality and application specific analysis functionality should be able to operate on such sets of data. New implementation-specific and analysis functions must be readily incorporated into the user interface to reflect the increasing functionality being made available by the GIS vendors and third parties.

6.4. DBA & CUSTOMIZER FUNCTIONALITY

Access to an embedded CASE tool via an icon in the normal UGIX interface is seen as advantageous. The ability to integrate functional (process, event) and data analysis, task design and user interface design within the target environment provides a unified system that may be extended as the application requirements change. The ability to develop tasks from lower level functions using graphical programming techniques permits extensibility of both function and interface. Changes should be immediately reflected on return to the UGIX environment.

7. Conclusions

This paper has attempted to summarize the recent developments in user interfaces for GIS and present a feasible solution to many user requirements. It can be said in conclusion that these developments have become central to the future of the "GIS revolution" of the late 1980s, since it is now clear that the massive (and desirable) growth of the use of GIS cannot be supported without improvements to the use environment experienced by the user. Already a variety of vendors are attempting to create user interfaces which go beyond GUIs, but none have appeared as yet.

A key assumption, however, which remains to be tested is whether and how a GIS user interface should condition GIS use. Whilst there are many who would argue that a measure of technical knowledge is desirable in those that use a GIS and a protection against the misuse of a powerful tool, it must now be established that maximum "achievable" use of a software system owes much to the creation of a structured use environment, with logic controls built in to the interface. The user is then free to choose the environment which best matches their use characteristics or which improves their aggregate efficiency measured in time, error or quality terms.

The use of a layered approach to design and development of the interface environment appears to offer a strategy for maximizing the benefits of existing and proposed standards. Proposals for enhancements to SQL provide the basis for spatial extensions to the language to allow the management and manipulation of spatial features. An iconic query language utilizing the functionality of the underlying command language and GIS will provide the user with a customizable and extensible environment for GIS interaction.

Finally, it is considered that the user interface question can also be interpreted as another expression of the data and conceptual modelling debate in GIS. User interfaces are not just pretty screen representations: as their use is extended they will come to express the whole nature of the system data model, and will probably become highly specialized as the interfaces move from function-oriented to task-oriented forms. Hence in the 1990s the user may not gain freedom of interaction with spatial data but become a captive of a particular design and control purpose. Therefore, the challenge for GIS user interface design is to gain credibility for generic forms of interface: it is this form of GIS use that will ensure wider and more informed use of GIS in the years ahead.

8. Acknowledgements

The authors of this paper acknowledge the help of ESRI in the funding of work leading to the prototyping of the first generation of UGIX. Apple Computer (UK) have also assisted in this work by establishing a laboratory of Macintosh computers which form the Apple Mapping Centre at Birkbeck College. Receipt of an ESRC/NERC grant from the collaborative programme in GIS is acknowledged by MSB. Smallworld Systems have assisted the project with hardware and software tools. We would also like to thank Elsa João for her efforts in reviewing and commenting on various drafts of this paper.

9. References

Abel, D. J. (1988) 'Relational Data Management Facilities for Spatial Information Systems', Proceedings Third International Symposium on Spatial Data Handling, Sydney, Australia, 9-18.

Abel, D. J. (1989) 'SIRO-DBMS: A Database Tool-kit for Geographical Information Systems', International Journal of Geographical Information Systems 3(2), 103-116.

ANSI (1986) American National Standards Institute X3.135-1986 Database Language SQL.

ANSI (1989) American National Standards Institute X3.135.1-1989 Database Language SQL.

Astrahan, M. M. et al. (1976) System R: Relational Approach to Database Management, ACM TODS 1, No 2.

Bundock, M. S. (1987) 'An Integrated DBMS Approach to Geographical Information Systems', Proceedings, International Symposium on Computer-Assisted Cartography (Auto-Carto 8), Baltimore, 292-301.

Cassel, R. A. and Parker, H. D. (1989) 'Design and user testing of a GIS user interface', Proceedings GIS/LIS '89, Orlando, 538-541.

Chamberlain, D. D. et al. (1981) 'A history and evaluation of System R', Communications of the ACM 24(10).

Charlwood, G. Moon, G. and Tulip, J. (1987) 'Developing a DBMS for Geographic Information: A Review', Proceedings, International Symposium on Computer-Assisted Cartography (Auto-Carto 8), Baltimore, 302-315.

Cleaveland, J. C. (1986) An Introduction to Data Types, Addison-Wesley, Reading, Massachusetts.

Codd, E. F. (1988) 'Fatal flaws in SQL', Datamation 34(16), 45-48.

Codd, E. F. (1985) 'Is Your Relational Database Management System Really Relational? An Evaluation Scheme', Proceedings, Oracle Users Conference, San Diego California.

Cowen, D. J. and Love, S.R. (1988) 'A Hypercard based workstation for a distributed GIS network', Proceedings, GIS/LIS '88, San Antonio, Texas, 285-294.

Date, C. J. (1986a) Relational Database: Selected Writings, Addison-Wesley, Reading Massachusetts.

Date, C. J. (1986b) An Introduction to Database Systems, V.1, 4th Edition Addison-Wesley, Reading.

Date, C. J. (1989) A Guide to the SQL Standard, Second Edition Addison-Wesley, Reading Massachusetts.

Easterfield, M. E., Newell, R. G. and Theriault, D. G. (1990) 'Version Management in GIS - Applications and Techniques', Proceedings, EGIS 90, Amsterdam, 288-297.

Egenhofer, M. J. (1987) 'An extended SQL syntax to treat spatial objects', Proceedings, 2nd International Seminar on Trends and Concerns of Spatial Sciences, New Brunswick, 83-95.

Egenhofer, M. J. (1989) 'Concepts of spatial objects in GIS user interfaces and query languages', Paper presented at GIS/LIS '89, Orlando, Florida.

Egenhofer, M. J. and Frank, A. U. (1988) 'Designing object-oriented query languages for GIS: human interface aspects', Proceedings, Third International Symposium on Spatial Data Handling, Sydney, Australia, 79-96.

Feuchtwanger, M. (1989) 'Geographic Logical Database Model Requirements', Proceedings, International Symposium on Computer-Assisted Cartography (Auto-Carto 9), Baltimore, 599-609.

Freeman, J. (1975) 'The Modelling of spatial relations', Computer Graphics and Image Processing 4, 156-171.

Gould, P. and White, R. (1974) Mental Maps. Hammondsworth, Penguin.

Güting, R. H. (1988) 'Geo-Relational Algebra: A Model and Query Language for Geometric Database Systems', Extending Database Technology, Lecture Notes in Computer Science, Springer-Verlag, 506-527.

Herring, J. R. Larsen, R. C. and Shivakumar, J. (1988) 'Extensions to the SQL Query Language to Support Spatial Analysis in a Topological Data Base', Proceedings, GIS/LIS '88, San Antonio, 741-750.

Ingram, K. J. and Phillips, W. W. (1987) 'Geographic Information Processing Using a SQL-Based Query Language', Proceedings, International Symposium on Computer-Assisted Cartography (Auto-Carto 8), Baltimore, 326-335.

ISO (1987) Report ISO 9075 Information Processing Systems-Database Language SQL

ISO (1989) Report ISO 9075 Information Processing Systems-Database Language SQL

ISO-ANSI (1989) Database Language SQL2 and SQL3 (ISO-ANSI working draft) X3H2-89-252 and ISO DBL FIR-3, J Melton (ed.), July 1989

Johnson, M. (1987) The Body in the Mind: the bodily basis of meaning and reason, University of Chicago Press,Chicago.

Kerridge, J. M. (1989) A Proposal to add User Defined Datatypes to SQL, ISO report ISO/TC97/SC21/WG3-DBL-FIR-3.

Mainguenaud, M. (1990) 'Is an Extended Relational Database Management System Powerful Enough to deal with Network Applications', Proceedings, EGIS '90, Amsterdam, 701-710.

Mark, D. M. (1989) 'Cognitive image schemata for geographic information: relations to user views and GIS interfaces', Proceedings, GIS/LIS '89, Orlando, 551-560.

Mark, D. M. and Frank, A. U. (1989) 'Concepts of space and spatial language', Proceedings International Symposium on Computer-Assisted Cartography (Auto-Carto 9), Baltimore, 538-556.

Moellering, H. (1986) 'Developing Digital Cartographic Data Standards for the United States', Proceedings Auto Carto London , London, 312-322.

Morrison, R. (1981) S-Algol Reference Manual, Internal Report CSR-80-81, Dept. of Computer Science, University of Edinburgh.

Newell R. G. and Easterfield M. E. (1990) 'Version Management - The Problem of the Long Transaction', Proceedings Mapping Awareness Conference, Oxford, England

Peuquet, D. J. (1988) 'Toward the Definition and use of Complex Spatial Relationships', Proceedings, Third International Symposium on Spatial Data Handling, Sydney, Australia, 211-224.

Pong-Chai, Goh (1989) 'A GQL for Cartographic and Land Information Systems', International Journal of Geographic Information Systems 3(3), 245-255.

Pullar, D. V. and Egenhofer, M. J. (1988) 'Toward Formal Definitions of Topological Relations among Spatial Objects', Proceedings Third International Symposium on Spatial Data Handling, Sydney, Australia, 225-242.

Raper, J. F. and Green, N. P. A. (1989) 'Development of a Hypertext Based Tutor for Geographical Information Systems', British Journal of Educational Technology 3, 164-172.

Rhind, D. W., Raper, J. F. and Green, N. P. A, (1989) 'First UNIX then UGIX', Proceedings International Symposium on Computer-Assisted Cartography (Auto-Carto 9), Baltimore, 735-744.

Sacks-Davis, R., McDonell, K. J. and Ooi, B. C. (1987) GEOQL - A Query Language for Geographic Information Systems, Internal Report No. 87/2, Dept of Computer Science, Royal Melbourne Institute of Technology.

Westwood, K. (1989) 'Toward the Successful Integration of Relational and Quadtree Structures in a Geographic Information System', Proceedings, National Conference - GIS - Challenge for the 1990's, Ottawa, Canada, 1026-1034.

DEFICIENCIES OF SQL AS A GIS QUERY LANGUAGE

MAX J. EGENHOFER
National Center for Geographic Information and Analysis
Department of Surveying Engineering
University of Maine
Orono, Maine 04469, USA

ABSTRACT. Numerous proposals have been made to extend the relational database query language SQL to serve as a spatial query language and currently efforts are under way to establish a standardized *spatial SQL*. Here it is argued that the SQL framework is inappropriate for an interactive spatial query language and an extended spatial SQL is at best a short-term solution. The serious deficiencies of any spatial SQL are divided into two groups:

- the severe difficulties to incorporate necessary spatial concepts into SQL such as graphical display and its specification and
- the lack of power within the relational framework with missing support for complex objects, object identity, meta queries, and non-first-order logic—all crucial when dealing with spatial data.

1. Introduction

The Structured Query Language SQL is enjoying much popularity in the database world and has become the standard for relational database management systems (ANSI, 1986). Although originally designed as a high-level interface to rcelational database management systems (Chamberlin *et al.*, 1976) to manipulate relational tables (Codd, 1970), the use of SQL has been proposed for a variety of applications, among them for spatial data handling (Lorie and Meier, 1984). Since SQL lacks spatial concepts, users are forced to simulate them in terms of the few, predefined non-spatial concepts (Westlake and Kleinschmidt, 1990). A number of SQL extensions to cope better with the properties of spatial data have been proposed (Roussopoulos and Leifker, 1985; Sikeler, 1985; Ingram and Phillips, 1987; Egenhofer, 1987; Egenhofer, 1988; Herring *et al.*, 1988; Roussopoulos *et al.*, 1988; Ooi *et al.*, 1989; Raper and Bundock, 1991), however, there is currently little consensus among researchers about the application of SQL for spatial data as demonstrated by the number of different SQL extensions. The SQL dialects tailored to spatial applications differ considerably in (1) the extent they cover spatial properties, (2) the degree at which they formally define the semantics of the extensions, (3) their syntactical implementations, and (4) the degree at which they comply with the standardized SQL structure. At the same time, they all struggle with incorporating spatial concepts into a framework designed for data modeled as 2-dimensional tables.

D. M. Mark and A. U. Frank (eds.), Cognitive and Linguistic Aspects of Geographic Space, 477–491.

This paper argues that these problems are due to SQL's inappropriateness as a framework for a high-level spatial query language. It is not discussed whether or not it is possible to express some spatial queries in SQL. Definitely, SQL can be used for querying any data modeled in a tabular format, and spatial data can be mapped onto such tables (Waugh and Healey, 1987; Abel, 1988; Westlake and Kleinschmidt, 1990); however, such a representation is not the most natural form for modeling spatial data and leads to unnecessarily complex queries. Comparing database query languages with programming languages, such a low-level treatment of spatial data is comparable to breaking a RECORD type in a high-level programming language into physical address assignments in Assembler language.

The scope of this paper is to underline this position with a number of arguments demonstrating severe shortcomings in the idea of extending the SQL framework to *the* query language for spatial data (Raper and Bundock, 1991). Our findings are based upon extensive studies of spatial query languages carried out by researchers at the University of Maine in requirements analysis (Frank, 1984a; Pullar, 1987; Egenhofer and Frank, 1988a, 1988b), query language design (Frank, 1982, 1984b; Egenhofer, 1991), and prototyping (Egenhofer, 1984, 1989b; Egenhofer and Frank, 1990). The focus will be exclusively on the role of SQL as a *query language* in its literate sense and any other roles, e.g., as a *manipulation language*, will not be discussed. To further restrict the focus, only one of the many interpretations of a query language will be considered. Currently, the use of SQL as a query language is overloaded and too many different tasks are performed with SQL. Some users consider it as a high-level interface language to tie a database management system into an application program; others use SQL as an interactive *ad hoc* query language; for another group of users, SQL is the data exchange language to transfer data from one database to another. Unfortunately, very different requirements exist for each of these roles of a query language. For example, the interface between database and application program must fit well into a high-level programming language, while an *ad hoc* query language must consider the interaction between humans and computers. In order to avoid further confusions of different requirements and expectations, the discussions in this paper are limited to the use of SQL as an interactive spatial query language.

The remainder of this paper is structured as follows: first, a brief description of SQL is given. Section 3 focuses on the role of a spatial query language within a geographic information system (GIS), a particular range of applications in the realm of spatial data handling, and typical concepts GIS users employ when they interact with geographic data. Section 4 reviews how various spatial extensions have been implemented in SQL dialects and stresses the diversity of the approaches. The impediments of the SQL framework for a high-level, interactive spatial query language are analyzed in Section 5. The conclusions propose the design of spatial database interface languages, in lieu of query languages, to better integrate the interaction between user and database.

2. SQL

SQL (Chamberlin *et al.*, 1976) is an implementation of the five relational operations selection, projection, Cartesian product, set union, and set difference (Codd, 1970)—plus a few useful extensions for operations upon tuples such as aggregate functions. The syntax framework is the SELECT-FROM-WHERE clause which corresponds to the relational operations of projection, Cartesian product, and selection, respectively. Set union and set difference can be explicitly formulated among several SQL queries. For example, given two relations **parcel** and **road** with the respective attributes **parcel.number**, **parcel.owner**, **parcel.roadname**, **road.name**, and **road.width**, the following SQL query is to retrieve the owners of all parcels which are located on roads narrower than 15 feet:

```
SELECT parcel.owner
FROM   parcel, road
WHERE  road.width < 15 and
       parcel.roadname = road.name
```

The query result is a relation which is in an interactive environment by default presented in form of a table on the screen. More extensive discussions of SQL can be found in database textbooks (Date 1986; Korth and Silberschatz 1986; Ullman, 1988) and SQL tutorials (Date, 1989).

3. Spatial Queries

SQL has been successfully used as a query language for applications from the commercial realm which can be easily expressed in terms of tables, frequently called *standard database applications*. So-called *non-standard applications* (Härder and Reuter, 1985), such as CAD/CAM, VLSI design, image databases, or geographic information systems, require a more complex data model than tables. A crucial criterion for evaluating the usability of a query language for a non-standard application domain is: "How useful are the database operations provided by the query language for the particular application?" (Schek, 1988). The following list of typical queries in a GIS environment will help to assess the usability of SQL as a spatial query language:

- "Display a map of Las Navas."
- "What is the shortest path from Castillo Magalia to the train station?"
- "Where is the nearest bus station?"
- "In which direction is El Escorial?"
- "What is this?"—and the user points to some place on a screen displaying some graphical rendering.

Obviously, spatial queries refer to particular spatial concepts (Egenhofer and Frank, 1988b) in a spatial data model (Frank, 1991a), e.g., geometric objects with a complex internal structure, spatial relationships to select objects of interest, complex graphical display, and selection by pointing. GIS users base their spatial queries upon such concepts. Therefore, they are critical criteria in a spatial query language; however, neither SQL nor the relational data model support them. Relations lack the proper set of fundamental operations to properly support the majority of geometric and pictorial operations (Joseph and Cardenas, 1988) and relational operations *per se* are insufficient to solve some typical GIS queries (Frank, 1982; Egenhofer and Frank, 1988b). For example, important problems, such as queries over combined maps, cannot be solved unless specific support of spatial operations is provided (Laurini and Milleret, 1988) and spatial data are frequently selected upon their spatial location, rather than by references to some values.

4. Spatial SQL Dialects

Several proposals have been made to turn SQL into a spatial query language, e.g., a spatial SQL dialect (Sikeler, 1985), the pictorial query language PSQL (Roussopoulos and Leifker, 1985; Roussopoulos *et al.*, 1988), Spatial SQL (Egenhofe, 1987, 1988, 1991), GEOQL (Ooi *et al.*, 1989; Ooi 1990), SQL-SX (Raper and Bundock, 1991), and the SQL-based GIS query languages for KGIS (Ingram and Phillips, 1987) and TIGRIS (Herring *et al.*, 1988). The most significant extensions will be reviewed subsequently to provide the reader an overview of what has been accomplished, but also to show how the individual extensions proposed differ.

4.1. SPATIAL DATA TYPE

Almost every spatial SQL extends the domains of the relational calculus with *spatial data types*. Although there have been attempts to design a spatial query language exclusively based on the standard domains in relational calculus (Go *et al.*, 1975; Berman and Stonebraker, 1977), i.e., on integers, reals, and character strings, there is a general agreement that users of spatial query languages need a high-level abstraction of spatial data. An attribute over such a spatial data type will be referred to as a *spatial attribute* and a relation with a spatial attribute will be called a *spatial relation*.

What varies among the different spatial SQL dialects is (1) the spatial data model for which the data types are used and (2) the degree at which this extension is integrated into the SQL environment. The variety of spatial data types proposed for spatial SQLs includes

- a universal spatial data type (Sikeler, 1985; Herring *et al.*, 1988; Ooi *et al.*, 1989);
- data types for each spatial dimension, i.e., points, lines, areas, and volumes (Raper and Bundock, 1991), and their generalizations to a dimension-independent spatial superclass (Egenhofer, 1988)—essentially, the universal spatial data type above;
- a number of data types for a multitude of spatial properties such as area, perimeter, and length (Ingram and Phillips, 1987) and the graphical representation as well; and
- a data type for bitmap graphics (Roussopoulos and Leifker, 1985).

The actual syntax of these extensions differs considerably. Three particular approaches have been pursued:

- Some query languages reserve specific attribute names for spatial attributes such as *loc* (Roussopoulos and Leifker, 1985; Sikeler, 1985) or *map*, *area*, and *perimeter* (Ingram and Phillips, 1987).
- Others assign no explicit spatial attribute to a spatial relation, but implicitly assume its existence (Herring *et al.*, 1988; Ooi *et al.*, 1989).
- Others define only the names of the spatial data types and with the data definition, spatial attributes are defined by associating an attribute name with a spatial data type (Egenhofer, 1988; Raper and Bundock, 1991).

4.2. SPATIAL RELATIONSHIPS

The extension with spatial data types is useless unless the pertinent operations and relationships are also defined. Spatial relationships are boolean operations to check whether or not a particular predicate holds true between tuples of two or more spatial relations. SQL tends to overload predicates, i.e., an operation may have multiple implementations and the system selects one depending on the type(s) of the argument(s) of the predicate. While such an overloading of SQL's standard predicates is sufficient for some applications, e.g., temporal relationships (Allen, 1983), it is insufficient for spatial relationships. The set of standard predicates in SQL is too small to cover all spatial relationships (Egenhofer, 1989a; Egenhofer and Herring, 1990). For example, SQL lacks predicates for topological concepts, such as *neighbor* and *intersect*; therefore, syntax extensions are necessary to allow users to formulate queries with spatial predicates.

Similar to the extension with a spatial data type, all spatial SQL dialects include such spatial operations and spatial relations. Four major differences in the use of spatial operations and relationships can be observed:

- The number of spatial relationships covered by the individual SQL extensions varies considerably and different terminology is used as well.
- With two exceptions (Herring *et al.*, 1988; Egenhofer, 1991), no formal definitions are given for the semantics of the relationships.
- Spatial relations are introduced in prefix (Ingram and Phillips, 1987) or infix form (Egenhofer, 1988; Ooi *et al.*, 1989; Raper and Bundock, 1991).
- Spatial predicates are used upon spatial attributes (Egenhofer, 1988; Raper and Bundock, 1991) or upon relations (Sikeler, 1985; Ingram and Phillips, 1987; Herring *et al.*, 1988; Ooi *et al.*, 1989).

4.3. GRAPHICAL DISPLAY

An *ad hoc* spatial query language requires that query results can be displayed graphically. The graphical display of query results has two separate components in a query language: (1) specifying that the query result (or parts of it) should be graphically displayed—as opposed to the implied tabular representation of alphanumeric data in SQL—and (2) describing how to represent the query result. For both issues, different solutions have been proposed.

- A qualifier for a spatial attribute in the SELECT clause specifies that a particular relation be graphically represented (Sikeler, 1985).
- All spatial attributes in the SELECT clause are displayed, while non-spatial parts of the query result are shown as alphanumeric tables (Ingram and Phillips, 1987).
- The query result is displayed according to the status in a display environment (Egenhofer, 1991).

The description of the graphical display of the query result has drawn only little attention. Most spatial query languages disregard this part or use only a default representation. Only PSQL and Spatial SQL propose solutions for this problem.

- PSQL displays query results according to predefined definitions, called ***picture lists*** (Roussopoulos and Leifker, 1985)—without providing a language to create or modify a picture list.
- Spatial SQL displays spatial query results according the definitions in the ***graphical representation environment*** (Egenhofer, 1991). A comprehensive display language as a superset of SQL allows users to describe the use of colors, patterns, symbols, etc. for spatial relations.

4.4. SELECTION BY POINTING

Each SQL query is a stand-alone instruction without any reference to the previously asked queries or their results. Likewise, query results are always represented as a single "rendering" and no interaction with the currently displayed result is possible. For a spatial query language with graphical representation of query results, this SQL feature is a major restriction, because the graphical representation of query results animates users to refer to the drawings when formulating further queries.

Since SQL has no provisons for input other than typed characters, some spatial SQL dialects include an operator to identify a spatial object by pointing to its spatial location on the screen. Most commonly, pointing is implemented as a keyword, e.g., MOUSE (Ingram and Phillips, 1987) or PICK (Egenhofer, 1988). Such references may occur in WHERE clauses when a user refers to ***this*** object and uses a pointing device to select the object from the screen drawing.

4.5. COMPLIANCE WITH THE SYNTAX FRAMEWORK

Any SQL extension suffers from the dilemma of preserving the standardized form of SQL while extending SQL's functionality. A variety of syntax modifications of the SELECT-FROM-WHERE framework have been proposed in order to enable users to query spatial data. Three fundamentally different trends can be found:

- the addition of new clauses in which particular spatial properties are addressed, such as WITH LOCATION (Sikeler, 1985) and AT (Roussopoulos and Leifker, 1985) for spatial conditions, and ON (Roussopoulos *et al.*, 1988) to specify an output format;
- the modification of the SQL framework by treating relations in lieu of attributes both in the SELECT and WHERE clauses and cancelling the FROM clause which becomes superfluous; and
- minimal extensions within the given framework to comply with standard SQL (Egenhofer, 1988; Ooi *et al.*, 1989) and the definition of a ***display environment*** outside of SQL (Egenhofer, 1991).

5. Impediments of the SQL Framework for Spatial Query Languages

Any SQL extension which attempts to comply with the SQL standard has to live with the conceptual and syntactical constraints set forth by the standardized framework. This section will show why it is so difficult to extend standard SQL so that it becomes a useful spatial query language. A number of shortcomings of the SQL framework for spatial data handling will be examined, some of which are due to the attempt of extending a given standardized query language. Two types of deficiencies are distinguished: (1) the lack of spatial concepts in the SQL framework and (2) the lack of expressive power to formulate types of (spatial) queries. All these shortcomings are crucial impediments for a spatial query language; however, it is not the lack of a particular functionality that makes the SQL framework inappropriate for a spatial query language. It is rather the fact that a great number of deficiencies exists for which no easy or simple remedies can be provided. Some of these problems may be fixed in SQL with considerable modifications of the initial SQL language, such as a recent proposal to integrate some syntax mechanisms for recursive query formulation into SQL (ANSI-ISO, 1989). For other problems, no such "solutions" are in sight. Of course, such extensions are only patches and do not help to make SQL a sound and coherent spatial query language.

5.1. DECOUPLED RETRIEVAL AND DISPLAY

Standard SQL is targeted almost entirely to retrieving data modeled in a tabular form. It stresses the functionality of formulating complex and powerful queries through the implementation of the relational algebra operations and, at the same time, disregards how to present the query result to the user. Only a single, default representation is provided—the display in the form of a table. This uneven balance between retrieval and display is characterized by two features of SQL:

- The SELECT clause is overloaded with the projection of the attributes onto the resulting relation *and* the implied tabular representation of each query result. Such a combination may be appropriate for those applications which always require the results to be presented as tables only; however, it is a major impediment for a query language with renderings other than tables, e.g., drawings, or with a choice of renderings (Egenhofer, 1987).

- The retrieval language is decoupled from the representation of query results. SQL neither memorizes what has been displayed nor does it allow the users to formulate queries with respect to the query results displayed.

These shortcomings are most apparent when users request to display the query results in a non-tabular format.

5.1.1. *Specification of Graphical Display.* The potentials for graphical variations of objects on drawings are much greater than those for manipulating the frame of a table, e.g., by adding headers over the columns or changing the sequence of columns. Graphical display involves the use of different colors, patterns, symbols, etc. (Bertin 1983).

What makes the *display specifications* more difficult is that a set of these graphical properties may be assigned not only to the entire result of a query, but also to specific spatial objects or classes of objects in the result. Especially for GIS applications with high-quality map output, these display specifications are too complex to be integrated into the actual query statement.

At a first glance, it may appear as a viable solution to separate a spatial query into two parts: (1) the instruction to retrieve the data wanted and (2) the *subsequent* command to display the query result. The query language—or better the *retrieval language*—would describe *what* to retrieve and a display language would specify *how* to graphically represent the query result previously retrieved; however, such a separation does not take into account that the query and specification of graphical representation frequently depend upon each other. For instance, given the following retrieval instruction:

```
SELECT road.geometry
FROM   road, district
WHERE  district.name = "Avila" and
       road.geometry INSIDE district.geometry
```

After its processing, further information will be necessary in order to draw the major roads with a different line style than the remaining roads. In essence, more *queries* are necessary to separate the intermediate result—the geometry of *all* roads—into two sets of roads so that the elements of each set can be represented with the same graphical attributes. Therefore, the demand for alternative displays is not solved by just drawing the spatial attributes in the SELECT clause. It must also be considered that the graphical display of query results may require more information from the database than what has been provided by the query result.

5.1.2. *Combinations of Graphically Displayed Query Results.* Spatial query results may be depicted as drawings at which users look; however, a more dynamic interaction with query results is also necessary (Egenhofer and Frank, 1988b). Users want to refer to the current drawing by asking further questions about it or they want to modify the current drawing, e.g., by adding further information to it or by removing information displayed (Egenhofer, 1990). Such a dynamic interaction may be richer than the interaction with tables and requires support from the query language. The SQL framework supports only the retrieval of data based upon input typed by the user and the presentation of the data retrieved for the user, and no provisions have been made for an alternative retrieval, i.e., one partially based on the currently displayed result. Extensions of the standardized SQL framework to incorporate multiple representation types are essentially impossible, unless changes are made to a degree that the extension is no longer compatible with the standard.

5.2. POWER OF THE LANGUAGE

The power of the SQL language to retrieve data is determined by the operations of the relational algebra plus some additional concepts such as aggregate functions. Although SQL is *relationally complete* (Ullman, 1982), i.e., any manipulation of tables is possible with a combination of the operations provided, it does not support some fundamental concepts, such as meta-data queries and higher-order queries. Furthermore, the retrieval part of SQL is inherently value-based, i.e., all operations are performed upon attributes and tuples can be only compared for equal values, but not for identity. These problems are general SQL problems which apply to non-spatial applications as well. While they may be less critical for traditional SQL applications, they impose serious restrictions upon the use of any spatial extension of SQL. The impact of these deficiencies on a spatial query language is discussed below.

5.2.1. *Meta-Data Queries.* The SQL query facilities are designed for retrieving data, but they are limited for asking queries about the structure of particular tuples. *Data queries* typically start with some knowledge about the structure of the tables, i.e., users remember the names of the tables and the corresponding attributes. The instructions are to find corresponding tuples the values of which fulfill certain constraints. While these queries may cover the majority of questions users ask against a database, there are other kinds of queries with which users request information about data, e.g., "To which relation(s) does a particular tuple belong?", "To which attribute(s) does a specific value belong?", "What is the domain of a specific attribute?", and "To which view does a particular relation belong?"

These ***meta-data queries*** are particularly important for spatial databases when information is graphically represented about which users request further information. Topographic and thematic maps, for example, typically contain symbols and labels for various features. Users may point to them asking "What is this?" and expecting an answer like "This is a castle" or "That is a train station." The answers to such queries are the names of relations or attributes, not particular values of attributes.

Answers to meta-queries can be combined from data and meta-data. For instance, an answer to a query about a label on a map may be "This is the mileage along the main road from Las Navas to Avila." This answer contains meta-data—the mileage is the attribute name of the label—but also data, namely that the label identifies the mileage for a road and that it is the value from Las Navas to Avila.

Another geographic application shows the use of meta-data queries about the domain of an attribute. The query "What are possible soil classifications?" refers to the domain of an attribute *soil type*. Note that this query is different from "What are all the values stored in the attribute *soil type*?" The result of the first query is independent of the actual values stored in a table and its definition is the subject of the data definition, while the result of the latter query may change with the storage, deletion, or update of any value.

5.2.2. *Knowledge Queries.* Queries of a nature similar to meta-queries are ***knowledge queries*** (Motro and Yuan, 1990), i.e., those used for explanations of the reasoning process of a query language. Spatial relationships, for example, are typically derived from a representation in a spatial data model, rather than being explicitly recorded (Davis 1986). The derivation is based upon rules which formalize the criteria for the individual relationships. Spatial information systems that allow users to tailor their spatial predicates (Herring *et al.*, 1988) must also include facilities to let them inquire about the rules used for a specific predicate. For example, the adjacency between two parcels may be defined such that they share at least one common boundary, but have no common interior (Egenhofer and Herring, 1990) and users may want to know from the system "Why did you identify the two objects as neighbors?"—"Because they share a common boundary." Such knowledge queries about the characteristics of operations may also be extended to include their properties. For example, the query "When Avila can be reached by train from Las Navas, can Las Navas be reached from Avila by train as well?" is a query about the ***symmetry*** of the relationship *reachable_by_train*.

In a similar way, users may ask queries about consistency constraints. For example, the query "What are the geometrical constraints about rivers?" may be answered by "They must not cross each other and a river cannot cross a road, unless there is a bridge." Such queries are of particular interest when users modify the database. Prior to making a change, they may want to get information what data they must provide, or after an unsuccessful attempt to update the database, they may want to find out more details about their—or the system's—failure.

5.2.3. *Qualitative Answers.* SQL allows users to ask queries with quantitative results, i.e., about the values of tuples; however, it does not support queries with qualitative results, e.g., about the kind of relationship between objects. For instance, while it is possible to ask for "all roads in the district of Avila which are wider than the road from Madrid to Las Navas," it is impossible to formulate an SQL query for "the relation between the two roads from Madrid to Las Navas and from Las Navas to Avila," with an expected answer like "The road from Las Navas to Avila is narrower." Likewise, relationships cannot be used as variables, for instance to ask a query like "Find all scenic places which are in the same direction as El Escorial is from Las Navas."

The conceptual problem behind this shortcoming is that SQL—like any other relational query language—is *first-order*, i.e., the predicates are not variable (Gallaire *et al.*, 1984). GIS users frequently ask qualitative queries, e.g., "What is the directional relationship between Las Navas and El Escorial?"

5.2.4. *Object Identity.* Since SQL is inherently value-based it provides only means to compare objects for the same value(s), but lacks the concept of identity. *Identity* is that property which distinguishes each object from all others (Khoshafian and Copeland, 1986); therefore, object identity allows for the comparison whether two objects are the same, independent of their particular attribute values. Object identity has been recently acknowledged as a major component of next-generation database systems (Atkinson *et al.*, 1989; Stonebraker *et al.*, 1990), because it is crucial for any application in which objects change over time (Al-Taha and Barrera, 1990). Geographic objects are prototypical examples thereof. A land parcel, for example, may change all its characteristics over time: the boundaries get re-surveyed and adjusted, new owners buy the land, the land may be used differently, the parcel may get a new postal address, etc. Nevertheless, the piece of land is still the same and in order to relate its state at one time in history with another state it is necessary that each parcel has an object identifier which is independent of the descriptive values of the parcel.

6. Conclusion

The intent of this paper was to bring a new perspective to the ongoing discussion about the use of the SQL framework for the standard spatial query language. While SQL has been fairly successfully applied as a query language for standard database applications, such as banking accounts, its success in the domain of spatial applications, such as geographic information systems or CAD/CAM systems, has been very limited. Despite numerous attempts to extend SQL with various spatial features, no satisfying solutions for an SQL-based interactive spatial query language have been found.

A spatial SQL can be considered only a short-term "solution" for an interactive spatial query language. Extensions of the SQL framework with a spatial abstract data type (ADT), including spatial relationships as predicates, are possible; however, prior to discussing syntax extensions, the semantics of the operations upon the spatial ADT must be formally defined. Any spatial query language which provides only a list of operation names and their parameters (Raper and Bundock, 1991) treats only syntactical problems and oversees the difficult, but crucial aspect of the semantics of spatial operations (Egenhofer, 1989b).

Informal sets of operations to be carried out by a spatial database have been collected (Joseph and Cardenas, 1988; Tomlin, 1990); however, only little effort has been put into the formalization of spatial data models (Güting, 1988; Dorenbeck and Egenhofer, 1991) and spatial relationships and operations such as cardinal directions (Peuquet, 1988; Chang *et al.*, 1989; Frank, 1991b; Hernández, 1991) and topological relationships (Egenhofer, 1989a; Egenhofer and Herring, 1990; Egenhofer and Franzosa, 1991).

Geographic databases need query languages which are more powerful and better-suited than SQL or an extension of it. SQL *per se* is already difficult to use (Reisner, 1981) and any addition to the SQL concepts increases its complexity. SQL lacks crucial concepts for a spatial query language, the most important deficit being the missing integration of data retrieval and representation of query results. The retrieval of data and its alphanumeric representation in a tabular format are insufficient for spatial data. Particularly the graphical display of query results in all its varieties cannot be handled with traditional query languages.

In lieu of a database query language, spatial applications need a *database interface language*, considering both the retrieval of data and the representation of query results. The design of such a language must start at the user level by investigating what kinds of operations users want to perform upon spatial databases and how they do it (Egenhofer and Frank, 1988b; Pizano *et al.*, 1989; Mainguenaud and Portier, 1990). Recent studies of the use of metaphors for the interaction with spatial data (Kuhn and Frank, 1991) provide a starting point for investigations into database interface languages.

7. Acknowledgements

This research was partially funded by grants from Digital Equipment Corporation under Sponsored Research Agreement No. 414 and TP-765536, and Intergraph Corporation. Additional support from NSF for the NCGIA under grant number SES 88-10917 is gratefully acknowledged. I also appreciate Andrew Frank's and Robert Cicogna's comments on an earlier version of this paper.

Author's electronic mail address: max@mecan1.maine.edu

8. References

Abel, D. (1988) 'Relational Data Management Facilities for Spatial Information Systems', In D. Marble (ed.), Third International Symposium on Spatial Data Handling, Sydney, Australia, 9–18.

Al-Taha, K. and Barrera, R. (1990) 'Temporal Data and GIS: An Overview', in GIS/LIS '90, Anaheim, CA, 244–254.

Allen, J. (1989) 'Maintaining Knowledge about Temporal Intervals", Communications of the ACM, 26(11), 832–843.

ANSI (1986) 'X3H2-86-2 American National Standard Database Language SQL', American National Standards Institute.

ANSI-ISO Committee (1989) 'Working Draft, Database Languages SQL 2 and SQL 3', American National Standards Institute and International Standards Organization.

Atkinson, M., Bancilhon, F., DeWitt, D., Dittrich, K., Maier, D., and Zdonik, S. (1989) 'The Object-Oriented Database System Manifesto', in Deductive and Object-Oriented

Databases, Kyoto, Japan.

Berman, R. and Stonebraker, M. (1977) 'GEO-QUEL, A System for the Manipulation and Display of Geographic Data', ACM Computer Graphics, 11(2), 186–191.

Bertin, J. (1983) Semiology of Graphics, The University of Wisconsin Press, Madison, WI.

Chamberlin, D., Astrahan, M., Eswaran, K., Griffiths, P., Lorie, R., Mehl, J., Reisner, P., and Wade, B. (1976) 'SEQUEL 2: A Unified Approach to Data Definition, Manipulation, and Control', IBM Journal of Research and Development 20(6), 560–575.

Chang, S.K., Jungert, E., and Li, Y. (1989) 'The Design of Pictorial Databases Based Upon the Theory of Symbolic Projections', in A. Buchmann, O. Günther, T. Smith, and Y. Wang (eds.), Symposium on the Design and Implementation of Large Spatial Databases, Lecture Notes in Computer Science 409, Springer-Verlag, New York, NY, 303–323.

Codd, E. (1970) 'A Relational Model for Large Shared Data Banks', Communications of the ACM 13(6), 377–387.

Date, C. (1986) An Introduction to Database Systems, Addison-Wesley Publishing Company, Reading, MA.

Date, C. (1989) A Guide to the SQL Standard, Addison-Wesley Publishing Company, Reading, MA.

Date, C. and White, C. (1989) A Guide to SQL/DS, Addison-Wesley Publishing Company, Reading, MA.

Davis, E. (1986) Representing and Acquiring Geographic Knowledge, Morgan Kaufmann Publishers, Inc., Los Altos, CA.

Dorenbeck, C. and Egenhofer, M. (1991) 'Optimization Strategies for Overlay Operations', in D. Mark and D. White (eds.), Auto-carto 10, Baltimore, MD, 296–312.

Egenhofer, M. (1984) 'Implementation of MAPQUERY, a Query Language for Land Information Systems' (in German), Technical Report 79, Institute for Geodesy and Photogrammetry, Swiss Federal Institute of Technology (ETH), Zurich, Switzerland.

Egenhofer, M. (1987) 'An Extended SQL Syntax To Treat Spatial Objects', in Y.C. Lee (ed.), Second International Seminar on Trends and Concerns of Spatial Sciences, Fredericton, New Brunswick, Canada, 83–95.

Egenhofer, M. (1988) 'A Spatial SQL Dialect', Technical Report 100, Department of Surveying Engineering, University of Maine, Orono, ME.

Egenhofer, M. (1989a) 'A Formal Definition of Binary Topological Relationships', in W. Litwin and H.-J. Schek (eds.), Third International Conference on Foundations of Data Organization and Algorithms (FODO), Paris, France, Lecture Notes in Computer Science 367, Springer-Verlag, New York, NY, 457–472.

Egenhofer, M. (1989b) 'Spatial Query Languages', PhD thesis, University of Maine, Orono, ME.

Egenhofer, M. (1990) 'Interaction with Geographic Information Systems via Spatial Queries', Journal of Visual Languages and Computing 1(4), 389–413.

Egenhofer, M. (1991) 'Extending SQL for Cartographic Display', Cartography and Geographic Information Systems (in press).

Egenhofer, M. and Frank, A. (1988a) 'Designing Object-Oriented Query Languages for GIS: Human Interface Aspects', in D. Marble (ed.), Third International Symposium on Spatial Data Handling, Sydney, Australia, 79–96.

Egenhofer, M. and Frank, A. (1988b) 'Towards a Spatial Query Language: User Interface Considerations', in D. DeWitt and F. Bancilhon (eds.), 14th International Conference on Very Large Data Bases, Los Angeles, CA, 124–133.

Egenhofer, M. and Frank, A. (1990) 'LOBSTER: Combining AI and Database Techniques for GIS', Photogrammetric Engineering & Remote Sensing 56(6), 919–926.

Egenhofer, M. and Franzosa, R. (1991) 'Point-Set Topological Spatial Relations', International Journal for Geographical Information Systems 5, (in press).

Egenhofer, M. and Herring, J. (1990) 'A Mathematical Framework for the Definition of Topological Relationships', in K. Brassel and H. Kishimoto (eds.), Fourth International Symposium on Spatial Data Handling, Zurich, Switzerland, 814–819.

Frank, A. (1982) 'MAPQUERY—Database Query Language for Retrieval of Geometric Data and its Graphical Representation', ACM Computer Graphics 16(3), 199–207.

Frank, A. (1984a) 'Requirements for Database Systems Suitable to Manage Large Spatial Databases', in International Symposium on Spatial Data Handling, Zurich, Switzerland, 38–60.

Frank, A. (1984b) 'Extending a Database with Prolog', in L. Kerschberg (ed.), First International Workshop on Expert Database Systems, Kiawah Island, SC, 665–676.

Frank, A. (1988) 'Requirements for a Database Management System for a GIS' Photogrammetric Engineering & Remote Sensing 54(11), 1557–1564.

Frank, A. (1991) 'Spatial Concepts, Geometric Data Models and Data Structures', Computers and Geo-Sciences (in press).

Gallaire, H., Minker, J., and Nicolas, J. (1984) 'Logic and Databases: A Deductive Approach', ACM Computing Surveys 16(2), 153–185.

Go, A., Stonebraker, M., and Williams, C. (1975) 'An Approach to Implementing a Geo-Data System', Technical Report, Memo ERL-M529, Electronics Research Laboratory, University of California, Berkeley, CA.

Güting, R. (1988) 'Geo-Relational Algebra: A Model and Query Language for Geometric Database Systems', in J. Schmidt, S. Ceri, and M. Missikoff (eds.), International Conference on Extending Database Technology, Venice, Italy, Lecture Notes in Computer Science 303, Springer-Verlag, New York, NY, 506–527.

Härder, T. and Reuter, A. (1985) 'Architecture of Database Systems for Non-Standard Applications' (in German), in A. Blaser and P. Pistor (eds.), Database Systems in Office, Engineering, and Science, Informatik Fachberichte 94, Springer Verlag, New York, NY, 253–286.

Hernández, D. (1991) 'Relative Representation of Spatial Knowledge: The 2-D Case', in

D. Mark and A. Frank (eds.) Cognitive and Linguistic Aspects of Geographic Space, Kluwer Academic Publishers, Dordrecht.

Herring, J., Larsen, R., and Shivakumar, J. (1988) 'Extensions to the SQL Language to Support Spatial Analysis in a Topological Data Base', in GIS/LIS '88, San Antonio, TX, 741–750.

Ingram, K. and Phillips, W. (1987) 'Geographic Information Processing Using a SQL-Based Query Language', in N.R. Chrisman (ed.), Auto-carto 8, Eighth International Symposium on Computer-Assisted Cartography, Baltimore, MD, 326–335.

Joseph, T. and Cardenas, A. (1988) 'PICQUERY: A High Level Query Language for Pictorial Database Management', IEEE Transactions on Software Engineering 14(5), 630–638.

Khoshafian, S. and Copeland, G. (1986) 'Object Identity', in N. Meyrowitz (ed.) Object-Oriented Programming Systems, Languages, and Applications—OOPSLA '86, Portland, OR, SIGPLAN Notices 21(11), 406–416.

Korth, H. and Silberschatz, A. (1986) Database System Concepts, McGraw-Hill Book Company, New York, NY.

Kuhn, W. and Frank, A. (1991) 'A Formalization of Metaphors and Image-Schemas in User Interfaces', in D. Mark and A. Frank (eds.) Cognitive and Linguistic Aspects of Geographic Space, Kluwer Academic Publishers, Dordrecht.

Laurini, R. and Milleret, F. (1988) 'Spatial Data Base Queries: Relational Algebra versus Computational Geometry', in M. Rafanelli, J. Klensin, and P. Svensson (eds.), Statistical and Scientific Database Management, 4th International Conference, Rome, Italy, Lecture Notes in Computer Science 339, Springer-Verlag, New York, NY, 291–313.

Lorie, R. and Meier, A (1984) 'Using Relational DBMS for Geographical Databases', Geo-Processing, 2, 243–257.

Mainguenaud, M. and Portier, M.-A. (1990) 'Cigales: A Graphical Query Language for Geographical Information Systems', in K. Brassel and H. Kishimoto (eds.), Fourth International Symposium on Spatial Data Handling, Zurich, Switzerland, 393–404.

Motro, A. and Yuan, Q. (1990) 'Querying Database Knowledge', in H. García-Molina and H. Jagadish (eds.), 1990 ACM SIGMOD, International Conference on Management of Data, Atlantic City, NJ, SIGMOD Record 19(2), 173–183.

Ooi, B.C., Sacks-Davis, R., and McDonell, K. (1989) 'Extending a DBMS for Geographic Applications', in IEEE Fifth International Conference on Data Engineering, Los Angeles, CA, 590–597.

Peuquet, D. (1988) 'Toward the Definition and Use of Complex Spatial Relationships', in D. Marble (ed.), Third International Symposium on Spatial Data Handling, Sydney, Australia, 211–223.

Pizano, A., Klinger, A., and Cardenas, A. (1989) 'Specification of Spatial Integrity Constraints in Pictorial Databases', Computer, 22(12), 59–71.

Pullar, D. (1987) 'Query Language for Spatial Relationships', in ASPRS-ACSM Annual Convention, Baltimore, MD, 180–192.

Reisner, P. (1981) 'Human Factors Studies of Database Query Languages: A Survey and Assessment', ACM Computing Surveys 13(1), 13–31.

Roussopoulos, N. and Leifker, D. (1985) 'Direct Spatial Search on Pictorial Databases Using Packed R-Trees', in S. Navathe (ed.), International Conference on Management of Data, Austin, TX, SIGMOD Record 14(4), 17–31.

Roussopoulos, N., Faloutsos, C., and Sellis, T. (1988) 'An Efficient Pictorial Database System for PSQL', IEEE Transactions on Software Engineering 14(5), 630–638.

Schek, H.-J. (1988) 'Perspectives of Future Database Systems—Extensibility and Object-Orientation' (in German), in AM/FM International, European Division, Regional Conference, Siegen, F.R. Germany, 2–14.

Sikeler, A. (1985) 'Examination of Storage Structures for 3-Dimensional Objects' (in German), Technical Report, University Kaiserslautern.

Stonebraker, M., Rowe, L., Lindsay, B., Gray, J., Carey, M., Brodie, M., Bernstein, P., and Beech, D. (1990) 'Third Generation Data Base System Manifesto', SIGMOD Record 19(3), 31–44.

Tomlin, C. (1990) Geographic Information Systems and Cartographic Modeling, Prentice-Hall, Englewood Cliffs, NJ.

Ullman, J. (1982) Principles of Database Systems, Computer Science Press, Rockville, MD.

Ullman, J. (1988) Principles of Database and Knowledgebase Systems, Computer Science Press, Rockville, MD.

Waugh, T. and Healey, R. (1987) 'The GEOVIEW Design: A Relational Database Approach to Geographical Data Handling', International Journal of Geographical Information Systems 1(2), 101–118.

Westlake, A. and Kleinschmidt, I. (1990) 'The Implementation of Area and Membership Retrievals in Point Geography using SQL', in Z. Michalewicz (ed.), Statistical and Scientific Database Management, Fifth International Conference, Charlotte, NC, Lecture Notes in Computer Science 420, Springer-Verlag, New York, NY, 200–218.

THE ROLE OF THE USER IN GENERALIZATION WITHIN GEOGRAPHIC INFORMATION SYSTEMS

ELSA MARIA JOÃO
Department of Geography
Birkbeck College
7-15 Gresse St..
London W1P 1PA United Kingdom

ABSTRACT. At present, most GIS fail to take adequate account of the problems of generalization. The reason for the difficulty in generalizing in an *automated* environment is the complexity of the generalization process and, to an extent, the subjectivity of the process as carried out on data before it is encoded. An ideal generalization procedure within a GIS would select the best available generalization process based on purpose, scale, data types and quality requirements. Also, it would keep track of the changes operated upon the data by the generalization procedures. The aim, ultimately, is to rely as much as possible on a knowledge-based system to carry out this process for the user. But at the moment, knowledge-based generalization is severely limited and so any successful generalization within a GIS will need to depend heavily on the input of the user. The user interaction with each function of such a generalization system is described from the point of view of the user. The flexibility of the system to support different user views and the importance of friendly user interfaces is stressed.

1. Introduction

One of the major challenges in GIS research is to perform map generalization automatically (João *et al..*, 1990). Until recently map generalization was a traditional task done by manual cartographers which took great skill and required adaptability to different circumstances. These different circumstances could arise either from a different geographical region that needed to be mapped, a different scale, a different size paper to draw the map in, a different purpose for which the map was going to be used, etc. All such factors would influence the cartographer in his selection of 'how to draw what', with what priority and detail, and which attributes to display as text. This would be accomplished using basic cartographic rules, using experience and natural intuition, and using the privileged human vision of being able to perceive a map as a whole and for what it represents. At present it is much too difficult for a computer to be able to mimic the cartographer (see Beard, 1987) and explains why, to date, all GIS fail to take adequate account of the problem of generalization.

Because of the difficulty of automating generalization, there is a real danger that the completed system may perform merely *what is currently possible* rather than *what is needed* or *what is desirable*. After some time, the existence of such systems set *de facto* standards which constrain and focus the user's perception of what is needed. This paper proposes the existence of a set of

D. M. Mark and A. U. Frank (eds.), Cognitive and Linguistic Aspects of Geographic Space, 493–506.

essential generalization functions, that are needed from the user's perspective, independent of the viability of their implementation in the short-term. This set of generalization functions describe what an automated system designed to perform generalization within a GIS would do for the users. The functional requirements of such an automated system may be represented using a special type of data flow diagram - a user concept diagram (Martin, 1988). User concept diagrams identify automated processes according to the user's view and trace the flow of data through the processes. As user concept diagrams give special attention to the identification of anticipated system uses and how the users will interact with the system, these diagrams were especially appropriate for the analyses presented in this paper.

It is the author's belief that the GIS user will always need to play an important role in GIS generalization if the process is to have satisfactory results. From the specification of user's needs to the solving of data conflicts the user should be given room to increase or control the system's performance. There might be some cases where fully automatic production can be possible and desirable in constrained map production, e.g. in national atlas mapping (see Richardson ,1989). In general though, it would not be possible for a knowledge-based system to carry out all of the generalization process and so some user interaction would always be needed. In addition, as long as the generalization techniques available within GIS continue to be severely limited, the user role will be even more crucial. Weibel and Buttenfield (1988) suggested that the automation of generalization should be subdivided among the user and the computer according to problem solving potential - the "amplified intelligence approach". The computer would implement some tasks for which it is good at solving but would rely on the user for control and knowledge. It is important therefore that the interaction of the user with the GIS be as efficient and friendly as possible.

This paper also proposes some specific advantages of using automated processes to perform generalization which could not be possible manually. Automated generalization should not just aspire to mimic manual processes but should also set up higher standards for the type of functions available. And so, the highest standard that automated generalization should aspire to is the control of quality of generalized cartographic products. This is not just a control of quality similar to that of the manual cartographer's check, but a *quantification* and *storage* of quality values that can then be carried forward and taken into account in subsequent GIS analysis.

This paper starts with a broad definition of generalization and of its operators. This is followed by a description of generalization facilities available within a raster-based GIS and a vector-based GIS available in the market. This description sets up the background for the new generalization functions proposed in the next section. The new generalization functions are presented using user concept diagrams. These diagrams have a hierarchical nature so that they start with a system overview and then the major functional area - map generalization - is decomposed to obtain a comprehensive view of the system. The paper concludes with some comments on ways of implementing such a system and on the importance of user interfaces.

2. Generalization Within GIS

In GIS, data generalization will be carried out 'on the fly' unless computing resources are inadequate or unless the same generalized data set will suffice for many purposes (João *et al.*, 1990). There are three main reasons for doing generalization: to create a level of detail appropriate for the scale of display or analysis, to permit analysis of data at multiple levels of resolution, and to minimize data storage and input/output operations. In relation to the generalization of different types of data, little difference exists, in principle, between the generalization applied to earth- and social-science data. Continuous and categorical data sets are found in both and, whilst some data-specific rules may be applied to take account of the peculiarities of one data set, the general absence of such knowledge suggests that both be treated identically - especially if they are to be combined inside a GIS and analyzed together (João *et al.*, 1990).

It can be argued that two distinctive classes of operators can be identified in the process of generalization: operators for raster-mode generalization and operators for vector-mode generalization. Table 1 shows a classification of generalization according to these two types of operators.

RASTER-MODE GENERALIZATION (McMaster and Monmonier, 1989)	**VECTOR-MODE GENERALIZATION**	
	Attribute data (McMaster, 1989 & Brassel and Weibel, 1988)	Spatial data (McMaster, 1989 & inferred from Tobler 1989)
Structural generalization		
simple structural reduction		
resampling	**Thematic data**	**Punctiform data**
tesselation change	classification	omit
vector-to-raster conversion	symbolization	aggregate
Numerical generalization	**Temporal data**	displace
low-pass filters		**Linear data**
high-pass filters		simplify
compass gradient masks		smooth
band-rationing		displace
vegetation indices		merge
Numerical categorization		enhance
minimum-distance-to-means classifier		omit
parallelepiped classifier		**Areal data**
maximum-likelihood classifier		omit
Categorical generalization		amalgamate
merging (of categories)		collapse
aggregation (of cells)		displace
non-weighted		**Volumetric**
category-weighted		**(surface) data**
neighborhood-weighted		smooth (trend surface)
attribute change		enhance
feature dissolve		simplify
minimum-size suppression		**Flow data**
categorical suppression		amalgamate
erode smoothing		omit
gap bridging		

Table 1. Classification of generalization according to raster and vector mode operators (João *et al.*, 1990).

For each type of operator more than one type of algorithm may exist. Although these algorithms were developed to perform the same kind of task, the effects that they can cause in the geometric properties of the generalized data can differ. For example McMaster (1987) showed how geometric properties of the data, such as direction and area displacement, varied for different line simplification algorithms. GIS that will base the selection of generalization algorithms on quality requirements should therefore have a wide range of these algorithms available. If, for example, a user requires the simplification of a line with minimum vector displacement, the GIS should have available an algorithm that could optimize this.

Independent of which of the operators displayed in table 1 are used, generalization should be a combination of both local and global factors. A cartographer, for example, when performing generalization needs to make simultaneous decisions about groups of features (local level) and about the layout of the map as a whole (global level) (Mackaness and Scott, 1988). The following examples indicate when local relationships could affect the application of global generalization rules:

- generalization of community centers does not depend solely on total population, since local density can influence whether small towns are retained or not.
- generalizing roads, rail lines and houses should ensure that the derived result preserves the correct relationship between the three entity types.

Current attempts at algorithmic-based generalization lack the flexibility of a cartographer and will usually apply the same rule to the map as a whole without being sensitive to local variations of features. The sequential application of the algorithms from feature to feature may also produce different results according to the sequence in which the features to be generalized are encountered. In order to optimize the application of generalization algorithms within a GIS not only should there be a wide range of algorithms available but new methodologies, e.g. data dependent, for applying these algorithms should also be developed.

Generalization is seen as a broader process than that normally conceived of by traditional cartographers, i.e., the creation of a data set with more desirable and usually less complex properties than that of its parent (João *et al.*, 1990). The grounds for defining 'desirable', according to the authors, may vary but must be explicit and - wherever possible - be quantitative : in essence, it must relate to the purpose for which the data set is to be used. It follows that at least some of the statistical properties and geometric measurements of the derived data set will differ from those of its parent.

On the basis of the above, generalization as will be considered in this paper can comprise at least the following (João *et al.*, 1990):

- the reduction of the number of areal units when data are aggregated to a higher level in a spatial hierarchy (e.g. counties to states or small pixels to large ones);
- the selection of one areal base or framework rather than another, different but similarly detailed, when there is a very large number of alternatives (e.g. see Openshaw, 1984);
- the deliberate modification of the initially recorded position, magnitude, shape or nature of discrete geographical objects or the change of their relationships with other discrete objects;
- the filtering of features of specified wavelengths where the phenomenon under study is continuous in nature;

- changes of measurement scale in the attribute data on aggregation (e.g. point data representing houses can be aggregated to census tracts with the individual data being replaced by census tract statistics; for instance, 0/1 variables may become ratio variables).

3. Generalization by Existing Systems

To set the background of the new GIS generalization functions proposed in this paper, a description is given below of how generalization is carried out within two existing and widely used GIS. The two GIS chosen are the vector-based system ARC/INFO (ESRI, 1989) and the raster-based system ERDAS (ERDAS, 1987). Both these systems are well known within the GIS user community and their limited generalization capabilities are symptomatic of the general situation observed in the present GIS market.

3.1. ARC/INFO

The well known ARC/INFO, produced by the Environmental Systems Research Institute, can perform three types of generalization:

- Line generalization
- Classification
- Elimination

Line generalization can be effected by the command GENERALIZE. This command simplifies lines using the Douglas-Peucker algorithm (Douglas and Peucker,1973) to reduce the amount of points used to represent each arc. The command can be applied either to the whole coverage or only to selected arcs. In both cases the user needs to specify the *generalize tolerance* on which the simplification is based. Depending on the geometric characteristics of the lines and their geographic distribution, the choice of this tolerance can be difficult. A tolerance too small might not simplify the lines as required and a tolerance too large might make the lines spiky and eventually overlapping. The user can only determine the best tolerance value by a trial and error procedure which can be time-consuming and frustrating, especially with very large coverages. No on-line help is provided in the choice of this tolerance as a function of the characteristics of the lines to be generalized.

Generalization by classification is often possible, depending on the type of data. This can be performed by recoding feature types, e.g. coniferous and deciduous trees are recoded as forest, or by dissolving borders between polygons based on a higher hierarchical classification, e.g. soils sub-groups are dissolved according to groups. Depending on the homogeneity of the distribution of the data types, the reclassification can simplify the coverage or not. There is no ready-available facility in ARC/INFO for the reclassification of geographical features based on its neighbors. A program (*macro*) could however be written, based on a polygon neighbor list, that would perform this operation.

Elimination of certain features, e.g. small polygons that would become imperceptible with scale change, can be performed using a set of selection commands. The selections can be made by size and/or shape. These type of commands are normally used for the elimination of sliver polygons but they can also be used for eliminating other types of polygons. However a major restriction exists on the elimination of small polygons that share a border with the outside polygon. For the border not to be destroyed those polygons cannot be eliminated unless recoded one by one. In cases where there are large numbers of small areas of coastal soils, for example, this can be a problem.

To analyze the changes created in the data, a visual check can be made by displaying the resulting coverage, and summary statistics can be computed on area of areal features and length of linear features. These are however the only facilities available and no generalization report on the changes made to the data is generated.

3.2. ERDAS

Being a raster-based system the facilities available for generalization resemble image processing techniques. Four types of generalization can be effected by ERDAS:

- Aggregation
- Smoothing
- Classification
- Elimination

These types of generalization are related with the raster-mode generalization procedures described by McMaster and Monmonier (1989). Cell aggregation can be performed by the GIS command AGGIE. During this process a structural generalization is performed by the reduction of the pixel size. Smoothing can be performed by the GIS command SCAN which can work as a low pass filter and therefore reduces the complexity of the GIS file by smoothing deviations. Generalization by classification can be performed by aggregating classes through recoding of more than one class to the same class number. Elimination of small groups of cells can be performed by using the command SIEVE. This command eliminates clumps of class values smaller than a user-specified minimum size. Due to the characteristics of the raster, all these commands are intrinsically easier to perform and the user can more easily control/understand what is happening to the data, than equivalent processes within a vector-based system.

As with ARC/INFO, visual checks must be made to analyze the changes made to the data by displaying the resulting coverage; summary statistics can be computed on area and number of cells per class values. The commands MATRIX and SUMMARY can also analyze the difference between two GIS input files before and after generalization. These are, however, the only facilities available and no generalization report on the changes made to the data is generated.

4. How Should GISs Generalize?

At present GISs have little capability to generalize. There are certain new functions that need to be built into the system. Some of these functions need to be included even before generalization can begin. Those include advice to let the user know whether generalization is required. The generalization process, in particular, should include a range of facilities that do not exist in commercially available GISs at the moment. These include functions that would allow the user to have a better control of the quality of the process. Once these functions have been outlined only then will it be possible to implement the generalization process in full.

4.1. BASIC FUNCTIONS PRIOR TO GIS GENERALIZATION

The basic functions that need to be available to the GIS user, before starting to generalize, are displayed in figure 1. The most elemental one is function 1, which allows display and query on the map content and on quality parameters such as scale, accuracy, source, etc. Table 2 shows a list of quality parameters for a general and cartographic classification of the data. This function, then, is the basis for map selection, whether for display or analysis purposes. Although it is already available in GIS, it usually lacks the listing of quality parameters integrated with map display.

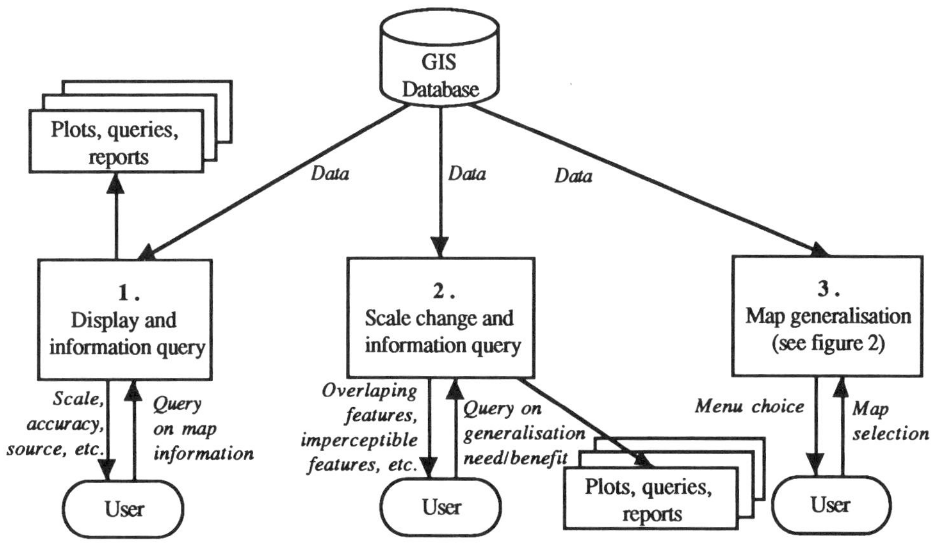

Figure 1. User concept diagram of generalization: top-level functional breakout.

Coupled with function 1, another important function would be to permit the user to query the need to generalize a particular map or the benefit likely to occur from so doing (function 2). Based on the change of scale required, the GIS should give the user a 'pre-generalization report' that would list the proportion of features that would become imperceptible on scale reduction, the number and type of features that would overlap, the areas of the map which would become cluttered, etc. A more intelligent system could then tell the user which type of generalization would be required (e.g. line simplification) and for which type of features (e.g. coast line). The user could then obtain information on these types of generalization and learn about the type of changes that could be made to the data (e.g. vector displacement). Based on the information given by functions 1 and 2, the user could then decide to select function 3 which would perform the generalization. The functions available within function 3 are explained in section 4.2 below.

General classification	Cartographic classification
• Format data was obtained	• Scale
• Date source-data was collected	• Coordinate system
• Origin of the source data	• Resolution
• Organization responsible for distribution of data in the final format	• Type of source map used to derive the digital data
	• Accuracy

Table 2 - Examples of quality parameters for data classification.

4.2. MAJOR FUNCTIONS OF GIS GENERALIZATION

4.2.1. *Purpose*. The user concept diagram for map generalization (function 3 of figure 1) is displayed in figure 2. When the user chooses to generalize a map he or she is first asked what the map is to be used for (function 3.1). The user chooses among a list of purposes already available and eventually can add new purposes if necessary. In a GIS environment, the two main types of purposes are display or analysis. Beside the choice between these two, the user will also be asked to tell for which specific purpose the map is going to be used, e.g. a map of where to watch birds in Scotland. This is important not only to determine the appropriate scale but also to help determine priority and importance of features, and which measures of generalization effects will be most appropriate.

4.2.2. *Scale*. Based on the purpose the system will then ask the user to retrieve the new scale required (function 3.2). At this point a mechanism will warn the user if the scale for the specified purpose will not be suitable (function 3.3). The determination of an appropriate scale for analysis can be a complex procedure as it can be dependent of the "uncertainty about the accuracy of individual observations and the need to lower random fluctuations and data noises" (Muller, 1989).

4.2.3. *Map Details*. Next, the user is asked to input map information relevant for the generalization (function 3.4). The classification parameters should be included in GIS attribute tables. Within this function users could possible classify spatial data differently according to different image-schemata (see Mark, 1989a). A preliminary classification of entity types was proposed in João *et al.*, (1990). The classification starts with a description of the geometric form and measurement level of the geographical features (which depends on scale), and of certain quality factors. This is followed by a classification of the geographical features according to its integration in a geographical region (i.e., measures of density, spacing and interaction) and finally each geographical feature is classified on how it relates to other types of geographical features, i.e., how rivers relate to contours.

4.2.4. *Quality Requirements*. Finally, the user inputs the quality requirements for the generalization, i.e., the maximum generalization effect that the user accepts per entity type (function 3.5). Generalization effects (João *et al.*, 1990), affecting single or multiple layers of the GIS

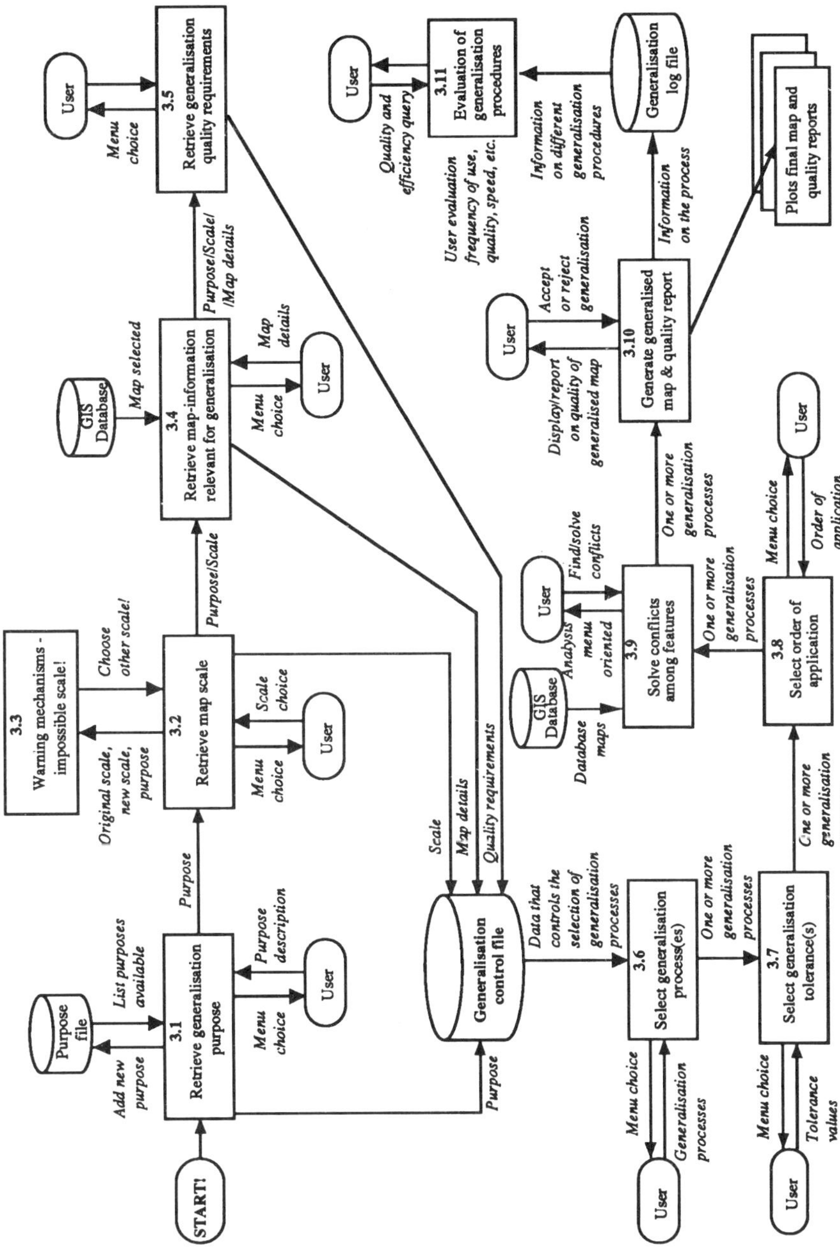

Figure 2. User concept diagram for function 3 - map generalisation.

database, can be geometry type problems (e.g. Steinhaus' paradox and feature shift) or statistically related problems (e.g. Modifiable Areal Unit Problem). The quality requirements can include details such as the maximum acceptable vector and area displacement, and the maximum acceptable change of feature length. Within this function the user should also identify an hierarchy of feature importance so conflicts between features (arising from a concentration of features in smaller scales) can be solved. These priorities shall also guide the sequence of application of the algorithms to each individual geographical feature for consistency purposes.

4.2.5. *Generalization Control File.* The information on purpose, scale, map details, and quality requirements together constitute the generalization control file. The generalization control file is the fulcrum of the system. This file will control the selection of generalization process(es) to be applied to the data (function 3.6). The way the system is designed, the algorithm(s) selected shall conform to the user's quality requirements, i.e., they will minimize certain generalization effects. The selection of the generalization algorithms is followed by the selection of the generalization tolerance(s) for the respective algorithms (function 3.7). This is followed by the selection of the order with which these processes will be applied (function 3.8). For example, if both a simplification and smoothing algorithm were selected, the simplification algorithm would be applied first.

At this point a function will report any possible conflicts between different types of geographical features (function 3.9). For example, if a forest map is being generalized the user needs to confirm that the amalgamation of forest areas is not going to overlap on top of a river running between the areas. It is very important that geographic features are not generalized in isolation or in the abstract (Mark, 1989b). In the case of a GIS in which different geographic features are stored in separate layers (e.g. ARC/INFO coverages), each layer is generalized independently of all other layers. However, mechanisms must be found to take into account the geographical relationship between features at the time of generalization.

The classification of entity types in function 3.4 can eventually help solve this problem but it is important that a conflict resolution function (function 3.9) exists. The user can be asked to solve specific conflicts and assign priority of importance to certain features. For example, if we need to produce a map for display and a rail line, a road, a river and a power line all run by the same narrow valley, which should be the feature to be displaced the least ? Finally, the generalized map is generated (function 3.10). This is coupled with the generation of a quality report that informs the user about the changes created in the data. Based on this quality report the user shall accept or reject the generalization.

In the situation that the generalization performed is rejected, the user shall be asked to input different quality requirements or, alternatively, the user could choose to re-initialize the whole process. If the generalization is accepted, the steps to perform the generalization process are stored in a *generalization log file.* By storing information describing the generalization processes performed for each specific situation it will be possible to repeat these processes for more than one map, e.g. for map series production. The generalization log file shall also store information on efficiency and quality of each generalization process, e.g. speed of the procedure, number of iterations until the user was satisfied with the result, amount and time of user interaction, etc. Based on the latter information, an additional function is available for evaluating the generalization procedures within their own terms (function 3.11). This function allows the user to compare different generalization procedures performed at different times or for different types of data.

4.2.6. *User Interface.* During the whole process the system should have a good user interface, i.e., the GIS interface should help the user make selections and decisions. For example, in the case of the selection of the generalization tolerance(s) for the selected algorithm the user could be asked to select the minimum distance between two buildings before they are aggregated into one. According to the buildings' distribution the user might want to choose a certain tolerance in order to coalesce only closer buildings. Given this situation the GIS could provide a user interface of the following types:

- color coded display of distance ranges between buildings that would indicate to the user which buildings would be aggregated and which would not for certain tolerance values;
- interactive selection by the user of two specific buildings which are representative of the distance that the user would like the tolerance to be;
- confirmation facility by letting the user appreciate the results - the generalized map could be displayed to the user using the original one as backdrop; the user could then say if he or she liked the result or not.

GIS interfaces, like the ones described above, for the selection of generalization tolerances are examples of the interfaces that should be developed for the whole system. Special concern should be given to the interface for quality requirements input, and for the presentation of the generalized map with the quality report. Better control of the quality of the generalization process may be possible if the user is able to customize their own user interface. In this way users could emphasize the control of particular quality parameters (e.g. preservation of angularity) that are most important for their applications. Error messages and warning indicatives could also be developed according to the user preferences. The customization of such a generalization-control system could be possible through the interface facilities provided by a system such as UGIX (Rhind *et al.*, 1989; see also Raper and Bundock, 1991).

Independent on how the ultimate user interface would look like, the interface should make clear to the user the cause-consequence connection between algorithm and generalization effects. The identification of the relationship between generalization algorithms and generalization effects for different classes of features and for different scale transitions is the key element within this automated process.

5. The Implementation of such a Generalization System

Knowledge-Based Systems (KBSs) could be used in selecting the particular type of generalization processes (e.g. simplifying lines) and the particular algorithm required (e.g. Douglas-Peucker algorithm). Such a procedure is shown in figure 2 as function 3.6. The particular generalization process chosen by the KBS would depend on the purpose, scale, map information, and quality requirements *input* by the user. Application of KBSs in the area of GIS has been proposed or reviewed by an extensive bibliography which includes Rhind (1988), Fisher and Mackaness (1987), Robinson and Frank (1987), and Mark and Buttenfield (1988).

To date, the use of KBSs has been investigated for specific cartographic functions such as name placement, e.g. AUTONAP (Freeman and Ahn, 1984) and for more general map design and scale transition applications, e.g. ASTRA (Leberl and Olson, 1986) and MAPAID (Mackaness *et al.*, 1986). Whether KBSs would be suitable for the generalization processes proposed in this paper remains to be determined. One of the advantages of KBSs is that they have been shown to be a viable option in many different areas of expertise. However, Robinson and Frank (1987) point out that the very nature of geographical problems do not make them ideal candidates for current KBSs. This is because KBSs tend to work better with more rigid and logical type problems that are easily definable. Certain aspects of GIS generalization tend to be less rigid, more of a subjective and commonsensical nature. It might then be more appropriate to investigate the potential of neural networks to deal with these more subjective areas. However, practical problems in implementing neural networks may mean that a pragmatic generalization system will have to rely on KBSs.

One important aspect of any new KBS is to make explicit the formalization of the rules used. These rules, gleaned from the human experts and literature, are to an extent permanent, and are independent of the particular computer language the KBS is embedded in. For example, AUTONAP used twenty five general rules as a basis for placing a name on a map in a sensible fashion. These rules then become the currency of any future KBS dealing with this topic. From the research at Birkbeck it seems likely that for the generalization problem the essential rules will be to relate the generalization procedure to the generalization effects, and subsequently the quality desired. By concentrating on these ground rules first, it limits the possibility of implementing a KBS in an *ad hoc* and piece-meal fashion.

6. Conclusions

It is very important that the generalization effects are controlled and taken into account in GIS analysis. "In general terms, spatial analysis can be considered to be the formal quantitative study of phenomena that manifest themselves in space. This implies a focus on location, area, distance and interaction" (Anselin, 1989). Generalization can affect all these variables to a lesser or greater extent, with a possible dramatic effect on the results of any GIS analysis. Despite the generalization problem being widely acknowledged among academic circles, commercial GIS have tended to sweep the problem under the carpet.

Before a system can be designed the effects of generalization need to be fully identified and measured. McMaster (1987) described mathematical measures for the evaluation of spatial accuracy of linear simplification; equivalent measures need to be defined for other aspects of generalization and tested empirically on data with known properties. In this way more efficient algorithms can be used to minimize any potential error introduced by generalization. Likewise, it will be important to define the limits of what is possible and sensible to attempt, and to educate the user about it.

Any system that is developed will rely crucially on the user to supply information about the purpose, scale required, quality tolerances, relevant map information and priority that should be taken into account during generalization. At the moment it will be assumed that the user will also make decisions about the algorithms that will be used to generalize. However, it is hoped that, eventually, as much as possible of the user's role will be replaced by KBSs. Any such KBS will be more general in its scope than the specialized cartographic KBSs developed to date.

Ultimately, it seems likely that the user will always play an instrumental role in a GIS generalization system. For such a system to be effective therefore, it is essential that the user interface is flexible and friendly enough to deal with a wide range of users and queries, from users of different backgrounds and nationalities. Whilst important in concept this can prove difficult to implement. Petchenik (1983), for instance, has pointed out how complex the hypothetical user can be, because even if we know how the response to certain directed tasks might be, we know little about the questions the user might ask spontaneously. However, by addressing this issue now, a user friendly approach can be built into the system at an early stage, rather than by merely tagging it on at the end.

7. Acknowledgements

Thanks are due to the UK Economic and Social and the Natural Environment Research Councils who are supporting the research described. I would also like to thank Prof. David Rhind, Andrew Thompson, Mike Bundock, Graeme Herbert and Jonathan Raper for their comments on the draft.

8. References

Anselin, L. (1989) 'What is special about spatial data?: Alternative perspectives on spatial data analysis', Technical Paper 89-4, National Center for Geographic Information and Analysis, University of California, Santa Barbara.

Beard, M. (1987) 'Data descriptions for automated generalization', First International Symposium on Geographic Information Systems 2, NASA, Crystal City, Virginia, 3-10.

Brassel, K. and R. Weibel (1988) 'A review and framework of automated map generalization', International Journal of Geographical Information Systems 2, 229-244.

Douglas, D. and T. Peucker (1973) 'Algorithms for the reduction of the number of points required to represent a digitized line or its caricature', Canadian Cartographer 10, 112-113.

ERDAS (1987) ERDAS user's guide, Version 7.2.XX, ERDAS, Inc.

ESRI (1989) ARC/INFO user's guide, Version 5.0. Environmental Systems Research Institute, Inc., Redlands, California.

Fisher, P. and W. Mackaness (1987) 'Are cartographic expert systems possible?', Proceedings, International Symposium on Computer-Assisted Cartography (Auto-Carto 8), Baltimore, 530-534.

Freeman, H. and J. Ahn (1984) 'AUTONAP - An expert system for automatic name placement', Proceedings, International Symposium on Spatial Data Handling, Zurich, 544-569.

João, E. , D. Rhind, S. Openshaw and B. Kelk (1990) 'Generalisation and GIS databases', Proceedings, First European Conference on GIS, Amsterdam.

Leberl, F. and D. Olson (1986) 'ASTRA—A system for automated scale transition', Photogrammetric Engineering and Remote Sensing 52, 251-258.

Mackaness, W., P. Fisher and G. Wilkinson (1986) 'Towards a cartographic expert system', Proceedings, Auto Carto London 1, 578-587.

Mackaness, W. and D. Scott (1988) 'The problems of operationally defining the map design process for cartographic expert systems', Proceedings, Austra Carto III, 7th Australian Cartographic Conference, Sydney, 715-723.

Mark, D. M. and B. P. Buttenfield (1988) 'Design criteria for a cartographic expert system', Proceedings, Eighth International Workshop "Expert systems and their applications", Avignon, 413-425.

Mark, D. M. (1989a) 'Cognitive image-schemata for geographic information: relations to user views and GIS interfaces', Proceedings, GIS/LIS, Orlando, 551-560.

Mark, D. M. (1989b) 'Conceptual basis for geographic line generalization', Proceedings, International Symposium on Computer-Assisted Cartography (Auto-Carto 9), 68-77.

Martin, C. (1988) User-Centered Requirements Analysis, Prentice-Hall, Englewood Cliffs, New Jersey.

McMaster, R. (1989) Introduction to 'Numerical Generalization in Cartography', Cartographica, Monograph 40, 26(1), 1-6.

McMaster, R. and M. Monmonier (1989) 'A conceptual framework for quantitative and qualitative raster-mode generalization', Proceedings, GIS/LIS'89, Orlando, 390-403.

McMaster, R. (1987) 'The geometric properties of numerical generalization', Geographical Analysis 19(4), 330-346.

Muller, J.-C. (1989) 'Theoretical considerations for automated map generalization', ITC Journal, 200-204.

Openshaw, S. (1984) The Modifiable Areal Unit Problem, CATMOG—Concepts and Techniques in Modern Geography 38, Geo Abstracts: Norwich.

Petchenik, B. (1983) 'A map maker's perspective on map design research 1950-1980', in D. Taylor (ed.), Progress in Contemporary Cartography 2, John Wiley & Sons Ltd., New York, 37-68.

Raper, J. F. and Bundock, M. S. (1991) 'UGIX: A Layer Based Model For A GIS User Interface', in D. M. Mark, D. M. and A. U. Frank (eds.), Cognitive and linguistic aspects of geographic space, Kluwer, Dordrecht.

Rhind, D. (1988) 'A GIS research agenda', International Journal of Geographical Information Systems 2(1), 23-28.

Rhind, D., J. Raper and N. Green (1989) 'First UNIX, then UGIX', Proceedings, International Symposium on Computer-Assisted Cartography (Auto-Carto 9), Baltimore, 735-744.

Richardson, D. (1989) 'Rule based generalization for map production', Proceedings, National Conference 'GIS - Challenge for the 1990s', Ottawa, Canada, 718-739.

Robinson, V. and A. Frank (1987) 'Expert systems applied to problems in GIS: Introduction, review and prospects', Proceedings, International Symposium on Computer-Assisted Cartography (Auto-Carto 8), Baltimore, 510-519.

Tobler, W. (1989) 'Frame independent spatial analysis', in M. Goodchild and S. Gopal (eds.), Accuracy of spatial data bases, Taylor and Francis, London, 115-122.

Weibel, R. and B. Buttenfield (1988) 'Map design for GIS', Proceedings, GIS/LIS '88, San Antonio, Texas, 350-359.

VIRTUAL WORLDS, INSIDE AND OUT

ROBERT JACOBSON
Human Interface Technology Laboratory
The Washington Technology Center
FU-20, University of Washington
Seattle, Washington 98195 USA

ABSTRACT. A place, physical or cognitive: the difference between simply being here, and *knowing here*, can be profound. This knowing, or *virtuality*, reflects one's ability to craft a dynamic internal —virtual— model of the world. The paper postulates that, being overly-dependent on media-supplied information, we are allowing our virtual sense to diminish. This renders us less capable of finding our way in both physical and conceptual space. However, a new technology, virtual world technology, permits us to understand virtuality and simulate its use through immersion in deliberately designed, computer-generated virtual worlds. Understanding this inside-outside relationship may provide the foundation for a new science of virtuality. This science may be able to unify the current potpourri of spatially oriented disciplines (environmental psychology, behavioral geography, sociology, urban design, and so forth), each with its own dogma and limitations, that attempt to define the human experience of space.

1. The Virtual World Within

A place, physical or cognitive: the difference between simply being here, and *knowing here*, can be profound. This knowing I call *virtuality*, the ability to craft a dynamic internal — *virtual* — model of the world. We rely upon our virtual world models to guide us not only in physical space but also in "idea space," where concepts stand in for things, places, and processes that we may never directly, personally experience.

The virtual world model, though it may include a cognitive map, is more than a cognitive map. A cognitive map is constructed over time to locate items and the map-holder in relation to one another, in both Euclidean and non-Euclidean space (Downs and Stea, 1973). The internal virtual world includes cognitive maps expressing the relations among entities, qualities, and the self. It also contains an "encyclopedia" and "glossary" describing the properties of these "objects" (including the self) and the processes that affect them — even suggested strategies for altering these alignments. The virtual world is to the cognitive map as the ecosystem is to the biostructure.

Like a virtual model contained within a computer's memory, the human being's internal virtual world model is dynamic and malleable — and only as efficient and reliable as the algorithms and data with which it is built. Thus, the development and maintenance of our virtual sense is essential to our coordination with the exterior world and our ability to act in it and upon it.

Take our knowledge of places. A small fraction of all human beings peregrinate around the world. Many more travel within one region or among several regions. Most people, however,

D. M. Mark and A. U. Frank (eds.), Cognitive and Linguistic Aspects of Geographic Space, 507–514.

leave the bounds of their day-to-day neighborhood, village, or town only on an infrequent basis. Human beings in contemporary society generally learn what they know of places — their own place, as well as others — from the media (the press, films, television, and radio) and from conversation heavily laden with media content (Meyrowitz, 1985). Today, the geographic component of most people's virtual worlds is built with media-supplied information. But information, even in its most textural form, is alienated experience. Information is produced by culling out discreet percepts from a rich mix of cues and clues that are available within a particular environment and then synthesizing these 'factoids' into a portrait of the environment (Sowa, 1984). Mediated information about the exterior world almost without exception is distributed as a definitive, objective, and complete *representation* of the environment, not as the peculiarly limited and highly subjective *interpretation* that it is . We construct our virtual worlds mainly with information supplied by others and filtered through the technological and organizational idiosyncrasies of our media messengers. Exposure to different types and mixes of media may explain the anomalous barriers to knowledge that exist among people who live in physical proximity (Gould and White, 1986).

It is unsurprising that popular surveys measuring knowledge of conventional geography in advanced industrial societies find that this knowledge is slightest where the local media is introverted, as in the United States, and only a little less so where an international perspective is more common (Kirby, 1899). The media are imperfect teachers, because they employ so little of our innate virtual sense. Quite the opposite: by requiring us to sit still, either in a darkened room opposite a screen or cathode-ray tube or under a reading light, the media reduce our getting about in the physical world. Simultaneously, they use only a fraction of our sensory apparatus, relegating the rest to somnolescence and retirement.

The paper postulates that, being over-dependent on media-supplied information, we are allowing our virtual sense to diminish. This renders us less capable of finding our way in both physical and conceptual space. However, a new technology, virtual world technology, permits us to understand virtuality and stimulate its use through immersion in deliberately designed external virtual worlds. Understanding this inside-outside relationship may provide the foundation for a new science of virtuality. This science may be able to unify the current potpourri of spatially oriented disciplines (environmental psychology, behavioral geography, sociology, urban design, and so forth), each with its own dogma and limitations , that attempt to define the human experience of space (Spencer and Blades, 1986; Strassoldo, 1986).

2. The Metasense of Virtuality

Virtuality is a metasense, a psychic function of the five physical senses and a sixth, nonphysical sense, the experience of time. Virtuality enables human beings to position themselves in space and time. The other senses sweep the environment, generating complex, multidimensional data for the mind to process. The virtual metasense synthesizes this rich milieu and produces a simplified, intuitive model of the environment.

The virtual metasense may exist, in a primitive form, among other higher forms of life, but it is most fully developed in human beings. For us, virtuality is essential to survival. Location-finding may not be instinctual among human beings in the same way it is among other migratory animals. Over time, however, virtuality has come to fill in for the location-finding genetic capabilities of other creatures, permitting humans to learn spatial and environmental conditions and then use them to guide our actions (Stea, 1974). The Australian Aborigines' "dreamtime" is perhaps the most evocative surviving expression of virtuality. In dreamtime, the Aborigine explores mythical realms that often correspond with places, events, and processes in the material world (Lynch, 1960). There is increasing agreement among cognitive scientists that the movement of a person through space helps create and refine mental maps (Gärling, Böök, and Lindberg, 1984). It is not difficult to believe that these maps are in some way metaphors, if not agents, for changes in the physical makeup of the brain. Our virtual sense, our ability to relate what's inside with what's outside, may be as central to our human identity as our capacity for thought: *Homo sapiens* may be better characterized as *Homo virtualis*.

Over the millennia, our internal virtual worlds have expanded to encompass the terrain of the the intangible world of ideas and emotion. In contemporary society, virtuality has an expanded function, to orient the person among the plethora of themes and memes that populate the conceptual world. Well-schooled individuals can traverse conceptual networks of great complexity, metaphorically portrayed in Hesse's *The Glass Bead Game* (1969) as a vast armature on which competing scholars skillfully arranged tokens of knowledge. When properly arrayed, these individual elements produced beautiful peals of music mightier than Bach's concertos. Gibson, in his immensely popular and influential *Neuromancer* series of novels (1984, 1986, 1988), wickedly transformed this metaphorical machinery. In Gibson's allegorical near-future world, disembodied individuals ("network cowboys") and intelligent, self-actuating computer programs slip their geospatial anchors to sail over intertwined computer networks — *cyberspace* — in pursuit of knowledge (represented by visible constellations of data). Those who use electronic networks today experience a weak manifestation of cyberspace: our conceptual breadth increases with our network journeys, although the engagement of our virtual sense is limited by the linear, textual nature of computer communications (Rheingold, 1987).

We can communicate in the virtual mode. Then our exchange is quick and complete, because virtuality synergistically draws upon many or all of our senses working in synchrony. Unfortunately, we can only communicate at a virtual level with true intimates since to do so requires closer contact of mind, body, and (some would say) spirit than our mores or our current technologies allow. In fact, civilization has gone in exactly the other direction, against virtuality. In order to speak to broader communities of interest, we have diminished our reliance on virtuality. We speak and listen to more people more loudly now, but with less meaning exchanged. Consequently, we are increasingly out of touch with our virtual sense and, in turn, with the external reality.

3. Technology: For and Against Virtuality

Spoken language, for example, enables us to share ideas and experiences but only after we have learned an abstract, symbolic code with arduous semantic and syntactic rules. These rules do not permit more than a smattering of nonverbal amplification, mostly in the form of body language. Written language is even more rigorous in what it will and will not allow. Though its powers to communicate meaning are vast, written language uses even less of our powers of virtuality.

So it has been with each successive innovation in "communication." Print, photography, film, telephony, recorded sound, radio, and video each requires its users to work at making sense from the symbols it presents in unnatural ways. Native peoples without prior exposure to these media often cannot make sense of the images and cues that they contain. That we can experience pleasurable internal states based on these limited modes of communication is a credit to our adaptability, not to the qualities of these media. If it is true that people "self-regulate" (choose) their environments to gain pleasure or at least reduce pain (Korpela, 1989), it may be that many are most comfortable in front of a television set because the TV medium does not challenge them where they are weakest, in their virtual selves. Whether computer networks and other media enhance or diminish the development of an individual's (and by extension, a society's) virtual capabilities depends on the extent to which they make the individual more cognizant of, and capable of exercising, his or her virtual metasense. Thus does virtuality become a design criterion for the emerging technologies of communication and learning (Johnson-Lenz and Johnson-Lenz, 1990).

That individuals have different virtual capabilities, in proportion to their personal and collective resources, is obvious. Some decry this situation as antidemocratic and even anti-human. Others assert that it merely expresses the distribution of knowledge in society and that those who are conceptually broad and integrated are more fit to occupy positions of power. The proposition that computer networks and other media of communication should be constructed to enable every person to develop his or her virtual capabilities to the fullest — to further empower individuals in a mass society — is a design criterion that, to this day, remains in dispute (Mandeville, 1983). Normative methodologies like *sense-making* challenge the powerful role exercised in contemporary society by conventional information media (Dervin, 1989b; Dervin and Clark, 1989). What they communicate, according to the sense-making theorists, is a meta-message: "be passive, be quiet, be accepting." The sense-makers argue that what changing societies require are new modes and structures of communicating experience that increase the ability of individuals to seek out, locate, and share knowledge with others similarly situated. With these new methods and structures, individuals could empower themselves to increase their useful knowledge about the physical and conceptual worlds they inhabit, take action to improve their own condition, and in so doing retrain those virtual learning capacities that have eroded with exposure to virtuality deadening, misinforming media.

Virtual worlds technology answers well the design parameters expressed by the sense-makers. This newly emergent technology enables an intimate sharing of knowledge and experience within the external virtual environments it creates and sustains. Instead of forcing people, as do other media, to learn new and peculiar methods of receiving information, virtual world technology connects with people "where they live" — in their virtual selves (Furness, 1987). A virtual world is successful to the degree that it plays on and with our sense of virtuality and incites us to deeper involvement with the virtual environment. Virtual world technology is both inclusive and responsive: it invites participation in the ongoing design of the dynamic worlds it presents. Although its communicative capabilities are profound, virtual world technology is simple in concept (M. Bricken, 1990). (Implementation is a bit more difficult.)

An external virtual world is a three-dimensional model programmed in a computer's memory and experienced by its "participants" as an all-encompassing environment of sights, sounds, and tactile cues. The participants in an external virtual environment don special headgear ("eyephones" and a 3-D sound system) and manipulative devices (for example, the dataglove) equipped with sensors for communicating the participants' position and movement to the computer. The computer constantly regenerates visual and aural "sense-surrounds" and adjusts these fields according to the participants' movements in space. The virtual world system allows its participants to dynamically rescale the virtual environment so that, individually or collectively, they can fly across large expanses of terrain (even the universe!) or dive into an atom. Being in such demanding environments will test our virtual sense and strengthen it. It is also possible for multiple participants to experience an external virtual world (W. Bricken, 1990). Multiple participants in televirtual worlds (shared via telecommunications networks) will have their virtual capabilities tested by groupwork situations or when simply enjoying a virtual party.

Virtual world theoreticians speak of a "hyper-reality" that will be "more real" than the world we commonly inhabit and in which we will be able to have experiences denied to us in the "real world." External virtual worlds can reveal processes that may be invisible to us because they are distant in time and space, occur too fast or too slow, or are too large or too small for our physical senses. A more prosaic use of virtual world technology (although revolutionary for geography) will be to create maps that accurately reflect the subjective capabilities and predispositions of their users (Edwards, 1990). In the process of constructing such maps we will certainly learn more about how people conceptualize the physical and cognitive worlds they inhabit and how their images differ from culture to culture. Ultimately, working and playing in external virtual worlds will cause us to better appreciate and "design" our corresponding internal virtual worlds.

4. A Science of Virtuality

The development of virtual world technology is proceeding at a prodigious rate. The commercial application of virtual world technology is probably only a few years away. Developers around the world are competing to perfect computational and display systems for application in 1995 or sooner (Ditlea, 1989; Rheingold, 1990).

The rapt fascination expressed by the popular press for virtual world technology and the "virtual realities" it creates is not balanced by a dispassionate, scientific critique of the consequences of this technology for human affairs. On the one hand, advocates argue that virtual world technology may revolutionize interpersonal communication, creating at long-last the "group mind" that science-fiction writers have proposed as the ultimate convergence of individual personalities (Leary, 1989). On the other hand, virtual world technology is castigated by critics who foresee it inducing the end of individuality and, even worse, the construction and distribution of artificial realities so compelling that truth no longer has meaning (see Winograd's statements in Paulson, 1990).

Despite the expectations placed on virtual world technology as an indispensable tool in professional, industrial, educational, and creative settings, virtual world technology is a theme without a theory. It is time to start building that theory, now that the technological means for assessing it arc almost ready.

A science of virtuality would be an amalgam of computer science, technology assessment, psychology, education, and the creative arts, this new science could eventually subsume the disparate fields that today poke and probe at the process of human social understanding without achieving any consensus. Virtual science could unite them in an holistic understanding of what it is to be a human being in a complex but malleable world.

What might the science of virtuality entail? Its central issue would be the relation of human beings' internal virtual worlds to the real world. The study of human beings' interaction with external virtual worlds created by technological systems should provide interesting findings in this regard. Phenomenology, sociology, and cognitive psychology would all be useful in developing theories and methodologies for this endeavor. The science of virtuality might also draw upon and influence such "harder" sciences as human factors, physiology, and computer science, because what we learn about human beings and information environments will require the thoughtful design of environment-creating systems.

Virtual scientists would attempt to understand how human beings identify and understand the various information environments they inhabit, how they situate themselves within this environment, and how they attempt to manipulate the environment to satisfy their perceived needs. Obviously, there is a continuous, dynamic interaction between the external mediated world and the world inside a person's head. Of course, the Heisenberg Principle applies here as it does to studies of other quantum phenomena — the observation may affect what is being observed. But the fact that reliable observation is difficult should not — cannot — itself be an argument against learning how virtuality works.

The methodology for a science of virtuality remains to be worked out. Initially, it may be fumbling and uncertain. Even at an early stage, however, there will be things that can be learned from shifting our paradigms and blurring the line that separates objective inquiry from subjective experience. For example, the nascent methodology of sense-making gives public information campaign managers a better idea of how they can serve their clients' information needs (Dervin, 1989a). Similarly, *wayfinding*, a discipline in the practical field of urban design, permits architects and urban planners to anticipate how people will perceive and act in the built space (Passini, 1984; Arthur and Arthur, 1988). No one will argue that these methodologies are complete sciences. Still, they have value and they point the way toward more robust constructs regarding the connection between the world outside and the virtual world within.

A theory of virtuality will be important to designers of virtual worlds. The technological components that will permit computers to create three-dimensional environments of convincing verisimilitude are exist today; their refinement and assemblage into systems is only a few years off. There is no clear indication of how these systems can best be designed and implemented for the enhancement of human understanding and performance. For the moment, everything known about virtual worlds is ad hoc or, at best, only partial. A science of virtuality will enable more beneficial directions to be taken in the design of virtual world technology and possibly prevent large-scale personal dislocations from occurring — as they occur for most people today, in the thrall of conventional media — once virtual worlds systems become pervasive and ubiquitous.

5. Conclusion

From the standpoint of geographers and others concerned about spatial knowledge, virtual worlds systems can be used positively to learn more about the virtual worlds within us, as well as to catalyze others' virtual senses so that they can be more active in the real world. The virtual world is a permeable boundary between the world inside and the world outside, joining as it separates the spaces inside and out. Acceptance of the virtual world paradigm — that experiences inside us are a function of the world outside, and vice versa — may help ease the archetypal fear of eternal separation that haunts our species and enable us to rejoin the natural world.

6. References

Bricken, M. (1990) 'No Interface to Design', Paper presented at First Cyberspace Conference, School of Architecture, University of Texas, Austin, Texas, May 5-6.

Bricken, W. (1990) 'Coordination of Multiple Patrons in Virtual Space: Preliminary Notes', Unpublished paper, Human Interface Technology Laboratory, University of Washington, Seattle, WA.

Dervin, B. (1989a) 'Audience as Listener and Learner, Teacher and Confidante: The Sense-Making Approach', in Rice, R., and Atkins, C. (eds.), Public Communication Campaigns, 2nd Ed., Newbury Park, CA., Sage, 67-85.

Dervin, B. (1989b) 'Users of Communication/Information Systems as Research Inventions: Perpetuating Myths, Reifying Inequities', in Journal of Communication, special edition.

Dervin, B. and Clark, K. D. (1989) 'Communication as Cultural Identity: The Invention Mandate', in Media Development, special edition.

Ditlea, S. (1989) 'Inside Artificial Reality,' PC Computing, November, 91-99+.

Downs, R. M. and Stea, D. (1973) 'Cognitive Maps and Spatial Behavior', in Downs, R.M., and Stea, D. (eds.), Image and Environment, Chicago: Aldine, 8-26.

Edwards, T. (1990) The Virtual Realization of Geographic Information, Seattle, University of Washington, Department of Geography, unpublished paper.

Gibson, W. (1988) Mona Lisa Overdrive, New York City, Bantam.

Gibson, W. (1986) Count Zero, New York City, Ultramarine.

Gibson, W. (1984) Neuromancer, New York City, Ace Books.

Gould, P. and White, R. (1986) Mental Maps, Boston, Allen and Unwyn.

Gärling, T., Böök, A., and Lindberg, E. (1986) 'Cognitive mapping of large-scale environments: The interrelationship of action plans, acquisition, and orientation', Environment & Behavior 16(1), 3-34.

Hall, E. T. (1966) The Hidden Dimension, Garden City, NY, Doubleday & Company.

Hesse, H. (1969) The Glass Bead Game, New York City, Holt, Rinehart & Winston. (Originally published in 1943 as Das Glasperkenspiel, Zurich, Fretz & Wasmuth.)

Johnson-Lenz, P. and Johnson-Lenz, T. (1990) Rhythms, Boundaries, and Containers: Creative Dynamics of Asynchronous Group Life, Lake Oswego, Oregon, Awakening Technology Research Report No. 4.

Kirby, A. (1988) 'Context, Common Sense and the Reality of Place: A Critical Reading of Meyrowitz', Journal for the Theory of Social Behavior 18(2), 239-250.

Korpela, K. M. (1989) 'Place Identity as a Product of Environmental Self-Regulation', Journal of Environmental Psychology 9, 241-256.

Leary, T. (1989) 'The Interpersonal, Interactive, Interdimensional Interface', Unpublished paper.

Mandeville, T. (1983) 'The Spatial Effects of Information Technology', Futures, February, 65-72.

Meyrowitz, J. (1985) No Sense of Place: The Impact of Electronic Media on Social Behavior, New York, Oxford University Press.

Passini, R. (1984) Wayfinding in Architecture, New York, Van Nostrand Reinhold.

Paulson, T. (1990) 'Virtual Reality — Next computer revolution asks: "Is it live, or is it cyberspace?"' Seattle Post-Intelligencer, May 8.

Rheingold, H. (1990) 'Travels in Virtual Reality', Whole Earth Review, Summer, 80-87.

Rheingold, H. (1987) 'Virtual Communities', Whole Earth Review, Winter, 78-80.

Spencer, C., and Blades, M. (1986) 'Pattern and Process: A Review Essay on the Relationship between Behavioural Geography and Environmental Psychology', Progress in Human Geography 10(2), 230-248.

Stea, D. (1974) Program Notes on a Spatial Fugue, Los Angeles: UCLA School of Architecture and Urban Planning, Monograph No. DP38.

Strassoldo, R. (1986) 'The Sociology of Space: A Typological Framework', Paper presented at International Sociological Association conference.

Wallace, R. (1989) 'Cognitive Mapping and the Origin of Language and Mind', Current Anthropology 30(4), 518-526.

Appendix: NATO Advanced Study Institute Participants

* author or coauthor of a chapter in this book

Directors and Lecturers:

Canada:

C. Grant Head*	Department of Geography, Wilfrid Laurier University, Waterloo Ontario, N2L 3C5
Zenon Pylyshyn	Center for Cognitive Science, University of Western Ontario, London, Ontario, N6A 1C2

France:

Jean-Gabriel Ganascia	LAFORIA, Tour 46-0, Université Pierre eb Marie Curie, 4 Place Jussieu, 75252 Paris, CEDEX 05

Spain:

Maria Cátedra Thomas*	Department of Anthropology, Universidade Complutense de Madrid, Madrid

United Kingdom:

Mark Blades*	Department of Psychology, The University of Sheffield, Sheffield S10 2TN

USA:

Andrew U. Frank, Co-Director*	NCGIA, Department of Surveying Engineering, University of Maine, Orono, Maine 04469
John R. Herring*	Intergraph Corporation, M.S. IW17A2, One Madison Industrial Park, Huntsville, Alabama 35807-4201
George Lakoff	Department of Linguistics, University of California, Berkeley, California
David Mark, Director	Department of Geography, State University of New York at Buffalo, Buffalo, New York 14261
Len Talmy	Cognitive Science Center, University of California, Berkeley, California
Current address:	Center for Cognitive Science, State University of New York at Buffalo, Buffalo, New York 14260

Other participants

Canada:

Elaine M. Bjorkland*	Department of Geography, University of Western Ontario, London, Ontario, N6A 5C2
Geoffrey Edwards*	Centre de Géomatique, Department des Sciences Géodesiques et Télédétection, Université Laval, Ste-Foy, Québec, G1K 7P4
Janet Mersey	Department of Geography, University of Guelph, Guelph, Ontario, N1G 2W1
Martin Mujica	Department of Sociology, Université de Moncton, Moncton, New Brunswick, E1A 3E9
Allan K.Philbrick*	Department of Geography, University of Western Ontario, London, Ontario, N6A 5C2
Hansgeorg Schlichtmann*	Department of Geography, University of Regina, Regina, Saskatchewan, S4S 0A2
David Zaidel	Department of Industrial Engineering, University of Toronto, Toronto, Ontario, M5S 1A4

Denmark:

Prof Ole Jacobi	Institute of Surveying and Photogrammetry, Technical University of Denmark, Landmaalervej 7, Dk - 2800 Lyngby
Jesper Kaae Petersen	Institute of Surveying and Photogrammetry, Technical University of Denmark, Landmaalervej 7, Dk - 2800 Lyngby

France:

Denis Blamont	UPR 299, CNRS, 34 nue de L'Orangerie, 78000 Versailles

Germany:

Christian Freksa*	Institut für Informatik, Technische Universität München, Arcisstr. 21, 8000 Munich 2
Daniel Hemández*	Institut für Informatik, Technische Universität München, Arcisstr. 21, 8000 Munich 2
Wolfgang Hoeppner	University of Koblenz, Computational Linguistics, Rheinau 3-4, D-5400 Koblenz

Greece:

Marinos Kavouras — Hellenic Mapping & Cadastral Organization, T. VASSOU 11-13, 115 21 Athens

Italy:

Irene Campari* — Instituto CNUCE, Consiglio Nazionale delle Recherche, v.s. Maria 36, 56100, Pisa

Paolo G. Franciosa — Instituto di Analisi dei Sistemi ed Informatica, Consiglio Nazionale delle Ricerche, Viale Manzoni 30, I-00185 Rome, ITALY

Roberto Scopigno — Instituto CNUCE, Consiglio Nazionale delle Ricerche, v.s. Maria 36, 56100, Pisa

Netherlands:

Peter Essens — TNO-Institute for Perception, Cognitive Psychology Group, P.O. Box 23, 3769 ZG Soesterberg

Anthony Williams — Geography, Universitaat van Amsterdam, Netherlands

Current address: Department of Geography, Pennsylvania State University, University Park, PA 16802 USA

Spain:

Jose Chantada — Professor Titular de Geografía, Universidade de Santiago de Compostela

Francisco Escobar — Departamento de Geografía, c/Colegios n:2, U. Alcalá de Henares, Alcalá De Henares, Madrid

Joan Nunes* — Departament de Geograpahia, Universitat Autonoma Barcelona, 08193 Bellaterra

Mauricio Ruiz-Pérez — Departament de Ciencies de la Terra, Universitat de les Illes Balears, Carretera de Valldemossa Km. 7.5, Palma de Mallorca - 07071, Baleares

Maria Jesus Vidal — Departamento de Geografía, Facultad de Filosof-a y Letras, Universidad Autonoma de Madrid, Cantoblanco, 28049-Madrid

Turkey:

Ahmet Arslan	Department of Computer Engineering and Information Sciences, Bilkent University, P.O. Box 8, Maltepe, 06572 Ankara
Selma Çapan	Faculty of Education, Department of Foreign Languages, Middle East Technical University, 06531 Ankara
Ali Kas*	Faculty of Education, Department of Foreign Languages, Middle East Technical University, 06531-ODTU Ankara

United Kingdom:

Michael Bundock*	Department of Geography, Birkbeck College, University of London, 7-15 Gresse Street, London W1P 1PA
Elsa Maria João*	Department of Geography, Birkbeck College, University of London, 7-15 Gresse Street, London W1P 1PA
David J. Medyckyj-Scott	Department of Computer Studies, University of Technology Loughborough, Loughborough, Leicestershire LE11 3TU
Ian Pratt*	Department of Computer Science, University of Manchester, Manchester, M13 9PL
Jonathan Raper*	Department of Geography, Birkbeck College, 7-15 Gresse Street, University of London, London W1P 1PA
Andrew Runnalls	Computer Science, University of Kent at Canterbury
Michael Worboys*	Department of Computing Studies, University of Leicester, Leicester, LE1 7RH

USA:

Kelly Chan*	Harvard Lab for Computer Graphics & Spatial Analysis, 48 Quincy Street, Cambridge, MA 02138
John Cloud	Department of Geography, University of California, Santa Barbara, California 93106
Max Egenhofer*	NCGIA, Department of Surveying Engineering, University of Maine, Orono, Maine 04469
Charles O. Frake	Department of Anthropology, State University of New York at Buffalo, Buffalo, New York 14260
Scott Freundschuh*	Department of Geography, State University of New York at Buffalo, Buffalo, New York 14261
Current address:	Department of Geography, Memorial University of Newfoundland, St. John's, Newfoundland, Canada

USA (continued):

Thomas A. Gentry*	Department of Psychology, California State University, Stanislaus, 801 Monte Vista Avenue, Turlock, California 95380
Myke Gluck*	School of Information Studies, 4-206 Center for Science & Technology, Syracuse University, Syracuse, NY 13244-4100
Michael Gould*	Department of Geography, State University of New York at Buffalo, Buffalo, New York 14261
Current address:	ESRI-España, Arapiles, 14, 28015 Madrid, España
Susan Haller	Department of Computer Science, State University of New York at Buffalo, Buffalo, New York 14260
Robert Jacobson*	Human Interface Technology Laboratory, University of Washington, FU-20, Seattle, Washington 98195
Werner Kuhn*	NCGIA, Department of Surveying and Engineering, University of Maine, Orono, Maine 04469
Merry Makela	Department of Biology, Texas A & M University, College Station, Texas 77843
Matthew McGranaghan*	Department of Geography, University of Hawaii, 2424 Maile Way, Honolulu, Hawaii 96822
Dalia Varanka*	Department of Geography, BOX 413 UWM, Milwaukee, WI 53201

Non-NATO countries:

Finland:

Tapani Sarjakoski	Finnish Geodetic Institute, Ilmalankatu 1 A, SF-00240 Helsinki

Taiwan (Republic of China):

Bor-Wen Tsai	Department of Geography, National Taiwan University, Taipei